DYNAMIC STRUCTURE OF NUCLEAR STATES

PROCEEDINGS OF 1971 MONT TREMBLANT INTERNATIONAL SUMMER SCHOOL

DYNAMIC STRUCTURE OF NUCLEAR STATES

PROCEEDINGS OF 1971 MONT TREMBLANT INTERNATIONAL SUMMER SCHOOL

Edited by

D.J. Rowe

L.E.H. Trainor

S.S.M. Wong

T.W. Donnelly

University of Toronto Press

Toronto and Buffalo
Printed in Canada

ISBN 0-8020-1868-8
Microfiche ISBN 0-8020-0199-8
LC 75-186282

1971 MONT TREMBLANT INTERNATIONAL SUMMER SCHOOL

ON THE DYNAMIC STRUCTURE OF NUCLEAR STATES

ORGANIZERS

The Theoretical Physics and Nuclear Physics Divisions
of the Canadian Association of Physicists

Director
D.J. Rowe (Toronto)

Committee
T.W. Donnelly (Toronto)
A.E. Litherland (Toronto)
M. Harvey (Chalk River)
Diane Schwartz (Toronto)
L.E.H. Trainor (Toronto)
S.S.M. Wong (Toronto)

SPONSORS

North Atlantic Treaty Organization
National Research Council of Canada
Atomic Energy of Canada Limited

ACKNOWLEDGEMENTS

The Canadian Association of Physicists is pleased to acknowledge financial support from the North Atlantic Treaty Organization, the National Research Council of Canada, the Atomic Energy of Canada Limited, and the provinces of Alberta and Ontario. Contributions to the school were also gratefully received from Hydro-Quebec, the City of Montreal, Imperial Oil Ltd., and the Bank of Montreal.

PREFACE

Summer schools in physics have come to play an important role in the life of the Canadian scientific community. The first summer school was organized by the Theoretical Physics Division of the Canadian Association of Physicists in the summer of 1957. It was a collaborative effort with the Canadian Mathematical Congress and lasted four weeks, three seminar weeks in Edmonton followed by a Congress week in Banff. Highlights of that summer school included a description of the new "BCS theory" of superconductivity by John Bardeen and an explanation by Eugene Wigner of the significance of the non-conservation of parity in the then current experiments by C.S. Wu and collaborators on ^{60}Co.

Since the "Edmonton experiment" many summer schools have been held in Canada, at first under the auspices of the Theoretical Physics Division of CAP, but more recently as collaborative efforts between that division and various other subject divisions of CAP. Over the years the character of the summer schools has evolved from four-week exposures of theoretical physics in-the-large to concentrated two-week exposures of specialized fields. The main intent of the CAP summer school series has been to provide an opportunity for young scientists, primarily at the senior graduate and post-doctoral levels, to study in depth areas of current interest in physics and to interact with senior scientists working in the forefront of these areas.

The Mont Tremblant International Summer School was primarily concerned with the dynamic structure of nuclear states. In selecting the topics we deliberately avoided the well-established approaches to nuclear physics and emphasized recent developments which seemed to us to have considerable promise for the future. One of the currently exciting prospects is the possibility of artificially synthesizing superheavy elements. Interest in this field is strongly coupled with a renewal of interest in several related fields, for example, the fission process, the cluster structure of nuclei, and heavy ion physics. The stimulus for this renewed interest comes partly from the developments of new experimental techniques, partly from the development of new probes of the nucleus represented by the latest generation of accelerators, and partly from recent developments in theoretical physics which have led to an improvement in the microscopic understanding of nuclear bulk properties. The lectures at the summer school strongly reflected these interests and presented many of the new developments.

It is noteworthy that some of the most spectacular successes in nuclear physics have been achieved with very simple semi-phenomenological, semiclassical models: for example, the shell-model, cluster models, the liquid drop model, and collective vibrational and rotational models. These models were devised to explain a wide variety of nuclear properties, and are basic to the general understanding of nuclear stabilities, radioactivity, fission phenomena, and many systematics of nuclear spectroscopy. Yet although these models give an immediate superficial description of collective phenomena, there has been a conspicuous absence, for the most part, of more complete microscopic theories to explain either the models or the phenomena. Significant advances have recently been made in this direction and some of the more representative developments are recorded in these proceedings. They include developments in the microscopic approaches to collective motion, cluster models, and fission theory and new theories of nuclear excitations and level distributions.

Every summer school faces the difficult decision of whether or not to publish the lectures, as publication easily doubles the work load both on the speakers and on the organizing committee. In retrospect we are pleased that our decision was in the affirmative, since the quality of the lectures was extremely high and the invited speakers have all cooperated magnificently to facili early publication.

The school was jointly sponsored by the Theoretical Physics and Nuclear Physics Divisions of CAP and every effort was made to bring the theoreticians and experimentalists together to discuss their common problems from their different viewpoints. It seemed to be a unanimous feeling on the part of those attending that the school achieved this objective and was, in modest terms, a success. Whether cordiality was stimulated by the French-Canadian cuisine, by the social and entertainment amenities of Mont Tremblant Lodge, and by the beauty of the Laurentian countryside, certainly it was founded on excellent lectures and resulted in enthusiastic discussion and good all-round scientific fellowship.

It is not possible to thank individually each and every one who contributed to the success of the summer school, since the list would include all the participants. We are, however, especially indebted to a number of individuals and organizations: to Monsieur Guy Landry, Manager of Mont Tremblant Lodge, who facilitated our stay in every way, and who with his staff was a most gracious and considerate host; to our publisher, the University of Toronto Press, for its patience in dealing with the whims and peculiarities of science and scientists; to our sponsors (listed elsewhere in this book), whose financial support made the summer school possible and whose continued faith in the ultimate value of science persists in these troubled times; to the invited speakers, whose expertise and willingness to share it were central to the success of the summer school; to Miss Kate Birkinshaw, who helped with many of the conference arrangements and who typed several of the manuscripts; and in particular to Mrs. Diane Schwartz for her tremendous efforts in every aspect of the conference arrangements, for typing or retyping most of the manuscripts, and for generally preventing the physicists from creating unbounded confusion.

Toronto
November 1971

D.J.R.
L.E.H.T.
S.S.M.W.
T.W.D.

CONTENTS

SECTION 5
NUCLEAR STRUCTURE AND INTERMEDIATE ENERGY PHYSICS

SECTION 6
CONTRIBUTED PAPERS

DYNAMIC STRUCTURE OF NUCLEAR STATES

PROCEEDINGS OF 1971 MONT TREMBLANT INTERNATIONAL SUMMER SCHOOL

HARTREE-FOCK THEORY AND COLLECTIVE MOTION

FELIX M.H. VILLARS
Department of Physics and Laboratory for Nuclear Science
Massachusetts Institute of Technology
Cambridge, Massachusetts

1. TIME DEPENDENT HARTREE-FOCK THEORY AND COLLECTIVE CANONICAL EQUATIONS

A. Wave Functions, Density Matrices and Hartree-Fock Equations

In Hartree-Fock theory, the Schroedinger equation for a system of A identical nucleons is approximately solved by means of a variational principle, with a trial function $\phi(1,2,\ldots A)$ of the form of a determinant of A orthogonal single-particle wave functions†

$$\phi(1,2,\ldots A) = \frac{1}{\sqrt{A!}} \det|\psi_\mu(i)| \; . \tag{1.1}$$

It will be assumed that the $\psi_\mu(i)$ are the A lowest energy members of a complete orthogonal set $\psi_\alpha(i)$ of single particle functions (orbitals). The indices $\mu,\nu\ldots$, will be reserved to designate the "occupied" orbitals (occurring in ϕ); indices $\sigma,\tau\ldots$ will designate "unoccupied" single particle states. These latter will play a role in formulating the variational problem.

The Hamiltonian of the A-nucleon system will be assumed to have the form

$$H = \sum_i \frac{p_i^2}{2m} + \sum_{i<j} V_{ij} \quad . \tag{1.2}$$

It is well known that, for H as given by (1.2), the mean energy is expressible in terms of the one- and two-body density matrices. Without yet restricting the wave function $\Psi(1,2,\ldots A)$ to the form (1.1), we may define the density matrices:

$$(1|\rho|1') = A \int d\tau_2 \ldots d\tau_A \; \Psi(1,2,\ldots A)\, \Psi^*(1',2,\ldots A) \tag{1.3}$$

† i stands for the position, spin and isospin coordinates of nucleon i, and μ for the set of all quantum numbers (e.g., n,l, j,m) needed to specify the state of the particle.

and

$$(12|\rho|1'2') = A(A-1) \int d\tau_3 .. d\tau_A \ \Psi(1,2,3...A) \ \Psi^*(1',2',3...A). \tag{1.4}$$

The kinetic energy operator $p_i^2/2m$ is represented by the matrix

$$(i|t|i'); \tag{1.5}$$

the potential energy V_{ij} between particles i and j by

$$(ij|V|i'j') \ . \tag{1.6}$$ †

With these definitions, one has (exactly)

$$\begin{aligned} E &= \int d\tau_1 .. d\tau_A \ (\Psi^* H \Psi) \\ &= \int d\tau_1 d\tau_{1'} \ (1|t|1')(1'|\rho|1) \\ &+ \int d\tau_1 d\tau_{2'} \ (12|V|1'2')(1'2'|\rho|12) . \end{aligned} \tag{1.7}$$

The choice of the Slater determinant ϕ, eq. (1.1), as an approximation for Ψ leads to the following reductions:

$$(1|\rho|1') = \sum_{\mu=1}^{A} \psi_\mu(1) \ \psi_\mu^*(1') \tag{1.8}$$

$$(12|\rho|1'2') = (1|\rho|1')(2|\rho|2') - (1|\rho|2')(2|\rho|1') \tag{1.9}$$

The structure (1.8) of the one-body density matrix leads to the additional properties

$$\int d\tau_{1''} \ (1|\rho|1'')(1''|\rho|1') = (1|\rho|1') \tag{1.10}$$

$$\int d\tau_1 \ (1|\rho|1) = A \tag{1.11}$$

for which we shall use the shorthand notation

$\rho^2 = \rho$ (1.10a) trace $\rho = A$. (1.11a)

† This seemingly fussy notation will have the advantage that all formulae which we develop will be <u>invariant</u> under a change of representation.

With (1.8) and (1.9), the mean energy E can be written as

$$E = \int d\tau_1 d\tau_{1'} \, (1|t|1')(1'|\rho|1) + \tfrac{1}{2}\int d\tau_1 .. d\tau_{2'} \, (12|V^A|1'2')(1'|\rho|1)(2'|\rho|2) \quad (1.12)$$

where we have defined an antisymmetrized matrix

$$(12|V^A|1'2') = (12|V|1'2') - (12|V|2'1') \, . \quad (1.13)$$

We are now in a position to state the variation problem for E: $\delta E=0$. The variation of ϕ is entirely expressible in terms of a variation $\delta\rho$ of the density matrix ρ. From (1.12), and using the symmetry property $(12|V^A|1'2') = (21|V^A|2'1')$ of the potential energy matrix, one has at once

$$\delta E = \int d\tau_1 d\tau_{1'} \left\{(1|t|1') + (1|U|1')\right\} (1'|\delta\rho|1)$$
$$= \int d\tau_1 d\tau_{1'} \, (1|h|1')(1'|\delta\rho|1) \equiv \text{trace } (h\delta\rho) \quad (1.14)$$

where

$$(1|U|1') = \int d\tau_2 d\tau_{2'} \, (12|V^A|1'2')(2'|\rho|2) \quad (1.15)$$

and

$$(1|h|1') = (1|t+U|1') \, . \quad (1.16)$$

h, as defined by (1.16) is the Hartree-Fock single particle Hamiltonian.

There remains to find an expression for $\delta\rho$. Variations which leave (1.10) and (1.11) invariant are unitary transformations

$$\rho' = \rho+\delta\rho = U^+\rho U = (1-if+..)\,\rho(1+if+..)$$

A first order variation of ρ is therefore expressed by

$$\delta\rho = i[\rho, f] \quad (1.17)$$

where f, the infinitesimal generator of U, is an arbitrary hermitian matrix. Inserting this into (1.14), one has

$$\delta E = +i \text{ trace } (h[\rho, f])$$
$$= i \text{ trace } (f[h,\rho]) \, . \quad (1.18)^{\dagger}$$

† The cyclic invariance of the trace, trace (abc) = trace (cab), has here been used.

The variational condition $\delta E=0$ therefore leads to

$$\text{trace}\ (f[h,\rho]) = 0 \tag{1.19}$$

for any f, from which one concludes that

$$[h,\rho] = 0\ . \tag{1.20}$$

In other words, this means that in a representation, in which ρ is diagonal, h may also be diagonal. Now, by (1.8), ρ is diagonal with regards to the quantum numbers α of the single-particle function ψ_α:

$$(\alpha|\rho|\beta) = \int d\tau_1 d\tau_{1'}\ \psi^*_\alpha(1)\ (1|\rho|1')\ \psi_\beta(1')$$

$$= \sum_{\mu=1}^{A} \delta_{\mu\alpha}\ \delta_{\mu\beta} = \begin{cases} \delta_{\alpha\beta} & : \alpha,\beta \le A \\ 0 & \text{otherwise}\ . \end{cases} \tag{1.21}$$

In the same representation, h (eq. (1.16)) is given by

$$(\alpha|h|\beta) = (\alpha|t|\beta) + (\alpha|U|\beta)$$

$$(\alpha|U|\beta) = \sum_{\mu=1}^{A} (\alpha\mu|V^A|\beta\mu)\ .$$

Hence, we satisfy (1.20) by the condition

$$(\alpha|h|\beta) = \varepsilon_\alpha\ \delta_{\alpha\beta}\ . \tag{1.22}$$

This is the Hartree-Fock equation for the states. In the original representation, this is equivalent to the equation

$$\int d\tau_{1'}\ (1|h|1')\ \psi_\beta(1') = \varepsilon_\beta\ \psi_\beta(1) \tag{1.22a}$$

or, in more detail:

$$-\frac{1}{2m}\nabla^2\ \psi_\beta(1) + \int d\tau_{1'}\ (1|U|1')\ \psi_\beta(1') = \varepsilon_\beta\ \psi_\beta(1) \tag{1.22b}$$

with

$$(1|U|1') = \int d\tau_2 d\tau'_2\ (12|U^A|1'2')\ \sum_\mu\ \psi_\mu(2')\ \psi^*_\mu(2)\ . \tag{1.23}$$

We close this section with a few supplementary results which will be useful later. Let

$$F = \sum_{i=1}^{A} f_i$$

(with matrices $(i|f|i')$) be an arbitrary hermitian one-particle operator. It is then clear that

$$\langle F\rangle = \int d\tau\ (\phi^*, F\phi) = \text{trace}\ (\rho f) \tag{1.24}$$

and it is easily established that

$$\langle [H,F]\rangle = \text{trace}\ (\rho[h,f]) \tag{1.25a}$$

$$= -\ \text{trace}\ (f[h,\rho])\ . \tag{1.25b}$$

This last relation shows that $\delta\rho$, as given by (1.17) corresponds to a variation of ϕ,

$$\delta\phi = iF\phi = i(\sum_i f_i)\phi\ ,$$

and that the condition $\delta\langle H\rangle = 0$, leading to (1.19) corresponds to

$$\langle [H,F]\rangle = 0 \tag{1.26}$$

for <u>any</u> one body operator F, as a comparison with (1.25b) shows.

B. Time Dependent Hartree-Fock Theory

Starting from the time dependent Schroedinger equation

$$i\frac{\partial}{\partial t}\Psi = H\Psi\ , \tag{1.27}$$

one has

$$\frac{d}{dt}\langle F\rangle = \frac{d}{dt}\int d\tau\ (\Psi^*, F\Psi) = i\langle [H,F]\rangle\ . \tag{1.28}$$

A time dependent Slater determinant will <u>not</u> satisfy (1.27), but we can ensure that (1.28) holds for any one-particle operator $F=\Sigma f_i$. In this case, $\langle F\rangle$ and $\langle [H,F]\rangle$ are given by eqs. (1.24) and (1.25), so that (1.28) reads

$$\frac{d}{dt}\ \text{trace}\ (\rho f) = -i\ \text{trace}\ (f[h,\rho])$$

or

$$\text{trace}\ (f\ i\frac{\partial\rho}{\partial t}) = \text{trace}\ (f[h,\rho])$$

for any f. This leads at once to

$$i\frac{\partial\rho}{\partial t} = [h,\rho]\ . \tag{1.29}$$

An alternative to establish (1.29) is to start from (1.28) and construct the time derivative of the exact one-particle density matrix:

$$i \frac{\partial}{\partial t} (1|\rho|1') = \int d\tau_{1''} \left\{ (1|t|1'')(1''|\rho|1') - (1|\rho|1'')(1''|t|1') \right\}$$

$$- \int d\tau_2 \iint d\tau_{1''} d\tau_{2''} \left\{ (12|V|1''2'')(1''2''|\rho|1'2) - (12|\rho|1''2'')(1''2''|V|1'2) \right\}$$

Upon imposing on $(12|\rho|1'2')$ the form (1.9), one reproduces (1.29).

C. Time Dependent H-F Theory of Collective Motion

In this section we describe the simplest version of a theory of collective motion which is possible within the framework of Hartree-Fock theory. Let

$$Q = \sum_{i=1}^{A} q_i \tag{1.30a}$$

be a one-particle operator which describes a collective aspect of a system (e.g. $Q = \Sigma r_i^2/A$, the mean square radius of the matter distribution). Choose a conjugate momentum variable

$$P = \sum_{i=1}^{A} p_i \tag{1.30b}$$

(in the example given for Q, p_i would be $\frac{1}{2}(\vec{r}_i \cdot \vec{p}_i)/r_i^2$ satisfying the commutation relation $i[P,Q] = 1$). Clearly, in general the equations of motion for P and Q:

$$\dot{Q} = i[H,Q] \qquad \dot{P} = i[H,P]$$

will couple these variables to other degrees of freedom of the system. [We shall come back to this point in II,B.] By averaging over these, an effective Hamiltonian for the P-Q motion alone can be imagined. The dynamical approximation embodied in time dependent H-F lends itself particularly well for that purpose.

Consider first the following general argument: Let us add to the Hamiltonian H of the system two "driving terms": $-\lambda Q$, and $-\mu P$, and define

$$\mathcal{G} = H - \lambda Q - \mu P . \tag{1.31}$$

The function of these driving terms is to control the mean values of the collective variables P and Q, and to find the (approximate) lowest state of the system subject to this constraint. In this way, a density matrix $\rho(\lambda,\mu)$ is found, which of course does not satisfy (1.20), but in its place, by applying the variational principle to $\mathcal{G}$, satisfies:

$$[g,\rho] = 0 \tag{1.32}$$

where

$$g = h - \lambda q - \mu p \tag{1.33}$$

and h is given, as before, by (1.15)(1.16). Once this is found, we consider again the system without constraints. For this system, governed by H rather than by G, $\rho(\lambda,\mu)$ represents a non stationary state, in which the collective variables are displaced off their equilibrium values, and will therefore change with time. So, as we remove the constraints, we let λ and μ be time dependent, and use this device as a parametrization for a time-dependent ρ.

Before doing this, consider some general relations evolving from (1.31), (1.32) and (1.33): Assume a solution of (1.32) to be found and let $\langle G \rangle$, $\langle Q \rangle$, $\langle P \rangle$ be the mean values of G, P, Q for the associated $\rho(\lambda,\mu)$. Since ϕ (and ρ) derive from a variational principle, the Feynman-Hellman theorem applies, and one has

$$\frac{\partial\langle G \rangle}{\partial\lambda} = -\langle Q \rangle , \qquad \frac{\partial\langle G \rangle}{\partial\mu} = -\langle P \rangle \tag{1.34}$$

Using the <u>same</u> ρ to determine the mean value $\langle H \rangle$ of H, we have upon a variation of λ and μ:

$$d\langle H \rangle = d(\langle G \rangle + \lambda\langle Q \rangle + \mu\langle P \rangle) = \frac{\partial\langle G \rangle}{\partial\lambda}d\lambda + \frac{\partial\langle G \rangle}{\partial\mu}d\mu$$

$$+ \langle Q \rangle d\lambda + \langle P \rangle d\mu + \lambda d\langle Q \rangle + \mu d\langle P \rangle ,$$

or using (1.34):

$$d\langle H \rangle = \lambda d\langle Q \rangle + \mu d\langle P \rangle . \tag{1.35}$$

Hence, $\langle H \rangle$, viewed as a function of $\langle P \rangle$ and $\langle Q \rangle$ satisfies

$$\frac{\partial\langle H \rangle}{\partial\langle Q \rangle} = \lambda, \qquad \frac{\partial\langle H \rangle}{\partial\langle P \rangle} = \mu . \tag{1.36}$$

(Mean values determined with a ρ that makes $\langle G \rangle$ stationary!) We are now in a position to go back to time dependent Hartree-Fock, for a system governed by the original Hamiltonian H. Let us, for explicitness' sake, write $\rho(\lambda,\mu)$ for the density matrix satisfying (1.32). As mentioned before, we now let these parameters be functions of time: $\rho(\lambda(t), \mu(t))$. This matrix will not satisfy exactly the time-dependent Hartree-Fock equation (1.29), since we have a simple parametrization of the time dependence of ρ; we will, however, not want to satisfy (1.28) for all F's but only for F=P and F=Q. This will determine the two parameters $\lambda(t)$ and $\mu(t)$. Since eq. (1.32) implies that

$$<[\mathcal{G},F]> = 0$$

we have, with $H = \mathcal{G} + \lambda Q + \mu P$:

$$\frac{d<P>}{dt} = i<[H,P]> = i\lambda<[Q,P]> = -\lambda \tag{1.37a}$$

and similarly

$$\frac{d<Q>}{dt} = i<[H,Q]> = i\mu<[P,Q]> = +\mu . \tag{1.37b}$$

Combining these results with (1.36) we have

$$\frac{d<P>}{dt} = -\frac{\partial<H>}{\partial<Q>} \tag{1.38a}$$

$$\frac{d<Q>}{dt} = +\frac{\partial<H>}{\partial<P>} \tag{1.38b}$$

that is, canonical equations of motion for the time dependent mean values $<P>$ and $<Q>$ of the collective variable pair. In addition, the Hartree-Fock equation for the Hamiltonian g supplies the values for $<g>$, $<Q>$ and $<P>$ as functions of λ and μ, from which the function $<H> = \mathcal{H}_{eff}(<P>, <Q>)$ can be constructed. This problem will be discussed in more detail in Section II.

As a final observation, we draw attention to the obvious generalization of this scheme: First, the case of several collective canonical pairs of variables; second, the case where collective variables define a Lie-algebra:

$$i[Q_r,Q_s] = \sum_t C^t_{rs} Q_t . \tag{1.39}$$

In this case, one would define an operator

$$\mathcal{G} = H - \sum_r \lambda_r Q_r .$$

In place of (1.34) and (1.36) one then has

$$\partial<\mathcal{G}>/\partial\lambda_r = -<Q_r> , \quad \partial<H>/\partial<Q_r> = +\lambda_r \tag{1.40}$$

and (1.37) is replaced by

$$\frac{d<Q_r>}{dt} = i<[H,Q_r]>$$
$$= i\sum_s \lambda_s <[Q_s,Q_r]> = \sum_{st} \lambda_s C^t_{sr} <Q_t>$$

and, using (1.40):

$$\frac{d\langle Q_r\rangle}{dt} = \sum_s \frac{\partial\langle H\rangle}{\partial\langle Q_s\rangle} \sum_t C^t_{sr} \langle Q_t\rangle \ . \tag{1.41}$$

Considering the $\langle Q_r\rangle$ as classical variables, and introducing an effective Hamiltonian by

$$\langle H\rangle = \mathcal{H}_{eff}(\langle Q_r\rangle)$$

the relations (1.41) are nothing but the classical Poisson-bracket relations

$$\frac{d\langle Q_r\rangle}{dt} = \left\{ \mathcal{H}_{eff}, \langle Q_r\rangle \right\} \tag{1.42}$$

Of course, the main problem is again the determination of the structure of $\mathcal{H}_{eff}$.

A particular example of the above case is that of the angular momentum operators

$$J_k = \sum_i (\vec{r}_i \times \vec{p}_i + \tfrac{1}{2}\vec{\sigma}_i)_k$$

which, however, will not be discussed in these notes.

D. Hartree-Fock Theory in 2nd Quantization

For the further development of our method, outlined so far, it will be of advantage (although by no means necessary) to introduce the language and notation of 2nd quantization. We will summarize it very briefly:

a) To each single particle orbital ψ_α is associated a creation operator a^+_α, and it adjoint, an annihilation operator a_α, describing the placement of a particle into (or removal of a particle from) a single particle state with quantum numbers **α**, as defined in (A).

b) The operators a^+_α, a_β, obey anticommutation relations

$$a_\alpha a^+_\beta + a^+_\beta a_\alpha \equiv \left\{ a_\alpha, a^+_\beta \right\} = \delta_{\alpha\beta} \tag{1.43}$$

$$\left\{ a_\alpha, a_\beta \right\} = \left\{ a^+_\alpha, a^+_\beta \right\} = 0 \ . \tag{1.44}$$

Equation (1.44) guarantees that the state

$$|\phi_{[\alpha]}\rangle = a^+_{\alpha_1} a^+_{\alpha_2} \ldots a^+_{\alpha_A} |0\rangle \ , \tag{1.45}$$

$|0\rangle$ being the vacuum state, is antisymmetric in the quantum labels, and thus expresses the Pauli-principle (notice that by (1.44), $(a^+_\alpha)^2 = 0$).

c) The vacuum state $|0>$ has the property

$$a_\alpha |0> = 0 \qquad (\text{all } \alpha) \ . \tag{1.46a}$$

The adjoint relation is

$$<0|a_\alpha^+ = 0 \qquad (\text{all } \alpha) \ . \tag{1.46b}$$

d) Each state of type (1.45) is in one-to-one correspondence with a Slater-determinant of single particle wave functions $\psi_{\alpha_1}, \psi_{\alpha_2}, \ldots \psi_{\alpha_A}$.

e) Any operator of type $a_{\alpha 2}^+ a_{\alpha 1}$ is a substitution operator; acting on $|\phi>$, it gives zero if α_1 does not occur in $|\phi>$; if α_1 occurs, it is replaced by α_2. (Upon "ordering" the resulting product of operators a^+, a change in signs may occur.) Now it is clear that any one particle operator $F = \Sigma f_i$ acting on a determinantal wave function can only carry out a single substitution, by virtue of

$$f_i \ \psi_\alpha(i) \equiv \int d\tau_i' (i|f|i') \ \psi_\alpha(i') = \sum_\beta \psi_\beta(i)(\beta|f|\alpha) \ .$$

Hence, the operator F, in the notation of 2nd quantization, will have the form†

$$\hat{F} = \sum_{\alpha\beta} (\alpha|f|\beta) \ a_\alpha^+ a_\beta \ . \tag{1.47a}$$

f) A two-particle operator (such as the potential energy)

$$F = \sum_{i<j} f_{ij}$$

is a double-substitution operator; acting on ϕ, it changes the state of two particles. With this reminder, we give, without further proof, the form of the Hamiltonian operator $\hat{H}$:

$$H = \sum (\alpha|t|\beta) \ a_\alpha^+ a_\beta + \tfrac{1}{4}\sum (\alpha\beta|V^A|\gamma\delta) \ a_\alpha^+ a_\beta^+ a_\delta a_\gamma \ . \tag{1.48}$$

g) To the ground-state determinant $\phi_0(1,2,..A)$ is associated the ground-state vector

$$|\phi_0> = \prod_{\mu=1}^{A} a_\mu^+|0> = a_1^+ a_2^+ \ldots a_A^+|0> \ . \tag{1.49}$$

Again we use labels μ, ν, ... to indicate single particle states occuring in ϕ_0 ("occupied states"), and labels σ, τ, .. to indicate unoccuplied states. By virtue of (1.44), the ground state $|\phi_0>$ has the property

† We shall always use the caret (^) to indicate the abstract operator in 2nd quantization.

$$a^+_\mu|\phi_0\rangle = 0 \qquad \text{all } \mu \leq A \tag{1.50a}$$

$$a_\sigma|\phi_0\rangle = 0 \qquad \text{all } \sigma > A\ . \tag{1.50b}$$

h) Equations (1.50) illustrate the fact that, relative to $|\phi_0\rangle$, the a^+_μ behave the same way as the annihilation operators relative to the true vacuum $|0\rangle$. A change in notation makes use of this property.

We introduce $|\phi_0\rangle$ as a "reference" state, and along with it, annihilation operators c_α, creation operators c^+_α defined by

$$\begin{aligned} c_\mu &= a^+_{\bar\mu} & c_\sigma &= a_\sigma \\ c^+_\mu &= a_{\bar\mu} & c^+_\sigma &= a^+_\sigma \end{aligned} \tag{1.51}$$

$\bar\mu$ being the state time reversed to μ (e.g. if $\mu=n\ell jm$, $\bar\mu=n\ell j,-m$). Then, for all α:

$$c_\alpha|\phi_0\rangle = 0 \tag{1.52}$$

($|\phi_0\rangle$ is the "c-vacuum".) Notice now that, among the operators $a^+_\alpha a_\beta$, one has

$$a^+_\sigma\, a_\tau = c^+_\sigma\, c_\sigma$$

$$a^+_\sigma\, a_\mu = c^+_\sigma\, c^+_{\bar\mu}\ , \quad a^+_\mu\, a_\sigma = c_{\bar\mu}\, c_\sigma$$

$$a^+_\mu\, a_\nu = c_{\bar\mu}\, c^+_{\bar\nu} = \delta_{\mu\nu} - c^+_{\bar\nu}\, c_{\bar\mu}\ . \tag{1.53}$$

This is an example of ordering; ordered operators, of type $c^+..c^+c\ ...c$ have zero expectation value with regards to $|\phi_0\rangle$. Equations (1.53) can be summarized as

$$a^+_\alpha\, a_\beta = \langle a^+_\alpha\, a_\beta\rangle_0 + :a^+_\alpha\, a_\beta:$$

where $\langle\ \rangle_0$ is the $|\phi_0\rangle$-expectation value† of $a^+_\alpha\, a_\beta$ and $:a^+_\alpha\, a_\beta:$ the c-ordered product (think of the a, a^+ written in terms of c, c^+ and anticommute all c^+ to the left of c). The general case of ordering is covered by Wick's theorem, of which we give the example relevant to us:

† Notice that $\langle a^+_\alpha\, a_\beta\rangle_0$ is nothing but the density matrix $(\beta|\rho|\alpha)$ defined in eq. (1.21).

$$a^+_\alpha a^+_\beta a_\delta a_\gamma = :a^+_\alpha a^+_\beta a_\delta a_\gamma:$$

$$+ \quad :a^+_\alpha a_\gamma: \langle a^+_\beta a_\delta\rangle_0 - (\alpha\rightleftarrows\beta) - (\gamma\rightleftarrows\delta) + (\alpha\rightleftarrows\beta,\ \gamma\rightleftarrows\beta)$$

$$+ \quad \langle a^+_\alpha a_\gamma\rangle_0 \langle a^+_\beta a_\delta\rangle_0 - (\gamma\rightleftarrows\delta). \tag{1.54}$$

The Hamiltonian $\hat{H}$ can now be written as

$$\hat{H} = E_0 + \sum (\alpha|h|\beta) : a^+_\alpha a_\beta :$$

$$+ \tfrac{1}{4}\sum (\alpha\beta|V^A|\gamma\delta) \quad : a^+_\alpha a^+_\beta a_\delta a_\gamma : \tag{1.55}$$

with $E_0 = \sum_\mu (\mu|t|\mu) + \tfrac{1}{2}\sum_{\mu\nu} (\mu\nu|V^A|\mu\nu)$

and $(\alpha|h|\beta)$ defined in eq. (1.16). We had seen already in section (A), eq. (1.26), that the condition for a stationary value of E_0 was that $\langle\phi_0|i[\hat{H},\hat{F}]|\phi_0\rangle = 0$ (based on the argument that $|\delta\phi_0\rangle = i\hat{F}|\phi_0\rangle$), $\hat{F}$ being an arbitrary one-particle operator. Since this will be used later extensively, consider the operator $\dot{\hat{F}} = i[\hat{H},\hat{F}]$ in general: $\dot{F}$ is usefully decomposed in the same manner as H in eq. (1.55), giving its mean value, one-particle and two-particle parts:

$$\dot{\hat{F}} = \dot{\hat{F}}^{(0)} + \dot{\hat{F}}^{(1)} + \dot{\hat{F}}^{(2)} \tag{1.56}$$

$$\dot{\hat{F}}^{(0)} = \langle\dot{\hat{F}}\rangle_0 = \sum_\mu i(\mu|[h,f]|\mu)$$

$$= i \text{ trace } (\rho[h,f]) = i \text{ trace } (f[\rho,h]) \tag{1.56a}$$

which simply reiterates an old result, e.g. (1.25). $\dot{\hat{F}}^{(1)}$ may be written, in a basis of Hartree-Fock states α:

$$\dot{\hat{F}}^{(1)} = \sum_{\alpha\beta} (\alpha|\dot{f}|\beta) :a^+_\alpha a_\beta :$$

where

$$- i\ (\alpha|\dot{f}|\beta) = (\varepsilon_\alpha - \varepsilon_\beta)(\alpha|f|\beta)$$

$$+ \sum_{\tau\nu} \left\{ (\nu\alpha|V^A|\tau\beta)(\tau|f|\nu) - (\nu|f|\tau)(\tau\alpha|V^A|\nu\beta)\right\} \tag{1.56b}$$

and finally

$$\dot{\hat{F}}^{(2)} = \tfrac{1}{2}\sum \left\{ \sum_\varepsilon (\alpha\beta|V^A|\varepsilon\delta)(\varepsilon|f|\gamma)\right.$$

$$\left. - \sum_\varepsilon (\alpha|f|\varepsilon)(\varepsilon\beta|V^A|\gamma\delta)\right\} : a^+_\alpha a^+_\beta a_\delta a_\gamma : \quad . \tag{1.56c}$$

Most important for us will be the one-particle part of $\dot{F}$: $\dot{F}^{(1)}$. Consider for instance the mean value of the commutator of $\dot{F}$ with another one-particle operator S:

$$C = \langle\phi_0|[\dot{\hat{F}},S]|\phi_0\rangle \ .$$

In this expression, only $\dot{\hat{F}}^{(1)}$ contributes since $\dot{F}^{(0)}$ is a constant, and $\dot{F}^{(2)}$ generates double substitutions in $|\phi_0\rangle$ which cannot be undone by S. Using (1.52) and (1.53), one has

$$C = \sum_{\sigma\mu} \Big\{ (\mu|\dot{f}|\sigma)(\tau|S|\nu) \langle\phi_0|c_{\bar{\mu}}c_{\sigma}c^+_{\tau}c^+_{\bar{\nu}}|\phi_0\rangle - (\nu|S|\tau)(\sigma|\dot{f}|\mu) \langle\phi_0|c_{\bar{\nu}}c_{\tau}c^+_{\sigma}c^+_{\bar{\mu}}|\phi_0\rangle \Big\}$$

or

$$\langle\phi_0|[\dot{\hat{F}},S]|\phi_0\rangle = \sum_{\sigma\mu} \Big\{ (\mu|\dot{f}|\sigma)(\sigma|S|\mu) - (\mu|S|\sigma)(\sigma|\dot{f}|\mu) \Big\}$$

$$= \sum_{\mu} (\mu|[\dot{f},S]|\mu) = \text{trace}\ (\rho[\dot{f},S]) \ . \qquad (1.57)$$

The matrix elements $(\sigma|\dot{f}|\mu)$ and $(\mu|\dot{f}|\sigma)$ will occur frequently in what follows, and we write them down for reference:

$$\begin{aligned} -i\ (\sigma|\dot{f}|\mu) &= (\varepsilon_\sigma - \varepsilon_\mu)(\sigma|f|\mu) \\ &+ \sum \Big\{ (\sigma\nu|V^A|\mu\tau)(\tau|f|\nu) - (\nu|f|\tau)(\sigma\tau|V^A|\mu\nu) \Big\} \\ +i\ (\mu|\dot{f}|\sigma) &= (\varepsilon_\sigma - \varepsilon_\mu)(\mu|f|\sigma) \\ &+ \sum \Big\{ (\nu|f|\tau)(\mu\tau|V^A|\sigma\nu) - (\mu\nu|V^A|\sigma\tau)(\tau|f|\nu) \Big\} \ . \end{aligned} \qquad (1.58)$$

The structure of these equations emerges more clearly if we consider the matrix elements

$$(\sigma|f|\mu) \equiv f_{\sigma\mu}, \qquad (\mu|f|\sigma) \equiv f^*_{\sigma\mu}$$

as components of a complex vector f; in this case we may write (using $(\sigma|\dot{f}|\mu) \equiv \dot{f}_{\sigma\mu}$):

$$\begin{aligned} -i\ \dot{f}_{\sigma\mu} &= \sum \mathcal{A}_{\sigma\mu;\tau\nu}\ f_{\tau\nu} - \sum \mathcal{B}_{\sigma\mu;\tau\nu}\ f^*_{\tau\nu} \\ +i\ \dot{f}^*_{\sigma\mu} &= \sum \mathcal{A}^*_{\sigma\mu;\tau\nu}\ f^*_{\tau\nu} - \sum \mathcal{B}^*_{\sigma\mu;\tau\nu}\ f_{\tau\nu} \end{aligned} \qquad (1.58a)$$

where

$$\mathcal{A}_{\sigma\mu;\tau\nu} = (\varepsilon_\sigma - \varepsilon_\mu)\ \delta_{\sigma\tau}\ \delta_{\mu\nu} + (\sigma\nu|V^A|\mu\tau) \qquad (1.59a)$$

$$\mathcal{B}_{\sigma\mu;\tau\nu} = (\sigma\tau|V^A|\mu\nu) \ . \qquad (1.59b)$$

As an example of a question involving the matrices $\mathcal{A}$, $\mathcal{B}$, consider the second variation of the mean energy: $\delta^{(2)}\langle H\rangle$. Let F be the infinitesimal generator of a variation of $|\phi\rangle$

$$|\delta\phi\rangle = e^{iF}|\phi\rangle = (1 + i\,\hat{F} - \tfrac{1}{2}\,\hat{F}^2 + \ldots)|\phi\rangle$$

Then

$$\delta^{(2)}\langle H\rangle = \frac{i}{2}\langle\phi|[[i\hat{H},\hat{F}],\hat{F}]|\phi\rangle = \frac{i}{2}\langle\phi|[\dot{\hat{F}}^{(1)},\hat{F}]|\phi\rangle \qquad (1.60)$$

This reduces to

$$\frac{i}{2}\sum_{\mu}(\mu|[\dot{f},f]|\mu) = \tfrac{1}{2}\sum_{\sigma\mu}\left\{f^*_{\sigma\mu}(-i\,\dot{f}_{\sigma\mu}) + f_{\sigma\mu}(i\,\dot{f}^*_{\sigma\mu})\right\}$$

$$= \tfrac{1}{2}f^*(\mathcal{A}f - \mathcal{B}f^*) + \tfrac{1}{2}f(\mathcal{A}^*f^* - \mathcal{B}^*f)$$

$$= \tfrac{1}{2}(f^*,f)\begin{pmatrix}\mathcal{A} & -\mathcal{B}\\ -\mathcal{B}^* & \mathcal{A}^*\end{pmatrix}\begin{pmatrix}f\\ f^*\end{pmatrix} \qquad (1.61)$$

where, in the last line, an obvious notation has been introduced.

The condition that the variational solution satisfying $\delta^{(1)}\langle H\rangle = 0$ be stable, that is, correspond to an actual minimum of the energy, is that $\delta^{(2)}\langle H\rangle$ be positive, or, that the matrix

$$\mathcal{M} = \begin{pmatrix}\mathcal{A} & -\mathcal{B}\\ -\mathcal{B}^* & \mathcal{A}^*\end{pmatrix}$$

be a positive matrix (having only positive eigenvalues). It is clear that the positiveness of the particle-hole single particle excitation energies ($\varepsilon_\sigma - \varepsilon_\mu > 0$) is not a sufficient condition for stability.

We observe that if $|\delta\phi\rangle$ is generated by a constant of the motion, as for example

$$|\delta\phi\rangle = i(\lambda\cdot\hat{P})|\phi\rangle \quad \text{(infinitesimal displacement)}$$

or

$$|\delta\phi\rangle = i(\omega\cdot\hat{J})|\phi\rangle \quad \text{(infinitesimal rotation)},$$

then, due to $[\hat{H},\hat{P}] = [\hat{H},\hat{J}] = 0$, $\delta^{(2)}\langle H\rangle$ is zero, the Hartree-Fock solution is not stable against such variations, and $\mathcal{M}$ is at best semi-positive definite.

2. CONSTRUCTION OF THE COLLECTIVE HAMILTONIAN

In Section 1 most of the preliminary methods and concepts were developed. The main business at hand now is the construction of the effective collective Hamiltonian $H_{eff}(\langle P\rangle,\langle Q\rangle)$ which describes the collective dynamics. In 1C, we had reduced this problem to the one of finding Hartree-Fock solutions to the Hamiltonian $G(\lambda,\mu) = H - \lambda Q - \mu P$. The eventual dynamic role of λ and μ is exhibited in eqs. (1.34) and (1.36). Clearly, our first aim is to determine how the solutions to the Hartree-Fock problem for G depend on λ and μ.

A. The Infinitesimal Generators $\hat{R}$ and $\hat{S}$

The Hartree-Fock states $|\phi(\lambda,\mu)\rangle$ satisfy the variational principle

$$\delta \langle\phi(\lambda\mu)|\,\hat{G}(\lambda,\mu)|\phi(\lambda\mu)\rangle = 0 \tag{2.1}$$

for all values of λ, μ. Let us now define the two infinitesimal generators $\hat{R}$ and $\hat{S}$. which describe the effect on $|\phi\rangle$ of an infinitesimal change of λ or μ:

$$|\phi(\lambda + \delta\lambda,\mu)\rangle = (1 + i\delta\lambda\hat{R})|\phi(\lambda,\mu)\rangle \tag{2.2}$$

$$|\phi(\lambda,\mu + \delta\mu)\rangle = (1 + i\delta\mu\hat{S})|\phi(\lambda,\mu)\rangle \quad . \tag{2.3}$$

The variational principle for $|\phi(\lambda + \delta\lambda,\mu)\rangle$ then takes the form

$$\delta \langle\phi(\lambda,\mu)|(1 - i\delta\lambda\hat{R})\ \hat{G}(\lambda + \delta\lambda,\mu)(1 + i\delta\lambda\hat{R})|\phi(\lambda,\mu)\rangle = 0$$

and by assuming (2.1) to hold, one has, to first order in $\delta\lambda$:

$$\delta \langle\phi(\lambda\mu)|i[\hat{G},\hat{R}] - \hat{Q}|\phi(\lambda\mu)\rangle = 0 \quad . \tag{2.4}$$

An analogous equation evolves from a variation of μ:

$$\delta \langle\phi(\lambda\mu)|i[\hat{G},\hat{S}] - \hat{P}|\phi(\lambda\mu)\rangle = 0 \quad . \tag{2.5}$$

Now $i[\hat{G},\hat{R}] = \dot{\hat{R}}$, $i[\hat{G},\hat{S}] = \dot{\hat{S}}$ (with S as Hamiltonian), and the variation equations (2.4), (2.5) lead again to

$$\langle\phi|[\dot{\hat{R}} - \hat{Q},\hat{F}]|\phi\rangle = 0 \tag{2.4a}$$

any $\hat{F}$

$$\langle\phi|[\dot{\hat{S}} - \hat{P},\hat{F}]|\phi\rangle = 0 \tag{2.5a}$$

or, using again the notation of eqs. (1.56), and using (1.57):

$$(\sigma|\dot{r}|\mu) - (\sigma|q|\mu) = 0 \tag{2.6}$$

$$(\mu|\dot{r}|\sigma) - (\mu|q|\sigma) = 0 \tag{2.7}$$

and a similar equation for $\dot{s}$ and p. Using the equations (1.58), (1.59) for the matrices $\dot{r}$ and $\dot{s}$, along with the vector notation $(\sigma|q|\mu) = q_{\sigma\mu}$, $(\mu|q|\sigma) = q^*_{\sigma\mu}$, defining the vector q, equations (2.6) and (2.7) may be written as

$$\begin{pmatrix} A & -\mathcal{B} \\ -\mathcal{B}^* & A^* \end{pmatrix} \begin{pmatrix} r \\ r^* \end{pmatrix} = \begin{pmatrix} -\ iq \\ +\ iq^* \end{pmatrix} \qquad (2.8)$$

and

$$\begin{pmatrix} A & -\mathcal{B} \\ -\mathcal{B}^* & A^* \end{pmatrix} \begin{pmatrix} s \\ s^* \end{pmatrix} = \begin{pmatrix} -\ ip \\ +\ ip^* \end{pmatrix} . \qquad (2.9)$$

Clearly, here the matrices A and $\mathcal{B}$ derive from $G(\lambda,\mu)$, and are functions of the parameters (λ,μ).

B. The Collective Mass Parameter

In selecting collective variables P and Q, it is understood that Q is a deformation or shape variable, and even under time reversal; by contrast, P will be odd, and we may assume that $\langle\phi|P|\phi\rangle_{\mu=0} = 0$. In this case, for small values of μ, one has, to lowest non-vanishing order:

$$\langle P\rangle = \mu \langle\phi(\lambda,0)|i[P,S]|\phi(\lambda,0)\rangle \qquad (2.10)$$

and

$$\begin{aligned} \langle G\rangle &= \langle\phi(\lambda,0)|\hat{H} - \lambda\hat{Q}|\phi(\lambda,0)\rangle \\ &\quad + \frac{i}{2}\mu^2 \langle\phi|[i[\hat{H} - \lambda\hat{Q},\hat{S}],\hat{S}]|\phi\rangle - i\mu^2 \langle\phi|[\hat{P},\hat{S}]|\phi\rangle \\ &= \langle\hat{G}\rangle_{\mu=0} - \frac{i}{2}\mu^2 \langle\phi(\lambda,0)|[\hat{P},\hat{S}]|\phi(\lambda,0)\rangle . \end{aligned} \qquad (2.11)$$

This shows that the relation (1.34) indeed holds:

$$\frac{\partial\langle G\rangle}{\partial\mu} = -\langle P\rangle .$$

Also, since by (1.37), the parameter μ is identified with $d\langle Q\rangle/dt$, equation (2.10) is the relation

$$\langle P\rangle = M \frac{d\langle Q\rangle}{dt}$$

and defines the collective mass M. This mass is therefore given by the expression

$$M = \langle\phi(\lambda,0)|[i\hat{P},\hat{S}]|\phi(\lambda,0)\rangle . \qquad (2.12)$$

The significance of this expression must now be explored a bit. In the course of this, we will also find alternative ways to write it:

a) M as given in (2.12), is exactly equal to

$$M = - i \sum_{\sigma\mu} (p_{\sigma\mu} s^*_{\sigma\mu} - p^*_{\sigma\mu} s_{\sigma\mu}) \tag{2.12a}$$

Equation (2.9) gives the relation of $p_{\sigma\mu}$ to $s_{\sigma\mu}$. In an approximation, where $\mathcal{A}_{\sigma\mu,\tau\nu} \approx (\varepsilon_\sigma - \varepsilon_\mu)\, \delta_{\sigma\tau}\, \delta_{\mu\nu}$ and $\mathcal{B}_{\sigma\mu,\tau\nu} \approx 0$, one has

$$s_{\sigma\mu} \cong - i\, p_{\sigma\mu}/(\varepsilon_\sigma - \varepsilon_\mu)$$

and

$$M \cong 2 \sum_{\sigma\mu} \frac{|p_{\sigma\mu}|^2}{\varepsilon_\sigma - \varepsilon_\mu} \tag{2.13}$$

In this "cranking approximation", M is always positive since in any event $\varepsilon_\sigma - \varepsilon_\mu > 0$ for the ground state.

b) More exactly, however, we may use eq. (2.5a) with $\hat{F} = \hat{S}$, to show that

$$M = i^2 \langle\phi|\, [[\hat{G},\hat{S}],\hat{S}]\, |\phi\rangle$$

that is, it is proportional to a second variation $\delta^{(2)}\langle G\rangle$, as may be seen from eq. (1.59). Hence M will be positive only if the associated energy $\langle G\rangle$ is stable with respect to a variation initiated by $\hat{S}$.

c) At this point, the question also arises about how P and Q are actually chosen. Clearly, if collective motion can at all be isolated in the way attempted here, they must in some sense represent "normal coordinates" of the system. As such, the method of the RPA equations would appear to be the appropriate means to identify them. However, the approach presented here is aimed at going beyond the situations that can be handled by the RPA (harmonic oscillations about a single equilibrium point); we have in mind a situation such as large amplitude collective phenomena such as occurring in the description of fission. In this case, the potential energy of deformation is described in terms of a few chosen variables Q_i, each with a clear geometric interpretation. In the same spirit, we choose here a variable Q, without at the moment being concerned whether, near its unconstrained equilibrium value Q_0, it is exactly a "normal coordinate" in the sense of the RPA.

The choice of the associated variable P is a separate question. In the developments of Section 1, it has been assumed that both P and Q are one-particle operators. An example for such a Q is the quadrupole moment. Even with the above restriction on P, its choice is ambiguous; as a matter of fact, it can easily be seen, going through Section 1, that all results reported there hold if only

$$i \langle\phi|[P,Q]|\phi\rangle = 1 \tag{2.14}$$

is satisfied. More stringent restrictions of course enter if several canonical pairs are introduced.

d) We can exploit the freedom to choose P, subject to (2.14), by linking it to the operator $\hat{R}$ defined by eq. (2.2). It follows from this equation that

$$\frac{\partial\langle Q\rangle}{\partial\lambda} = \langle\phi|i[\hat{Q},\hat{R}]|\phi\rangle . \tag{2.15}$$

Therefore, a possible choice for $\hat{P}$ is:

$$\hat{P} = -\hat{R}\Big/\left(\frac{\partial\langle Q\rangle}{\partial\lambda}\right) \tag{2.16}$$

which, because of (2.15), satisfies (2.14).† R can be chosen such that, for all λ, $\langle\phi|R|\phi\rangle_{\mu=0} = 0$, so that $\langle\phi|P|\phi\rangle_{\mu=0} = 0$. A special feature of this choice is that

$$\frac{\partial\langle P\rangle}{\partial\lambda} = \langle\phi|i[\hat{P},\hat{R}]|\phi\rangle = 0 \tag{2.17‡}$$

which in turn implies, because of (1.34), that

$$\frac{\partial\langle Q\rangle}{\partial\mu} = \langle\phi|i[Q,S]|\phi\rangle = 0 . \tag{2.18}$$

It thus appears that the mass $M = \partial\langle P\rangle/\partial\mu$ is independent of λ.

e) We can now reproduce yet another, if inaccurate, expression for the mass M, based on the choice (2.16) of the operator $\hat{P}$. The matrix elements $p_{\sigma\mu}$ occurring in the cranking approximation (2.13) are

$$p_{\sigma\mu} = (\sigma|p|\mu) = i\int d\tau\, \psi_\sigma^* \frac{\partial\psi_\mu}{\partial\lambda}\Big/\left(\frac{\partial\langle Q\rangle}{\partial\lambda}\right)$$
$$= i\int d\tau\, \psi_\sigma \frac{\delta\psi_\mu}{\delta\langle Q\rangle} \equiv i\,(\sigma|\frac{\delta}{\delta\langle Q\rangle}|\mu)$$

so that

$$M \cong 2\sum_{\sigma\mu} |(\sigma|\frac{\delta}{\delta\langle Q\rangle}|\mu)|^2 \Big/ (\varepsilon_\sigma - \varepsilon_\mu) . \tag{2.19}$$

† This choice makes P the operator of an infinitesimal displacement of $\langle Q\rangle$.

‡ The right hand side of (2.17) is zero; this only establishes that $\partial\langle P\rangle/\partial\lambda = \langle\partial P/\partial\lambda\rangle$; the conclusion that M is independent of λ is incorrect.

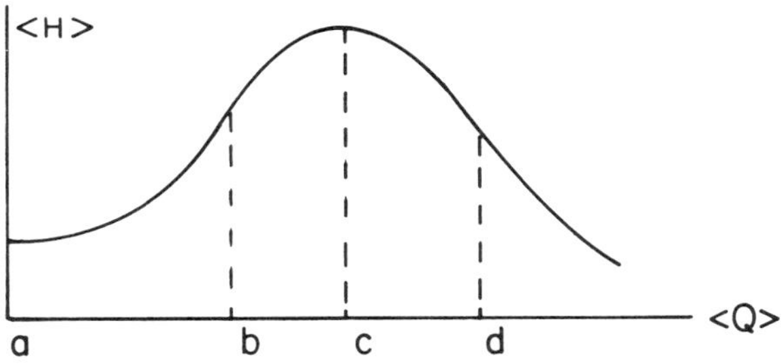

Figure 1
Mean energy $\langle H \rangle$ plotted as a function of mean value $\langle Q \rangle$ of Q.

C. Singular Points

We come here to a question which implicitly is raised already in Section 1C, and explicitly in the last section, 2Bd: Can the relation between (λ,μ) and $\langle P \rangle$, $\langle Q \rangle$ be inverted all the time? The solution of the Hartree-Fock problem for $\mathcal{G}(\lambda,\mu)$ gives us $\langle \mathcal{G} \rangle$, $\langle P \rangle$ and $\langle Q \rangle$ as functions of λ, μ: to go from there to $\langle H \rangle = \mathcal{H}_{eff}(\langle P \rangle, \langle Q \rangle)$ we need $\lambda = \lambda(\langle P \rangle, \langle Q \rangle)$, and $\mu(\langle P \rangle, \langle Q \rangle)$.

To see what problems may arise, let us plot a tentative curve of $\langle H \rangle$ vs $\langle Q \rangle$ (Figure 1). Using the relation (1.36): $\partial\langle H \rangle/\partial\langle Q \rangle=\lambda$, we can also plot a curve of λ vs $\langle Q \rangle$ (Figure 2): This shows that at points b and d, $d\langle Q \rangle/d\lambda = \infty$ and that $\langle Q \rangle$ is a multivalued function of λ. It thus appears that our definition (2.16) of $\hat{P}$ becomes useless as we approach point b. Also, in the interval b-d, we are on a different branch of the relation $\lambda = \lambda(\langle Q \rangle)$; it remains to be seen whether our methods hold in this domain (where $\langle H \rangle$ goes through a maximum), and how to make the connections with the interval a-b at point b. The physics of the feature that at point c in Figure 2, we can have a non-zero value of $\langle Q \rangle$ for zero λ lies in a feature of the two body interaction (having, say a quadrupole-quadrupole term), which at sufficient value of $\langle Q \rangle$ sustains its own deformation without external forcing. We can analyze this point with a model, in which we separate such a term from the $\mathcal{A}$, $\mathcal{B}$ matrices:

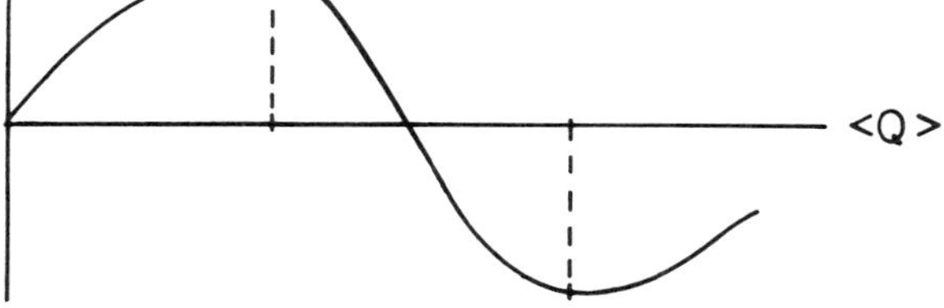

Figure 2
Value of the constraint parameter λ as a function of $\langle Q \rangle$.

$$A_{\sigma\mu,\tau\nu} = A^{0}_{\sigma\mu,\tau\nu} - k_0\, q_{\sigma\mu}\, q^{*}_{\tau\nu}$$

$$\mathcal{B}_{\sigma\mu,\tau\nu} = \mathcal{B}^{0}_{\sigma\mu,\tau\nu} - k_0\, q_{\sigma\mu}\, q_{\tau\nu} \,. \tag{2.20}$$

The equations (2.8), (2.9) for $r_{\sigma\mu}$ and $s_{\sigma\mu}$ can now be written as

$$\begin{pmatrix} A^0 & -\mathcal{B}^0 \\ -\mathcal{B}^*_0 & A^*_0 \end{pmatrix}\begin{pmatrix} r \\ r^* \end{pmatrix} = \left\{ - i + k_0 (q^*\cdot r - q\cdot r^*)\right\} \begin{pmatrix} q \\ -q^* \end{pmatrix} . \tag{2.21}$$

In the corresponding equation for $s_{\sigma\mu}$, the term

$$q^*\cdot s - q\cdot s^* = \langle [\hat{S},\hat{Q}] \rangle$$

drops out because of (2.18). Hence

$$\begin{pmatrix} A^0 & -\mathcal{B}^0 \\ -\mathcal{B}^*_0 & A^*_0 \end{pmatrix}\begin{pmatrix} s \\ s^* \end{pmatrix} = \begin{pmatrix} - i\, p \\ + i\, p^* \end{pmatrix} . \tag{2.22}$$

Assuming that the matrix

$$M^0 = \begin{pmatrix} A^0 & -\mathcal{B}^0 \\ -\mathcal{B}^*_0 & A^*_0 \end{pmatrix}$$

has an inverse, we can easily solve (2.21) for the quantity

$$(q^*\cdot r - q\cdot r^*) = \langle [\hat{Q},\hat{R}] \rangle = - i \,\frac{\partial\langle Q\rangle}{\partial\lambda} : \tag{2.15}$$

$$\frac{\partial\langle Q\rangle}{\partial\lambda} = \frac{C_0}{1 - k_0\, C_0} \tag{2.23}$$

where

$$C_0 = (q^*, -q)\,(M^0)^{-1} \begin{pmatrix} q \\ -q^* \end{pmatrix} . \tag{2.23a}$$

The vector $\begin{pmatrix} r \\ r^* \end{pmatrix}$ itself is given by

$$\begin{pmatrix} r \\ r^* \end{pmatrix} = - i\Big(1 + k_0 \frac{\partial\langle Q\rangle}{\partial\lambda}\Big)(M^0)^{-1} \begin{pmatrix} q \\ -q^* \end{pmatrix} .$$

To get to $\begin{pmatrix} p \\ p^* \end{pmatrix}$, we multiply with $-1/\frac{\partial\langle Q\rangle}{\partial\lambda}$; now

$$\Big(1 + k_0 \frac{\partial\langle Q\rangle}{\partial\lambda}\Big) \frac{\partial\langle Q\rangle}{\partial\lambda} = C_0^{-1} ,$$

so that

$$\begin{pmatrix} p \\ p^* \end{pmatrix} = \frac{i}{C_0} (M^0)^{-1} \begin{pmatrix} q \\ -q^* \end{pmatrix} . \qquad (2.24)$$

We see that $\partial\langle Q\rangle/\partial\lambda$ with its possible infinity has dropped out. Similarly, eq. (2.22) for (s,s^*) also involves only M^0:

$$\begin{pmatrix} s \\ s^* \end{pmatrix} = (M^0)^{-1} \begin{pmatrix} p \\ -p^* \end{pmatrix} \qquad (2.25)$$

and we can now calculate the mass from (2.24) and (2.25) according to (2.12a):

$$M = -i\,(p\cdot s^* - p^*\cdot s) .$$

D. Large Deformations

In this section we discuss the additional measures that have to be taken to carry the previously outlined procedures to the range of deformations (as measured by $\langle Q\rangle$ corresponding to b to d in Figure 1; in this range $\langle H\rangle$ goes through a maximum.

Disregarding for a moment the parameter μ, we dealt so far with the operator

$$\hat{G}(\lambda) = \hat{H} - \lambda\hat{Q}$$

and its self-consistent mean value

$$\mathcal{G}(\lambda) = \langle \hat{G}(\lambda)\rangle_\lambda \equiv \langle\phi(\lambda)|\hat{H} - \lambda\hat{Q}|\phi(\lambda)\rangle . \qquad (2.26)$$

Choosing a fixed value λ_0 of λ, a particular variation of $G(\lambda_0)$ is to consider

$$\langle \hat{G}(\lambda_0)\rangle_{\lambda_0+\delta\lambda} = \langle\phi(\lambda_0+\delta\lambda)|\,\hat{G}(\lambda_0)\,|\phi(\lambda_0+\delta\lambda)\rangle .$$

Now

$$\langle \hat{G}(\lambda_0)\rangle_\lambda = \mathcal{G}(\lambda) + (\lambda - \lambda_0)\,\langle\hat{Q}\rangle_\lambda$$

and hence

$$\frac{\partial}{\partial\lambda}\langle \hat{G}(\lambda_0)\rangle_\lambda = \frac{\partial\mathcal{G}}{\partial\lambda} + \langle Q\rangle_\lambda + (\lambda - \lambda_0)\frac{\partial}{\partial\lambda}\langle Q\rangle_\lambda$$

$$= (\lambda - \lambda_0)\frac{\partial}{\partial\lambda}\langle Q\rangle_\lambda \qquad (2.27)$$

since, by (1.34), $\partial\mathcal{G}/\partial\lambda = -<Q>_\lambda$. Hence, the variation of $<G(\lambda_0)>_\lambda$ with respect to λ (in $\phi(\lambda)$) is zero for $\lambda=\lambda_0$; the second variation gives

$$\left.\frac{\partial^2<\hat{G}(\lambda_0)>_\lambda}{\partial\lambda^2}\right|_{\lambda=\lambda_0} = + \left.\frac{\partial<Q>}{\partial\lambda}\right|_{\lambda=\lambda_0} . \tag{2.28}$$

Looking at Figure 2, this is positive between a and b, but negative between b and d; in this latter domain $<G(\lambda_0)>_\lambda$ has a maximum at $\lambda=\lambda_0$, making the Hartree-Fock iteration procedure unstable.

This situation can be remedied by the simple device of redefining $\mathcal{G}$ in the following manner:

$$\mathcal{G}(\lambda') = <\hat{H} - \lambda'\hat{Q}>_{\lambda'} + \tfrac{1}{2}\gamma_0 <\hat{Q}>^2_{\lambda'} \tag{2.29}$$

where γ_0 is a sufficiently large positive constant. The variational principle for $\mathcal{G}$ ($\delta\mathcal{G} = 0$) now takes the form

$$\delta\mathcal{G} = \text{trace }(\delta\rho(h - \lambda'q + \gamma_0<Q>_{\lambda'}q)) = 0 \tag{2.30}$$

or

$$\text{trace }(\delta\rho(h - \lambda q)) = 0 \tag{2.30'}$$

with

$$\begin{aligned}\lambda &= \lambda' - \gamma_0<Q>_{\lambda'} \\ &= \lambda' - \gamma_0 \text{ trace }(\rho(\lambda')q) .\end{aligned} \tag{2.31}$$

One still has eq. (1.34)

$$\frac{\partial\mathcal{G}}{\partial\lambda'} = - <Q> + \text{trace}\left(\frac{\partial\rho}{\partial\lambda'}(h - \lambda q)\right) = - <Q> \tag{2.32}$$

but eq. (1.36) is changed into

$$\frac{\partial<H>}{\partial<Q>} = + \lambda \quad (\text{not } \lambda') , \tag{2.33}$$

since

$$\begin{aligned}d<H> &= \frac{\partial\mathcal{G}}{\partial\lambda'}\, d\lambda' + d\lambda'<Q> + \lambda'd<Q> - \gamma_0<Q>d<Q> \\ &= (\lambda' - \gamma_0<Q>)\, d<Q> = \lambda d<Q> .\end{aligned}$$

Furthermore, if we define again

$$<G(\lambda_0')>_{\lambda'} = <\hat{H} - \lambda_0'\hat{Q}>_{\lambda'} + \tfrac{1}{2}\gamma_0<Q>^2_{\lambda'}$$

then it follows that

$$\frac{\partial \langle G(\lambda_0')\rangle_{\lambda'}}{\partial \lambda'} = (\lambda' - \lambda_0') \frac{\partial \langle Q\rangle}{\partial \lambda'} \xrightarrow[\lambda' \to \lambda_0']{} 0 \quad . \tag{2.34a}$$

The second derivative is

$$\frac{\partial^2 \langle G(\lambda_0')\rangle_{\lambda'}}{\partial \lambda'^2} = \frac{\partial \langle Q\rangle}{\partial \lambda'} \quad . \tag{2.34b}$$

Now, from (2.33) one has

$$\frac{\partial^2 \langle H\rangle}{\partial \langle Q\rangle^2} = \frac{\partial \lambda}{\partial \langle Q\rangle}$$

which, combined with (2.31) gives

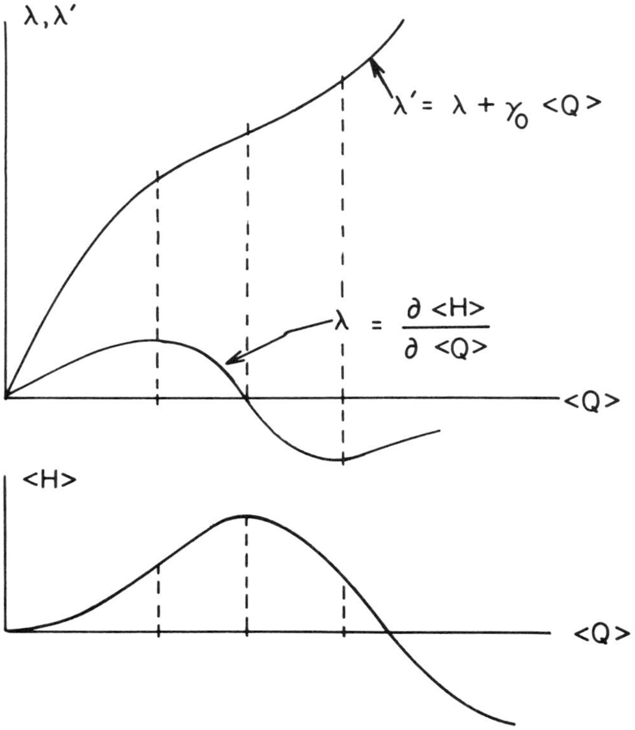

Figure 3
Values of λ, λ' and <H> as functions of the mean value <Q> of Q.

$$\frac{\partial\langle Q\rangle}{\partial\lambda'} = \left\{\gamma_0 + \frac{\partial^2\langle H\rangle}{\partial\langle Q\rangle^2}\right\}^{-1} > 0 \qquad (2.35)$$

provided γ_0 is sufficiently large. Hence, according to (2.34a,b) $\langle G(\lambda_0')\rangle_{\lambda'}$ has an actual minimum at $\lambda'=\lambda_0'$ with respect to the variation $\phi(\lambda_0')\rightarrow\phi(\lambda_0'+\delta\lambda_0')$. The relations between $\langle H\rangle$, λ and λ' are illustrated in Figure 3. It becomes clear that with γ_0 large enough so as to make the right hand side of (2.35) positive, $\langle Q\rangle$ becomes a one-valued function of λ', in contrast to the relation between $\langle Q\rangle$ and λ.

In what manner does this modification of the method affect the mass parameter? Not at all, if we proceed with the following understanding: Equations (2.2) and (2.3) remain valid in the sense that $\hat{R}$ generates now a displacement in λ':

$$|\phi(\lambda'+\delta\lambda',\mu)\rangle = (1 + i\delta\lambda'\hat{R})|\phi(\lambda',\mu)\rangle \qquad (2.3a)$$

$$|\phi(\lambda',\mu + \delta\mu)\rangle = (1 + i\delta\mu\hat{S})|\phi(\lambda',\mu)\rangle \qquad (2.3b)$$

then, the momentum operator $\hat{P}$ is defined as in eq. (2.16);

$$\hat{P} = -\hat{R}\Big/\frac{\partial\langle Q\rangle}{\partial\lambda'} \qquad (2.16a)$$

and the mass M is still given by (2.12).

The effect of the term $\frac{1}{2}\gamma_0\langle Q\rangle^2$ on the A, B matrices is best exemplified in the discussion starting only with eq. (2.20). With the extra term at hand, the only change in eq. (2.20) is to replace the constant k_0 by $k_0-\gamma_0$; in consequence, one has, in place of (2.23):

$$\frac{\partial\langle Q\rangle}{\partial\lambda'} = \frac{C_0}{1 + (\gamma_0-k_0)C_0} \qquad (2.36)$$

and with a proper choice of γ_0, this will be finite for all values of C_0. The final argument of Section C then also makes it clear that the value of the mass parameter is unaffected by the presence of the term $\frac{1}{2}\gamma_0\langle Q\rangle^2$ and is independent of γ_0.

3. BCS THEORY

Nuclear fission of heavy elements is the best known example of a nucleus performing large amplitude collective oscillations. It is well known that a single collective parameter is not sufficient, and two or three variables are needed to describe this motion. This kind of generalization is straightforward in principle, and will not be discussed here. More important, in the context of a general approach to collective dynamics as outlined here, is the fact that, as the collective variable $\langle Q\rangle$ increases, the minimum of $\langle H\rangle$ will be associated with different configurations (that is, different single-particle occupation patterns). See Figure 4. The switchover between configurations, effected by the residual

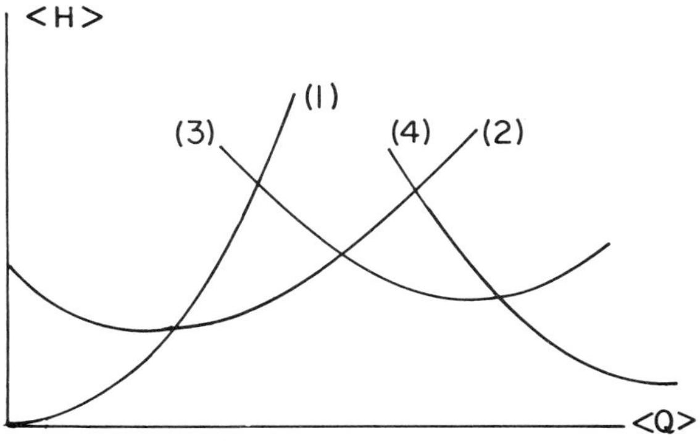

Figure 4a
Energies <H> for four different H-F configurations, as functions of <Q>.

interaction, cannot be easily incorporated in a straight Hartree-Fock scheme. By contrast, the BCS theory, in which the density matrix ρ has the more general form

$$(\alpha|\rho|\beta) = \delta_{\alpha\beta}\, v_\alpha^2$$

with v_α^2 representing the fractional occupation of the single particle state α, describes configurations with a "diffuse" Fermi-surface, and the distinct H.F. configurations of Figure 4a gradually and continuously melt into each other, by a continuous adjustment of the occupation probabilities v_α^2 to the value of <Q>. In this way, a continuous minimum energy surface, labelled "M" in Figure 4b, is generated.

In what follows, the elements of the BCS method will be sketched, and its close relationship to Hartree-Fock emphasized. Particularly relevant for us is its aspect as a variational principle, since so much depends on that particular property of the approximation scheme.

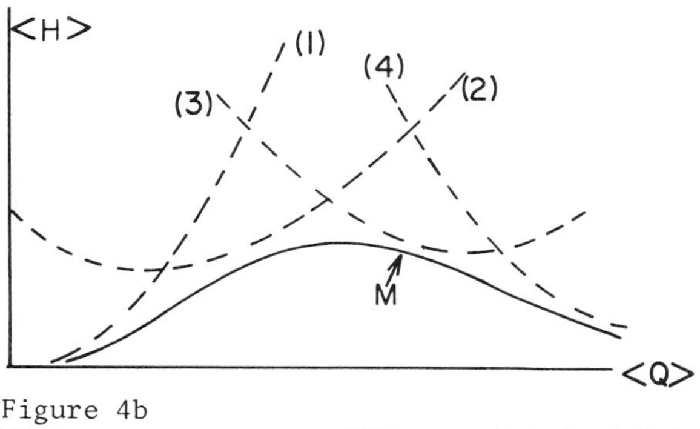

Figure 4b
Minimum energy curve "M" associated with lowest BCS configuration.

We first present BCS theory as an approximation for stationary states, and then introduce time dependent BCS.

A. Stationary BCS

Let ψ_α be a complete set of single particle wave functions, and a_α, a_α^+ the associated annihilation and creation operators. We had already, in (1.51), introduced the operators c and c^+ by

$$a_\mu = c_{\bar{\mu}}^+ \quad (\mu \leq A)$$
$$a_\sigma = c_\sigma \quad (\sigma \geq A) \tag{3.1}$$

giving $c_{\bar{\mu}}^+$ the role of <u>creating</u> a hole in the Fermi sea. A diffuse Fermi surface is introduced by the generalization of (3.1):

$$a_\alpha = u_\alpha c_\alpha + v_\alpha c_{\bar{\alpha}}^+ \tag{3.2}$$

with u_α, v_α being real coefficients, whose α-dependence is sketched in Figure 5.

c_α^+, c_α will be called creation- and annihilation-operators of quasi-particles (quasi particles are holes for $\varepsilon_\alpha \ll \eta$, particles for $\varepsilon_\alpha \gg \eta$, but hybrid for $\varepsilon_\alpha \approx \eta$).

The approximate ground state, $|0\rangle$ which we search for, and which minimizes the energy, is to be the <u>quasi-particle vacuum</u>:

$$c_\alpha |0\rangle = 0 \qquad \langle 0| c_\alpha^+ = 0 . \tag{3.3}$$

For the quasi-particle operators to have the same anticommutation relations as the a_α, a_α^+, we must supplement (3.2) by the statements

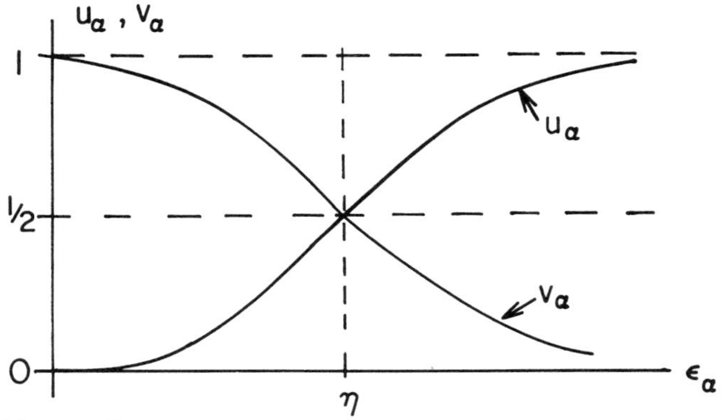

Figure 5
Dependence of the occupation amplitudes u_α, v_α on the single particle energies ε_α. η is the Fermi-energy.

$$a_{\bar\alpha} = u_\alpha c_{\bar\alpha} - v_\alpha c_\alpha^+$$

$$a_{\bar\alpha}^+ = u_\alpha c_{\bar\alpha}^+ - v_\alpha c_\alpha \tag{3.2a}$$

(that is, for any conjugate pair α, $\bar\alpha$ of single particle states, one has $u_{\bar\alpha} = u_\alpha$, $v_{\bar\alpha} = - v_\alpha$, and

$$u_\alpha^2 + v_\alpha^2 = 1 \quad . \tag{3.4}$$

With (3.2) and (3.4) one verifies that the anticommutation relations (1.43), (1.44) lead to the isomorphic relations

$$\left\{ c_\alpha, c_\beta^+ \right\} = \delta_{\alpha\beta} \, , \quad \left\{ c_\alpha, c_\beta \right\} = \left\{ c_\alpha^+, c_\alpha^+ \right\} = 0 \quad . \tag{3.5}$$

The ground state density matrix $\rho_{\alpha\beta}$ is defined as before by (see footnote following (1.53))

$$\rho_{\alpha\beta} = \langle 0 | a_\beta^+ \, a_\alpha | 0 \rangle$$

where $|0\rangle$ satisfies (3.3); thus:

$$\rho_{\alpha\beta} = \langle 0 | (u_\beta c_\beta^+ + v_\beta c_{\bar\beta})(u_\alpha c_\alpha + v_\alpha c_{\bar\alpha}^+) | 0 \rangle = v_\alpha^2 \, \delta_{\alpha\beta} \tag{3.6}$$

[this reduces to ρ of the Hartree-Fock theory if u and v become step-functions: $u_\sigma = v_\mu = 1$; $u_\mu = v_\sigma = 0$]. The choice of (3.3) as the characterization of the ground state means that it is not an eigenstate of the number operator $\hat N$:

$$\hat N = \sum_\alpha a_\alpha^+ a_\alpha = \sum (u_\alpha c_\alpha^+ + v_\alpha c_{\bar\alpha})(u_\alpha c_\alpha + v_\alpha c_{\bar\alpha}^+)$$

$$= \sum v_\alpha^2 + \sum (u_\alpha^2 - v_\alpha^2) \, c_\alpha^+ c_\alpha$$

$$+ \sum u_\alpha v_\alpha (c_\alpha^+ c_{\bar\alpha}^+ + c_{\bar\alpha} c_\alpha) \quad . \tag{3.7}$$

This equation shows that the mean number is

$$\langle \hat N \rangle = \sum_\alpha v_\alpha^2 = \text{trace } \rho \tag{3.8a}$$

and

$$\langle \hat N^2 \rangle - \langle N \rangle^2 = \sum_\alpha (u_\alpha v_\alpha)^2 > 0 \quad . \tag{3.8b}$$

Using eqs. (3.2), (3.2a), the Hamiltonian $\hat H$, as given by (1.48), may be expressed in terms of the quasi-particle operators, and then ordered in c^+, c by using Wick's theorem. In this process, the three following contractions occur:

$$<\phi|a_\beta^+ a_\alpha|\phi> \equiv (\alpha|\rho|\beta) \tag{3.9}$$

$$<\phi|a_\beta a_\alpha|\phi> \equiv (\alpha\beta|\chi)$$

$$<\phi|a_\alpha^+ a_\beta^+|\phi> \equiv (\chi|\alpha\beta) \ . \tag{3.10}$$

For a state $|\phi>$ satisfying eq. (3.3), one has the more specific results

$$(\alpha|\rho|\beta) = v_\alpha^2 \ \delta_{\alpha\beta} \tag{3.9a}$$

$$(\alpha\beta|\chi) = (\chi|\alpha\beta) = u_\alpha \ v_\alpha \ \delta_{\beta,\bar{\alpha}} \ . \tag{3.10a}$$

The mean value of $\hat{H}$ now takes the form

$$\begin{aligned}<\hat{H}> &= \sum (\alpha|t|\beta)(\beta|\rho|\alpha) \\ &+ \tfrac{1}{2}\sum (\alpha\beta|V^A|\gamma\delta) \left\{(\gamma|\rho|\alpha)(\delta|\rho|\beta) + \tfrac{1}{2} (\chi|\alpha\beta)(\gamma\delta|\chi)\right\} \\ &= \sum_\alpha (\alpha|t|\alpha) \ v_\alpha^2 + \tfrac{1}{2}\sum (\alpha\beta|V^A|\alpha\beta) \ v_\alpha^2 \ v_\beta^2 \\ &+ \tfrac{1}{4}\sum (\alpha\bar{\alpha}|V^A|\beta\bar{\beta})(u_\alpha \ v_\alpha)(u_\beta \ v_\beta)\end{aligned} \tag{3.11}$$

and $\hat{H}$ itself can now be expressed as

$$\begin{aligned}\hat{H} &= <\hat{H}> + \sum (\alpha|h|\beta) \ :a_\alpha^+ a_\beta: \\ &+ \tfrac{1}{2}\sum \left\{(\alpha\beta|\Delta) \ :a_\alpha^+ a_\beta^+: + (\Delta|\alpha\beta) \ :a_\beta a_\alpha:\right\} \\ &+ \tfrac{1}{4}\sum (\alpha\beta|V^A|\gamma\delta) \ :a_\alpha^+ a_\beta^+ a_\delta a_\gamma: \ .\end{aligned} \tag{3.12}$$

The one-quasi-particle part of this operator contains the matrix elements

$$(\alpha|h|\beta) = (\alpha|t|\beta) + \sum_\gamma (\alpha\gamma|V^A|\beta\gamma) \ v_\gamma^2 \tag{3.13a}$$

$$(\alpha\beta|\Delta) = \tfrac{1}{2}\sum (\alpha\beta|V^A|\gamma\bar{\gamma}) \ u_\gamma \ v_\gamma \tag{3.13b}$$

$$(\Delta|\alpha\beta) = \tfrac{1}{2}\sum u_\gamma \ v_\gamma \ (\gamma\bar{\gamma}|V^A|\alpha\beta) = (\alpha\beta|\Delta)^* \ .$$

This last equation determines the pairing potential matrix Δ. We shall at this point introduce the commonly used dynamic approximation† that the pairing potential is diagonal and real, in the

† This approximation is not mandatory, but should be made unless at the same time the transformation (3.2)(3.2a) is generalized too.

sense that

$$(\alpha\beta|\Delta) = (\Delta|\alpha\beta) = \delta_{\beta,\bar{\alpha}}\ \Delta_\alpha \tag{3.14}$$

where

$$\Delta_\alpha = \sum_\gamma\ (\alpha\bar{\alpha}|V^A|\gamma\bar{\gamma})(u_\gamma\ v_\gamma)\ .$$

With these preliminaries, let us now look at the variational principle for <H>. By contrast to simple Hartree-Fock, this variation has now two objectives:
a) To determine the self consistent single particle basis of functions ψ_α.
b) To determine the coefficients u_α, v_α.
Consider a) first, and introduce the following variation of the ground state:

$$|0'\rangle = e^{i\sum\ (\alpha|f|\beta)\ a^+_\alpha\ a_\beta}\ |0\rangle\ .$$

The generator of this transformation commutes with the number operator $\hat{N}$, and $|0'\rangle$ has the same mean N as $|0\rangle$. To first order in f, one has

$$\begin{aligned} -\ i\ \delta\langle H\rangle &= \text{trace}\ ([f,\rho]h) + \sum\ \Big\{ (\chi|\alpha\gamma)(\gamma|f|\delta)(\delta\alpha|\Delta) \\ &\qquad - (\Delta|\alpha\gamma)(\gamma|f|\delta)(\delta\alpha|\chi) \Big\} \\ &= \text{trace}\ (f[\rho,h])\ . \end{aligned}$$

The other terms cancel by (3.13b), (3.14). Hence h may be diagonal in a representation, in which ρ is:

$$(\alpha|h|\beta) = (\alpha|t|\beta) + \sum_\gamma\ (\alpha\gamma|V^A|\beta\gamma)\ v^2_\gamma = \varepsilon_\alpha\ \delta_{\alpha\beta}\ . \tag{3.15}$$

This done, we can now proceed to b); variation of u_α, v_α. This variation will affect $\langle\hat{N}\rangle = \Sigma v^2_\alpha$, a number which we want to hold constant. This subsidiary condition is handled by a Lagrange multiplier η: We formulate the condition

$$\delta_{u,v}\ \langle\hat{H} - \eta\ \hat{N}\rangle = 0$$

using (3.11) for $\langle\hat{H}\rangle$, and assuming (3.14) and (3.15) to hold. This gives

$$2\ v_\alpha\ \delta v_\alpha(\varepsilon_\alpha - \eta) + \Delta_\alpha\ (v_\alpha\ \delta u_\alpha + u_\alpha\ \delta v_\alpha) = 0\ .$$

Using (3.4), $u_\alpha\delta u_\alpha + v_\alpha\delta v_\alpha = 0$, one has

$$2\ u_\alpha\ v_\alpha\ (\varepsilon_\alpha - \eta) + \Delta_\alpha\ (u^2_\alpha - v^2_\alpha) = 0\ . \tag{3.16}$$

From (3.16) and (3.4), one has, by simple algebra:

$$2\,u_\alpha v_\alpha = -\frac{\Delta_\alpha}{E_\alpha}\,,\quad u_\alpha^2 - v_\alpha^2 = \frac{\varepsilon_\alpha - \eta}{E_\alpha}\,, \tag{3.17a}$$

where

$$E_\alpha = +\sqrt{(\varepsilon_\alpha - \eta)^2 + \Delta_\alpha^2}\,. \tag{3.17b}$$

The occupation probability v_α^2 is thus given by

$$v_\alpha^2 = \tfrac{1}{2}(1 - (\varepsilon_\alpha - \eta)/E_\alpha)\,, \tag{3.18}$$

showing that the Fermi-level η marks the energy, at which v_α^2 has dropped to half its peak value of 1. (See also Figure 5.) It is determined numerically by prescribing the value of $\langle\hat{N}\rangle = \Sigma v_\alpha^2$.

Let us now turn our attention to the one-particle part of H, the middle terms in (3.12). With (3.14) and (3.15) it follows that

$$(\hat{H} - \eta\,\hat{N})^{(1)} = \sum (\varepsilon_\alpha - \eta)\ :a_\alpha^+ a_\alpha:$$
$$+ \tfrac{1}{2}\sum \left\{ \Delta_\alpha\ :a_\alpha^+ a_{\bar{\alpha}}^+: + \Delta_\alpha\ :a_{\bar{\alpha}} a_\alpha: \right\}\,.$$

Writing these ordered operators in terms of the quasi-particle operators c^+, c one has

$$(\hat{H} - \eta\hat{N})^{(1)} = \sum \left\{ (\varepsilon_\alpha - \eta)(u_\alpha^2 - v_\alpha^2) - 2\,\Delta_\alpha u_\alpha v_\alpha \right\}\ c_\alpha^+ c_\alpha$$
$$+ \tfrac{1}{2}\sum \left\{ (\varepsilon_\alpha - \eta)\ 2\,u_\alpha v_\alpha + \Delta_\alpha\ (u_\alpha^2 - v_\alpha^2) \right\}\ (c_\alpha^+ c_{\bar{\alpha}}^+ + c_{\bar{\alpha}} c_\alpha)\,.$$

Application of eqs. (3.17a) now reveals that the coefficient of the diagonal part $c_\alpha^+ c_\alpha$ reduces to E_α, and the coefficient of the non-diagonal part $(c^+ c^+ + c\ c)$ to zero:

$$(\hat{H} - \eta\,\hat{H})^{(1)} = \sum_\alpha E_\alpha\, c_\alpha^+ c_\alpha\,. \tag{3.19}$$

In ordinary Hartree-Fock theory, the analogous result for the one-particle part of H is contained in eq. (1.55):

$$\hat{H}^{(1)} = \sum_\alpha \varepsilon_\alpha\ :a_\alpha^+ a_\alpha:$$
$$= \sum_{\sigma>A} \varepsilon_\sigma\, c_\sigma^+ c_\sigma - \sum_{\mu<A} \varepsilon_\mu\, c_\mu^+ c_\mu\,.$$

The diagonality of the one particle part of $\hat{H}$, (or $(\hat{H} - \eta\,\hat{N})$ respectively), that is, the absence of terms of type $c^+ c^+$, c c, in both Hartree-Fock and BCS, finds its general expression in the property

$$<\phi_0|[\hat{H},\hat{F}]|\phi_0> = 0 \qquad \text{(H.F.)} \qquad (3.20a)$$

$$<0|[\hat{H} - \eta\,\hat{N},\hat{F}]|0> = 0 \qquad \text{(BCS)} \qquad (3.20b)$$

for any one particle operator $\hat{F}$. In this context it may be worth pointing out that the concept of a one-particle operator in BCS is somewhat more general than it is in H.F. In this latter context, it is defined as

$$F = \sum_{i=1}^{A} f_i \quad \text{or equivalently} \quad \hat{F} = \sum (\alpha|f|\beta)\, a^+_\alpha\, a_\beta \, .$$

Written in terms of quasi-particle operators, this takes the form

$$\hat{F} = \sum (\alpha|f|\alpha)\, v^2_\alpha + \sum \left\{u_\alpha\, u_\beta\, (\alpha|f|\beta) - v_\alpha\, v_\beta\, (\bar{\beta}|f|\bar{\alpha})\right\} c^+_\alpha\, c_\beta + \sum u_\alpha\, v_\beta\, (\alpha|f|\beta)\, c^+_\alpha\, c^+_{\bar{\beta}} + (\beta|f|\alpha)\, u_\alpha\, v_\beta\, c_{\bar{\beta}}\, c_\alpha \, . \qquad (3.21)$$

This operator commutes with the number operator N. By analogy, we may define a general 1-particle operator F, in BCS, directly through its matrix elements in the quasi-particle representation:

$$\hat{F} = F_0 + \sum ([\alpha|f|\beta]\, c^+_\alpha\, c_\beta + \sum f_{\alpha\beta}\, c^+_\alpha\, c^+_\beta + f^*_{\alpha\beta}\, c_\beta\, c_\alpha), \qquad (3.22)$$

F_0 being a constant. The complex amplitudes $f_{\alpha\beta}$, $f^*_{\alpha\beta}$ are the BCS analogs of the vector components $f_{\sigma\mu}$, $f^*_{\sigma\mu}$ defined in eq. (1.57). Such an operator as defined by eq. (3.22) will not in general commute with the number operator N; yet it is natural to define for instance the BCS analog of the displacement operators $\hat{R}$ and $\hat{S}$, as given by (2.2) and (2.3), in a quasi-particle representation only; in fact, to fulfil their role, they may be chosen to be of the form

$$\hat{R} = \sum_{\alpha\beta} (r_{\alpha\beta}\, c^+_\alpha\, c^+_\beta + r^*_{\alpha\beta}\, c_\beta\, c_\alpha) \qquad (3.23)$$

and a similar form for $\hat{S}$.

We have not yet given the complete form of $\hat{H}$, eq. (3.12), in terms of the quasi-particle operators c^+, c; one has

$$\hat{H} - \eta\,\hat{N} = <\hat{H} - \eta\,\hat{N}> + \sum E_\alpha\, c^+_\alpha\, c_\alpha + \tfrac{1}{4}\sum [\alpha\beta|V|\gamma\delta]\, c^+_\alpha\, c^+_\beta\, c_\delta\, c_\gamma + \frac{1}{12}\sum W^*_{\alpha\beta\gamma\delta}\, c^+_\alpha\, c^+_\beta\, c^+_\gamma\, c^+_\delta + \frac{1}{12}\sum W_{\alpha\beta\gamma\delta}\, c_\delta\, c_\gamma\, c_\beta\, c_\alpha \, , \qquad (3.24)$$

where $[\alpha\beta|V|\gamma\delta]$ and $W_{\alpha\beta\gamma\delta}$ contain matrix elements of V and u and v coefficients and where we have omitted terms in c^+ccc and

with $\mathring{r}$ etc., given in terms of r by eqs. (3.26), relations which generalize eqs. (1.58).

The operators $\hat{R}$ and $\hat{S}$ so defined are of the form (3.23), and will not commute with N; in fact $<[R,N]>=0$ for instance requires that

$$\sum_\gamma u_\gamma v_\gamma (r_{\gamma\bar{\gamma}} - r^*_{\gamma\bar{\gamma}}) = 0 . \qquad (3.35)$$

But the $r_{\gamma\bar{\gamma}}$ following from (3.33) are not real and there is no reason for (3.35) to hold. We must therefore generalize the definition of $\hat{P}$ as given in (2.16). The proper choice is

$$\hat{P} = c_1 \hat{R} + c_2 \hat{Z}$$

with c_1, c_2 determined by

$$i <[\hat{P},\hat{Q}]> = c_1 <i[\hat{R},\hat{Q}]> + c_2 <i[\hat{Z},\hat{Q}]>$$

$$= - c_1 \frac{\partial<Q>}{\partial\lambda} - c_2 \frac{\partial<Q>}{\partial\eta} = 1 \qquad (3.36a)$$

$$i <[\hat{P},\hat{N}]> = - c_1 \frac{\partial<N>}{\partial\lambda} - c_2 \frac{\partial<N>}{\partial\eta} = 0 . \qquad (3.36b)$$

This gives

$$c_2 = - c_1 \frac{\dfrac{\partial<N>}{\partial\lambda}}{\dfrac{\partial<N>}{\partial\eta}} = + c_1 \left(\frac{\delta\eta}{\delta\lambda}\right)_{<N>}$$

so that

$$\hat{P} = - \left(\hat{R} + \left(\frac{\delta\eta}{\delta\lambda}\right)_{<N>} \hat{Z}\right) \Big/ \left(\frac{\partial<Q>}{\partial\lambda} + \left(\frac{\delta\eta}{\delta\lambda}\right)_{<N>} \frac{\partial<Q>}{\partial\eta}\right) . \qquad (3.37)$$

For Q we need not have this problem; we may choose it to be of the form $Q = \Sigma q_i$, so that $\hat{Q}$ is of the form (3.22). In the notation defined in (3.22), one has then

$$q_{\alpha\beta} = - q_{\beta\alpha} = \tfrac{1}{2} \left\{ v_\alpha u_\beta (\beta|q|\bar{\alpha}) - u_\alpha v_\beta (\alpha|q|\bar{\beta}) \right\} . \qquad (3.38)$$

With this, one can procced to solve eqs. (3.34).

The determination of the collective mass M, defined by

$$<\hat{P}>_{\lambda,\mu} = \mu <i[\hat{P},\hat{S}]>_{\substack{\lambda \\ \mu=0}} = \mu M$$

as in (2.12), is now straightforward, the generalization of (3.12a) is now

$$M = -i \sum_{\alpha<\beta} (p_{\alpha\beta} s^{*}_{\alpha\beta} - p^{*}_{\alpha\beta} s_{\alpha\beta}) \quad . \tag{3.39}$$

Again, the equivalent of the cranking approximation (2.13) is possible:

$$\mathcal{A}_{\alpha\beta;\gamma\delta} \cong (E_\alpha + E_\beta)\, \delta_{\alpha\gamma}\, \delta_{\beta\delta} \quad , \quad \mathcal{B}_{\alpha\beta,\gamma\delta} \cong 0$$

in which case

$$-i \overset{\circ}{s}_{\alpha\beta} \cong (E_\alpha + E_\beta)\, s_{\alpha\beta} = -i\, p_{\alpha\beta}$$

and

$$M \cong 2 \sum_{\alpha<\beta} \frac{|p_{\alpha\beta}|^2}{E_\alpha + E_\beta} \quad . \tag{3.40}$$

This concludes the discussion of the BCS approximation.

REFERENCES

For general information on Hartree-Fock theory and microscopic theories of collective motion, the following references may be found useful.

Baranger, M. "Theory of Finite Nuclei" in "1962 Cargese Lectures in Theoretical Physics", ed. M. Levi, W.A. Benjamin, Inc., N.Y. (1963).

Rowe, D.J. "Nuclear Collective Motion; Models and Theory", Methuen and Co., Ltd. London (1970).

Belyaev, S.T. "Collective Excitations in Nuclei", Gordon and Breach, Inc., N.Y. (1968).

Baumgaertner, G. and Schuck, P. "Kernmodelle B·I. Hochschultaschenbücher", Vol. 203/203a*, Bibliographishes Institut, Mannheim/ Zurich (1968).

Lane, A.M. "Nuclear Theory", W.A. Benjamin, Inc., N.Y. (1964).

Brown, G.E. "Unified Theory of Nuclear Models", North Holland Publishing Co., Amsterdam (1964).

THEORIES OF COLLECTIVE MOTION

ABRAHAM KLEIN
Department of Physics
University of Pennsylvania

During the past decade, numerous methods, both phenomenological and microscopic, of describing collective phenomena in nuclear physics have been put forward. An effort is made in these lectures to establish such relationships as exist among the various proposals, including in many instances proofs of identity and approximate equality. This is done in part I of the work. In order to cover so much ground conceptually, we have simplified formulas by inventing a monopole vibration (breathing mode) for the nucleus which, for the most part, we suppose to be uncoupled from other degrees of freedom. In part II, we study a model system which exhibits such a monopole vibration in order to illustrate some of the microscopic approaches.

Several features of these notes should be stressed in advance. Though we believe that we have gone quite far in producing a synthesis of existing ideas, certain bases have been left completely untouched. As an example we have not discussed any method based on the explicit definition of collective variables as functions of the coordinates and momenta of individual nucleons. It is also likely that every reader will find some of the discussion too terse for his liking. Finally, in addition to a review of the old (if not always well-known), we have used this occasion to put forward some new ideas which will be developed more fully in future publications.

I. GENERAL METHODS IN THE COLLECTIVE DESCRIPTION OF A MONOPOLE DEGREE OF FREEDOM

A. Conditions for Collective Motion

Let us imagine (since it doesn't occur) that in some regions of the periodic table even-even nuclei exhibit a spectrum of relatively low-lying 0^+ states which we label $n=0,1,2,3,\ldots$, with $n=0$ the ground state of the system, and that these "bands" have the following properties: (i) The spacing is nearly harmonic. Thus the excitation energy, E_n, is represented by the formula

$$E_n = a_1 n + a_2 n(n-1) + a_3 n(n-1)(n-2) + \ldots, \qquad (1.1)$$

where

$$a_1 >> a_2 >> a_3 >> \ldots \qquad (1.2)$$

(ii) Enhanced monopole transitions are observed to take place between neighbouring states in the band and to depend on n in a characteristic way. Thus if M is the monopole operator (e.g.

$$M = \sum_i (r^{(i)})^2 \rightarrow \hat{M} = \int \psi^\dagger(x) r^2 \psi(x) dx \ , \qquad (1.3)$$

where the caret indicates a second quantized operator and $\psi^\dagger(x)$ creates a nucleon at $x=(r,\sigma=\pm 1,\tau=p \text{ or } n)$), we have

$$<n+1|\hat{M}|n> = M_o\sqrt{n+1}\ [1+q_1 n+q_2 n(n-1)+\ldots] \ , \qquad (1.4)$$

where

$$1 >> q_1 >> q_2 >> \ldots \qquad (1.5)$$

Of course, this is approximately the transition rule obeyed by the coordinate of a simple harmonic oscillator. (iii) Crossover intraband transitions are suppressed compared to the direct transitions described in (ii), i.e.,

$$|<n+1|\hat{M}|n>| >> |<n+2|\hat{M}|n>| >> \ldots \qquad (1.6)$$

(iv) The change of the average mean-square radius in going from state to state in the band is small:

$$\left| \frac{<n+1|\hat{M}|n+1> - <n|\hat{M}|n>}{<n|\hat{M}|n>} \right| << 1 \ . \qquad (1.7)$$

(v) Finally, interband transitions are not much enhanced if at all over single-particle values.

It should be clear that we have sought to delineate the conditions for the nucleus to sustain a well-developed breathing mode of oscillation. In the sections which follow immediately we shall try to describe a number of useful phenomenologies to represent this situation. In some instances these suggest many-body approaches, but the latter will be taken up separately in succeeding sections. Some methods will be described in general terms in the later sections of this part, but some will only be described directly for the simple model which underlies all the illustrations of part II.

B. Number Diagonal ("Physical Boson") Representation

The most obvious representation of the "data" given above is in terms of a one-dimensional nearly harmonic oscillator. In this connection the elementary fact of representation up to unitary equivalence is often overlooked, or at least not sufficiently stressed. The interest of a representation in which the number operator is diagonal has only recently been appreciated.[1,2]

For our case, we introduce boson creation and annihilation operators $B^\dagger, B$,

$$[B, B^\dagger] = 1 \quad , \tag{1.8}$$

and represent the exact eigenstates of the nucleus as uncoupled harmonic oscillator eigenstates,

$$|n\rangle \to |n\rangle_B = (n!)^{-\frac{1}{2}} (B^\dagger)^n \; |0\rangle_B \tag{1.9}$$

$$B|0\rangle_B = 0 \; . \tag{1.10}$$

Equation (1.9) should be understood as an imaging or projection, but we shall henceforth drop the subscript B on the image state. Since the states (1.9) are eigenstates of $B^\dagger B$ with eigenvalue n, if we introduce a collective phenomenological Hamiltonian, H,

$$H = a_1 \; B^\dagger B + a_2 (B^\dagger)^2 B^2 + a_3 (B^\dagger)^3 B^3 + \ldots \; , \tag{1.11}$$

it is easily seen that

$$H|n\rangle = E_n |n\rangle \; , \tag{1.12}$$

with E_n given by (1.1). Thus we are dealing with a representation in which the collective Hamiltonian is a power series in the number operator: Near harmonicity means rapid convergence of this series.

In the same representation, the data (1.4) is reproduced by the operator

$$\hat{M} = \hat{M}_0 B^\dagger [1 + q_1 B^\dagger B + q_2 (B^\dagger)^2 B^2 + \ldots]$$
$$+ \text{h.c.} + q_{11} B^\dagger B + \ldots + q_{20}[(B^\dagger)^2 + B^2] + \ldots \quad , \tag{1.13}$$

where h.c. means Hermitian conjugate, and the additional parameters q_{11}, ... are included in order to describe a possible change in mean square radius with excitation, and the additional parameters q_{20}, ... to represent cross-over transitions. In contrast to the Hamiltonian (1.11) which contains the minimum number of parameters possible for any phenomenological choice, namely, one for each excitation energy, the transition operator is as complex as it can be in terms of number of parameters, since in principle it must contain a parameter for every pair of states.

The number-diagonal representation does not suggest directly any microscopic approach to the description of collective motion. The boson variables which can be immediately useful in such a microscopic program can be variously chosen, but these do not in general render the Hamiltonian diagonal and a subsequent diagonalization must be performed. There is a method in part IIB — the algebraic approach — which makes no direct use of boson variables, which can however be interpreted as the direct construction of the physical boson representation.

Aside from its convenience as a representation of the solution of the problem of nearly-harmonic vibrations, emphasis on the power series (1.1) suggests that if we move toward a region of nuclei which exhibit anharmonicities large enough so that series such as (1.1) cease to be useful, we should still expect that any improved formula for the excitation energy should represent an analytic continuation of this series, i.e., there should be a range of values of the parameters and of the variable n for which it can be expanded in a series of the form (1.1).

C. Transition Operator Representation

In our definition of collective motion we have emphasized two operators, the Hamiltonian and a characteristic transition operator. Our first choice of boson emphasized simplicity in the representation of the former. Another "extreme" representation suggests itself, namely one in which the transition operator has the simplest possible representation. We define boson operators $\beta^\dagger,\beta$ by requiring

$$\hat{M} \to M_\beta = M_o(\beta^\dagger+\beta) \quad . \tag{1.14}$$

In the basis

$$|n;\beta\rangle = (n!)^{-\frac{1}{2}} (\beta^\dagger)^n \; |0;\beta\rangle \tag{1.15}$$

the collective Hamiltonian, H, will no longer be diagonal, but will be a general polynomial in β and $\beta^\dagger$. The eigenstates

$$|n\rangle = \sum_p C_{np} \; |p;\beta\rangle \tag{1.16}$$

in this representation can, in principle, be determined from sufficient data on the matrix elements of $\hat{M}$ together with the assumed form, (1.14). Given the state vectors, the Hamiltonian, if desired could be constructed. These points can be verified by the reader.

This representation has perhaps equal claim to the appelation "physical boson representation" as does that of the previous section. Though in principle equivalent to the previous representation, it appears to us more cumbersome as a phenomenology (for vibrations!). It does have the distinct advantage of being more closely tied to a possibly useful version of the microscopic theory, but even here, as we shall see by example, the most useful choices are not suggested by the phenomenology. Indeed, for the purposes of phenomenology no further boson representations commend themselves particularly, but we shall see that the microscopic theory is quite rich in alternatives.

D. Concept of Ideal Collective Coordinate and Formal Hill-Wheeler Projection

In this section we shall move a little closer to making a connection between the phenomenological manifestation of a collective degree of freedom and general properties of the dynamics. Here we admit that we are looking for a representation of (1.12) in the form

$$\mathcal{H}(X,P)u_n(X) = E_n u_n(X) \ , \qquad (1.17)$$

and that as far as our microscopic ambitions are concerned two cases are of paramount interest: Either $\mathcal{H}$ is slightly anharmonic or we can make an adiabatic approximation, i.e., the kinetic energy is small compared to the potential energy of "deformation". In what follows we shall gain some insight into both situations.

We start by idealizing the physical situation, assuming the spectrum of states $|n\rangle$ to be limitless. This permits us to define an ideal many-body intrinsic state $|X\rangle$,[3,4]

$$|X\rangle = \sum_n |n\rangle\, u_n^*(X) \ , \qquad (1.18)$$

$$|n\rangle = \int dX\, u_n(X)\,|X\rangle \ , \qquad (1.19)$$

which establishes a correspondence between the many-body states $|n\rangle$ and the one-dimensional eigenfunctions $u_n(X)$,

$$u_n^*(X) = \langle n|X\rangle \ . \qquad (1.20)$$

The states $|X\rangle$ are sharply localized,

$$\langle X|X'\rangle = \delta(X-X') \ . \qquad (1.21)$$

The definition (1.18) is purely formal. Dynamics enters with the assumption that $\hat{H}$ and $\hat{\mathcal{M}}$ are collective in the sense that the overlaps

$$\langle X|\hat{\Theta}|X'\rangle \equiv \Theta(X|X') \ , \qquad (1.22)$$

where $\hat{\Theta}$ is any of the operators of interest, remain strongly peaked in the variable $X-X'$. Let us write

$$X = Y + \tfrac{1}{2}\xi \ ,$$

$$X' = Y - \tfrac{1}{2}\xi \ , \qquad (1.23)$$

$$\Theta(X|X') = \tilde{\Theta}(Y,\xi) \ , \qquad (1.24)$$

and $\tilde{\Theta}$ is assumed to be strongly peaked in the variable ξ.

Such a peaked function can be expanded in terms of the δ function and its derivatives (dropping the tilde)

$$\Theta(Y,\xi) = \sum_{m} \Theta^{(m)}(Y)(p_\xi)^m \delta(\xi) \ , \tag{1.25}$$

where $p_\xi = -i\partial_\xi$

$$\Theta^{(m)}(Y) = \int d\xi \frac{(-i\xi)^m}{m!} \ \Theta(Y,\xi) \ . \tag{1.26}$$

Equations (1.25) and (1.26) express the basic content of the idea of collectivity in the ideal limit. This limit is implicitly assumed in almost all treatments of collective motion.

Thus by applying these formulas to the Hamiltonian, the following simple computation yields an expression for the collective Hamiltonian. We have, using Eq. (1.19), (W_o is the ground state energy)

$$\begin{aligned}\langle n|H|n'\rangle &= (W_o + E_n)\delta_{nn'} \\ &= \int dXdX' \ u_n^*(X) \ H(X,X') \ u_{n'}(X') \\ &= \int dY \ d\xi \ u_n^*(Y+\tfrac{1}{2}\xi) \ H(Y,\xi) \ u_{n'}(Y-\tfrac{1}{2}\xi) \\ &= \int dX \ u_n^*(X)[\mathcal{H}(X,P)+W_o]u_{n'}(X) \ ,\end{aligned} \tag{1.27}$$

where it follows from (1.25) by integration by parts that

$$\begin{aligned}\mathcal{H}(X,P) + W_o &= H^{(0)}(X) + \tfrac{1}{2}\{P,H^{(1)}(X)\} \\ &+ (\tfrac{1}{2})^2\{P,\{P,H^{(2)}(X)\}\} + \ldots\end{aligned} \tag{1.28}$$

We may interpret the result as follows. We have inserted into the "computer" the assumption that we can construct a sharply localized (strongly deformed) intrinsic state that hangs together in time (the action of H does not disperse it), and we have concluded that the collective Hamiltonian must possess a convergent expansion in the collective kinetic energy, or equivalently the effective inertial parameter associated with the collective motion must be large (see below).

E. Near Harmonic and Adiabatic Limits for Ideal Collective Coordinates

To carry this development further, we consider the two interesting limiting cases. The near harmonic limit is obtained by assuming that the various moment functions (1.26) are slowly varying functions of the collective coordinate, and can therefore be expanded in Taylor series in the latter:

$$H^{(m)}(X) = \sum_{r=0}^{\infty} H^{(mr)} \ X^r/r! \tag{1.29}$$

From (1.28) and (1.29), we have

$$\begin{aligned} \mathcal{H}(X,P) + W_o &= H^{(00)} + H^{(01)}X + H^{(10)}P \\ &+ [H^{(20)}P^2 + \frac{1}{2} H^{(11)} \{P,X\} + \frac{1}{2} H^{(02)}X^2] \\ &+ [H^{(30)}P^3 + \frac{1}{4} H^{(21)} \{P,\{P,X\}\} + \frac{1}{4} H^{(12)} \{P,X^2\} \\ &+ \frac{1}{3!} H^{(03)}X^3] + \dots \end{aligned} \tag{1.30}$$

This expression can be simplified. We may set $H^{(01)} = 0$, implying that $X = 0$ is a position of equilibrium. We further assume X and P can be chosen so that X is even, P odd under time reversal. Since the $H^{(m)}(X)$ are real functions, it follows that they must vanish for m odd. We thus obtain to fourth order

$$\begin{aligned} \mathcal{H}(X,P) + W_o &= H^{(00)} + [H^{(20)}P^2 + \frac{1}{2} H^{(02)}X^2] \\ &+ [\frac{1}{4} H^{(21)} \{P,\{P,X\}\} + \frac{1}{3!} H^{(03)}X^3] \\ &+ [H^{(40)}P^4 + \frac{1}{4} \cdot \frac{1}{2} H^{(22)} \{P,\{P,X^2\}\} + \frac{1}{4!} H^{(04)}X^4] + \dots \end{aligned} \tag{1.31}$$

To lowest order we must have

$$H^{(00)} = W_o \tag{1.32}$$

$$H^{(20)} = (2B)^{-1} \tag{1.33}$$

$$H^{(02)} = C \tag{1.34}$$

where B is the inertial parameter of the harmonic motion and C the stiffness parameter.

We must now consider conditions for the convergence of the series (1.31). We take the basic excitation energy to be of order unity,

$$\omega \equiv \sqrt{C/B} = O(1) \ , \tag{1.35}$$

which can be taken as definition of the energy scale. We assume that $B = O(N)$, where N is some measure of the number of particles taking part in the collective motion. Thus from (1.35),

$$C \sim B \sim O(N) \ . \tag{1.36}$$

Since for the harmonic oscillator

$$\omega \sim \frac{P^2}{2B} \sim \frac{1}{2} CX^2 \sim O(1) \ , \tag{1.37}$$

it follows that

$$P = O(N^{\frac{1}{2}}), \ X = O(N^{-\frac{1}{2}}) \ . \tag{1.38}$$

Examination of (1.32) - (1.34), remembering that $W_0 = H^{(00)} = O(N)$, suggests that

$$H^{(mr)} = O(N^{-m+1}) \tag{1.39}$$

and is independent of r.

If these assumptions are valid, examination of (1.31) establishes that $\mathcal{H}(X,P)$, the collective Hamiltonian is formally a power series in $N^{-\frac{1}{2}}$. It can be shown, however, that if one computes in perturbation theory about the harmonic limit, third and fourth order terms contribute the same order of magnitude to the energy shifts because they are off-diagonal and diagonal respectively. (Notice also that the X,P coordinates of this section could under suitable further restrictions correspond to either of the bosons previously discussed on more phenomenological grounds).

The adiabatic limit differs from the near-harmonic limit in that we suppose ω to become so small that (1.35) cannot be maintained. In contrast to (1.38), we assume that

$$X \sim P \sim O(1) \tag{1.40}$$

corresponding to larger amplitudes and slower oscillations. The expansion (1.29) may no longer converge. Equation (1.39) can, however, be taken in the form

$$H^{(m)}(X) = O(N^{-m+1}) \ . \tag{1.41}$$

Returning to (1.28), we see that the potential energy $H^{(0)}(X)$ dominates and the kinetic energy is a small correction. In this limit we would study the Hamiltonian

$$\mathcal{H}(X,P)+W_0 = H^{(0)}(X) + \frac{1}{4}\{P,\{P,H^{(2)}(X)\}\} \ , \tag{1.42}$$

and in general demand that the microscopic theory allow us to calculate the two functions $H^{(0)}(X)$ and $H^{(2)}(X)$. We shall show by example in part II how this can be done.

We leave it as an exercise to the reader to compute formal expressions for the matrix elements of the operator $\hat{M}$ using the methods of this section.

F. Operator Formulation of the Ideal Collective Coordinate and Some Relations to Microscopic Theories

From Eq. (1.18) we have the following two equations:

$$X|X> = \sum_{nn'} |n>[\int u_n^*(Y)Yu_{n'}(Y)dY]u_{n'}(X) \ , \tag{1.43}$$

$$-i\ \partial_X|X> = \sum_{nn'} |X>[\int u_n^*(Y)(i\partial_Y)u_n(Y)dY]u_{n'}(X) \ . \tag{1.44}$$

If we define many-body operators $\hat{X}$, $\hat{P}$ by the equations

$$\langle n|\hat{X}|n'\rangle = \int u_n{}^*(X)X\,u_{n'}(X)dX\ , \tag{1.45}$$

$$\langle n|\hat{P}|n'\rangle = \int u_n{}^*(X)(-i\partial_X)\,u_{n'}(X)dX\ , \tag{1.46}$$

so that these many-body operators are defined strictly only within the subspace of collective states, we have, as expected from (1.43) and (1.44),

$$\hat{X}|X\rangle = X|X\rangle\ , \tag{1.47}$$

$$\hat{P}|X\rangle = i\partial_X|X\rangle\ , \tag{1.48}$$

which guarantees the canonical commutation relations between X and P.

These relations mean that in the space of states $|n\rangle$, that is, as a projection of the total many-body space, we can replace the relation (1.31), on which we focus temporarily, by the operator equation

$$H_{\text{projected}} = \sum_{\substack{k,r=0\\ k\ \text{even}}}^{\infty} H^{(kr)}\ \{\underline{\hat{P}},(\hat{X})^r\}^{(k)}(\tfrac{1}{2})^k\frac{1}{r!}\ , \tag{1.49}$$

where

$$\{\underline{\hat{P}},\hat{\Theta}\}^{(k)} = \underbrace{\{\hat{P},\{\hat{P}\ldots\{\hat{P}}_{k\ \text{times}},\hat{\Theta}\}\ldots\}\}\ , \tag{1.50}$$

and a similar definition will hold below with the commutator replacing the anticommutator. The subscript "projected" will henceforth be understood.

It is now a straightforward algebraic exercise to "solve" Eq. (1.49) for $H^{(kr)}$ by forming multiple commutators, first with $\hat{X}$, then with $\hat{P}$. We find

$$H^{(kr)} = \frac{(-i)^k(i)^r}{k!}\ [\underline{\hat{P}},[\underline{\hat{X}},\hat{H}]^{(k)}]^{(r)}$$

$$- \sum_{\substack{\ell\geq k,s\geq k\\ \ell+s>k+r}} \frac{1}{k!}(\tfrac{1}{2})^{(\ell-k)}\ \frac{\ell!}{(\ell-k)!(s-r)!}\ H^{(\ell s)}\{\underline{P},(X)^{s-r}\}^{(\ell-k)}\ . \tag{1.51}$$

Equations (1.49) and (1.51) constitute a version of the phenomenological part of Villars' theory of collective motion.[5,6,7,8,9] We have thus established the formal equivalence between this theory and that of the ideal collective coordinate. It is amusing to note that according to (1.26) and (1.29), we have

$$H^{(kr)} = [(\partial_Y)^{(r)} \int d\xi \, \frac{(-i\xi)^k}{k!} \, \langle Y+\tfrac{1}{2}\xi|\hat{H}|Y-\tfrac{1}{2}\xi\rangle_{Y=0}$$

$$= \int d\xi \, \frac{(-i)^k(i)^r}{k!} \, \langle\tfrac{1}{2}\xi|[\underline{\hat{P}},[\underline{\hat{X}},\hat{H}]^{(k)}]^{(r)}|-\tfrac{1}{2}\xi\rangle \; . \qquad (1.52)$$

We leave it as an exercise to the reader that (1.51) satisfies (1.52).

In order not to leave matters in so abstract a form, let us show by a simple example how this formulation can be utilized for the microscopic theory. For this purpose let us consider the harmonic limit $H^{(kr)} = 0$, $k + r > 2$. We then have the following equations, remembering that $H^{(kr)} = 0$ for k odd,

$$H^{(10)} = 0 = -i[\hat{X},\hat{H}] - 2H^{(20)}\hat{P} \; , \qquad (1.53)$$

$$H^{(01)} = 0 = i[\hat{P},\hat{H}] - H^{(02)}\hat{X} \; , \qquad (1.54)$$

or with the definitions (1.33), (1.34),

$$[\hat{X},\hat{H}] = i\,\hat{P}/B = i\,\dot{\hat{X}} \; , \qquad (1.55)$$

$$[\hat{P},\hat{H}] = -iC\,\hat{X} = i\,\dot{\hat{P}} \; . \qquad (1.56)$$

These last equations are clearly recognizable as the equations of motion for the canonical (boson) collective variables X, P. They can be utilized in at least three distinguishable ways:

(i) We consider Eqs. (1.55) and (1.56) strictly in the space of the canonical variables. Our task is then to find the part of H depending on X and P using one or more of the boson transformation methods described in part II of these lectures. The resulting Hamiltonian $H=H_B$ may be diagonalized directly using any convenient basis or may be studied by means of the equations of motion of which (1.55), (1.56) represent only the harmonic approximation.

(ii) Equations (1.55), (1.56) may be viewed as equations in the original fermion space. We then keep the original Hamiltonian H but assume suitable expansions for the operators X, P in terms of a complete (or overcomplete) set of fermion operators. The study of the matrix elements of (1.55), (1.56) between a complete set of fermion states, together with the matrix elements of the canonical commutation relation, then constitutes one of the standard approaches to the random phase approximation (RPA) and its generalizations.

(iii) A method related to either (i) or (ii) insofar as its detailed working out can be done either in the boson or in the fermion space is obtained either by using a different pair of the relations (1.51) or more directly from (1.55), (1.56) by a further commutation. We have

$$[\hat{X},[\hat{X},\hat{H}]] = -B^{-1} = i[\hat{X},\hat{P}]B^{-1} \; , \qquad (1.57)$$

$$[\hat{P},[\hat{P},\hat{H}]] = -C = i[\hat{X},\hat{P}]C \ . \qquad (1.58)$$

By taking the expectation values of (1.57) and (1.58) in any convenient state $|\Phi\rangle$, we can form from the combination of (1.57) and (1.58) a homogeneous form for the square of the excitation energy, $\omega^2 = (C/B)$, namely,

$$\omega^2[\langle\Phi|[\hat{X},\hat{P}]|\Phi\rangle]^2 = \langle\Phi|[\hat{X},[\hat{X},\hat{H}]]|\Phi\rangle \langle\Phi|[\hat{P},[\hat{P},\hat{H}]]|\Phi\rangle \ . \quad (1.59)$$

We appear, superficially, to have gone from operator identities in (1.57), (1.58) to a restricted matrix element. But we must remember two facts: The equations to which (1.57), (1.58) represent a first approximation are themselves only a projection of the full many-body space on to a restricted subspace. Secondly because of the further (harmonic) approximation, instead of being "true" in the entire projected space, they may be true, if at all, near the ground state of the projected space. It seems most sensible then to consider $|\Phi\rangle$ to be some candidate for the ground-state vector emerging from an antecedent variational consideration.

With the ground state assumed known, we may now consider (1.59) as a variational expression for the energy of the first excited state. In this form it is equivalent, in the fermion space, to an approximate form of the equation-of-motion method due to Rowe[10,11,12] and in a certain approximation, then, to the RPA.

The brief discussion given above is an inadequate account of the use of the equations of motion for the study of the excited states. In section IIB,C we shall present what we consider to be the most powerful use of equations of motion for the study of collective motion and there it is most convenient to work with fermion pair operators directly. On the other hand, as discussed in section IID,E there is some indication that the most powerful boson expansions can be constructed with <u>little</u> or no use of the equations of motion, and that once this is done, one should simply diagonalize the resulting Hamiltonian in the boson space. Therefore, for our purposes, it would not appear to be fruitful to go more deeply into the direct use of the formalism of this section. For the sake of improved clarity, however, we have chosen to give a further discussion in section IIF of various aspects of equations-of-motion methods within the context of the model studied in part II.

For many readers, it would probably be useful at this stage to delay further reading of part I, until the corresponding sections A-F of part II have been studied.

G. Operator Formulation in the Adiabatic Case

We give a brief discussion of the adiabatic limit, which generalizes the considerations of the previous section. In this case, we write for the partial sum of (1.49)

$$H = \sum_{0}^{\infty} (\tfrac{1}{2})^k \{\underline{\hat{P}}, H^{(k)}(\hat{X})\}^{(k)} . \tag{1.60}$$

From this equation, we derive the identity

$$[\underline{\hat{P}}, [\underline{\hat{X}}, \hat{H}]^{(k)}]^{(r)} = (i)^k (-i)^r k! H^{(kr)}(\hat{X})$$

$$+ \sum_{p=1}^{\infty} (\tfrac{1}{2})^p (i)^k (-i)^r \frac{(k+p)!}{p!} \{\underline{\hat{P}}, H^{(k+p,r)}(\hat{X})\}^{(p)} , \tag{1.61}$$

which becomes identical with (1.51) upon expansion. For the special case of Eq. (1.42) these equations become

$$i\dot{\hat{P}} = [\hat{P}, \hat{H}] = -i\, \partial_X H^{(0)}(\hat{X}) , \tag{1.62}$$

$$i\dot{\hat{X}} = [\hat{X}, \hat{H}] = i\, \{\hat{P}, H^{(2)}(\hat{X})\} \tag{1.63}$$

which are recognized as the equations of motion associated with the Hamiltonian (1.42).

H. Equivalence to Conventional Hill-Wheeler-Peierls-Yoccoz Projection

The previous considerations revert ultimately to the definition of an ideal generating state given by Eq. (1.18). This is a state which we cannot, in reality, construct until after we have solved the problem of collective motion as derived from the concept. We should like to contrast this viewpoint with the long-established one[13-17] which starts (deus ex machina or the result of previous work) with an intrinsic state, $|\Phi(a_0)\rangle$, depending on certain values a_0 of a set of parameters. The state $|\Phi(a_0)\rangle$ is supposed to be a superposition of the members of the ground-state band. With $|n\rangle \equiv |\Psi_n\rangle$, we write

$$|\Phi(a_0)\rangle = \sum_n |\Psi_n\rangle f_n^*(a_0) . \tag{1.64}$$

Needless to say, to obtain an intrinsic state as accurate as (1.64) requires again a solution of the problem of collective motion, but for the sake of discussion we allow that it may be possible to find such a solution outside the framework of the formalism which has been developed so far. We therefore suppose $\Phi(a_0)$ to be known. It follows that the expansion coefficients $f_n(a_0)$ can also be obtained by the solution of the problem as previously formulated. For we know the eigenstates $|\Psi_n\rangle$ in the following sense: We have

$$\begin{aligned} f_n^*(a_0) &= \langle\Psi_n|\Phi(a_0)\rangle \\ &= \int dX \sum_{n'} \langle\Psi_n|X\rangle\langle X|\Phi_{n'}\rangle\langle\Phi_{n'}|\Phi(a_0)\rangle \\ &= \int dX \sum_{n'} u_n^*(X) \langle X|\Phi_{n'}\rangle\langle\Phi_{n'}|\Phi(a_0)\rangle . \end{aligned} \tag{1.65}$$

Here we have introduced a new set of states $|\Phi_n\rangle$. These are supposed to be a set of simple states which span the collective fermion space and which map onto the collective boson space in a known, preassigned manner, e.g., as the states of a harmonic oscillator which we have used as a basis for the diagonalization of the full collective Hamiltonian $H(X,P)$. We may therefore assume that the amplitudes

$$u_n^{(0)}(X) = \langle X|\Phi_n\rangle \tag{1.66}$$

have been defined and that the amplitudes

$$f_n^{(0)*}(a_0) \equiv \langle\Phi_n|\Phi(a_0)\rangle \tag{1.67}$$

can be calculated from the knowledge of $\Phi(a_0)$. Thus we have

$$\begin{aligned} f_n^*(a_0) &= \int dX \sum_{n'} u_n^*(X)u_{n'}^{(0)}(X)f_{n'}^{(0)*}(a_0) \\ &\equiv \sum_{n'} C_{n'}(n)\, f_{n'}^{(0)*}(a_o) \ , \end{aligned} \tag{1.68}$$

where we might have more readily available the expansion coefficient $C_{n'}(n)$ of the exact projected boson states on to the harmonic ones. In the last form, the evaluation of $f_n^*(a_0)$ is most clearly related to known or calculable quantities.

Notice that even if $|\Phi(a_0)\rangle$ is not an exact generating state, i.e., even if it has a non-vanishing overlap with the non-collective part of the space, the formulas above give the overlap with the collective space.

As we shall see in the succeeding sections the concept of a normalizable generating state has been used in the literature in several different ways, of which we presently consider only one. In this approach[18-21] we choose one parameter (for our monopole example) $a_0 \to a$ to be a variable, of the same physical nature, let us suppose, as the previous variable X. In the expansion (1.64), the functions $f_n^*(a)$ can then be viewed as a set of single-particle functions which contain the same physical information as the $u_n^*(X)$. We have changed the name of the variable in order to emphasize that the representation is by non-orthogonal functions. Thus for the overlap we have

$$\begin{aligned} N(a|a') &= \langle\Phi(a)|\Phi(a')\rangle = \sum_a f_n(a)f_n^*(a') \ , \\ N(a|a) &= 1 \ . \end{aligned} \tag{1.69}$$

In applications $N(a|a')$ is often approximated as a gaussian function in $(a-a')$. We return to this point below.

In order to invert (1.64), which is necessary in order to compute the energies and other observables from $|\Phi(a)\rangle$, we introduce a set of functions $\tilde{f}_n(a)$, bi-orthogonal to the $f_n(a)$,

$$(\tilde{f}_n, f_{n'}) = \int \tilde{f}_n^*(a) f_{n'}(a) da = \delta_{nn'} \quad , \tag{1.70}$$

and write

$$|\Psi_n\rangle = \int da\ \tilde{f}_n(a)\ |\Phi(a)\rangle \quad . \tag{1.71}$$

We then have formally

$$W_n = \langle n|H|n\rangle = \frac{\int \tilde{f}_n^*(a)\ H(a|a')\tilde{f}_n(a')dada'}{\int \tilde{f}_n^*(a)\ N(a|a')\tilde{f}_n(a')dada'} \quad , \tag{1.72}$$

where

$$\langle\Phi(a)|H|\Phi(a')\rangle = H(a|a') = \sum_n f_n(a) W_n f_n^*(a') \quad . \tag{1.73}$$

If we remember that $W_n = W_0 + \omega_n$ and that

$$\delta_{nn'}\omega_n = \int u_n^*(X) \mathcal{H}(X,P) u_{n'}(X) dX \quad , \tag{1.74}$$

Eq. (1.73) can be rewritten in a very suggestive form, namely

$$\begin{aligned} H(a|a') = W_0 N(a|a') &+ \frac{1}{2} \int K(a|a'')N(a''|a')da'' \\ &+ \frac{1}{2} \int N(a|a'')K^\dagger(a''|a')da'' \quad , \end{aligned} \tag{1.75}$$

where $K^\dagger$ is the hermitian conjugate of K and

$$K(a|a') = \int [\sum_n f_n(a) u_n^*(X)] \mathcal{H}(X,P) [\sum_{n'} u_{n'}(X) \tilde{f}_{n'}^*(a')] dX \quad . \tag{1.76}$$

Equation (1.75) then is seen to be equivalent to (1.73), if we notice the equations which follow from (1.70) and (1.76), namely

$$\int K(a|a')\ f_n(a')da' = \omega_n f_n(a) \tag{1.77}$$

and its complex conjugate. This is equivalent to the usual form of Hill-Wheeler integral equation, derivable directly from (1.72) as variational equation if we notice the relationship between the f_n and the $\tilde{f}_n$, which can be written

$$f_n(a) = \int N(a|a')\tilde{f}_n(a')da' \quad . \tag{1.78}$$

In the above we have transformed our solution based on the ideal collective coordinate into the Hill-Wheeler form. In practice, if we are simply given $|\Phi(a)\rangle$, we must calculate $N(a|a')$ and $H(a|a')$ and then solve the integral equation (1.75) for $K(a|a')$. This is often done approximately starting from a gaussian approximation to $N(a|a')$. Our considerations make it plain that this is inadequate if one wishes to go beyond the harmonic approximation, since the overlap is a function both of (a-a') and of ½(a+a'). An improved treatment would be based on the assumption of a peaked behaviour in (a-a') and of a slowly varying behaviour in ½(a+a'). This is considered in the next section.

Let us finally remark that under the condition of small anharmonicity $K(a|a')$ can be replaced, except for the lack of Hermiticity, by the analogue of (1.30). We show this in a simple case by writing

$$K(a|a') \to \delta(a-a')\, K(a,p_a) \quad , \tag{1.79}$$

$$K(a,p_a) = C + \alpha a^2 + \beta p_a{}^2 + i\gamma a\, p_a + \dots \, , \tag{1.80}$$

$$K^\dagger(a,p_a) = C + \alpha a^2 + \beta\, p_a{}^2 - i\gamma\, p_a a + \dots \, . \tag{1.81}$$

The real constants C, α, β, γ can be determined from the (assumed) knowledge of the $f_n(a)$ and the ω_n, i.e. from the previous solution, using Eq. (1.75) and taking the scalar product left and right with all possible choices of $\tilde{f}_0, \tilde{f}_1$, thus giving four conditions, the required number. This procedure can be continued to arbitrary order in the expansion of (1.79).

I. Alternative Treatments of the Relationship Between the Hamiltonian Overlap and the Unit Overlap: Generalized Peierls-Yoccoz (P.Y.) Method

We refer here to Eq. (1.75). We shall carry through our considerations within the level of approximation indicated by Eqs. (1.79) - (1.81), but there is no difficulty of principle obstructing extension beyond the harmonic approximation. The reader should have developed some suspicion by now that the various forms of (1.75) contain the essence of the theory of collective motion.[22,23] We shall illustrate three ways of treating this relationship so as to obtain acceptable approximate theories of collective motion, which could become exact theories in a more complete treatment. We shall also be able to understand theoretically their differences.

We write

$$H(a|a') = h_o N(a|a') + \tfrac{1}{2}(\alpha a^2+\beta p_a{}^2+i\gamma a p_a)N(a|a') + \tfrac{1}{2}N(a|a')(\alpha a'^2+\beta p_{a'}{}^2-i\gamma p_{a'} a') \, . \tag{1.82}$$

We consider the case that either the intrinsic state $|\Phi(a)\rangle$ is given or some procedure for its determination is given and that we can actually calculate the unit overlap N(a|a') and the Hamiltonian overlap H(a|a'). We then introduce the variables X,ξ,

$$a = X + \tfrac{1}{2}\xi \, , \quad a' = X - \tfrac{1}{2}\xi \, ,$$

$$X = \tfrac{1}{2}(a+a') \, , \quad \xi = a-a' \quad , \tag{1.83}$$

and write

$$N(a|a') = \mathcal{N}(X,\xi) \ , \ H(a|a') = \mathcal{H}(X,\xi) \quad . \tag{1.84}$$

Acting on these functions

$$\partial_a = \tfrac{1}{2}\partial_X + \partial_\xi \quad , \quad \partial_{a'} = \tfrac{1}{2}\partial_X - \partial_\xi \ . \tag{1.85}$$

If we assume that $N(a|a')$ is real, it is then also symmetric, and we have

$$\mathcal{N}(X,\xi) = \mathcal{N}(X,-\xi) \quad . \tag{1.86}$$

We now find that (1.82) is transformed into the expression

$$\mathcal{H}(X,\xi) = h_o\mathcal{N}(X,\xi) + [-\tfrac{1}{4}\beta(\partial_X)^2 + \alpha X^2 + \gamma X\partial_X]\mathcal{N}(X,\xi)$$
$$+ [-\beta(\partial_\xi)^2 + \tfrac{1}{4}\alpha\xi^2 + \gamma\partial_\xi\xi]\mathcal{N}(X,\xi) \ . \tag{1.87}$$

All further progress is based on the assumptions that the overlaps are peaked in ξ (the sine qua non of collective motion) and slowly varying in X (for near harmonic motion). This allows us in a first approximation to drop the terms depending on $\beta(\partial_X)^2$ and $\alpha\xi^2$. In (1.87) effectively $X \sim \partial_X \sim O(1)$, $\xi \sim O(N^{-\frac{1}{2}})$, $\partial_\xi \sim O(N^{\frac{1}{2}})$; we also know that $\beta \sim O(N^{-1})$, $\alpha \sim O(N)$, but γ is a mystery (probably O(1) making the terms in which it occurs intermediate in size between the dominant ones and those we drop). We retain the terms in γ for the time being.

We now procced by computing moments with respect to ξ of both sides of (1.87) utilizing the definition (1.26) (and later (1.29)). For the zeroth and second moments (the first vanishes by (1.86)) we find

$$\mathcal{H}^{(0)}(X) \cong h_o\mathcal{N}^{(0)}(X) + \alpha X^2\mathcal{N}^{(0)}(X) + \gamma X\partial_X\mathcal{N}^{(0)}(X) \ , \tag{1.88}$$

$$\mathcal{H}^{(2)}(X) \cong h_o\mathcal{N}^{(2)}(X) + \alpha X^2\mathcal{N}^{(2)}(X) + \gamma X\partial_X\mathcal{N}^{(2)}(X)$$
$$+ \beta\mathcal{N}^{(0)}(X) - 2\gamma\mathcal{N}^{(2)}(X) \quad . \tag{1.89}$$

A sufficient number of equations to determine the constants h_o, α, β, γ is obtained from the leading terms of the Taylor series for (1.88), (1.89) around X = 0, three terms from (1.88) and one from (1.89),

$$\mathcal{H}^{(00)} = h_o\mathcal{N}^{(00)} \quad , \tag{1.90}$$

$$\mathcal{H}^{(01)} = h_o\mathcal{N}^{(01)} + \gamma\mathcal{N}^{(01)} \quad , \tag{1.91}$$

$$\mathcal{H}^{(02)} = h_o\mathcal{N}^{(02)} + \gamma\mathcal{N}^{(02)} + 2\alpha\mathcal{N}^{(01)} \quad , \tag{1.92}$$

$$\mathcal{H}^{(20)} = h_o\mathcal{N}^{(20)} - 2\gamma\mathcal{N}^{(20)} + \beta\mathcal{N}^{(00)} \quad . \tag{1.93}$$

The determinant of the coefficients is $2(N^{(00)})^2(N^{(01)})^2$. Thus if $N^{(01)}$, which measures the skewness of the unit overlap, is small it is best to drop Eq. (1.91) and set $\gamma = 0$. The equations then reduce to those given by Peierls and Yoccoz.[16]

The theory based on (1.90) - (1.93) was in disfavour for many years simply because examples could be given in which it did not correctly describe the exactly known translational motion of the centre-of-mass. From the point of view of these lectures, however, the fault lies with the intrinsic function $|\Phi(a)\rangle$ used in these examples since it can be proved that the specialization of Eqs. (1.90) - (1.93) to the case of translation ($\gamma = \alpha = 0$), gives $\beta = (2M)^{-1}$, where M is exactly the total mass, if $|\Phi(a)\rangle$ lies entirely within the collective band. The non-exactness of an intrinsic state, which is serious in principle, may not be so in practice, since the error in the total mass may be small. Corresponding errors for other collective parameters may be entirely tolerable since they may be no larger than that given by alternative methods which do give the exact mass even if the intrinsic state is not exact (the methods of section F are known to have this property, see the lectures by Professor Villars). The accuracy of these latter methods, which will be considered from our current standpoint below, depends on the physical possibility of constructing an intrinsic state of indefinitely large "deformation". This is precisely what one can do for translations and less so for rotations and vibrations.

Returning to (1.90) - (1.93), once the parameters are known, the resulting second-order differential equation can be recognized as the equation of a simple harmonic oscillator in the non-orthogonal representation and easily solved. We do not enter into such details.[18]

J. Strong Deformation and Skyrme-Levinson Theories

We consider the limit $\gamma = 0$ to produce a slight simplification in the following. We now make the assumption that $|\Phi(a)\rangle$ is sufficiently deformed so that Eqs. (1.47) and (1.48) are applicable, but that nevertheless $|\Phi(a)\rangle$ remains normalizable. Under these (contradictory) assumptions, we have, with the notation $O(X,\xi) = \langle X + \tfrac{1}{2}\xi|\hat{O}|X - \tfrac{1}{2}\xi\rangle$,

$$\lim_{X,\xi\to 0} \xi\, O(X,\xi) = \langle\Phi|[\hat{X},\hat{O}]|\Phi\rangle , \tag{1.94}$$

$$\lim \partial_X\, O(X,\xi) = i\langle\Phi|[\hat{P},\hat{O}]|\Phi\rangle , \tag{1.95}$$

$$\lim X\, O(X,\xi) = \tfrac{1}{2}\langle\Phi|\{\hat{X},\hat{O}\}|\Phi\rangle , \tag{1.96}$$

$$\lim \partial_\xi\, O(X,\xi) = \tfrac{1}{2}i\langle\Phi|\{\hat{P},\hat{O}\}|\Phi\rangle . \tag{1.97}$$

With these relations, we notice first that $\gamma = 0$, $\alpha = \tfrac{1}{2}C$, $\beta = (2B)^{-1}$, (1.87) becomes (strong deformation approximation)

$$H(X,\xi) = h_o\ 1(X,\xi) + (2B)^{-1}\ P^2(X,\xi) + \tfrac{1}{2}\ CX^2(X,\xi). \qquad (1.98)$$

Then, using the operations (1.94) and (1.95) twice each we reach the equations

$$B^{-1} = -\ \langle\Phi|[\hat{X},[\hat{X},\hat{H}]]|\Phi\rangle\ , \qquad (1.99)$$

$$C\ = -\ \langle\Phi|[\hat{P},[\hat{P},\hat{H}]]|\Phi\rangle\ , \qquad (1.100)$$

which coincide essentially with (1.57) and (1.58) and therefore represent correctly the leading terms of the equations of motion.

Though perhaps more circuitous than the above considerations, those of Sec. F are more valuable because they are formally exact. Nevertheless, one does derive important qualitative understanding from the present approach, based on the following two theorems which we state without proof, (the first is easy, the second trivial to prove): Theorem 1: Let $|\Phi(a)\rangle$ be an exact intrinsic state. Then the equations (1.90) - (1.93) of the P.Y. theory give correctly to the leading order in N the ground state energy (h_o) and the first excitation energy (as defined by the collective parameters B and C). Theorem 2: Let there exist an approximate projection

$$\hat{H} = \hat{H}_{in} + (\hat{P}^2/2B) + \tfrac{1}{2}\ C\hat{X}^2 + \ldots\ , \qquad (1.101)$$

where $\hat{X}$ and $\hat{P}$ commute with $\hat{H}_{in}$, the "intrinsic" Hamiltonian, and are canonical variables. Then (1.99) and (1.100) are correct for "any" $|\Phi\rangle$. We call this the insensitivity to the intrinsic state in the strong deformation limit. However, this limit gives us no information about the ground state itself, at least not superficially. (In practice it is quite otherwise as we shall see from our study of cranking theory.)

The moral here is that we should use the P.Y. theory when we believe in the intrinsic state (for rotations this means light nuclei) and the strong deformation theory for the rotations of heavy nuclei.

For the sake of completeness, we mention yet a third formalism which requires both a good intrinsic state and strong deformation to be "exact". Starting again from (1.98), we apply twice each the so-far neglected two operations (1.96) and (1.97). We thus obtain, remembering Eq. (1.101),

$$h_o = \langle\Phi|\hat{H}_{in}|\Phi\rangle = \langle\hat{H}_{in}\rangle\ , \qquad (1.102)$$

$$h_o\langle\hat{P}^2\rangle = \tfrac{1}{4}\langle\{\hat{P},\{\hat{P},\hat{H}_{in}\}\}\rangle\ , \qquad (1.103)$$

$$h_o\langle\hat{X}^2\rangle = \tfrac{1}{4}\langle\{\hat{X},\{\hat{X},\hat{H}_{in}\}\}\rangle\ . \qquad (1.104)$$

As stated above the "exactness" of this formulation which we call the Skyrme-Levinson method (see below) requires the conditions of both previous theorems. In the case of rotations where $\hat{X}$ is ignorable and the $\hat{P}$'s are known operators (components of the angular momentum), this method accords with one proposed by Skyrme,[24] and applied by Levinson in the s-d shell[25] (see also[26]). Though a method for the application of (1.102) - (1.104) could be devised, this method for theoretical and practical reasons does not commend itself as a theory of collective vibrations.

K. Cranking Theory and its Significance

All the results obtained up to now have been based on the use of two types of generating function (= intrinsic state), the non-normalizable ideal or infinitely deformed generating state on the one hand and any convenient or physically given or computed generating state on the other (Hartree-Fock or BCS or as much of an improvement on these as we can muster). In the latter case, the theory of the next few sections shows that the accuracy of our results is strongly dependent on the accuracy of this function. In this and the succeeding sections, we shall exhibit additional special generating functions which can be exceedingly useful in practice. We shall first show that the significance and utility of the cranking method, which is by far the most widely applied method in the theory of collective motion, especially in the theory of rotations, can be understood in terms of a special generating function. The result which we shall obtain, though previously understood in various special forms[22,23,27] and even understood to be general by the writer[28] was first proved in essentially general form by Marshalek and Weneser.[9] What follows is probably a simplification of their proof.

Consider the generating function

$$|\Phi(\alpha,p_\alpha)\rangle = \exp(iXp_\alpha)\exp(-iP\alpha)|\Phi\rangle \quad , \tag{1.105}$$

where we assume that $|\Phi\rangle$ is an exact intrinsic state for the collective band. Since the operators X and P carry any state within the band into well-defined states also in the band, we must admit that $|\Phi(\alpha,p_\alpha)\rangle$ is an exact intrinsic state if $|\Phi\rangle$ is. Therefore if we calculate the expectation value of H with respect to $|\Phi(\alpha,p_\alpha)\rangle$, we can do so with the projection, Eq. (1.49), or later Eq. (1.60). Using the well-known displacement properties of the exponential operators, we obtain, using (1.49),

$$\langle\Phi(\alpha,p_\alpha)|H|\Phi(\alpha,p_\alpha)\rangle = \langle\Phi|H(\hat{P}+p_\alpha,\hat{X}+\alpha)|\Phi\rangle$$

$$= \sum_{k,r=0}^{\infty} (H^{(kr)}+\ldots)\,\frac{(p_\alpha)^k(\alpha)^r}{r!} \quad , \tag{1.106}$$

where ... are, under normal conditions, terms which are smaller by one order of N than the leading term. For instance for $k = r = 0$, the parenthesis consists of

$$H^{(00)}+H^{(20)}\langle\hat{P}^2\rangle + \tfrac{1}{2}H^{(02)}\langle\hat{X}^2\rangle + \text{smaller terms yet.} \tag{1.107}$$

In this case, our contention is satisfied if

$$<P^2> \sim O(N) \;,\; <X^2> \sim O(N^{-1}) \;, \tag{1.108}$$

which corresponds to a strongly deformed intrinsic state. Thus we have the result that in the expectation value (1.106) the coefficient of $[(p_\alpha)^k(\alpha)^r/r!]$ is precisely the desired collective parameter $H^{(kr)}$ up to terms of relative order N^{-1}.

Of course, one obvious way of constructing $|\Phi(\alpha,p_\alpha)>$ is to construct separately the intrinsic function $|\Phi>$, which is an exact eigenstate of H_{in}, and then construct the operators $\hat{X}$ and $\hat{P}$. This latter is precisely what is accomplished by the methods of section F and of the preceding section.

The form of (1.105) suggests however a new procedure for calculating $|\Phi(\alpha,p_\alpha)>$. The latter belongs to the <u>class</u> of states for which

$$<\Phi(\alpha,p_\alpha)|\hat{P}|\Phi(\alpha,p_\alpha)> = p_\alpha \;, \tag{1.109}$$

$$<\Phi(\alpha,p_\alpha)|\hat{X}|\Phi(\alpha,p_\alpha)> = \alpha \tag{1.110}$$

(if $<P> = <X> = 0$), and should be a correct one of the class if it also minimizes the expectation of H_{in} under the prescribed constraints. Introducing Lagrange multipliers λ, μ and writing

$$|\Phi(\alpha,p_\alpha)> = |\Psi(\lambda,\mu)> \;, \tag{1.111}$$

we require therefore

$$\delta<\Psi(\lambda,\mu)|[\hat{H}_{in}-\lambda\hat{P}-\mu\hat{X}]|\Psi(\lambda,\mu)> = 0. \tag{1.112}$$

In practice, one often replaces $\hat{H}_{in}$ by $\hat{H}$ in (1.112), at least in the first approximation. This can be corrected in higher approximation since the method does in fact generate the operators $\hat{P}$ and $\hat{X}$. In fact, it is more convenient to use $\hat{H}$ itself and there seems to be no theoretical objection to this in the present case.

We discuss briefly the method of solution. For the anharmonic case, it is to expand $|\Psi(\lambda\mu)>$ in a double power series about $\lambda = \mu = 0$,

$$<\hat{H}>_{\lambda\mu} - h_{oo} = \sum_{q=0}^{\infty} h_{0q}\,\mu^{2+q} + \sum_{p=1,q=0}^{\infty} h_{pq}\,\lambda^{2p}\mu^{q} \;, \tag{1.113}$$

$$<\hat{P}>_{\lambda\mu} = p_\alpha = \sum_{p=0,q=0}^{\infty} a_{pq}\,\lambda^{2p+1}\mu^{q} \;, \tag{1.114}$$

$$<\hat{X}>_{\lambda\mu} = \alpha = \sum_{p=0,q=0}^{\infty} b_{pq}\,\lambda^{2p}\mu^{q+1} \;, \tag{1.115}$$

the general forms following from time-reversal invariance. In principle, one has only then to invert the series (1.114), (1.115), substitute into (1.113) and identify the coefficient of $(p_\alpha)^k(\alpha)^r$ as in (1.106). There are, however, a number of relationships among the coefficients in the above series which should be recognized.

From the variational principle (with H), we can derive the following equations for the expectation value with respect to $|\Psi(\lambda,\mu)>$:

$$\partial_\lambda <\hat{H}> = \lambda \partial_\lambda <\hat{P}> + \mu \, \partial_\lambda <\hat{X}> , \tag{1.116}$$

$$\partial_\mu <\hat{H}> = \lambda \partial_\mu <\hat{P}> + \mu \, \partial_\mu <\hat{X}> , \tag{1.117}$$

$$\partial_\mu <\hat{P}> = \partial_\mu <\hat{X}> . \tag{1.118}$$

Studying Eq. (1.118) first, we find no condition on the a_{po} or b_{oq} (naturally), but we must have

$$a_{p1} = 0 \tag{1.119}$$

and

$$(q+2)\ a_{p,q+2} = 2(p+1)\ b_{p+1,q} , \tag{1.120}$$

for p and q running from zero. From (1.116), using (1.118), we find

$$2(p+1)h_{p+1,q} = (2p+1+q)\ a_{pq} , \tag{1.121}$$

which also implies $h_{p+1,1} = 0$, and from (1.117)

$$(q+2)h_{p,q+2} = (2p+q+1)b_{p,q} . \tag{1.122}$$

One checks that (1.120) - (1.122) are compatible.

Equations (1.116) and (1.117) also state the relations

$$\lambda = \partial<\hat{H}>/\partial_{p_\alpha}\big|_\alpha , \tag{1.123}$$

$$\mu = \partial<\hat{H}>/\partial_\alpha\big|_{p_\alpha} , \tag{1.124}$$

which can be quite useful in the inversion of (1.114), (1.115), but again only in the case that we replace $\hat{H}_{in}$ by $\hat{H}$ in the variational principle.

L. Cranking and the Coherent Generating State

It is somewhat more transparent and leads to new insights if we replace the canonical variables with which we have been working by boson variables. We then take the projected Hamiltonian to be of the form

$$\hat{H} = \hat{H}_{in} + \sum_{r+s\geq 2} \omega_{rs} (b^{\dagger})^{r}(b)^{s} = \hat{H}_{in} + \hat{H}_{c} \quad . \tag{1.125}$$

This has then the advantage of being in normal form with respect to the ground state, $|\Phi_0>$, of the boson b, defined by

$$b|\Phi_0> = 0 \quad . \tag{1.126}$$

We now consider a very special form of generating state, namely

$$\begin{aligned} |\Phi(\beta,\beta^*)> &= \exp(ib\beta^* + ib^{\dagger}\beta)|\Phi_0> \\ &= \exp(ib\beta^*)\exp(ib^{\dagger}\beta)\exp(\tfrac{1}{2}|\beta|^2)|\Phi_0> \\ &= \exp(ib^{\dagger}\beta)\exp(ib\beta^*)\exp(-\tfrac{1}{2}|\beta|^2)|\Phi_0> \\ &= U(\beta,\beta^*)|\Phi_0> \quad , \end{aligned} \tag{1.127}$$

in which the exponential operator is essentially equivalent to that used in the previous section, since we can imagine that (1.125) has been obtained by a transformation such as

$$b = \frac{1}{2}\,[(BC)^{\frac{1}{4}}\hat{X} + i(BC)^{-\frac{1}{4}}\,\hat{P}] \tag{1.128}$$

and subsequent rearrangement into normal form. But now we insist on a very special intrinsic state, namely (1.126).

This has quite a remarkable consequence. Since

$$\begin{aligned} U^{-1}bU &= b + i\beta \quad , \\ U^{-1}b^{\dagger}U &= b^{\dagger} - i\beta^* \quad , \end{aligned} \tag{1.129}$$

and since

$$<\Phi_0|(b^{\dagger})^{r}(b)^{s}|\Phi_0> = 0 \quad , \tag{1.130}$$

we find

$$<\Phi(\beta,\beta^*),\hat{H}\Phi(\beta,\beta^*)> = <\Phi_0|\hat{H}_{in}|\Phi_0> + \sum_{r+s\geq 2} \omega_{rs}(\beta^*)^{r}\beta^{s} \quad . \tag{1.131}$$

Thus the generating function (1.127) allows us to construct the exact collective Hamiltonian.

We shall consider in a moment the cranking method associated with the present formulation, but we must first stop to examine more closely the difference between the procedures of the last two sections. In the first formulation of the cranking principle we were willing to accept any good intrinsic state $|\Phi>$ and did not insist on any special relationship between the collective variables and the intrinsic state. In the present formulation, we add the harmless looking condition (1.126), which then has very powerful consequences.

But to gain these consequences a price must be paid. For example, when we set out to construct b and $b^\dagger$ as expansions in fermion operators, we must continually readjust the definition of $|\Phi_0\rangle$ as we go to higher and higher terms and naturally this entails more labour than if such a requirement is ignored.

The generating function (1.127) can be obtained as the solution of the cranking problem

$$\delta\langle\Psi(\mu,\mu^*)|[H_{in}-\mu^* b-\mu\, b^\dagger]|\Psi(\mu,\mu^*)\rangle = 0 \quad , \tag{1.132}$$

$$\langle b\rangle = \langle b^\dagger\rangle^* = i\beta \quad , \tag{1.133}$$

$$b|\Psi(\mu=\mu^*=0)\rangle = b|\Phi(\beta=\beta^*=0)\rangle = 0 \quad . \tag{1.134}$$

One can solve by expanding in a power series in μ and μ^*.

We notice in passing that the generating state is a right eigenstate of b with eigenvalue $i\beta$, as follows from (1.129). We have furthermore

$$\Phi(\beta,\beta^*) = \sum_n C_n|\Phi_n\rangle \quad , \tag{1.135}$$

where

$$\Phi_n = \frac{(b^\dagger)^n}{\sqrt{n!}}|\Phi_0\rangle \quad , \tag{1.136}$$

$$C_n = \frac{(i\beta)^n}{\sqrt{n!}} e^{-\frac{1}{2}|\beta|^2} \quad . \tag{1.137}$$

This is the well-known coherent state,[29] which, if propagated in time, according to the Hamiltonian $b^\dagger b$, represents a wave-packet performing simple harmonic motion without dispersing in time.

Let us finally remark that the equivalent operator formulation of section F can be recast in the present notation as the problem of constructing operators b, $b^\dagger$, such that H takes the form (1.125). These operators are to satisfy the equations

$$[b,b^\dagger] = 1, \quad b|\Phi_0\rangle = 0 \quad , \tag{1.138}$$

and

$$[(b \text{ or } b^\dagger), H - H_c] = 0 \quad . \tag{1.139}$$

M. Cranking the Exact Ground State

We shall now go a step further and try to crank the exact ground state! We shall thereby arrive at a formulation which is of considerable phenomenological interest and leads back ultimately to the method of Hill-Wheeler projection.

We seek to replace the pair of operators b, $b^\dagger$ by the number operator $\hat{n} = b^\dagger b$ and its "conjugate" phase operator $\hat{\theta}$, formally satisfying

$$[\hat{\theta},\hat{n}] = i \quad . \tag{1.140}$$

A natural definition is

$$b^\dagger = \hat{n}^{\frac{1}{2}} e^{i\hat{\theta}} \quad , \tag{1.141}$$

$$b = e^{-i\hat{\theta}} \hat{n}^{\frac{1}{2}} \quad . \tag{1.142}$$

This definition satisfies the boson commutation relation if

$$[e^{-i\hat{\theta}},\hat{n}] = 1 \quad , \tag{1.143}$$

but there are difficulties. First of all, (1.140) does not follow from (1.143). As is well known, the commutation relations (1.140) require $\hat{\theta}$ and $\hat{n}$ to have continuous spectra, whereas in reality $\hat{n}$ is integral and $\hat{\theta}$ is required to be periodic with period 2π. Nevertheless one does not get into any trouble if one uses (1.140) in a domain of states consisting of sharp wave-packets in θ confined to the neighbourhood of the origin.[30] Under these conditions we shall assume that the relation (1.43), which can also be written for any integral ν as

$$e^{-i\nu\hat{\theta}} \hat{n}\, e^{i\nu\hat{\theta}} = \hat{n} + \nu \quad , \tag{1.144}$$

can be continued to non-integral ν for any ν and in any domain of states for which

$$\nu\langle\hat{\theta}^2\rangle^{\frac{1}{2}} << \pi \quad . \tag{1.145}$$

Another difficulty is that according to (1.142) $\exp(-i\hat{\theta}) = b\hat{n}^{-\frac{1}{2}}$ is undefined when acting on the state $|\Phi_0\rangle$. According to (1.141) and (1.142), we have

$$e^{i\hat{\theta}}|\Phi_n\rangle = |\Phi_{n+1}\rangle \quad ,$$

$$e^{-i\hat{\theta}}|\Phi_n\rangle = |\Phi_{n-1}\rangle \;, \quad n \neq 0 \quad . \tag{1.146}$$

In what follows, however, we shall avoid the undefined operation. The difficulties to which we have briefly alluded have been exhaustively discussed in the literature.[30]

Next we consider the generating state

$$|\Psi(\nu)\rangle = \exp(i\nu\Theta)|\Psi_0\rangle \quad , \tag{1.147}$$

where we have changed notation in order to emphasize that we want to insist that $|\Psi_0\rangle$ is the exact ground state. The equation $B|\Psi_0\rangle = 0$ now means that $B^\dagger$ is the physical boson defined in the first section. The implication is that the projected Hamiltonian has the form

$$\hat{H} = W_o + \sum_{k=1} a_k (\hat{n}) (\hat{n}-1) \ldots (\hat{n}-k) \quad . \tag{1.148}$$

Formally at least, then we have

$$<\Psi(\nu)|\hat{H}|\Psi(\nu)> = W_o + \sum_{k=1} a_k \nu(\nu-1) \ldots (\nu-k) \quad , \tag{1.149}$$

which gives us directly all the energies. This is not surprising since for ν = an integer, $|\Psi(\nu)>$ takes the form, according to (1.141),

$$|\Psi(\nu)>|_{\nu=n} = (n^{-\frac{1}{2}} B^{\dagger})^n |\Psi_o>$$

$$= \frac{(B^{\dagger})^n}{\sqrt{n!}} |\Psi_o> \quad , \tag{1.150}$$

the exact eigenstate in the physical boson representation.

We shall proceed further, despite the formal aspect, because of analogies that exist with recent work in the phenomenological description of rotational bands[31-36] and seek a cranking formulation. For this purpose we prescribe

$$<\Psi(\lambda^2)|\sqrt{\hat{n}}|\Psi(\lambda^2)> = \lambda \quad . \tag{1.151}$$

This choice is made in analogy with the case of two-dimensional rotations where the cranking operator is just the square root of the lowest power of the angular momentum which occurs in the excitation energy. We require

$$\delta<\Psi(\lambda^2)|[\hat{H}-\rho\sqrt{\hat{n}}]|\Psi(\lambda^2)> = 0 \quad , \tag{1.152}$$

where here we insist on the exact Hamiltonian H rather than the intrinsic one (why?). From (1.152), we conclude in the usual manner

$$\rho = \frac{d<\hat{H}>}{d<\sqrt{\hat{n}}>} \quad . \tag{1.153}$$

This equation is satisfied by the following formal power series solutions to (1.152):

$$<H> = W_o + \tfrac{1}{2}\rho^2 \left[\frac{1}{2\omega_o} + 3c\rho^2 + 5d\rho^4 + 7f\rho^6 + \ldots\right] \quad , \tag{1.154}$$

$$<\sqrt{n}> = \rho\left[\frac{1}{2\omega_o} + 2c\rho^2 + 3d\rho^4 + 4f\rho^6 + \ldots\right] \quad , \tag{1.155}$$

where $\omega_0 = a_1$ (cf. (1.148)). If we replace $\rho \to \omega$, $(2\omega_0)^{-1} \to I_0$, $\langle\sqrt{\hat{n}}\rangle \to \langle\hat{J}\rangle = \sqrt{I(I+1)}$, these are precisely the formulas given by Harris[31] for the phenomenological description of the ground-state rotational bands.

N. Further Analysis of Cranking: Cranking and Phenomenology

The implication of the formalism of the previous section is that there exists a one-parameter cranking formulation, such that when the parameter ρ goes through certain values the solution goes through the exact eigenstates. It is therefore amusing to show that for solutions of (1.152) which do pass through the eigenstates (it is obvious that inter alia such solutions must exist), ρ is not a continuous variable and so (1.151) cannot be satisfied. On the other hand, the variational principle (1.151), (1.152), divorced from its origins can have solutions for which ρ is a continuous variable, but these will not pass through the exact eigenstates. This inconsistency, however, as we then show, disappears in the semi-classical limit.

Of the three points mentioned in the previous paragraph, the first is established as follows: We look for a solution within the ground state band,

$$|\Psi(\lambda^2)\rangle = \sum_n C_n|\Psi_n\rangle \quad , \tag{1.156}$$

$$\sum_n |C_n|^2 = 1 \quad , \tag{1.157}$$

$$\sum_n \sqrt{n}\,|C_n|^2 = \lambda \quad . \tag{1.158}$$

Then (1.152) becomes

$$\delta \sum_n |C_n|^2\,[W_n - \rho\sqrt{n} - R] = 0 \quad , \tag{1.159}$$

where R is the Lagrange multiplier for Eq. (1.157). If the C_n are now freely variable, we have

$$[W_n - \rho\sqrt{n} - R]C_n = 0 \quad . \tag{1.160}$$

Thus at most two $C_n \neq 0$ can be chosen. From (1.158) there are in general an infinite number of choices n_1, n_2, as long as the condition $n_1 < \lambda^2 < n_2$ is satisfied. The most obvious choice, which we now make, is to take those values which are contiguous to λ, for then, clearly, we pass, as λ grows, through every eigenstate. Right at each eigenstate, we must anticipate a discontinuity, since at that point (1.158) becomes irrelevant, or only the combination $(\rho\sqrt{n} + R)$ is determined (as the exact energy W_n), and we shift thereafter to a different choice of n_1 and n_2. In fact from (1.160), one finds easily that

$$\rho = \frac{W_{n2} - W_{n1}}{\sqrt{n_2} - \sqrt{n_1}} \equiv \frac{\Delta E}{\Delta\sqrt{n}} \quad , \tag{1.161}$$

which is a step function.

If, on the other hand, we insist on a solution of the variational principle with a smooth behaviour for ρ, this can be achieved by imposing a form on the solution, that is, by choosing a dependence of the C_n upon a set of parameters $\alpha_i, C_n = C_n(\alpha_i)$. The variational equations (1.160) are then replaced by the smaller set

$$\sum_{n'} (W_{n'} - \rho \sqrt{n'} - R) \frac{\partial |C_{n'}(\alpha_i)|^2}{\partial \alpha_i} = 0 \quad . \tag{1.162}$$

These are now in general non-linear equations without the trivial solutions previously found. For trial ρ,R we shall then have solutions $\alpha_i = \alpha_i(\rho, R)$, the correct parameters ρ and R to be determined by satisfying (1.157), (1.158). Under these conditions (1.153) will be satisfied as a continuous function, but it is clear, or should be, that as λ^2 passes through the physical values n, the present solution can be far from an exact eigenstate.

Let us then finally consider a specialization in which C_n is centered around a large value n, and we write

$$n' = n + \delta \tag{1.163}$$

and suppose that the effective maximum value of δ, δ_m satisfies

$$|\delta_m/n| << 1 \quad . \tag{1.164}$$

With $W_n = W_0 + E_n$, we next expand the parenthetical expression in (1.162) and find

$$[W_0 + E_n - \rho(n)\sqrt{n} - R] \sum_{n'} \frac{\partial |C_{n'}(\alpha_i)|^2}{\partial \alpha_i}$$

$$+ \sum_{n'} \left[\left(\frac{\partial E_n}{\partial \sqrt{n}} - \rho(n) \right) (\Delta\sqrt{n}) + \frac{1}{2} \frac{\partial^2 E_n}{\partial^2 \sqrt{n}} (\Delta\sqrt{n})^2 \right] \frac{\partial |C_{n'}(\alpha_i)|^2}{\partial \alpha_i} \quad . \tag{1.165}$$

Here $\rho(n)$ is defined as

$$\rho(n) = \partial \sum_{n'} E_{n'} |C_{n'}|^2 / \partial\lambda |_{\lambda = \sqrt{n}} \quad . \tag{1.166}$$

The parentheses containing $\rho(n)$ in the second set of square brackets of (1.165) should then vanish with an error of the order of the remaining term within the brackets. This term is of the order $(\partial\rho/\partial\sqrt{n})(\delta^2/n)$. For fixed δ this term will, for typical classical behaviour, go to zero for large n. It will do so even for a harmonic oscillator and will go to zero even more rapidly if there is compression of the levels at high excitation. The remaining first bracket just characterizes the exact eigenstate.

In any practical calculation, necessary approximations will, in general, lead to a mixing with states outside the band, and this has been omitted from the present analysis.

It may be helpful to summarize the significance of the present discussion, in view of the fact that formally three different versions of cranking theory have been presented. The first two are practical and when fully implemented are completely equivalent to previously described methods of constructing the collective Hamiltonian, either approximately (in the first instance) or exactly (in the second). In addition to being suspect from the viewpoint of principle, the third method is also not practical because it requires evaluation of the square root of the exact number operator. If we insist on constructing directly the physical boson representation, we should use the algebraic method described for an example in part II, section B, C.

The significance of the present formulation (and the corresponding formulation for rotations) is the following: The formula (1.1), the power series in n, will not converge for large n even for moderate anharmonicity. One needs then an analytic continuation of the series. One method of analytic continuation is precisely through the introduction of a new variable which is a function of the old and single-valued in the physical domain. The series (1.154), (1.155) then represent such a possibility (satisfying Eq. (1.153), consistent with the physics of the situation for large n, precisely where it is needed. (One must be be cautioned, however, against taking the series as such too literally since the convergence of the new series may also be in question for large n.) Let us indeed emphasize that it is the relation

$$\rho = d\, E_n / d\, \sqrt{n} \tag{1.167}$$

to which we can asymptotically give credence, and (1.154) and (1.155) represent possible solutions at any polynomial level.

O. The Variable-Frequency Model

In this section, we shall show that by a suitable application of the method of H.W. projection, once more conceived in purely theoretical terms, one can ultimately reach a sounder basis for the relation (1.167). By yet another choice of intrinsic function we suppose that we have in hand a normalizable intrinsic state

$$|\Phi(X)\rangle = \sum_n C_n |n\rangle\, u_n^*(X) \quad , \tag{1.168}$$

where the $u_n^*(X)$ are the exact solutions of the collective

Hamiltonian in any convenient orthogonal representation. The expansion (1.168) can be thought of as a special case of the development of section H in which the functions f_n and $\tilde{f}_n$ differ only in scale, since $f_n = u_n C_n{}^*$. Using the method leading to (1.75), it is in fact straightforward to show that

$$H(X|X') = W_o N(X|X') + \tfrac{1}{2}\, \mathcal{H}(X,P) N(X|X') + \tfrac{1}{2}\, N(X|X')\, \mathcal{H}(X',P') \quad , \tag{1.169}$$

where $\mathcal{H}(X,P)$ is precisely the collective Hamiltonian defining the $u_n(X)$. We must therefore find

$$W_n = \frac{\int u_n{}^*(X)\; H(X|X')\; u_n(X')\; dXdX'}{\int u_n{}^*(X)\; N(X|X')\; u_n(X')\; dXdX'} = W_o + E_n \; , \tag{1.170}$$

conceptually separated into the residual term when n=0 and the term depending explicitly on n, here (but not later) the ground and excitation energies respectively, the latter in the form (1.1)

To obtain (1.170) for a given n, we do not in fact need an intrinsic state as precise and complete as (1.168). It is sufficient in practice to have a state of the form

$$|\Phi_n(X)> = C_n |n> u_n{}^*(X) + |\chi_n(X)> \quad , \tag{1.171}$$

where the state $|\chi_n(X)>$ satisfies

$$\int dX\; u_n(X) |\chi_n(X)> = 0 \quad , \tag{1.172}$$

which implies that $|\chi_n>$ is orthogonal to $|n>$. This can be turned into a flexible requirement in that $|\Phi_n(X)>$ can be constructed in such a way as to yield the result (1.171) for every n of interest by means of a slow and continuous variation with n of the intrinsic function.

We frame this as a hypothesis, by assuming the existence of a variable intrinsic state $|\Phi(\alpha_i,X)>$, where the α_i are those physical parameters (deformation parameters, for instance) whose slow variation with n will bring about the desired result. We then project from $|\Phi(\alpha_i,X)>$, leaving the precise value of α_i so far undetermined, assuming only that the form of the intrinsic state is sufficiently flexible so that by a proper (and slowly varying) choice of the α_i the desired result can be achieved. Projecting from $|\Phi(\alpha_i,X)>$ on to the state $|n>$, it must still be possible to write

$$W_n(\alpha_i) = \tilde{W}_o(\alpha_i) + \tilde{E}_n(\alpha_i) \quad , \tag{1.173}$$

which is here a separation of functions.

How shall we find the functions $\alpha_i(n)$? We use the method of variation after projection, i.e., we treat the expression for each n separately, at least to start with. Before doing so we complete the list of our assumptions. We write for the formal excitation function

$$\tilde{E}_n(\alpha_i) \equiv n\omega(n,\alpha_i) \tag{1.174}$$

and refer to ω as the variable frequency. We assume that ω is a monotonic function of n, which could be checked "experimentally" a posteriori. The essential role of this hypothesis will be seen imminently.

We turn first to the ground state variational principle which yields

$$\partial W_o/\partial\alpha_{io} = 0 \quad , \tag{1.175}$$

together with the condition that the matrix $\partial^2 W_o/\partial\alpha_{io}\partial\alpha_{jo}$ be positive definite. For the excited states, it is then convenient to rewrite (1.173) as

$$\begin{aligned} E_n &\equiv W_n(\alpha_i(n)) - W_o(\alpha_{io}) \\ &= \tfrac{1}{2} \sum(\partial^2 W/\partial\alpha_{io}\partial\alpha_{jo})(\alpha_i-\alpha_{io})(\alpha_j-\alpha_{jo}) + \ldots \\ &\quad + n\, \omega(n,\alpha_i) \quad . \end{aligned} \tag{1.176}$$

This can be transformed and simplified by the postulate of monotonicity, since this means there is a single-valued function

$$n = n(\omega) \quad , \tag{1.177}$$

and the assumption that the α_i are slowly varying functions of n can be translated into the assumption that the α_i are slowly varying functions of ω. Thus it is permissible to write

$$\alpha_i - \alpha_{io} = \kappa_i(\omega-\omega_o) + \ldots \tag{1.178}$$

and with

$$C_2 = \sum \kappa_i\kappa_j\, \partial^2 W_o/\partial\alpha_{io}\partial\alpha_{jo} > 0 \quad , \tag{1.179}$$

we have

$$E_n = \frac{1}{2!}\, C_2(\omega-\omega_o)^2 + \frac{1}{3!}\, C_3(\omega-\omega_o)^3 + \ldots + n\,\omega \quad . \tag{1.180}$$

Thus, because of the postulate that the variable ω serves just as well as n to label the states and the assumption of analyticity about zero energy, we have managed to replace the dependence on all the variables α_i by that upon a single variable ω. Therefore all that remains of the variational principle is the statement

$$(\partial E_n/\partial\omega)|_n = 0 \quad . \tag{1.181}$$

Formally, this equation defines ω_n as a function of n and the parameters,

$$\omega_n = \omega_n(n,\omega_o,C_2,C_3, \ldots) \quad . \tag{1.182}$$

Together with (1.180), this constitutes again an analytic continuation away from the region of small anharmonicity.

P. Equivalence of the Cranking and Variable-Frequency Analytic Continuations. Additional Formulations

Though the two modes of analytic continuation that have been presented are superficially quite distinct, we now prove their mathematical equivalence. We start from the cranking formulation and transform it into the other form. The required relationship between ρ and ω as one can guess from (1.154), (1.155) and (1.174) is

$$\sqrt{n} = \rho/2\omega \quad . \tag{1.183}$$

Equation (1.155) now shows that

$$\frac{1}{2\omega} - \frac{1}{2\omega_o} = \phi(\rho^2) \quad , \tag{1.184}$$

or

$$\omega = \omega_o = \Phi(\rho^2) \quad , \tag{1.185}$$

and we assume that this relation can be inverted.

Equation (1.183) permits us to write in place of (1.154)

$$E_n = \frac{\rho^2}{4\omega} + \frac{1}{2}\rho^2[C\rho^2 + 2d\rho^4 + 3f\rho^6 + \ldots]$$

$$= n\omega + F(\omega-\omega_o) \quad , \tag{1.186}$$

which is of the form (1.180). We have furthermore

$$\frac{dE}{d\rho} = \omega\frac{dn}{d\rho} + \frac{\partial E}{\partial\omega}\bigg|_n \frac{d\omega}{d\rho} \quad . \tag{1.187}$$

The first term is $\omega\cdot 2\cdot\sqrt{n}(d\sqrt{n}/d\rho = \rho(d\sqrt{n}/d\rho)$, which, according to (1.167), precisely cancels the left hand side. Thus we find that the second term must vanish separately, and this is the remaining condition (1.181) of the variable-frequency model.

It has turned out therefore that the two methods of analytic continuation that we have given are in reality one. Nevertheless a little consideration convinces one that it is not difficult to invent additional techniques of analytic continuation.

We give just one more whose analogue for rotations is interesting.[37] Consider Eq. (1.176) once more, and just for the sake of simplicity, a single parameter α, though the argument goes through for any number as before. With obvious definitions, we then have

$$E_n = \frac{1}{2} C'(\alpha-\alpha_0)^2 + n\,\omega$$
$$\cong \frac{1}{2} C'(\alpha-\alpha_0)^2 + n\,\omega_0 + n\,\frac{\partial\omega}{\partial\omega_0}\,(\alpha-\alpha_0) \quad . \tag{1.188}$$

We now first apply the variational principle, which gives

$$C(\alpha-\alpha_0) = -\,n\,\frac{\partial\omega}{\partial\alpha_0} \quad , \tag{1.189}$$

and (in one possible form)

$$E_n = -\,\frac{1}{2} C'(\alpha-\alpha_0)^2 + n\,\omega_0 \quad . \tag{1.190}$$

Strictly the first term should be replaced by a term in n^2, since $(\alpha-\alpha_0) \sim n$, according to (1.189). But instead of n we now choose for continuation purposes

$$(\alpha-\alpha_0) = \sigma\,E_n + \ldots \quad . \tag{1.191}$$

Again with a suitable definition, we now have

$$E_n = -\,\frac{1}{2} C\,E_n^{\,2} + n\,\omega_0 \quad , \tag{1.192}$$

which can be viewed first as a simple harmonic oscillator whose frequency diminishes with excitation energy (C>0)

$$E_n = n\omega_0/(1 + \frac{1}{2} C\,E_n) \quad , \tag{1.193}$$

and on the other hand yields the explicit solution

$$C\,E_n = \sqrt{4 + 8\,Cn\omega_0} - 1 \quad . \tag{1.194}$$

Q. Other Proposals for Intrinsic Functions to be Used in H.W. Projection

The generating functions already used have been sufficient to give us formulations of the problem of collective motion as precise and general as one may wish. Nevertheless additional suggestions have appeared in the past when the formal solution was not yet so clear and will continue undoubtedly to appear in the futute until the knowledge that the problem has been solved becomes sufficiently widespread. For example, several authors[28,21,20,18], differently motivated, have successfully utilized a method of double projection, using two parameters for every collective degree of freedom. Thus they consider projections in a form equivalent to the equation

$$|\Psi_n\rangle = \int d\alpha dp_\alpha \; f_n(\alpha,p_\alpha)|\Phi(\alpha,p_\alpha)\rangle \quad , \tag{1.195}$$

which can be viewed effectively as projection from some form of the cranking intrinsic state. One can also use a complex form instead of the real form shown and indeed it is only in the latter form that the problem has been carried out in any generality. Using the strong-deformation approximation (steepest descents) the results of these lectures can be regained. We should also refer at this point to one of the first papers[39,40] to recognize the importance of utilizing an intrinsic state which varied slowly with excitation, though the motivation was strictly to correct the occasional inadequacies of the P.Y. method.

R. Cranking and the Adiabatic Approximation

A generalization of the cranking theorem of section K to the adiabatic case can be made. Let $|\Phi(\alpha)\rangle$ be an exact intrinsic state with mean value α for X. We define the intrinsic state

$$|\Phi(\alpha,p_\alpha)\rangle = \exp[i\hat{X}\;p_\alpha]\;|\Phi(\alpha)\rangle \quad . \tag{1.196}$$

The analogy with (1.105) has dictated the notation. Using Eq. (1.60) for $\hat{H}$, we then find

$$\begin{aligned}\langle\Phi(\alpha,p_\alpha)|\hat{H}|\Phi(\alpha,p_\alpha)\rangle &= \langle\Phi(\alpha)|H(\hat{P} + p_\alpha,\hat{X})|\Phi(\alpha)\rangle \\ &= \sum(p_\alpha)^k \;\langle\Phi(\alpha)|[H^{(k)}(\hat{X}) + \ldots]|\Phi(\alpha)\rangle \quad .\end{aligned} \tag{1.197}$$

Here the continuation dots indicate the omission of corrections of relative order N^{-1}, as has previously been demonstrated (the arguments of section K continue to be applicable in this regard).

To render the present formulation useful, we must make a further expansion of (1.197), which amounts to assuming that we are near the strong deformation limit so that the state $|\Phi(\alpha)\rangle$ is sharply localized in X. In particular we assume that

$$|\langle(X-\alpha)^2\rangle H^{(k2)}(\alpha)| \ll |H^{(k)}(\alpha)| \quad . \tag{1.198}$$

Under these circumstances (1.197) becomes

$$\langle\Phi(\alpha,p_\alpha)|\hat{H}|\Phi(\alpha,p_\alpha)\rangle \cong \sum\;(p_\alpha)^k\;H^{(k)}(\alpha) \quad , \tag{1.199}$$

and again, if we can construct the intrinsic state (1.196), we can obtain a good approximate construction of the collective Hamiltonian.

In accordance with previous ideas, we shall use cranking to construct the state $|\Phi(\alpha)\rangle$. Here we believe it is particularly important to stress that it is the intrinsic Hamiltonian that must be cranked. To really obtain (1.196), we must compute $|\Phi(\alpha)\rangle$ for a relatively large range of values of α and insist that it be a "good" intrinsic state over this range. Thus in looking for solutions of

$$\delta<[\hat{H}_{in} - \mu\hat{X}]> = 0 \quad ,$$

$$<\hat{X}> = \alpha \quad , \tag{1.200}$$

a true solution will correspond to $\mu \equiv 0$, and a good solution to one which is relatively flat in μ about some minimum. Using the statement $\mu = 0$ gives us that $<\hat{H}_{in}>$ is independent of α, or

$$\frac{\partial}{\partial\alpha} <\Phi(\alpha)|\hat{H}|\Phi(\alpha)> \cong \frac{\partial}{\partial\alpha} <\Phi(\alpha)|H^{(0)}(X)|\Phi(\alpha)>$$

$$\cong \frac{\partial}{\partial\alpha} H^{(0)}(\alpha) \quad . \tag{1.201}$$

In this equation we have assumed that $H^{(0)}(X)$ is the main part of the collective Hamiltonian. (A correction for the kinetic energy can later be made if desired.) Thus, the cranking problem as thus defined determines the collective potential energy.

In this form of the problem, it is certainly advantageous to prescribe the operator X on physical grounds, because then to a reasonable approximation we have

$$<\Phi(\alpha)|[\hat{X},[\hat{X},\hat{H}]]|\Phi(\alpha)>$$

$$\cong - 2 H^{(2)}(\alpha) \quad , \tag{1.202}$$

and we need not explicitly construct the form of $\hat{P}$ in this case.

Do we in fact need an explicit version of P under any circumstance? This should not be necessary at all in the strong deformation approximation where we replace the action of P by the appropriate differential operator with respect to α.

Under these circumstances, as an alternative to (1.202) and as a check of consistency, we can study the cranking problem

$$\delta<H_{in} - \lambda\hat{P}> = 0 \quad ,$$

$$<\hat{P}> = p_{\alpha} \quad , \tag{1.203}$$

from which $H^{(2)}(\alpha)$ can be obtained for each α by a power series in λ. We leave details to the reader.

In the adiabatic limit, the procedure just described is not necessarily the only useful one. Suppose we have a strong deformation and an intrinsic state $|\Phi(\alpha_0)>$ determined at the equilibrium position. Suppose also that it is the operator P that is suggested by physical considerations. We then have directly

$$|\Phi(\alpha)> = \exp[-i\hat{P}(\alpha-\alpha_0)]|\Phi(\alpha_0)> \tag{1.204}$$

and

$$<\Phi(\alpha)|\hat{H}|\Phi(\alpha)> \cong H^{(0)}(\alpha) + \ldots \quad . \tag{1.205}$$

According to (1.202) we have now to construct X. This can be done by solving the equation of motion

$$[\hat{X},\hat{H}] = i\ \{\hat{P},H^{(20)}(X)\}\quad . \tag{1.206}$$

In this case $\hat{P}$ is known explicitly as a fermion operator, and we study matrix elements between members of a complete set of which $|\Phi(\alpha)\rangle$ is the lowest member. Here $H^{(20)}(\alpha)$ will have to be calculated self-consistently between (1.206) and (1.202).

The two methods we have given must in fact be connected by a unitary transformation, but we shall not enter here into any detailed discussion of this matter.

The concepts of this section need clarification in practice. For this the reader must turn to part II.H.

II. ILLUSTRATION OF VARIOUS MICROSCOPIC METHODS FOR AN EXACTLY SOLUBLE MODEL

A. The model of Meshkov-Glick-Lipkin (MGL), Qualitative Discussion

Since its introduction,[41-45] the model of MGL has done yeoman service as a testing ground for approximations in the theory of collective motion. We shall review some of this work, with obvious emphasis on our own contributions. We shall, however, also introduce some new suggestions, some of which are being carried out now. In many, but not all, cases these are related to the general descriptions of part I. We shall also indicate how to go beyond the study of a single degree of freedom and introduce coupling to other degrees of freedom in a consistent way. This will take us beyond our original program and a step closer to the real world.

The model to be studied consists of N identical fermions in two N-fold degenerate levels separated by the single-particle energy ε and undergoing a highly simplified residual interaction. The Hamiltonian for this system is

$$H = \tfrac{1}{2}\varepsilon \sum_{p,\sigma} \sigma a^{\dagger}_{p\sigma} a_{p\sigma} - \tfrac{1}{2} V \sum_{pp'\sigma} a^{\dagger}_{p\sigma} a^{\dagger}_{p'\sigma} a_{p'-\sigma}\, a_{p-\sigma}\quad , \tag{2.1}$$

where V is positive. The index σ, $\sigma = \pm 1$, distinguishes the upper from the lower level, and $p = 1,2,\ldots N$. Definition of the operators

$$J_+ = (J_-)^{\dagger} = \sum_p a^{\dagger}_{p+} a_{p-}\quad , \tag{2.2}$$

$$J_o = \tfrac{1}{2} \sum_p [a^{\dagger}_{p+} a_{p+} - a^{\dagger}_{p-} a_{p-}]\quad , \tag{2.3}$$

allows us to write the Hamiltonian as

$$H = \varepsilon J_0 - \tfrac{1}{2} V(J_+^{\,2} + J_-^{\,2}) \quad , \tag{2.4}$$

wherein the operators introduced obey the familiar commutation relations

$$[J_+, J_-] = 2J_0 \quad , \tag{2.5}$$

$$[J_0, J_\pm] = \pm J_\pm \quad , \tag{2.6}$$

that identify the generators of the group SU(2). We further see that the Casimir operator of the group,

$$J^2 = J_+ J_- + J_0^{\,2} - J_0 \quad , \tag{2.7}$$

commutes with the Hamiltonian, and we may thus consider separately those states which span a representation of the group. It is this decomposition of the entire shell model space into non-interacting "bands" which renders the problem exactly soluble - which only means here that we can carry out the diagonalization of the Hamiltonian for any reasonable value of N.

For the most part we shall study the ground state band for a system with particle number equal to the degeneracy of either level. Since when V = 0 the ground state consists of the lower level fully occupied, and it is thus the fully "stretched" state with $J_0 = -\tfrac{1}{2}N$, annihilated by J_-, we identify the value of J as $\tfrac{1}{2}N$ and find for (2.7)

$$J_+J_- + J_0^{\,2} - J_0 = \tfrac{1}{2}N\,(\tfrac{1}{2}N + 1) \quad . \tag{2.8}$$

We are thus dealing with a space of 2J+1 = N+1 states which we label $|n\rangle$, n = 0,1,...N. Because the Hamiltonian contains only the square of the raising or lowering operators, $J_\pm$, it does not connect states with even and odd n. This reduces by a factor of two the dimensionality of the largest matrix that need be diagonalized. In fact, it is the existence of this extra (discrete) symmetry of the Hamiltonian, which we may consider as a reflection symmetry, that renders the problem interesting, with some relation, conceptually, to the real world.

We shall justify this last remark by a qualitative discussion of the spectrum as a function of the interaction strength V, preceded by a few remarks about notation.

Because we can choose all matrix elements, we have

$$\langle n|J_-|n'\rangle = \langle n'|J_+|n\rangle \equiv J_-(nn')(n'>n) \equiv J_+(n'n)(n>n') \quad , \tag{2.9}$$

$$\langle n|J_0|n'\rangle = \langle n'|J_0|n\rangle = J_0(nn') \quad . \tag{2.10}$$

As $V \to 0$, we have by transcription of the well-known angular momentum matrices,

$$J_0(nn) = -\tfrac{1}{2}N + n \quad ,$$

$$J_-(n,n+1) = [(n+1)(N-n)]^{\frac{1}{2}} \quad , \qquad (2.11)$$

all others vanishing. At this limit the spectrum is purely harmonic, energy separation ε. If we apply perturbation theory to the computation of the energy change, it is easy to see from (2.11) that for $n << N$ the expansion parameter is $\delta = (NV/\varepsilon)$. As long as $\delta << 1$, there will be a near harmonic spectrum and the unperturbed ($V = 0$) states will only be weakly mixed. As δ approaches and exceeds unity, one should observe a rapid change of wave function with interaction strength, each true eigenstate becoming a strong mixture of unperturbed eigenstates. Detailed calculations show that this change occurs over a very small interval of δ, an interval of order 1. Nuclear physicists have borrowed the phrase "phase transition" from the field of statistical mechanics, though some workers from the latter are (unjustifiably) horrified.

When we reach the situation $\delta >> 1$, what indeed must be the structure of the ground state? The single-particle energy difference can no longer matter. The average occupancy of the upper level must equal that of the lower level in the exact ground state. This means that the relation

$$\langle\Psi_0|J_0|\Psi_0\rangle = 0 \qquad (2.12)$$

holds. (For the remainder of this discussion let us distinguish exact and corresponding free states by $|\Psi_n\rangle$ and $|\Phi_n\rangle$, respectively.) In fact the simplest way that (2.12) can be satisfied is by writing for the ground state its expansion in terms of unperturbed states,

$$|\Psi_n\rangle = \sum_{n,\text{even}} C_{on}\,|\Phi_n\rangle \quad , \qquad (2.13)$$

and anticipating that C_{on} defines a smooth distribution peaked about $n \fallingdotseq N/2$.

We must now remember that the state $|\Psi_1\rangle$ is the ground state of the level system with opposite parity. It is intuitively clear that the interaction will try to lower its energy as much as possible, and remembering that there is no repulsion between states of different symmetry, we expect that its state vector will be approximately

$$|\Psi_1\rangle = \sum_n C_{on}\,|\Phi_{n+1}\rangle = \sum_n C_{on}\,\frac{J_+}{\sqrt{(n+1)(N-n)}}\,|\Phi_n\rangle \quad , \qquad (2.14)$$

and that its energy will be nearly that of the ground state. In the limit $V \to \infty$, the distribution C_n should be rather strongly peaked about its maximum, so that we can approximately write $\sqrt{(n+1)(N-n)} \approx \sqrt{\frac{N}{2}(N - \frac{N}{2})} = \frac{1}{2}N$. We then have from (2.14)

$$\tfrac{1}{2}N|\Psi_1\rangle \cong J_+|\Psi_o\rangle \quad , \tag{2.15}$$

or

$$\langle\Psi_1|J_+|\Psi_o\rangle = \tfrac{1}{2}N \quad . \tag{2.16}$$

It turns out that the corrections to this result are O(1) and the rest of the argument to follow is qualitatively but not quantitatively correct. Remembering (2.12), (2.8), and assuming (incorrectly as just noted) that to the required order,

$$\langle\Psi_o|J_-J_+|\Psi_o\rangle \cong |\langle\Psi_1|J_+|\Psi_o\rangle|^2 = \tfrac{1}{4}N^2 \quad , \tag{2.17}$$

we obtain from (2.8)

$$\langle\Psi_o|J_o{}^2|\Psi_o\rangle = \tfrac{1}{2}N \quad . \tag{2.18}$$

A simple wave function[45] incorporating the results (2.12) and (2.18) is the Poisson distribution

$$C_{on} = \frac{\beta^n}{\sqrt{n!}}\ e^{-\frac{1}{2}\beta^2} \quad , \quad \beta^2 = \tfrac{1}{2}N \quad . \tag{2.19}$$

To summarize: In the weak coupling limit we have a near-harmonic spectrum. In the strong coupling limit, we can extend the previous argument to arrive at the conclusion that as long as $n \ll N$, we must expect the degeneracy or near degeneracy of pairs (0,1), (2,3), etc., with a gradual spreading out as $n \to N$, where our qualitative arguments break down. Where the argument relating to (2.17) and (2.18) goes wrong in detail is that we now have, in consequence of the doublet structure,

$$|J_+(30)| \sim |J_+(21)| \sim \sqrt{N} \quad . \tag{2.20}$$

Most theories of collective motion are special approximations which provide more or less accurate solutions in one or the other limit. We shall develop and review some of these approximations, but we start with an extremely powerful method, described in the next section, capable, without changing the approximation, of taking us through from the weak to the strong-coupling limit. It yields for all coupling strengths a consistent theory and at a suitable level of approximation an accurate one.

B. The Algebraic or Sum-Rule Approach as a Complete Dynamical Theory[46,47]

In the shell model and particularly in the present model, all physical operators are polynomials in the generators of a certain Lie group. Matrix elements of such quantities between physical states can then be expressed by sum-rule techniques in terms of the matrix elements between physical states of the generators themselves. If G_1 and G_2 are two such generators and $|A\rangle$, $|A'\rangle$ are actual eigenstates of H, we write

$$\langle A|G_1 G_2|A'\rangle = \sum_{A''} \langle A|G_1|A''\rangle \langle A''|G_2|A'\rangle \tag{2.21}$$

and assume that the sum rule is sensibly satisfied by a small number of intermediate states. Any assumptions as to order of magnitude can be checked for consistency from the results of the calculation that ultimately follows.

We now ask: To what conditions are such sum rules as (2.21) to be applied so as to give a closed set of equations determining a finite set of matrix elements? (It goes without saying that with some conceivable exceptions, we always start with ground-state matrix elements and study then the lowest energy states of the system which are collective with respect to the ground state and to each other.) The conditions divide themselves into two sets, kinematical and dynamical, to which we assign equal weight per relation in the present approach.

The kinematical relations are all the algebraically independent statements which define the representation of the Lie algebra under study. This means: (i) The commutation relations (here (2.5), (2.6)). (ii) The Casimir operators (here only (2.7)). One exploits these fully (see below) and then adds conditions from the dynamics. There are also two subdivisions for the latter: (iii) We want the Hamiltonian to be diagonal. The statement $\langle A|H|A'\rangle = 0$, $A \neq A'$, as one sees from (2.4), provides further usable sum rules. (iv) Finally we want the diagonal matrix elements $\langle A|H|A\rangle$ to yield the correct energies W_A. These relations can be utilized, as we have found, in two equivalent forms: We study the non-trivial matrix elements of the equations-of-motion for all the generators

$$[G_1, H] \equiv P_1[G] \quad , \tag{2.22}$$

where $P_1[G]$ means the appropriately-named polynomial in G. This brings into the equations and therefore into the set of quantities to be determined the energy differences $\omega_{AA'} = W_A - W_{A'}$. Alternatively, it turns out to be sufficient to study only the Hermitian polynomials in the generators, Q[G], for which the diagonal matrix elements

$$\langle A|[Q[G],H]|A\rangle = 0 \tag{2.23}$$

yield non-trivial equations. In fact (2.23) is just an operator form of the Rayleigh-Ritz variational principle. We see this by noting that (2.23) is equivalent to the requirements

$$\delta\langle A|H|A\rangle = 0 \quad , \tag{2.24}$$

$$\delta\langle A|A\rangle = 0 \quad , \tag{2.25}$$

since the second of these relations may be eliminated by writing

$$\delta|A\rangle = iqQ\,|A\rangle \quad , \tag{2.26}$$

where Q is an arbitrary Hermitian operator and q an infinitesimal real parameter. Substitution of (2.26) into (2.24) yields (2.23). Finally it is itself a consequence of completeness that if all possible relations (2.22), diagonal and off-diagonal, are satisfied then so are all possible relations (2.23), assuming that Q is a polynomial in the generators.

Such a program has been carried out with great success for the model of MGL[47] exactly as described, using (2.22) instead of (2.23), but only because the relationship between them was an afterthought of the actual work. In fact, the effort was not completely trivial and not without great educational value. One learns: (a) The set of approximate algebraic relations one finds at a certain level (defined by the number of exact eigenstates being described) normally exceeds the number of matrix elements being studied. In general then, there are consistency checks. (b) On the other hand, some of the equations become identical in the extreme weak or strong coupling limit, and therefore one has to examine these limits carefully and understand them in order to make a proper selection of equations to solve. (c) The limiting cases also determine whether a given level is a consistent one or not. Thus for the model of MGL, a consistent level must, because of the degeneracy in the strong coupling limit, add two states at a time. Thus sensible schemes are possible which include states 0 and 1 or 0-3 only. (d) Finally, then, we depend on a plausibility argument that if we have a scheme which makes sense in both extremes, it will extrapolate smoothly through the transition region. In this we were not disappointed.

C. Application of the Algebraic Approach to the Model

If we wish to describe the first excited degenerate doublet in the strong coupling limit, we need equations at a level which includes 4 states. Our philosophy, justified by calculation, is that the elements $J_\pm(n,n')$, $J_0(n,n')$ go rapidly to zero as $\Delta n = |n'-n|$ increases. Consequently for maximum accuracy we should try to use matrix elements involving states as near to the ground state as possible so as to exhaust the sum rules within the allowed space. In writing the relations which follow, we make the definite assumption that all sum rules break off within or at the edge of the allowed space.

From the commutator (2.5) we find the equations

$$2J_0(00) = J_+(01)^2 - J_-(01)^2 + J_+(03)^2 - J_-(03)^2 \ , \quad \text{(I)}$$

$$2J_0(11) = J_-(01)^2 - J_+(01)^2 + J_+(12)^2 - J_-(12)^2 \ , \quad \text{(II)}$$

$$2J_0(02) = J_+(01)J_-(12) - J_-(01)J_+(12) + J_+(03)J_+(23) - J_-(03)J_-(23) \quad , \tag{2.27}$$

$$2J_o(13) = J_-(01)J_-(03) - J_+(01)J_+(03) + J_+(12)J_-(23) - J_-(12)J_+(23) \quad , \qquad \text{(III)}$$

where the left hand side tells us which matrix element has been taken, and where we have not made a mistake in numbering. The reason for differentiation of these equations will be discussed below. From the commutator (2.6), we obtain the two relations

$$[J_o(11)-J_o(00)-1]J_-(01) = J_o(02)J_+(12) - J_o(13)J_-(03) \quad , \qquad \text{(IV)}$$

$$[J_o(11)-J_o(00)+1]J_+(01) = J_o(02)J_-(12) - J_o(13)J_+(03) \quad , \qquad \text{(V)}$$

from the matrix elements $<0|,|1>$ and $<1|,|0>$ respectively. From the Casimir operator, relation (2.8), we extract the lowest diagonal elements in the two relations

$$\tfrac{1}{2}N(\tfrac{1}{2}N+1) = J_o(00)[J_o(00)-1] + J_o^2(02)+J_+^2(01)+J_+^2(03) \quad , \qquad \text{(VI)}$$

$$\tfrac{1}{2}N(\tfrac{1}{2}N+1) = J_o(11)[J_o(11)-1] + J_o^2(13)+J_-^2(01)+J_+^2(12) \quad . \qquad \text{(VII)}$$

There are, of course, still additional kinematical relations obtained from taking matrix elements involving the upper states in our set, which we shall mention again below, but these are intrinsically less accurate because of the omission of contributions to the sum rules from states outside the original set.

Following along in the same spirit, we next consider the dynamics. From the conditions $<0|H|2> = <1|H|3> = 0$, we find the equations

$$0 = \varepsilon J_o(02) - \tfrac{1}{2}V[J_-(01)J_-(12) + J_+(01)J_+(12) + J_-(03)J_+(23) + J_+(03)J_-(23)] \quad , \qquad \text{(VIII)}$$

$$0 = \varepsilon J_o(12) - \tfrac{1}{2}V[J_+(01)J_-(03) + J_-(01)J_+(03) + J_-(12)J_-(23) + J_+(12)J_+(23)] \quad . \qquad \text{(IX)}$$

Finally, from the commutators

$$[J_\pm,H] = \mp \varepsilon J_\pm - VJ_\mp + 2VJ_oJ_\mp \quad , \qquad (2.28)$$

$$[J_o,H] = -V(J_+{}^2 - J_-{}^2) \quad , \qquad (2.29)$$

we derive five more dynamical relations,

$$\omega_{10}J_-(01) = \varepsilon J_-(01) - V[1-2J_o(00)]J_+(01) + 2VJ_o(02)J_-(12) \quad , \qquad \text{(X)}$$

$$\omega_{10}J_-(01) = -\varepsilon J_+(01) - V[1+2J_o(00)]J_-(01) - 2VJ_o(02)J_+(12) \quad , \tag{XI}$$

$$\omega_{21}J_-(12) = \varepsilon J_-(12) - V[1-2J_o(11)]J_+(12) + 2VJ_o(13)J_-(23) \quad , \tag{XII}$$

$$\omega_{21}J_+(12) = -\varepsilon J_+(12) - V[1+2J_o(11)]J_-(12) - 2VJ_o(13)J_+(23) \quad , \tag{XIII}$$

and finally

$$\omega_{20}J_o(02) = V[J_-(01)J_-(12) - J_+(01)J_+(12) + J_-(03)J_+(23) - J_+(03)J_-(23)] \quad . \tag{XIV}$$

If we now examine Eqs. (I) - (XIV), we see that they constitute a set of non-linear equations for the fourteen dynamical quantities $J_0(00)$, $J_0(11)$, $J_0(02)$, $J_0(13)$, $J_\mp(01)$, $J_\mp(12)$, $J_\mp(23)$, $J_\mp(03)$ and in addition the excitation energies ω_{10}, ω_{21}.

If instead of the equations of motion we wished to use the Rayleigh-Ritz principle, we leave it as an exercise to the reader to show that the last five equations can be replaced by the three equations

$$\langle 0|[\Theta_1,H]|0\rangle = \langle 1|[\Theta_1,H]|1\rangle = \langle 0|[\Theta_2,H]|0\rangle = 0 \quad , \tag{2.30}$$

where

$$\Theta_1 = i[J_+{}^2 - J_-{}^2] = \Theta_1{}^\dagger \quad , \tag{2.31}$$

$$\Theta_2 = i[J_+{}^2J_o - J_oJ_-{}^2] = \Theta_2{}^\dagger \quad . \tag{2.32}$$

These equations can be derived from (X) to (XIV) by appropriate combination and elimination of the energies. For a satisfactory analytical derivation, one must add to the set (X) - (XIV) the equations of motion

$$\omega_{30}J_\mp(03) = \pm\, \varepsilon J_\mp(03) + V[1\mp 2J_o(00)]J_\pm(03) \mp 2VJ_o(02)J_\pm(23) \tag{2.33}$$

Numerically it turns out that these last equations contribute little to the solution, and so (2.30) - (2.32) are for all intents and purposes numerically equivalent to the set which we have used.

It should also be clear by now that there are more relations available than we can use, a first example having been exhibited in (2.27). This presents us both with a problem and an opportunity. The problem is naturally one of choice. It is resolved by examining (analytically, if possible) the limiting cases of weak and strong coupling, where there is simplification, and demanding that the equations chosen give a unique solution which can be continued into the region of intermediate coupling. This was the reason, by no means evident on superficial examination, for rejecting Eq. (2.27) in favour of Eq. (III).

The fourteen equations were solved analytically for $V \to 0$ and $V \to \infty$ and numerically in between. For details we refer to the original paper. The analytic solution found in the strong-coupling limit, for example, represents a definite improvement over the remarks found in section A. Results, in our view, are quite exceptional. We present a table for N = 20 for various values of $NV(\varepsilon=1)$. It will be noted that, of the quantities contained explicitly in the set of equations, those least well given are $J_{\pm}(23)$ and $J_{\pm}(03)$. These involve the boundary state n=3, and so are the least well constrained. We also show some quantities not included among the fourteen variables. We find the absolute energies of the ground and first-excited states, W_0 and W_1, in the obvious way, by evaluating the expectation values of H,

$$W_0 = \varepsilon J_0(00) - V[J_+(01)J_-(01) + J_+(03)J_-(03)] \quad , \qquad (2.34)$$

$$W_1 = \varepsilon J_0(11) - V[J_-(01)J_+(01) + J_+(12)J_-(12)] \quad . \qquad (2.35)$$

Their difference is identical to the directly calculated ω_{10}. The quantity ω_{30} follows from Eq. (2.33), both versions yielding the same number. The absolute energies, W_2 and W_3, come from

Table 1

Approximate and exact results for the model of MGL with $N = 20$

NV	1.0		1.8		3.0		20.0	
	approx.	exact	approx.	exact	approx.	exact	approx.	exact
ω_{10}	0.5809	0.5817	0	0.0613	0	0.0021	0	0.0000
ω_{21}	0.8158	0.8112	1.6581	1.3721	3.5173	3.5362	25.5682	26.5975
ω_{30}	2.1168	2.3178	1.6581	2.1203	3.5173	3.6454	25.5682	26.5975
W_0	−10.2915	−10.2915	−11.6573	−11.6910	−16.1870	−16.1896	−96.1702	−96.1753
W_1	−9.7106	−9.7099	−11.6573	−11.6297	−16.1870	−16.1875	−96.1702	−96.1753
W_2	−8.8948	−8.8987	−9.9992	−10.2576	−12.6697	−12.6534	−70.6020	−69.5777
W_3	−8.1747	−7.9737	−9.9992	−9.5707	−12.6697	−12.5442	−70.6020	−69.5768
$J_0(00)$	−9.5641	−9.5644	−6.1586	−6.3765	−3.6176	−3.6257	− 0.5391	− 0.5391
$J_-(01)$	5.1886	5.1900	7.7418	7.8898	9.2708	9.2687	9.9527	9.9492
$J_+(01)$	2.8130	2.8115	7.7418	7.5292	9.2708	9.2581	9.9527	9.9492
$J_0(12)$	−8.1680	−8.1560	−6.1586	−5.9142	−3.6176	−3.6111	− 0.5931	− 0.5931
$J_-(12)$	6.4564	6.4690	3.5250	4.2241	2.7977	2.8253	2.0016	2.0077
$J_+(12)$	2.5181	2.5598	0.3294	1.1000	−0.7694	−0.6216	1.7111	− 1.7031
$J_0(02)$	1.0165	1.0137	2.4226	2.2774	2.5990	2.5126	2.6807	2.6771
$J_0(13)$	1.3617	1.3326	2.4226	2.2732	2.5990	2.5492	2.6807	2.6771
$J_-(03)$	0.3703	0.3422	3.5250	2.6956	2.7977	2.7859	2.0016	2.0076
$J_+(03)$	−0.1303	−0.1393	0.3294	−0.1074	−0.7694	−0.7085	− 1.7111	− 1.7032
$J_-(23)$	7.2963	7.4426	6.2255	6.7176	7.8136	7.7559	8.5086	8.8520
$J_+(23)$	2.7782	2.3922	6.2255	3.7947	7.8136	7.7559	8.5086	8.8512
$J_0(22)$	−6.947	−7.185	−4.952	−7.610	−3.049	−4.859	− 0.461	− 0.649
	−7.025		−6.200		−3.971		− 1.473	
$J_0(33)$	−5.735	−6.215	−4.952	−5.668	−3.049	−4.233	− 0.461	− 0.648
	−7.615		−6.200		−3.971		− 1.473	

combining properly ω_{21}, ω_{30}, W_o and W_1. Finally we show two approximations to $J_o(22)$ and $J_o(33)$, the only ones of the matrix elements within the allowed space not included in the equations solved. For the calculation of these, there remain several relations following from (2.6), namely

$$\pm J_{\pm}(12) = J_o(11)J_{\pm}(12) + J_o(13)J_{\mp}(23) - J_{\pm}(12)J_o(22) + J_{\mp}(01)J_o(02) \quad , \qquad (2.36\pm)$$

$$\pm J_{\pm}(03) = J_o(00)J_{\pm}(03) + J_o(02)J_{\pm}(23) - J_{\pm}(03)J_o(33) - J_{\pm}(01)J_o(13) \quad , \qquad (2.37\pm)$$

and several off-diagonal elements of the Casimir operator,

$$0 = J_+(01)J_-(12) + J_+(03)J_+(23) + J_o(00)J_o(02) + J_o(02)J_o(22) - J_o(02) \quad , \qquad (2.38)$$

$$0 = J_+(12)J_-(23) + J_-(01)J_-(03) + J_o(11)J_o(13) + J_o(13)J_o(33) - J_o(13) \quad . \qquad (2.39)$$

The two versions of $J_o(22)$ shown in the table follow from (2.36-) and (2.38), whereas the values for $J_o(33)$ follow from (2.37-) and (2.39). The remaining equations turn out to be inaccurate except as $V \to 0$.

Many of our results are indistinguishable from the exact ones except in the transition region. It is to be expected, of course, that fluctuations are most severe in this region.

We confine our final remarks on this section to the observation that the algebraic approach under study can be viewed as the natural method for the direct construction of the physical boson representation as it was defined in part I.B. Thus in the expansion (1.11), we can from the results of this section obtain the coefficients $a_{1,2,3}$ but we can also obtain the coefficients in the expansion,

$$J_+ = (J_-)^\dagger = \sum_{ik} j_{ik}(B^\dagger)^i(B)^k \quad , \qquad (2.40)$$

with J_o obtained from the commutator $J_o = \frac{1}{2}[J_+,J_-]$, and where $(i,k) \le N$ and $|i-k|$ odd. For example, we have

$$J_+(n,n+1) = (n+1)^{\frac{1}{2}}[j_{01}+nj_{12}+n(n-1)j_{23}+\ldots\ n!\ j_{n,n+1}] \quad , \qquad (2.41)$$

and as with the energy we would expect rapid convergence for a nearly harmonic situation.

Table 2

Excitation energy of the four-phonon state, ω_{40}, calculated by the physical boson method, (a) using approximate energies for the lower states, (b) using exact energies for the lower states, compared with the exact value for the model of MGL with $N = 20$

NV	Approx. ω_{40}, (a)	Approx. ω_{40}, (b)	Exact ω_{40}
0.2	3.9789	3.9995	3.9657
0.4	3.9263	4.0119	3.8665
0.6	3.8799	4.0811	3.7134
0.8	3.9153	4.2856	3.5251
1.0	4.1509	4.7336	3.3283

We illustrate the situation for the energy by predicting ω_{40} from the previous calculation and the physical boson idea. The results are shown in Table 2, where as expected the physical boson representation clearly shows the breakdown of the idea of small vibrations as we approach the transition region in coupling strength.

It should be emphasized that our algebraic solution is, however, general enough to include not only the vibrational picture in the weak coupling limit, but also the deformed picture in the strong coupling limit.

Some remarks about the connection with the usual RPA should be made, but we postpone them to section II.F, where a special discussion of the RPA will be carried out.

D. Transition Operator Boson Representation

In the following two sections we shall describe methods of solving the problem of collective motion by the specific transcription of fermion operators into functions of boson variables. We shall describe two general approaches. The first method developed in this section is new, its genesis barely predating the period of preparation of these lecture notes. It is, however, so obviously a powerful approach that though detailed calculation remains to be carried out, we shall describe the mathematics and the qualitative physics.

The search for such a method was stimulated by the discussion of section I.C, namely we look for a representation in which the collective monopole operator is directly the coordinate X of the simple harmonic oscillator. We must then ask which is the collective operator in the model of MGL. For an attractive interaction, we see that it is $J_x = (J_+ + J_-)/2$ by rewriting the Hamiltonian in the form

$$H = \varepsilon J_0 - |V|[J_x^2 - J_y^2] \quad . \tag{2.42}$$

In the weak coupling limit, either J_x or J_y can be considered, but as $|V| \to \infty$, in order to minimize the energy, $\langle J_x^2 \rangle$ will grow at the expense of $\langle J_y^2 \rangle$.

If we rewrite the commutation relations in Cartesian form,

$$[J_x,J_y] = iJ_o \quad , \tag{2.43}$$

$$[J_o,J_y] = -iJ_x \quad , \tag{2.44}$$

$$[J_o,J_x] = iJ_y \quad , \tag{2.45}$$

then one sees immediately that the ansatz

$$J_x = N\hat{X},\ J_y = -\hat{P}\ ,\ J_o = -N \quad , \tag{2.46}$$

where N is a constant, will satisfy (2.43), but J_o will have to be amplified in order to satisfy (2.44), (2.45). When this is done in a minimal way (one must add terms proportional to $\hat{P}^2$ and $\hat{X}^2$), then (2.43) is no longer satisfied. But here one sees upon examination that there are many ways of proceeding. At the extremes, one can keep <u>either</u> the simple form for J_x or the simple form for J_y. As it turns out (and as we shall see), the choice $J_y = -\hat{P}$ appears to be the best to make from a physical point of view, since it corresponds directly to the diminishing importance of the collective momentum in the limit of large interaction. Finally, it was noticed after some manipulation that a closed form of solution was possible.

Thus, the transformation

$$J_x = N\{\sin X,\ \emptyset(P^2)\} \quad ,$$

$$J_y = -P \quad ,$$

$$J_z = -N\{\cos X,\ \emptyset(P^2)\} \quad , \tag{2.47}$$

is an exact solution of (2.43) and (2.44). The remaining commutator (2.45) is evaluated most easily by converting trigonometric functions to exponential ones and using the displacement operator property

$$\exp[-i\lambda X]\emptyset(P^2)\exp[i\lambda X] = \emptyset((P+\lambda)^2) \equiv \emptyset_\lambda \quad . \tag{2.48}$$

We then find straightforwardly the difference equation

$$P = \tfrac{1}{2}N^2(\emptyset_{-1}-\emptyset_1)(\emptyset_{-1}+2\emptyset+\emptyset_1) \quad . \tag{2.49}$$

Before considering this equation in detail, we note that the constant N, which is the only handle we have for insisting that we are dealing with fermions, must then be determined from the Casimir operator. Evaluation of the latter yields the additional equation

$$\tfrac{1}{2}N(\tfrac{1}{2}N+1) = P^2 + \tfrac{1}{2}N^2[2\emptyset^2+2\emptyset(\emptyset_1+\emptyset_{-1})+(\emptyset_1)^2+(\emptyset_{-1})^2] \quad . \tag{2.50}$$

We solve Eqs. (2.49) and (2.50) by assuming that $\emptyset$ can be expanded in a power series,

$$\emptyset = 1 + ap^2 + bp^4 + \dots \quad , \tag{2.51}$$

in which we shall find consistently that $b \sim a^2$. To terms of order a^2, we then find

$$b = - \tfrac{1}{2}a^2 \quad , \tag{2.52}$$

and

$$4a^2N^2 - 8aN^2 - 1 = 0 \quad , \tag{2.53}$$

whence

$$a = 1 - \sqrt{1 + \frac{1}{4N^2}} \cong - \frac{1}{8N^2} + \frac{1}{128N^4} \quad , \tag{2.54}$$

$$b \cong - \frac{1}{128N^4} \quad . \tag{2.55}$$

From the constant term on the right hand isde of (2.50) (all P dependence must in fact cancel), we find, using Eq. (2.53)

$$N^2 + 2N + 2 \equiv M = 16N^2 + 4a^2N^2 \quad , \tag{2.56}$$

or

$$N^2 \cong \frac{M}{16} - \frac{1}{16M} + \dots \quad . \tag{2.57}$$

Thus $N \sim \tfrac{1}{4}N$. We see that the series (2.51) converges for all coupling strengths when N is large. In the strong-coupling limit, however, where $P \sim 1$, the convergence is extremely rapid for large N.

We note in passing that other formal solutions of the algebra are obtained by cyclic permutation of the angular momentum operators and/or the canonical transformation $X \leftrightarrow - P$. Among these possibilities is the original suggestion of the transition operator representation.

We turn next to the Hamiltonian and consider it only up to terms of order P^4. We have

$$\begin{aligned} H = &- 2\varepsilon N\cos X - 4|V|N^2\sin^2X \\ &- a\varepsilon N\{\cos X,P^2\} + \tfrac{1}{2} a^2\varepsilon N\{\cos X,P^4\} \\ &- 2|V|N^2a\{\sin X,\{\sin X,P^2\}\} \\ &+ |V| a^2N^2\{\sin X,\{\sin X,P^4\}\} \\ &- |V|N^2a^2 [\{\sin X,P^2\}]^2 + |V|P^2 + \dots \quad . \end{aligned} \tag{2.58}$$

This is, with only trivial modifications, of the general form given in Eq. (1.28), so that the first two terms constitute in all essentials the collective energy. We search for its minimum and find the condition

$$\varepsilon \sin X = 4|V|N \sin X \cos X \quad , \tag{2.59}$$

which gives either

$$\sin X_0 = 0 \quad , \tag{2.60}$$

the weak coupling solution, or

$$\cos X_0 = \varepsilon/4|V|N \quad , \tag{2.61}$$

which is possible for V large enough ($\delta \gtrsim 1$).

In the regime defined by (2.61), the potential energy instead of having a single minimum at Z = 0, develops a double minimum symmetrically disposed with respect to the origin, with a maximum in between. This then provides another natural explanation of the parity doubling for those levels contained in the well between the minima and the maximum.

Previous treatments based on "deformed" Hartree-Fock[45,51-53,11] have produced various versions of Eqs. (2.59), but in (2.58) we have the possibility of studying oscillations of the system for any value of X_0. To see this, let us expand (2.58) about $X_0 = 0$ to quartic terms. We find

$$H = H_0 + H_2 + H_4 \quad , \tag{2.62}$$

$$H_0 = -2\varepsilon N \quad , \tag{2.63}$$

$$H_2 = [\varepsilon N - 4|V|N^2]X^2 + [|V|-2a\varepsilon N]P^2 \quad , \tag{2.64}$$

$$H_4 = [\frac{-\varepsilon N}{12} + \frac{4}{3}|V|N^2]X^4 - 2|V|N^2aXP^2X$$
$$+ \{X^2,P^2\}[\tfrac{1}{2}a\varepsilon N - 2|V|N^2a] + a^2\varepsilon NP^4 \quad . \tag{2.65}$$

The main qualitative point to be extracted from these expressions concerns the behaviour of the potential energy. We see from H_2 that the spring constant vanishes just at the point where (2.61) begins to be satisfied. On the other hand the coefficient of X^4 is positive and growing rapidly. One's normal inclination would be to take H_2 as unperturbed Hamiltonian (RPA approximation) and treat H_4 as perturbation. This would fail near $\delta \sim 1$ and is not recommended even if a complete diagonalization is carried out. On the other hand, the zero-order Hamiltonian

$$H_{20} = \varepsilon NX^2 + 2|a|\varepsilon NP^2 \tag{2.66}$$

corresponds to a frequency which is fixed and nearly unity. The problem of diagonalizing H in a basis defined by H_{20} is <u>very</u>

nearly a problem which has previously been treated by the writer and his colleagues[54] (see next section) and on the basis of this previous work we can anticipate that such diagonalizations in a basis of the correct size, namely that of the original fermion basis (again see next section), will give excellent numerical results well beyond the so-called transition point defined by (2.61). In fact, one then should be able to tie on to results obtainable by starting from the strong coupling limit, $X_0 = \frac{\pi}{2}$ in a smooth way. This is in process of verification.

E. Shell-Model Boson Representation

Here we come to a second boson method, whose technical aspects have been quite adequately described in the literature[54] and of which we shall give here only the briefest mention. Our reason for going into this development (aside from the fact that it is closer to conventional approaches) is to alter the psychological emphasis as to the nature of this method compared to the implications in our previous work. In our paper three deriviations of the results are given. We take the shortest way round.

We start from the observation that in the unperturbed fermion basis, J_0 has the matrix elements

$$J_0(nn) = -\tfrac{1}{2}N + n \quad . \tag{2.67}$$

We thereupon immediately introduce a boson basis

$$|n)_B = \frac{1}{\sqrt{n!}} (A^\dagger)^n |0)_B \tag{2.68}$$

and interpret n as the eigenvalue of $A^\dagger A$, or

$$(J_0)_B = -\tfrac{1}{2}N + A^\dagger A \quad , \tag{2.69}$$

i.e. $(J_0)_B$ is that boson operator which with respect to the basis (2.68) has the same matrix elements as the original fermion operator has with respect to the fermion basis. We can then derive also closed forms for J_+, since Eq. (2.11) is satisfied with respect to the states (2.68) if

$$(J_+)_B = \sqrt{N}\, A^\dagger \, [1-(A^\dagger A/N)]^{\frac{1}{2}} \quad . \tag{2.70}$$

This form shows most clearly that we must restrict the basis in the boson space directly to that subspace which is in one-to-one correspondence to the fermion space. In principle one can replace (2.69) and (2.70) by operators which annihilate the unwanted states, but it is easier to remember to omit them.

Equation (2.70) can be expanded and rearranged in normal form, then the expressions for $J_\pm$, J_0 introduced into the Hamiltonian and the latter also so arranged. We shall not carry this through here, since the representation (2.68) is closely that of (2.66).

Two remarks only are important. First, the quartic approximation shows no sign, up to many orders of magnitude in $|V|$ beyond the phase transition, of failing to yield a good approximation to the lowest-lying levels. In seeking the analogue of such a powerful tool when we turn to realistic problems, we should realize that what we have constructed is the shell model or Tamm-Dancoff boson. We emphasize this point only because in the present example it is so irrelevant and in a more realistic situation so vital. For in the present instance the boson $A^\dagger$ can also be viewed as a kinematical boson, i.e. it is obtained by a correspondence between states in the $V = 0$ limit. But for our model the unperturbed states are also the Tamm-Dancoff states in the band. That is, if we look for a linear combination of single-particle, single-hole states (1p-1h) in the space to be an approximate eigenstate, we find a unique choice, the state proportional to $J_+|0\rangle$ and similarly for 2p-2h states, etc.

In the more general case, we will have a choice between kinematical and shell model bosons. Only with the proper choice of shell model bosons can one hope to do physics. (By choosing kinematical bosons, the best one can hope to do is write long papers. In this respect guilt is to be shared among all writers in the field, including the present one, though degrees of culpability can be distinguished.)

We should try to justify the rather rigid assertions made above. This is so easy to do that it is embarrassing. Let us suppose therefore that we have fermion pair operators $Q_0, Q_0^\dagger$ which we desire to represent as primary bosons, that is,

$$Q_0^\dagger \propto B_0^\dagger + \text{"corrections"} \quad , \tag{2.71}$$

where the representation is with respect to some prescribed subset of states called "collective". We can think of three conditions which will guarantee the success of this enterprise:
(i) We write the commutator between Q_0 and $Q_0^\dagger$ as

$$[Q_0, Q_0^\dagger] = P_0 + R_0 \quad , \tag{2.72}$$

and if $|\Psi\rangle$ is any state in the subspace in which the boson expansion is required to converge rapidly, we require

$$\langle\Psi|P_0|\Psi\rangle = N(1 + O(N^{-1})) \quad , \tag{2.73}$$

where $N >> 1$ is a number characterizing the collective aspect, and where

$$\langle\Psi|R_0|\Psi\rangle = O(1) \quad . \tag{2.74}$$

If we write

$$P_0 = N + \delta P_0 \quad , \tag{2.75}$$

with these observations, (2.72) becomes

$$[Q_0/\sqrt{N}\ ,\ Q_0^\dagger/\sqrt{N}] = 1 + \frac{1}{N}[\delta P_0 + R_0] \quad , \tag{2.76}$$

and thus we may start our expansion with $Q_0^\dagger = \sqrt{N}B^\dagger + \ldots$.

The analogy to the model of MGL should be obvious at this point. It is only to be emphasized that the introduction of the boson has been based on dynamics in that both (2.73) and (2.74) are dynamical statements referring to properties of physical states in some regime of coupling strength. It is correct to say that the boson "expansion" (2.70) is a priori appropriate to the weak coupling regime. (ii) If other bosons are to be introduced for additional fermion-pair operators, e.g., $Q_1, Q_1^\dagger$. we demand that these commute with Q_0, $Q_0^\dagger$ except for small terms,

$$[Q_0, Q_0^\dagger] = R_{01} \quad , \tag{2.77}$$

$$\langle\Psi|R_{01}|\Psi\rangle = O(1) \quad . \tag{2.78}$$

(iii) This requirement usually follows if the first is satisfied, but we state it separately. It is the condition that the individual fermion degrees of freedom decouple from the collective motion. We only mention it because since boson expansions have been introduced which violate (i), they will also violate (iii). Now condition (i) cannot be satisfied unless $Q_0^\dagger$ is a combination of many elementary pair operators, unless, i.e., it is formally a coherent operator. Under these circumstances, we may anticipate that if $a_\alpha^\dagger$ is a fermion mode creation operator (cf. again MGL), we have something like

$$[Q_0^\dagger, a_\alpha^\dagger] = a_\alpha^\dagger \quad . \tag{2.79}$$

Then equations (2.73), (2.74) guarantee that the formal coherence built into the definition of $Q_0^\dagger$ is relevant to the dynamics being described. Under these circumstances the equation

$$\left[\frac{Q_0^\dagger}{\sqrt{N}}\ ,\ a_\alpha^\dagger\right] = \frac{a_\alpha^\dagger}{\sqrt{N}} \tag{2.80}$$

may be viewed as the statement that the boson almost commutes with the individual fermion degrees of freedom. (These remarks will be illustrated in detail in section J.)

F. The Several Faces of the RPA

We have now illustrated several of the most powerful ways of solving the problem of collective motion in practice. In so doing, we have barely mentioned the RPA, which has been, for vibrations, the most widely used method of all. In fact both the algebraic approach and the previously given boson expansions can be viewed as generalizations of the RPA, as we shall indicate, and indeed these are not the only ways of carrying out the generalization. Two other ways make use of the formalism of section I.F, which should indeed be illustrated, and a third, the method of Rowe,[10] far from being a theory of collective motion, generalizes the shell model into an alternative to the standard shell model diagonalization!

To get off the ground, we define the RPA for present purposes as the harmonic approximation to the physical boson representation, i.e., for our model we seek such functions, B, $B^\dagger$, of the fermion operators as satisfy the equations

$$[B,H] = \omega B \quad , \tag{2.81}$$

$$[B,B^\dagger] = 1 \quad , \tag{2.82}$$

$$B|\Psi_0\rangle = 0 \quad , \tag{2.83}$$

the latter being the definition of the ground state. From the present point of view, the only equation which is wrong is (2.81), for according to (1.11) the correct equation of motion is

$$[B,H] = a_1 B + 2a_2\, B^\dagger B^2 + 3a_3 (B^\dagger)^2 B^3 + \ldots \quad . \tag{2.84}$$

Therefore to get a consistent result from (2.81) - (2.83), we must stick to a simple assumption for $B, B^\dagger$. With the understanding that we are only trying to represent the first excited state and include some ground state correlations, we write

$$B^\dagger = (\psi_+ J_+ - \psi_- J_-)/\sqrt{N} \quad , \tag{2.85}$$

and find

$$[B,B^\dagger] = -\frac{2J_0}{N}(\psi_+^{\,2} - \psi_-^{\,2}) \quad . \tag{2.86}$$

Therefore for consistency at this stage one must replace J_0 by a c-number and one usually starts with $J_0 = \frac{1}{2}N$, the unperturbed value in the ground state, which fixes the norm

$$\psi_+^{\,2} - \psi_-^{\,2} = 1 \quad . \tag{2.87}$$

The usual RPA equations are obtained then by taking the matrix elements $\langle 0|$, $|1\rangle$, $\langle 1|$, $|0\rangle$ of Eq. (2.81) where $|0\rangle$, $|1\rangle$, ... are here the unperturbed states. We obtain the equations

$$\omega\,\psi_+ = \varepsilon\psi_+ - V(N+1)\psi_- \tag{2.88}$$

$$\omega\,\psi_- = -\,\varepsilon\psi_- + V(N-1)\psi_+ \quad , \tag{2.89}$$

or

$$\omega = \varepsilon\,\sqrt{1-\delta^2(1-N^{-2})} \quad , \tag{2.90}$$

which goes through zero for $\delta = (1-N^{-2})^{-\frac{1}{2}}$.

The situation can be saved, if we replace J_0 by $\langle J_o\rangle = \langle\psi_o|J_o|\psi_o\rangle$ calculated self-consistently and if we drop unity compared to $\langle J_o\rangle$ in the equations of motion. The resulting scheme should then almost be equivalent to the two level scheme defined in Ref. 47 which shows no mathematical breakdown up to infinite coupling, but we have no numerical results to show whether it shares all these desirable properties since there is still a difference, compared to the method of Ref. 47. There we do not try to construct the physical boson; the equations we use are in this order equivalent to (2.81), (2.82), but in place of (2.83), which leads to an actual construction of $|\Psi_0\rangle$, we calculate $\langle J_o\rangle$ from an approximate form of (VI). We have the identities $\psi_\pm = J_+(01)/\sqrt{N}$.

Thus as a first result we can view our lowest approximation in the algebraic method as a renormalized first RPA and our entire approach as one possible logical extension of RPA. As we have already shown in section II.C, it can be viewed as the construction of the physical boson representation where the latter is applicable.

A second road to generalization of the RPA is to recognize it as an approximate expansion of the physical boson in terms of the "kinematical" (rather T.D. boson) of section II.E, namely

$$B^\dagger \cong \psi_+ A^\dagger - \psi_- A \quad . \tag{2.91}$$

Here, then, H must be computed to quadratic terms in $A, A^\dagger$. For this we need the correct equation for J_0, (2.69), and if we keep this also in the commutator (2.86), we have again the self-consistent RPA. We get the correct equations of motion by taking matrix elements between the boson states defining the $A^\dagger$ operator. The generalization to all orders consists precisely in inserting the expansions of section II.E in the Hamiltonian and of assuming that

$$B^\dagger = \sum_{ik} \ell_{ik} (A^\dagger)^i A^k, \quad (i,k)<N, \quad |i-k| \text{ odd} \tag{2.92}$$

(compare (2.40)) and of demanding that (2.84), (2.82) and (2.83) be satisfied. To find conditions to determine the ℓ_{ik} and the energies, we take all non-vanishing matrix elements of these equations in the $A^\dagger$ basis.

It is, however, clear that technically it should be easier to diagonalize the boson Hamiltonian once it has been found. The logical equivalence to the above procedure should be recognized and we should perhaps even keep a partially open mind about the simpler procedure when many degrees of freedom are involved.

Yet another equivalent procedure once the boson Hamiltonian has been found is to study the equations

$$\langle\Psi_n|[A,H]|\Psi_n'\rangle = \omega_{n'n} \langle n|A|n'\rangle \quad , \tag{2.93}$$

$$\langle n|[A,A^\dagger]|n'\rangle = \delta_{nn'} \quad . \tag{2.94}$$

This is precisely the algebraic method of sections B and C in the boson space and is another generalization of the RPA.

Still a fourth generalization of the RPA is to solve the originally posed problem (2.82), (2.83), (2.84), in the fermion space by generalizing (2.85) in a manner suggested by (2.92) namely,

$$B^\dagger = \sum_{ik} d_{ik}(J_+)^i(J_-)^k \quad , \tag{2.95}$$

where in fact J_0 is not needed because it can be thought of as a function of J_+ through the Casimir operator relation. The comparison of the form (2.95) with (2.92) implies that in consequence of the last statement the boson expansions of $J_\pm$, J_0 in terms of A, $A^\dagger$ can be inverted. That this is so, we leave as an exercise to the reader. The problem posed by (2.82) - (2.84) and (2.95) is solved by taking matrix elements with respect to a fermion basis, even the free one.

All of the above generalizations of the RPA are more or less equivalent theories of collective motion, at least in the limit of exact solution. We finally consider one last generalization,[10] which is, rather, a shell-model calculation. In generalizing the RPA previously, except for the current algebra method, which needs no such explicit assumption, we did so by introducing the two-phonon state, then a three-phonon state, etc., of the same excitation. In the present method we formally introduce an independent harmonic oscillator for each new level we wish to describe. Thus we write formally

$$H = \sum_i \omega_i B_i^\dagger B_i \quad , \tag{2.96}$$

$$[B_i, B_j^\dagger] = \delta_{ij} \quad , \tag{2.97}$$

$$B_i|\Psi_0\rangle = 0 \ , \quad \text{all } i \tag{2.98}$$

and

$$|\Psi_i\rangle = B_i^\dagger|\Psi_0\rangle \quad . \tag{2.99}$$

We solve this problem by assuming

$$B_i^\dagger = \sum_k [f_{ik}(J_+)^k - g_{ik}(J_-)^k] \quad . \tag{2.100}$$

Why the structure (2.100)? We know that to construct any state in our space from the unperturbed ground state, the set of operators $(J_+)^k$ is already complete. Therefore in (2.100) we have doubled the size of the operator basis. We then understand Eq. (2.98) as a set of supplementary conditions to undo the ambiguity caused by this doubling. We may use the equations of motion either directly as above or in a form described by Rowe.[10]

In either event, we are doing a form of shell model, albeit one with some beautiful properties, which are the direct generalization of those enjoyed by the RPA.

G. Remarks on Generator-Coordinate Calculations

We have proved that generator-coordinate calculations, if done correctly, represent the equivalent of the previous methods, carried out in a coordinate system in which the collective eigenfunctions are not orthogonal. There have been several interesting applications of this method to the exactly-soluble model studied in these lectures.[55-58] We shall, however, find the strength (barely) to resist giving any detailed account of this work. On the other hand, the structure of the method may suggest ingenious approximations, whose immediate relation to the general developments in part I is not yet clear to the writer.[59-63]

H. Adiabatic Cranking

We shall now try to illustrate the method of part I.R. According to Eq. (2.58), in the adiabatic limit, the Hamiltonian of the MGL model for large N and $|V|$ should be given by the approximate form

$$\mathcal{H}(X,P) + W_o = H^{(0)}(X) + \tfrac{1}{4}\{P,\{P,H^{(2)}(X)\}\} \quad , \qquad (2.101)$$

where

$$H^{(0)}(X) = -\tfrac{1}{2}N\varepsilon\cos X - |V|\tfrac{1}{4}N^2\sin^2 X \quad , \qquad (2.102)$$

$$2H^{(2)}(X) \equiv B^{-1}(X) = (2\varepsilon/N)\cos X + 2|V|(1+\sin^2 X) \ . \qquad (2.103)$$

The correction to each of these forms is of relative order N^{-1}. The general theory of section I.R tells us that we can identify these forms, if we can construct any exact intrinsic state $|\Phi(\alpha,p_\alpha)\rangle$ such that $\langle X\rangle = \alpha$, $\langle P\rangle \doteq p_\alpha$.

We look for such an intrinsic state in the form of a Slater determinant

$$|\Phi(\alpha,p_\alpha)\rangle = \prod_p c^\dagger_{p_-}|0,c\rangle \quad , \qquad (2.104)$$

$$c_{p_-}|0,c\rangle = 0 \quad , \qquad (2.105)$$

where

$$c^\dagger_{p_-} = \mu_{--}a^\dagger_{p_-} + \mu_{-+}a^\dagger_{p_+} \quad , \qquad (2.106)$$

$$c^\dagger_{p+} = \mu_{+-}a^\dagger_{p-} + \mu_{++}a^\dagger_{p+} \quad , \qquad (2.107)$$

and

$$U(\alpha,p_\alpha) = \begin{pmatrix} \mu_{--} & \mu_{-+} \\ \mu_{+-} & \mu_{++} \end{pmatrix} \tag{2.108}$$

is a unitary transformation. This induces an orthogonal transformation among the components of angular momentum

$$J_\alpha = \sum_i a_{\alpha i} J_i \;, \quad \alpha = x, y, z; \quad i = 1, 2, 3 \;. \tag{2.109}$$

Here U is the unit matrix when $\alpha = p_\alpha = 0$. The elements $a_{\alpha i}$ are defined so that

$$<\Phi(\alpha,p_\alpha)|J_3|\Phi(\alpha,p_\alpha)> = -\frac{N}{2} \;, \tag{2.110}$$

$$<\Phi(\alpha,p_\alpha)|J_{1,2}|\Phi(\alpha,p_\alpha)> = 0 \;, \tag{2.111}$$

or

$$(J_1 - iJ_2)|\Phi(\alpha,p_\alpha)> = 0 \;. \tag{2.112}$$

From these requirements we find easily

$$<J_3{}^2>_{\alpha,p_\alpha} = \tfrac{1}{4} N^2 \;, \tag{2.113}$$

$$<J_{1,2}{}^2>_{\alpha,p_\alpha} = \tfrac{1}{4} N \;. \tag{2.114}$$

With the help of the Hamiltonian (2.42), and of Eqs. (2.109) - (2.114), we find

$$<\Phi(\alpha,p_\alpha)|H|\Phi(\alpha,p_\alpha)> = -\tfrac{1}{2} N\varepsilon a_{z3} - |V|[(a_{x3}{}^2 - a_{y3}{}^2)\,\tfrac{1}{4}\, N(N-1)] \;. \tag{2.115}$$

In deriving (2.115), we have also used the normalization condition

$$\sum_\alpha a_{\alpha 3}{}^2 = 1 \;. \tag{2.116}$$

In order to make further progress, we must commit ourselves concerning the relation between the collective variables $\hat{X}$, $\hat{P}$ and the operators occurring in the Hamiltonian. Let us first consider the potential energy and assume that the condition $p_\alpha = 0$ means $<J_y>_{\alpha,0} = 0$. To satisfy this condition it is sufficient (but not necessary) to choose the rotation (2.109) to be about the y axis. Thus we write ($p_\alpha = 0$)

$$J_x = \cos\beta J_1 - \sin\beta J_3 \;, \tag{2.117}$$

$$J_y = J_2 \;, \tag{2.118}$$

$$J_z = \sin\beta J_1 + \cos\beta J_3 \quad . \tag{2.119}$$

Applied to (2.115) this yields the potential energy

$$H^{(0)}(\alpha) = \tfrac{1}{2}\, \varepsilon N \cos\alpha - V\tfrac{1}{4}N(N-1)\sin^2\alpha \quad , \tag{2.120}$$

which agrees with (2.102) to the required order. (The extra term in (2.120) arises from a careful treatment of exchange, but is not strictly warranted by the present approximation.)

To obtain the kinetic energy correctly is now but a short step if we use the requirement $J_y = -\hat{P}$ as part of the definition of collective variables. From (2.109) we then have

$$a_{y3} = \frac{2}{N}\, p_\alpha \quad , \tag{2.121}$$

and in order to satisfy (2.116) to order $p_\alpha{}^2$, it is sufficient to take

$$a_{z3} = \cos\beta\, [1 - \tfrac{1}{2}\left(\frac{2}{N}\right)^2 p_\alpha{}^2] \quad , \tag{2.122}$$

$$a_{x3} = -\sin\beta\, [1 - \tfrac{1}{2}\left(\frac{2}{N}\right)^2 p_\alpha{}^2] \quad . \tag{2.123}$$

Insertion into (2.115) yields, in addition to (2.120), the relation

$$B^{-1}(\alpha) = 2(\varepsilon/N)\cos\alpha + |V|(1 + \sin^2\alpha)(1 + N^{-1}) \quad , \tag{2.124}$$

which is again correct to the required order.

If the procedure of this section has seemed ad hoc to the reader, we shall now show how the limit considered can be reached in a systematic fashion.

I. Systematic Approach to the Semi-Classical Limit

A general procedure for obtaining the collective Hamiltonian in the semi-classical limit is not difficult to develop if we utilize the idea of generating functions for the various moments of any overlap. Thus remembering the notation introduced in Eqs. (1.22) - (1.24), we introduce a Fourier transform of the overlap[64]

$$\Theta(Y,P) = \int d\xi\, e^{-i\xi P}\, \tilde{\Theta}(Y,\xi)$$
$$= \sum_{q=0}^{\infty} \Theta^{(q)}(Y) P^q \quad , \tag{2.125}$$

and thus see that its various moments can be identified from a formal power series in the variable P. The utility of (2.125) can first be recognized if we note a connection between a theorem for the product of two operators and the physics of collective motion. Thus if L = JK, i.e.

$$L(X|X') = \int dX''J(X|X'')K(X''|X') \quad , \tag{2.126}$$

it is a straightforward exercise in change of variables to derive the expression

$$L(X,P) = \sum_{q=0}^{\infty} \sum_{r=0}^{\infty} (\tfrac{1}{2})^{q+r} \frac{1}{q!} \frac{1}{r!} (i)^q (-i)^r J_{qr}(X,P) K_{rq}(X,P) \tag{2.127}$$

where

$$J_{qr}(X,P) = (\partial_X)^q (\partial_P)^r J(X,P) \quad . \tag{2.128}$$

Up to first order terms, we have

$$L(X,P) = J(X,P)K(X,P) + \tfrac{1}{2}i[J,K]_{X,P} + \cdots \tag{2.129}$$

where

$$[J,K]_{X,P} = \partial_X J(X,P)\partial_P K(X,P) - \partial_P J \partial_X K \tag{2.130}$$

is the classical Poisson bracket of the two dynamical variables.

Arguments for the rapid convergence of the series (2.125) in the adiabatic limit have been elaborated previously and need not be repeated. On this basis the rapid convergence of the series (2.127) or (2.129) for the product may be inferred. Thus to calculate the leading contribution (in powers of N) to the generating function of a product, we have from (2.129) the correspondence

$$JK \to J(X,P)K(X,P) \quad , \tag{2.131}$$

but for a commutator we must go one step further to the famous result

$$[J,K] \to i[J,K]_{X,P} \quad . \tag{2.132}$$

We now apply this correspondence to the model under study. For the Hamiltonian, the commutators, and the Casimir operator, we have respectively

$$H(X,P) = \varepsilon J_z(X,P) - |V|[J_x^2(X,P) - J_y^2(X,P)] \quad , \tag{2.133}$$

$$[J_x, J_y]_{X,P} = J_z(X,P) \qquad \text{(and cyclic permutation)} \; , \tag{2.134}$$

$$J_x^2(X,P) + J_y^2(X,P) + J_z^2(X,P) = \tfrac{1}{4}N^2 \quad . \tag{2.135}$$

A further specialization of great practical importance is to note that it is consistent with the reality properties of the angular momentum operators to choose J_x and J_z even in P and J_y odd in P (cf. Eqs. (2.47)).

It is straightforward to verify that Eqs. (2.134), (2.135) have the solution (which can be generated systematically)

$$J_x = \tfrac{1}{2}N \sin X\, f(P^2) \tag{2.136}$$

$$J_y = - P \quad , \tag{2.137}$$

$$J_z = - \tfrac{1}{2}N \cos X\, f(P^2) \quad , \tag{2.138}$$

$$f(P^2) = \sqrt{1 - (2P/N)^2} \quad . \tag{2.139}$$

These results lead immediately to the collective Hamiltonian (2.101), (2.102), (2.103).

The considerations of the last two sections may appear rather far removed from the usual apparatus of cranking, namely cranked Hartree-Fock equations. The appropriate equations of this type for the present model have been derived in ref. 64 and their equivalence to the previous formulations established. They may be used conveniently to construct the collective Hamiltonian when a power series in X and P is considered, but appear to offer no advantages for obtaining the "exact" collective parameters over the procedures already described, at least for the present model. We see no profit in recording yet another derivation of the results of this section.

J. Coupling Between Vibrational and Quasi-Particle Degrees of Freedom

In this section, we shall take a tentative step, narrowing slightly the gap between the model of MGL and the real world by introducing some coupling between the otherwise independent bands of the model. So far we have dealt only with the ground-state band. We can, equally well, deal with any of the low-lying excited bands, as we proceed to show.

Starting from the unperturbed ground state, there are N particle-hole operators creating excitations at energy ε when $V = 0$, namely

$$J_{p+} = a_{p+}^{\dagger}\, a_{p-} \quad . \tag{2.140}$$

One linear combination, namely $\Sigma J_{p+} = J_+$ belongs to the ground-state band. We need N-1 other linear combinations, $O_+(q)$, to generate the unperturbed ground states of the N-1 lowest excited bands with $J = \tfrac{1}{2}N-1$. We define the operators $O_+(q)$ by the requirements

$$|0,q\rangle = O_+(q)\,|0,0\rangle \quad , \tag{2.141}$$

where with $q = 1, \ldots, N-1$,

$$\langle 0,0|J_- O_+(q)|0,0\rangle = 0 \quad , \tag{2.142}$$

$$\langle 0,0|O_-(q)O_+(q)|0,0\rangle = 1 \quad . \tag{2.143}$$

The following set of operators satisfy this condition (exercise for the reader)

$$O_+(q) = [(1 - \frac{q-1}{N})(1 - \frac{q}{N})]^{-\frac{1}{2}} D_+(q) \quad , \tag{2.144}$$

$$D_+(q) = \frac{1}{N} [\sum_{p=1}^{q-1} C_+(p)] + (1 - \frac{q-1}{N})C_+(q) \quad , \tag{2.145}$$

$$C_+(q) = J_{q+} - N^{-1} J_+ \quad . \tag{2.146}$$

We next show how the boson operators representing J_0, $J_\pm$, as given in Eqs. (2.69) and (2.70), which are correct within the ground-state band, may be generalized so as to be correct for either the ground-state band or for any one of the N-1 excited bands now under discussion. This can only be done if we extend the previous definition of the boson basis in a convenient manner. Previously we considered the image

$$|0,n\rangle \to \frac{(A^\dagger)^n}{\sqrt{n!}} |0)_B \quad . \tag{2.147}$$

To this we append the set of images

$$|q,n\rangle \to \frac{(A^\dagger)^n}{n!} B_{q+}|0)_B \quad , \tag{2.148}$$

where

$$B_{q+} = b_{+q}{}^\dagger b_{-q}, \; B_{q-} = (B_{q+})^\dagger \tag{2.149}$$

and $b_{\pm q}{}^\dagger$ are a set of 2(N-1) "ideal" fermion operators which are assumed to commute with the A, $A^\dagger$. The idea behind the transformation is to represent the interaction of the original system as a coupling between a vibrational degree of freedom and the remaining independent fermion degrees of freedom.

To see how to extend the imaging procedure for operators, we start with J_0 and the relations

$$\langle 0,n|J_0|0,n\rangle = -\tfrac{1}{2} N + n \quad , \tag{2.150}$$

$$\langle q,n|J_0|q,n\rangle = -\tfrac{1}{2} N + 1 + n \quad . \tag{2.151}$$

It is simple to write an operator function of the bosons and ideal p-h operator (2.149) with the matrix elements (2.150), (2.151), namely

$$(J_0)_{Boson} = -\tfrac{1}{2}N + A^\dagger A + \sum_q B_{q+} B_{q-} \quad . \tag{2.152}$$

Similarly from the relations

$$\langle 0,n+1|J_+|0,n\rangle = [(N-n)(n+1)]^{\frac{1}{2}} \quad , \tag{2.153}$$

$$\langle q,n+1|J_+|q,n\rangle = [(N-n-2)(n+1)]^{\frac{1}{2}} \quad , \tag{2.154}$$

we derive the operator

$$(J_+)_{Boson} = A^{\dagger}\sqrt{N}\,[1 - \frac{A^{\dagger}A}{N}]^{\frac{1}{2}}$$

$$+ \sum_q B_{q+}B_{q-}\; A^{\dagger}\sqrt{N}\,\{[1 - \frac{(A^{\dagger}A)+2}{N}]^{\frac{1}{2}} - [1 - \frac{A^{\dagger}A}{N}]^{\frac{1}{2}}\} \quad , \tag{2.155}$$

which can be expanded in a series in $(A^{\dagger}A/N)$ and N^{-1} and rearranged in normal form.

In addition to the generators it is now possible to represent operators which couple the bands. For example, taking cognizance of the normalizations involved we can show that

$$\langle q,n|O_+(q)|0,n\rangle = \sqrt{1 - \frac{n}{N}}\;\sqrt{1 - \frac{n}{N-1}} \quad . \tag{2.156}$$

This result suggests the representation

$$(O_+(q))_{Boson} = B_{q+}\sqrt{1 - \frac{A^{\dagger}A}{N}}\;\sqrt{1 - \frac{A^{\dagger}A}{N-1}} \quad . \tag{2.157}$$

As an example of a perturbation which couples the bands, we add to the previous Hamiltonian of the model the operator

$$H' = \sum_q V_q\;(O_+(q)J_+ + h.c.) \quad . \tag{2.158}$$

The problem of the coupled bands can then be solved in the representation (2.147), (2.148) by introducing into the total Hamiltonian the expressions given by (2.152), (2.155) and (2.158). The expressions we have given are adequate for the consideration of the alteration of the properties of the ground state band due to its mixing with the restricted set of higher bands considered here, all this provided that the mixing is not too great. From (2.155), (2.157) and (2.158), this will be the case as long as

$$|V_q|\; N/\varepsilon = \delta_q \ll 1 \quad . \tag{2.159}$$

Further analysis and numerical results will be presented in an independent publication.

To summarize: We have postulated a coupling between a collective mode and single-particle modes and shown how to introduce a representation in which this can be viewed as the coupling of a boson to kinematically independent fermions. In the new representation, the conditions for this coupling not to destroy our picture of basically uncoupled vibrational bands are easily derived.

REFERENCES

1. Brink, D.M., de Toledo Piza, A.F.R. and Kerman, A.K. Phys. Letters 19, 413 (1965).
2. Das, T.K., Dreizler, R.M. and Klein, A. Phys. Rev. C2, 632 (1970).
3. Klein, A., Dreizler, R.M. and Johnson, R.E. Phys. Rev. 171, B1216 (1968).
4. Klein, A. and Johnson, R.E. Phys. Rev. 171, B1224 (1968).
5. Villars, F.M.H. Nucl. Phys. 74, 353 (1965).
6. Villars, F.M.H., in Many-Body Description of Nuclear Structure and Reactions, edited by C. Bloch (Academic Press Inc., New York, 1966).
7. Kammuri, T. Progr. Theoret. Phys. (Kyoto) 37, 1131 (1967).
8. Marshalek, E.R. and Weneser, J. Ann. Phys. (N.Y.) 53, 569 (1969).
9. Marshalek, E.R. and Weneser, J. Phys. Rev. C2, 1682 (1970).
10. Rowe, D.J. Rev. Mod. Phys. 40, 153 (1968).
11. Parikh, J.C. and Rowe, D.J. Phys. Rev. 175, 1293 (1968).
12. Rowe, D.J., Ullah, N., Wong, S.S.M., Parikh, J.C. and Castel, B. Phys. Rev. C3, 73 (1971).
13. Hill, D.L. and Wheeler, J.A. Phys. Rev. 89, 1102 (1953).
14. Griffin, J.J. and Wheeler, J.A. Phys. Rev. 108, 311 (1957).
15. Griffin, J.J. Phys. Rev. 108, 328 (1957).
16. Peierls, R.E. and Yoccoz, J. Proc. Phys. Soc. (London) A70, 381 (1957).
17. Yoccoz, J. Proc. Phys. Soc. (London) A70, 388 (1957).
18. Wong, C.W. Nucl. Phys. A147, 545 (1970).
19. Ui, H. and Biedenharn, L.C. Phys. Letters 26B, 608 (1968).
20. Brink, D.M. and Weiguny, A. Nucl. Phys. A120, 58 (1968).
21. Jancovici, B. and Schiff, D.H. Nucl. Phys. 58, 678 (1964).
22. Kamlah, A. Thesis, Göttingen (1968).
23. Kamlah, A. Zeits. f. Phys. 216, 52 (1968).
24. Skyrme, T.H.R. Proc. Phys. Soc. (London) A70, 433 (1957).
25. Levinson, C.A. Phys. Rev. 132, 2184 (1963).
26. Hu, Chi-Yu. Nucl. Phys. 66, 449 (1965).
27. Klein, A. Lectures in Theoretical Physics, Proc. Boulder Summer Institute for Theoretical Physics, Vol.XIB (Gordon & Breach, New York), p.1.
28. Johnson, R.E., Klein, A. and Dreizler, R.M. Ann. of Phys. (N.Y.) 49, 496 (1968).
29. Glauber, R.J. Phys. Rev. 131, 2766 (1963).
30. Carruthers, P. and Nieto, M.M. Rev. Mod. Phys. 40, 411 (1968).
31. Harris, S.M. Phys. Rev. 138, B509 (1965).
32. Mariscotti, M.A.J., Scharf-Goldhaber, G. and Buck, B. Phys. Rev. 178, 1864 (1969).
33. Klein, A., Dreizler, R.M. and Das, T.K. Phys. Letters 31B, 333 (1970).
34. Das, T.K., Dreizler, R.M. and Klein, A. Phys. Rev. Letters, 25, 1626 (1970).
35. Das, T.K., Dreizler, R.M. and Klein, A. Phys. Letters 34B, 235 (1971).

36. Das, T.K. Thesis, Univ. of Pennsylvania (1971).
37. Holmberg, P. and Lipas, P.O. Nucl. Phys. A117, 552 (1968).
38. Peierls, R.E. and Thouless, D.J. Nucl. Phys. 38, 154 (1962).
39. Yoccoz, J., in Many-Body Description of Nuclear Structure and Reactions, edited by C. Bloch (Academic Press Inc., New York, 1966).
40. Rouhaninejad, H. and Yoccoz, J. Nucl. Phys. 78, 353 (1966).
41. Lipkin, H.J., Meshkov, N. and Glick, A.J. Nucl. Phys. 62, 188 (1965).
42. Meshkov, N., Glick, A.J. and Lipkin, H.J. Nucl. Phys. 62, 199 (1965).
43. Glick, A.J., Lipkin, H.J. and Meshkov, N. Nucl. Phys. 62, 211 (1965).
44. Volkov, A.B. Nucl. Phys. 43, 1 (1963).
45. Agassi, D., Lipkin, H.J. and Meshkov, N. Nucl. Phys. 86, 321 (1966).
46. Dreiss, G.J., Klein, A. and Pang, S.C. Phys. Letters 29B, 461 (1969).
47. Dreiss, G.J. and Klein, A. Nucl. Phys. A139., 81 (1969).
48. Dreizler, R.M. and Klein, A. Phys. Letters 30B, 236 (1969).
49. Gross, D.H.E. and Yamamura, M. Nucl. Phys. A140, 625 (1970).
50. Mihailovic, M.V. and Rosina, M. Nucl. Phys. A130, 386 (1969).
51. Dreizler, R.M., Klein, A., Krejs, F.R. and Dreiss, G.J. Nucl. Phys. A166, 624 (1971).
52. Faessler, A. and Plastino, A. Z. Phys. 220, 88 (1969).
53. Bleuler, K., Frederick, A. and Schütte, D. Nucl. Phys. A126, 628 (1969).
54. Pang, S.C., Klein, A. and Dreizler, R.M. Ann. of Phys. (N.Y.) 49, 477 (1968).
55. Mann, A., Nissimov, H. and Unna, I. Nucl. Phys. A139, 673 (1969).
56. da Providencia, J. Nucl. Phys. A157, 358 (1970).
57. da Providencia, J., Urbano, J.N. and Ferreira, L. Corrections to the Gaussian Overlap Approximation: A New Boson Expansion (to be published).
58. Schütte, D. (Private communication, Univ. of Bonn.)
59. Holtzwarth, G. Nucl. Phys. A113, 448 (1968).
60. Holtzwarth, G. Nucl. Phys. A133, 161 (1969).
61. Holtzwarth, G. Nucl. Phys. A156, 511 (1970).
62. Justin, D., Mihailovic, M.V. and Rosina, M. Phys. Letters 29B, 458 (1969).
63. Mihailovic, M.V., Kujawski, E. and Lesjak, J. Nucl. Phys. A161, 252 (1971).
64. Johnson, R.E., Klein, A. and Dreizler, R.M. Ann. of Phys. (N.Y. 49, 496 (1968).

EQUATIONS OF MOTION APPROACH TO NUCLEAR SPECTROSCOPY

D.J. ROWE†
Department of Physics
University of Toronto

INTRODUCTION

There are many general approaches to nuclear spectroscopy, and to other many-body problems. Some are primarily static in character while others are primarily dynamic. The most familiar approach is probably the direct application of the time-independent Schrodinger equation. Now the time-independent Schrodinger equation is the equation of motion for the stationary states of the system and, as such, may be regarded as a static equation. Of course a knowledge of the stationary states enables one to compute all the dynamic properties, such as the response of the system to electromagnetic radiation or some other external probe. But in a sense this is a trivial observation since all exact and complete formalisms are ultimately equivalent.

Formalisms differ in practice because in their application they naturally suggest different approximation procedures and lead to different models. And models are not universally applicable. For example, the independent-particle model may be excellent for a description of single-particle phenomena such as magnetic and quadrupole moments or ground state density distributions. But it cannot be expected to describe phenomena involving two or more particle correlations as evidenced, for example, in high energy photon or pion absorption. It is essential therefore to optimize one's efforts by selecting a formalism which describes directly those properties of the system which are of immediate concern.

If our concern is with dynamic properties of the nucleus, such as excitation energies, electromagnetic transitions, inelastic scattering cross sections, etc., we should select a dynamic formalism which aims directly at these properties. For example, we might attempt to calculate the appropriate Green function, whose poles are the excitation energies and whose residues at the poles are the relevant transition intensities or spectroscopic factors. Or we might attempt a direct application

†Alfred P. Sloan Fellow

of the time-dependent Schrodinger equation which, in one approximation, gives the time-dependent Hartree-Fock theory.

In recent years there has been considerable activity in the development of new dynamic formalisms employing, for example, equations of motion for transition amplitudes or boson expansions. Some of these developments are discussed in the lectures of Dr. Klein who has been at the forefront of much of this activity.

In this series I want to describe a very practical formalism based on equations of motion for excitation operators. This is not a new idea. Equations of motion were used many years ago to derive the Hartree-Fock and Hartree-Bogolyubov equations[1] and the Random Phase Approximation[2]. The major recent advance is to remove the arbitrary linearization procedures of the early methods which restricted their use to relatively simple situations. The new equations of motion are formally exact and applicable to a whole new range of problems.

THE EQUATIONS OF MOTION FOR EXCITATION OPERATORS[3]

Derivation of the Equations of Motion

The equations of motion we employ derive from the well-known equations obeyed by the boson step-up and step-down operators of a harmonic oscillator Hamiltonian:

$$[H_{HO}, O_\lambda^\dagger] = \omega_\lambda O_\lambda^\dagger \tag{1}$$

$$[H_{HO}, O_\lambda] = -\omega_\lambda O_\lambda \quad .$$

Similar equations are obeyed by the fermion creation and destruction operators of an independent-particle Hamiltonian:

$$[H_{IP}, a_\nu^\dagger] = \varepsilon_\nu a_\nu^\dagger \tag{2}$$

$$[H_{IP}, a_\nu] = -\varepsilon_\nu a_\nu$$

By a suitable linearization prescription (which we return to later), eq. (2) was formerly used[1] with the nuclear Hamiltonian to derive the Hartree-Fock Independent-Particle Model; and eq. (1) was used[2] to derive the RPA (Random Phase Approximation) theory of nuclear vibrations. But of course the nucleus is neither an independent particle system nor a harmonic oscillator and the linearization prescriptions, which extracted such approximate solutions, were rather arbitrary and restricted to a few simple situations.

We therefore need more general equations of motion which are applicable to any Hamiltonian and to more general classes of excitation.

By analogy with the harmonic oscillator and independent particle problems, we define excitation operators $O_\lambda^\dagger$ for all the states $|\lambda\rangle$ by the equations

$$O_\lambda^\dagger |o\rangle = |\lambda\rangle \qquad (3)$$

$$O_\lambda |o\rangle = 0 \quad ,$$

where $|o\rangle$ is the ground state, and proceed to derive equations of motion to obtain these excitation operators.

This approach is particularly appropriate when the excitation process is relatively simple compared to the stationary states themselves. For example, although Pb^{208} and Pb^{209} may be complicated many-body systems, it may nevertheless be a good approximation to describe the low-lying states of Pb^{209} in terms of a neutron added to a Pb^{208} core. It would then be appropriate to seek creation operators for the low-lying states of Pb^{209} of the form

$$O^\dagger \sim a^\dagger + a^\dagger a^\dagger a + \ldots. \quad , \qquad (4)$$

Where the second and higher terms allow for core polarization. If weak coupling is valid, it will be a good approximation to truncate this expansion at the first or second term.

Operators of this type, which are constructed from odd numbers of nucleon creation and destruction operators, will be described as fermi-like operators.

Other simple excitations, of what may be complicated nuclei, are the collective vibrations. For example, all nuclei exhibit giant dipole resonances and octupole vibrations. And we have learned to interpret these motions as rather simple coherent density oscillations. We therefore expect that the corresponding excitations of the nucleus will be generated by simple one-body operators. Thus, for a collective vibrational excitation, we expect to truncate the series expansion

$$O^\dagger \sim a^\dagger a + a^\dagger a^\dagger a\, a + \ldots \quad , \qquad (5)$$

at the first term and get reasonable results.

Operators of this latter type, which are constructed from even numbers of nucleon creation and destruction operators, will be described as bose-like operators. Other bose-like operators are the two-particle transfer operators

$$O^\dagger \sim a^\dagger a^\dagger + a^\dagger a^\dagger a^\dagger a + \ldots. \quad .$$

These are appropriate to the description of two-nucleon transfer reactions and pairing vibrations, which unfortunately we will not have time to discuss in this series.

It will be noted that the above expansions all conserve the nucleon number as a good quantum number. Sometimes, however, it is convenient to give up nucleon number as a constant of motion, for the sake of other benefits, and consider operators of the form

$$0^{\dagger} \sim a^{\dagger} + a + \ldots .$$

Equations of motion for such operators lead to the B.C.S. and Hartree-Bogolyubov theories of super-conductivity, which again time will prevent us from discussing in this series.

Now, regardless of the character of the excitation, it is clear from eq. (3) that $0_\lambda^\dagger$ and 0_λ obey the equations of motion

$$[H, 0_\lambda^\dagger] \; |o\rangle = \omega_\lambda \, 0_\lambda^\dagger \; |o\rangle \tag{6}$$

$$[H, 0_\lambda] \; |o\rangle = - \omega_\lambda \, 0_\lambda \; |o\rangle \quad ,$$

where ω_λ is the excitation energy

$$\omega_\lambda = E_\lambda - E_o \quad . \tag{7}$$

Eq. (6) can be reduced to matrix form by a simple pre-multiplication and integration: viz.

$$\langle o| \, 0_\kappa \, [H, 0_\lambda^\dagger] \; |o\rangle = \omega_\lambda \; \langle o| \, 0_\kappa \, 0_\lambda^\dagger \; |o\rangle = \delta_{\kappa\lambda} \, \omega_\lambda \tag{8}$$

$$\langle o| \, 0_\kappa^\dagger \, [H, 0_\lambda] \; |o\rangle = -\omega_\lambda \; \langle o| \, 0_\kappa^\dagger \, 0_\lambda \; |o\rangle = 0 \tag{9}$$

If we now take the complex conjugate of (9), we obtain

$$\langle o| \, [H, 0_\lambda^\dagger] \, 0_\kappa \; |o\rangle = \omega_\lambda \; \langle o| \, 0_\lambda^\dagger \, 0_\kappa \; |o\rangle = 0 \tag{10}$$

which can be added or subtracted to (8) to give

$$\langle o| \, [0_\kappa, [H, 0_\lambda^\dagger]]_\pm \; |o\rangle = \omega_\lambda \; \langle o| \, [0_\kappa, 0_\lambda^\dagger]_\pm \; |o\rangle = \delta_{\kappa\lambda} \, \omega_\lambda . \tag{11}$$

Similarly, by working with the adjoint equations, we could have obtained the equation

$$\langle o| \, [[0_\kappa, H], 0_\lambda^\dagger]_\pm \; |o\rangle = \omega_\lambda \; \langle o| \, [0_\kappa, 0_\lambda^\dagger]_\pm \; |o\rangle = \delta_{\kappa\lambda} \, \omega_\lambda . \tag{12}$$

Eqs. (11) and (12) are completely equivalent. But to maintain symmetry it is convenient to take their average and write

$$\langle o| \, [0_\kappa, H, 0_\lambda^\dagger]_\pm \; |o\rangle = \omega_\lambda \langle o| \, [0_\kappa, 0_\lambda^\dagger]_\pm \; |o\rangle \tag{13}$$

where the double commutator is defined

$$[0_\kappa, H, 0_\lambda^\dagger]_\pm \equiv 1/2([0_\kappa, [H, 0_\lambda^\dagger]]_\pm + [[0_\kappa, H], 0_\lambda^\dagger]_\pm) \tag{14}$$

Eq. (13) is the equation of motion that we shall use to determine the excitation operators $O_\lambda^\dagger$. It will be used with normal commutators (the negative sign) for bose-like excitations and with anti-commutators (the positive sign) for fermi-like excitations.

It is immediately obvious from the definition (3) of the excitation operators that they also obey the equation of motion

$$<o|\ O_\kappa\ H\ O_\lambda^\dagger\ |o> = E_\lambda\ <o|\ O_\kappa O_\lambda^\dagger\ |o> = \delta_{\kappa\lambda} E_\lambda$$

which is equivalent to the usual Schrodinger equation in matrix form:

$$<\kappa|\ H\ |\lambda> = E_\lambda\ <\kappa|\lambda> = \delta_{\kappa\lambda}\ E_\lambda. \qquad (15)$$

One may wonder therefore why we have gone to all the trouble of getting the double commutator into (13). The answer is that the double commutator $[O_\kappa, H, O_\lambda^\dagger]$ is much simpler than the straightforward product $O_\kappa H O_\lambda^\dagger$. This can be seen by evaluating it for the harmonic oscillator and independent particle Hamiltonians. Since boson operators obey commutation relations

$$[O_\kappa, O_\lambda^\dagger] \equiv [O_\kappa, O_\lambda^\dagger]_- = \delta_{\kappa\lambda} \qquad (16)$$

we have

$$[O_\kappa, H_{HO}, O_\lambda^\dagger]_- = \delta_{\kappa\lambda}\ \omega_\lambda\ .$$

and since fermion operators obey anti-commutation relations

$$\{a_\mu, a_\nu^\dagger\} \equiv [a_\mu, a_\nu]_+ = \delta_{\mu\nu} \qquad (17)$$

we have

$$[a_\mu, H_{IP}, a_\nu^\dagger]_+ = \delta_{\mu\nu}\ \varepsilon_\nu\ .$$

Thus in both cases the double commutator is a pure number. For general bose-like and fermi-like excitations the double commutator is not normally a pure number but it is always an operator of particle-rank two lower than the straightforward product. For example, suppose O_κ and $O_\lambda^\dagger$ were one-body bose-like operators. In other words suppose we truncated $O^\dagger$ to the first term in eq. (5). Then for H an arbitrary two-body Hamiltonian we would find that the double commutator $[O_\kappa, H, O_\lambda^\dagger]$ is at most a two-body operator whereas the product $O_\kappa H O_\lambda^\dagger$ contains four-body components. This reduction in the particle rank of the operators is of enormous practical advantage when it comes to evaluating their ground state expectations. This is not only for the obvious reason that two-body operators are easier to manipulate than four-body operators but also because we must inevitably make some model approximation for the ground state wave function and we would have little confidence in its accuracy at predicting four-body correlations.

Ground state transition matrix elements for any operator W are given by

$$<\lambda|\ W\ |o> \ = \ <o|\ O_\lambda\ W\ |o>\ .$$

Again we make use of eq. (3) to express this in the more useful form

$$<\lambda|\ W\ |o> \ = \ <o|\ [O_\lambda,\ W]|o>\ . \qquad (18)$$

Solution of the Equations of Motion

In a practical solution of the Schrodinger equation the fundamental approximation consists of selecting a finite dimensional space of basis states. Similarly in the equations-of-motion formalism one must select a finite set of basis operators. The truncation may be made on energetic or physical grounds.

In addition, in the equations-of-motion approach one must specify the ground state. This seems like a disadvantage, but in fact it is the opposite; for recall that in the shell model solution of the Schrodinger equation, it is necessary to isolate a closed-shell core in order to limit the number of active particles. The equations-of-motion formalism simply allows one to work with more general cores which need not be closed-shell.

In principle the ground state is defined a posteriori, by eq. (3), as the vacuum of the excitation operators. But the equations of motion were not designed for predicting stationary states and, after a number of investigations, we concluded that it is more reliable to derive the ground state from an approach designed specifically for that purpose. For the nuclear ground state we therefore employ a shell model approximation.

To solve the Schrodinger equation, (15), one first calculates the Hamiltonian matrix $<\phi_\alpha|H|\phi_\beta>$ for a finite set of basis states. If this is an orthonormal set, i.e.

$$<\phi_\alpha|\phi_\beta> \ = \ \delta_{\alpha\beta}\ , \qquad (19)$$

one proceeds to find the unitary transformation X which diagonalizes the Hamiltonian matrix;

$$\sum_{\alpha\beta} X^*_{\alpha\kappa} <\phi_\alpha|\ H\ |\phi_\beta> X_{\beta\lambda} = \delta_{\kappa\lambda}\ E_\lambda . \qquad (20)$$

The columns of the unitary matrix X are then the eigenvectors and give the eigenstates

$$|\lambda> \ = \ \sum_\beta X_{\beta\lambda}\ |\phi_\beta> \qquad (21)$$

within the selected Hilbert space. With the usual normalization for the eigenvectors of a Hermitian matrix,

$$\sum_{\alpha} X^{*}_{\alpha\kappa} X_{\alpha\lambda} = \delta_{\kappa\lambda}, \tag{22}$$

the eigenstates form an orthonormal set

$$\langle\kappa|\lambda\rangle = \sum_{\alpha\beta} X^{*}_{\alpha\kappa} \langle\phi_{\alpha}|\phi_{\beta}\rangle X_{\beta\lambda} = \delta_{\kappa\lambda}. \tag{23}$$

Eq. (23) relies on the orthonormality (19) of the basis vectors. However it is perfectly legitimate to work with a non-orthonormal basis.[4] Suppose we are given the non-orthonormal basis states $|\psi_i\rangle$. We then begin by diagonalizing the metric matrix $\langle\psi_i|\psi_j\rangle$. This is a Hermitian matrix which can be diagonalized by a unitary transformation:

$$\sum_{ij} C^{*}_{i\alpha} \langle\psi_i|\psi_j\rangle C_{j\beta} = \delta_{\alpha\beta}\, n_{\alpha}.$$

If the basis states are linearly independent, the metric matrix is necessarily positive definite and hence the n_{α} are positive real numbers. We may therefore define a new and orthonormal set of states

$$|\phi_{\beta}\rangle = \frac{1}{n_{\beta}^{1/2}} \sum_{j} C_{j\beta}|\psi_j\rangle. \tag{24}$$

If the basis states were not linearly independent, one or more of the n_{α} would vanish, indicating that the corresponding new basis vectors had vanishing norms. These states would therefore be the spurious states and should be eliminated from the space to leave a linearly independent orthonormal set. Transforming the Hamiltonian matrix to the new orthonormal basis, eq. (24), one proceeds as before.

To solve the equations of motion for excitation operators, one similarly calculates the Hamiltonian matrix $\langle o|\,[\eta_{\alpha}, H, \eta^{\dagger}_{\beta}]\,|o\rangle$ for a finite set of basis operators. Assume for the moment that this is an orthonormal basis, according to the definition

$$\langle o|\,[\eta_{\alpha}, \eta^{\dagger}_{\beta}]\,|o\rangle = \delta_{\alpha\beta}. \tag{25}$$

We may then proceed to find the unitary transformation X which diagonalizes the Hamiltonian matrix,

$$\sum_{\alpha\beta} X^{*}_{\alpha\kappa} \langle o|\,[\eta_{\alpha}, H, \eta^{\dagger}_{\beta}]\,|o\rangle X_{\beta\lambda} = \delta_{\kappa\lambda}\, \omega_{\lambda}, \tag{26}$$

and gives the excitation operators

$$O^{\dagger}_{\lambda} = \sum_{\beta} X_{\beta\lambda} \eta^{\dagger}_{\beta} . \tag{27}$$

Again, with the normalization of eq. (22), these excitation operators form an orthonormal set

$$<o| [O_{\kappa}, O_{\lambda}^{\dagger}]|o> = \sum_{\alpha\beta} X_{\alpha\kappa}^{*} <o| [\eta_{\alpha}, \eta_{\beta}^{\dagger}] |o> X_{\beta\lambda} = \delta_{\kappa\lambda} . \tag{28}$$

As with the Schrodinger equation, it is perfectly legitimate to work with a non-orthonormal basis. What is more, it frequently happens, in the equations of motion formalism, that the metric matrix is not positive definite. Diagonalization of the metric matrix can then only yield basis operators with the orthonormality relations

$$<o| [\eta_{\alpha}, \eta_{\beta}^{\dagger}] |o> = \pm \delta_{\alpha\beta} . \tag{29}$$

This is perfectly legitimate however and happens, for example, in the RPA (Random Phase Approximation) where it leads to the special orthonormality relations associated with the RPA amplitudes. We shall discuss this situation when we come to it.

HARTREE-FOCK SELF-CONSISTENT FIELD THEORY

HF (Hartree-Fock) theory may be derived in many ways, each of which yields different insights. Frequently it is represented as a variational theory for the ground state of a many-fermion system. Here we express it in the equations-of-motion formalism where it is primarily a theory of single-particle states in odd-mass nuclei.

We suppose, that, to a first approximation, the lowlying states of an odd-mass nucleus can be generated by the addition of a single nucleon to an even-nucleus core, which remains unperturbed in its ground state. Thus we seek nucleon creation operators $a_{\mu}^{\dagger}$ which satisfy the equations of motion

$$<o| \{a_{\mu}, H, a_{\mu'}^{\dagger}\} |o> = \varepsilon_{\mu} <o| \{a_{\mu}, a_{\mu'}^{\dagger}\} |o> = \delta_{\mu\mu'} \varepsilon_{\mu} . \tag{30}$$

Since <u>a priori</u> we do not know the required single-nucleon wave functions, we begin by expressing the Hamiltonian, in second quantized form, in terms of arbitrary single-particle wavefunctions ϕ_i. These wave functions might be, for example, harmonic oscillator wave functions. Let $\alpha_i^{\dagger}$ be the creation operator for a nucleon with wave function ϕ_i. The Hamiltonian then takes the form

$$H = \sum_{ik} T_{ik} \alpha_i^{\dagger} \alpha_k + 1/4 \sum_{ijkl} V_{ijkl} \alpha_i^{\dagger} \alpha_j^{\dagger} \alpha_l \alpha_k \tag{31}$$

where

$$T_{ik} = \int \phi_i^* \; T \; \phi_k \, d\tau \qquad (32)$$

is a matrix element of the one-body part of H (including the kinetic energy, the spin-orbit force, etc.) and V_{ijkl} is an antisymmetrized matrix element of the two-body interaction:

$$V_{ijkl} = \iint \phi_i^*(1)\, \phi_j^*(2)\, V(12)\{\phi_k(1)\, \phi_l(2) - \phi_l(1)\, \phi_k(2)\} \, d\tau_1 \, d\tau_2 \, . \qquad (33)$$

The next step is to evaluate the double commutators of the equations of motion. Performing the simple commutations among the fermion operators, we obtain

$$\langle o| \{\alpha_i, H, \alpha_k^\dagger\} |o\rangle = T_{ik} + \sum_{jl} V_{ijkl} \langle o| \alpha_j^\dagger \alpha_l |o\rangle \, . \qquad (34)$$

Thus, if we are given the ground state $|o\rangle$, or at least its one-body densities, we can evaluate the Hamiltonian matrix. Diagonalizing it with a unitary transformation X, then gives the required nucleon creation operators,

$$a_\mu^\dagger = \sum_i X_{i\mu} \alpha_i^\dagger \, , \qquad (35)$$

which satisfy eq. (30). The corresponding single-particle wave functions are given by

$$\psi_\mu = \sum_i X_{i\mu} \phi_i$$

and have single-particle energies ε_μ, which are the eigenvalues of the Hamiltonian matrix (34).

Closed-Shell Nuclei

In the HF self-consistent field theory of a closed-shell nucleus (i.e. standard HF theory) we further suppose that the ground state $|o\rangle$ of the A-particle nucleus is an independent-particle state made up of A nucleons in the A lowest energy single-particle states, as defined above. Thus $|o\rangle$ is taken to be the product independent particle state

$$|o\rangle = \prod_{\nu=1}^{A} a_\nu^\dagger \, |-\rangle \qquad (36)$$

where $|-\rangle$ describes the bare-particle vacuum.

Now since the single-particle states, defined by eq. (30), depend on the many-particle ground state, eq. (36), which in turn depends on the single-particle states which go to form it, it is clear that the HF equations must be solved iteratively. One starts from a first guess for the single-particle wave functions, say the harmonic oscillator wave functions ϕ_i, and the corresponding product of these single-particle states for the nuclear ground state. One then calculates the Hamiltonian matrix (34) and diagonalizes it to give a second approximation for the single-particle wave functions and correspondingly a second approximation for the ground state. Iterating the cycle in this way until there is no longer any change, one eventually achieves self-consistency.

The HF single-particle wave functions ψ_μ are then the eigenfunctions of the single-particle Hamiltonian

$$H_o = T + u \tag{37}$$

where u is the self-consistent field, whose matrix elements are defined

$$\begin{aligned} u_{\mu\mu'} &= \sum_{\nu\nu'} V_{\mu\nu\mu'\nu'} \langle o|\, a_\nu^\dagger\, a_{\nu'}\, |o\rangle \\ &= \sum_{\nu=1}^{A} V_{\mu\nu\mu'\nu}. \end{aligned} \tag{38}$$

Thus the field u represents the interaction of the single nucleon with all the other nucleons in their ground state density distribution.

Open-Shell Nuclei

In the self-consistent field theory of an open-shell nucleus, it is possible to persist with the product form (36) for the nuclear many-particle wave function. However if some subshell of spherical single-particle states remains fractionally occupied, as is generally the case, such a many-particle wave function cannot in general have good angular momentum. Thus it cannot represent the ground state of the nucleus. Of course it may be employed as the intrinsic state of a rotational model or for a Peierls-Yoccoz theory of angular momentum projection and such methods give valuable insights; particularly in understanding nuclear rotational spectra (see the lectures of Villars). But deformed HF theory (as the above procedure is sometimes called) is clearly not an end in itself.

Here we consider a generalization of the spherical HF theory, to open-shell nuclei, which conserves angular momentum as a good quantum number. But it should be emphasized at the outset that the open-shell HF theory is in no way a substitute for the more familiar deformed HF theory. The latter is concerned primarily with understanding rotation-like correlations and, in particular with the orbital component of the single-particle wave function.

It is much less concerned with the radial part which, in practice, is often fixed at the outset. On the other hand, the open-shell HF theory is concerned only with the single-particle radial wave functions since the orbital components are characterized by good quantum numbers and are therefore uninteresting.

Exactly the same equations-of-motion theory is employed for the open-shell as for the closed-shell HF equations. But, instead of computing the HF Hamiltonian matrix, (34), for a product ground state, we employ a more general ground state derived by diagonalizing the full Hamiltonian in a finite many-particle configuration space. For example, the ground state might be a conventional shell-model ground state in which a closed-shell component is considered inert and the Hamiltonian diagonalized for the valence particles in a subspace of single-particle subshells. If v is the number of valence particles the ground state would then be a homogeneous polynomial, of degree v in the creation operators for particles in the valence space, acting on the closed-shell state; viz.

$$|o\rangle = f(a_\nu^\dagger)\,|cs\rangle\,. \qquad (39)$$

Again the single-particle states defined by (30) depend on the ground state (39) which in turn depends on the single-particle states. And again the HF equations must be solved iteratively. One starts from a first guess for the single-particle wave functions, e.g. harmonic oscillator wave functions, and selects a finite dimensional configuration space of many-particle states of good angular momentum and parity. Diagonalizing the Hamiltonian in this space gives the first approximation for the ground state $|o\rangle$ and, in particular, its single-particle occupation numbers. We will assume for simplicity, although it is not necessary, that no two single-particle states in the valence space have the same spin and parity. The single-particle densities are then diagonal and the occupancies, n_j are defined

$$\langle o|\, \alpha_j^\dagger\, \alpha_l |o\rangle = \delta_{jl}\, n_j\;. \qquad (40)$$

With these occupancies, one calculates the HF matrix (34),

$$\langle o|\, \{\alpha_i, H, \alpha_k^\dagger\}\, |o\rangle = T_{ik} + \sum_j V_{ijkj}\, n_j\;, \qquad (41)$$

and diagonalizes it to give the second approximation for the single-particle wave functions and hence for the ground state. Iterating in this way one eventually achieves self-consistency.

The open-shell HF single-particle wave functions are ultimately eigenfunctions of the single-particle Hamiltonian

$$H_o = T + u \qquad (42)$$

where u is the self-consistent field with matrix elements

$$u_{\mu\mu'} = \sum_{\nu} V_{\mu\nu\mu'\nu} \, n_{\nu} \quad . \tag{43}$$

Given the occupation numbers, n_{ν}, this HF Hamiltonian is no more difficult to handle than for a closed-shell nucleus. The extra complication comes in evaluating the occupation numbers by the intermediate shell-model calculation. However, since two-body matrix elements are not particularly sensitive to details of radial wave functions, it is unlikely that the occupancies will change significantly from one iteration to the next. Therefore it should not be necessary to recalculate them at each iteration; maybe only during the first and final iterations.

In concluding we remark that there are many possible definitions of single-particle wave functions and the equations-of-motion (self-consistent field) wave functions are not necessarily the 'best'. For the closed-shell case the self-consistent wave functions are the best in the sense that the many-nucleon product state (36) satisfies the variational equation

$$\delta \langle o| \; (H - E_o) \; |o\rangle = o$$

This is not true for the open-shell problem. Variation of a wave function of the given form (39) leads to an alternative choice of single-particle wave functions.[5]

Relationship to Other Dynamic Approaches

It is instructive to relate the above equations-of-motion description of single-particle structure to that of other dynamic approaches.

Let us first consider the general significance of the single-particle energies ε_{ν}. Expanding the double commutator of eq. (30) and inserting complete sets of odd-mass nucleus intermediate states, we obtain

$$\begin{aligned} \varepsilon_{\nu} &= \langle o| \; \{a_{\nu}, H, a_{\nu}^{\dagger}\} \; |o\rangle \\ &= \sum_{\alpha} |\langle\alpha|a_{\nu}^{\dagger}|o\rangle|^2 \; (E_{\alpha} - E_o) + \sum_{\beta} |\langle\beta|a_{\nu}|o\rangle|^2 \; (E_o - E_{\beta}) \end{aligned} \tag{44}$$

where the states $|\alpha\rangle$ and $|\beta\rangle$ belong respectively to the A±1 particle nuclei.

In a pure independent-particle nucleus, as envisaged in the closed-shell HF theory, there would be a single A±1 nucleus state corresponding to a given single particle state ν. For this state the spectroscopic factor would be unity and for all other states it would vanish. It would be at energy $E_o \pm \varepsilon_{\nu}$ (E_o is the even nucleus ground state energy) and would belong to the A±1 nucleus according as the single-particle state were vacant or occupied in the A nucleus ground state.

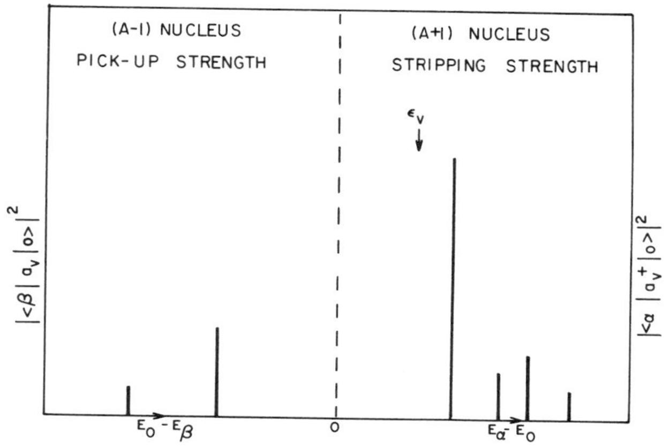

Figure 1
Single-particle strength distribution for a hypothetical nucleus

However, in general the single-particle strength is distributed over a number of states. What is more, if the state is fractionally occupied, there will be both stripping and pick-up strength. Plotting this strength as illustrated in fig. 1, ε_ν is the centroid energy of the distribution.

To the extent that the double commutator in eq. (30) is evaluated for the correct ground state, the equations of motion give the correct centroid energy. However, since the expectation of the double commutator depends only on the single-particle density of the ground state (cf. eq. (34)), there is reason to believe that the value obtained for ε_ν will be somewhat insensitive to details of the ground state. This suggests that HF single-particle energies are likely to be better estimates of pick-up and stripping centroid energies than the energies of particular nuclear states.

Where then does all the structure in the strength distribution come from? The answer is that nucleons are not independent and if we add a nucleon to a nucleus it will interact with the other nucleons and thus involve the core in its motion. In other words, there is a dynamic particle-core interaction which introduces many-particle correlations.

If we want to calculate the spreading effect of the particle-core interaction in the equations-of-motion formalism, we must include terms of the second type in the expansion

$$O^\dagger \sim a^\dagger + a^\dagger a^\dagger a + \ldots \qquad (45)$$

for the fermi-like creation operators. Such terms admit core-excitation by the added nucleon. With the additional terms there will clearly be many more solutions to the equations of motion and the distribution of the one-nucleon strength among these solutions should begin to resemble the gross structure of the true spectroscopic distribution. With the admission of third and higher terms in the expansion (45), more fine structure is obtained and ultimately the complete distribution. Thus the equations-of-motion give a natural description of the dissipation of single-particle strength into 2 particle-1 hole states (the doorway states), 3 particle-2 hole states, etc.

An alternative means of learning more about spectroscopic distributions is to calculate higher moments, as in French's spectral distribution theory. To get the second moment we have simply to calculate

$$\langle o| \{a_\nu, H, H, a_\nu^\dagger\} |o\rangle$$
$$= \sum_\alpha |\langle\alpha|a_\nu^\dagger|o\rangle|^2 (E_\alpha - E_o)^2 + \sum_\beta |\langle\beta|a_\nu|o\rangle|^2 (E_\beta - E_o)^2$$

where $\{a_\nu, H, H, a_\nu^\dagger\}$ is the average of $\{a_\nu,[H,[H, a_\nu^\dagger]]\}$ and $\{[[a_\nu, H],H], a_\nu^\dagger\}$. Similarly we get third moments from

$$\langle o|\{a_\nu, H, H, H, a_\nu^\dagger\} |o\rangle$$
$$= \sum_\alpha |\langle\alpha|a_\nu^\dagger|o\rangle|^2 (E_\alpha - E_o)^3 + \sum_\beta |\langle\beta|a_\nu|o\rangle|^2 (E_o - E_\beta)^3$$

etc. Of course it gets increasingly more difficult to calculate higher moments just as it gets increasingly difficult to include higher terms in the expansion (45).

In principle the advanced and retarded Green functions[6] †

$$G_1(\nu,\tau) = \langle o|\ a_\nu\ e^{-i(H-E_o)\tau}\ a_\nu^\dagger\ |o\rangle$$
$$G_2(\nu,\tau) = \langle o|\ a_\nu^\dagger\ e^{-i(H-E_o)\tau}\ a_\nu\ |o\rangle$$

contain complete information about the spectroscopic distribution. For example, the centroid energy is given by

$$\lim_{\tau\to 0+} i\frac{\partial}{\partial\tau}(G_1 + G_2) = \langle o|\ \{a_\nu, H, a_\nu^\dagger\}\ |o\rangle$$

and similarly higher moments by

$$\lim_{\tau\to 0+} \left(i\frac{\partial}{\partial\tau}\right)^n (G_1 + G_2) = \langle o|\{a_\nu, H,\dots H, a_\nu^\dagger\}|o\rangle\ .$$

However, the information available in practice is limited by one's ability to calculate these Green functions. Approximate Green functions are sometimes obtained via integral equations and sometimes via perturbation expansions and diagrammatic

†For convenience we use an unconventional retarded Green function. The conventional function is obtained from our expression by reversing the time; i.e. $G_2(\nu,-\tau)$.

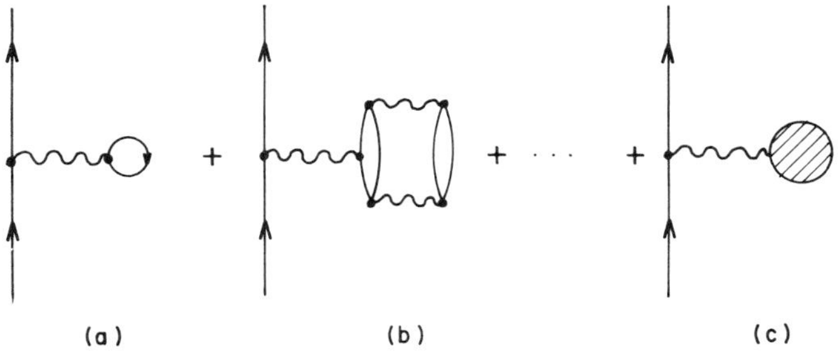

Figure 2
Diagrams which are summed to all orders in the self-consistent field definition of single-particle energies.

methods. For example, it can be shown that the Green functions predict a single line spectroscopic distribution at the centroid energy if the set of diagrams shown in fig. 2 is summed to all orders. If the core intermediate states are restricted to the valence space of the open-shell HF approximation, they give the open-shell HF results. And if the core states are restricted still further to the particle-hole vacuum of the closed-shell HF approximation, as shown in fig. 2(a), they give the closed-shell HF results. If one wishes to take into account the coupling of the single-particle to 2 particle-1 hole configurations, diagrams of the type shown in fig. 3 must be included.

The main disadvantage of the Green function, compared to the equations-of-motion, approach is that it is generally difficult to know which diagrams should be selected to give any desired result. Apart from a few simple cases, like the examples given above, the interesting relationships between the Green function and equations-of-motion formalisms have yet to be explored.

Finally, to put our equations-of-motion theory into perspective, we comment on other equations-of-motion approaches.

The old-fashioned linearization technique for closed-shell HF theory[7] was to consider the expansion

$$[H,\alpha_k^\dagger] = \sum_i T_{ik}\,\alpha_i^\dagger + 1/2\sum_{ijl} V_{ijkl}\,\alpha_i^\dagger\,\alpha_j^\dagger\,\alpha_l \qquad (46)$$

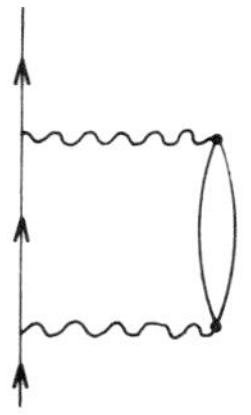

Figure 3
Core polarization corrections to a single-particle energy.

for some initial guess for the single-particle wave functions. Corresponding to this guess one defines an A-nucleus product state. This state is then regarded as the particle-hole vacuum for a normal ordering of eq. (46):

$$[H,\alpha_k^\dagger] = \sum_i \{T_{ik} + \sum_{j=1}^{A} V_{ijkj}\} \alpha_i^\dagger + 1/2 \sum_{ijl} V_{ijkl} :\alpha_i^\dagger \alpha_j^\dagger \alpha_l : \quad . \tag{47}$$

Disregarding the third term on the right hand side of (47), a unitary transformation on the single-particle basis is made such as to diagonalize the term in brackets - regarded as a single-particle matrix element $<i|H_o|k>$. Iterating in this way, which is clearly the HF procedure, one eventually achieves self-consistency and the solution

$$[H, a_\nu^\dagger] = \varepsilon_\nu a_\nu^\dagger + 1/2\sum_{\mu\mu'\nu'} V_{\mu\mu'\nu\nu'} : a_\mu^\dagger a_{\mu'}^\dagger a_{\nu'} : \tag{48}$$

The main disadvantage of the linearization prescription is that it does not readily extend to a nucleus with a non-independent-particle ground state. In particular, there is no obvious way to extend it to give open-shell HF theory, which was almost trivial with our general equations of motion.

An alternative procedure which grapples with the limitations of the linearization prescription is given by the core-particle coupling model (see lectures of Klein). In this approach, one starts from the equations of motion

$$<\alpha| [H,a_\nu^\dagger] |o> = (E_\alpha - E_o) <\alpha|a_\nu^\dagger |o> \tag{49}$$

for the spectroscopic amplitudes $<\alpha|a_\nu^\dagger|o>$. Expansion of the left hand side and insertion of a complete set of A-nucleus states $|\lambda>$, gives

$$\sum_\mu T_{\mu\nu} <\alpha|a_\mu^\dagger|o> + 1/2\sum_\lambda \sum_{\mu\mu'\nu'} V_{\mu\mu'\nu\nu'} <\alpha|a_\mu^\dagger|\lambda><\lambda|a_{\mu'}^\dagger a_{\nu'}|o>$$

$$= (E_\alpha - E_o) <\alpha| a_\nu^\dagger |o> . \tag{50}$$

This set of equations is solved, together with the coupled equations for the amplitudes $<\alpha|a_\mu^\dagger|\lambda>$ and $<\lambda|a_\mu^\dagger a_{\mu'}|o>$. to give the single particle strength distribution. Like our equations of motion, eq. (50) is formally exact. The approximations enter through truncation of the single-particle space and the number of intermediate states.

PARTICLE-HOLE EXCITATIONS

Tamm-Dancoff Approximation (TDA)

In the IPM (independent particle model) of a closed-shell N=Z nucleus, excited states are generated by the creation of a particle-hole excitation, i.e. by promoting a nucleon from below the fermi surface to above, as illustrated in fig. 4.

In this simple model the corresponding excitation energies are simply

$$\omega_{ph} = \varepsilon_p - \varepsilon_h$$

i.e. the difference in single-particle energy of the nucleon's final and initial states. The energy spectrum and the dipole strength of the five particle-hole configurations which can be excited by a dipole photon are shown for O^{16} in the upper half of fig. 5; note that in this calculation and in all calculations described in the latter half of this series, HF wave functions are approximated by harmonic oscillator wave functions and single-particle energies are taken from experiment.

For some time the IPM description of the giant dipole resonance was taken seriously, in spite of its major defect of predicting the excitation energy at approximately half its observed value. Clearly it is a naive model; for a nucleon in the particle state p, for example, has energy ε_p only if there is a full complement of nucleons below the fermi surface. With a hole below the fermi surface there is an interaction between the particle and the hole and, as a result, they scatter into other particle-hole configurations.

In the TDA this interaction is taken into account by expressing excited states as a finite sum of particle hole states:

$$|\lambda\Gamma\rangle = \sum_{ph} Y_{ph}(\lambda\Gamma) \; |(ph^{-1})\Gamma\rangle, \tag{51}$$

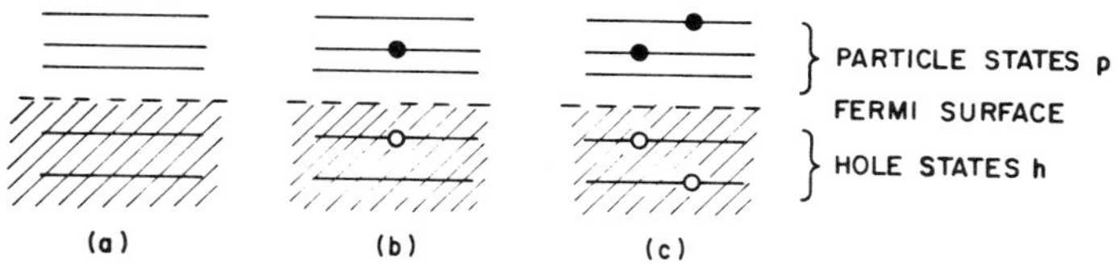

Figure 4
Schematic representation of (a) the particle-hole vacuum - the IPM ground state for a closed-shell nucleus - (b) a particle-hole excitation, and (c) a more complicated two particle-hole excitation.

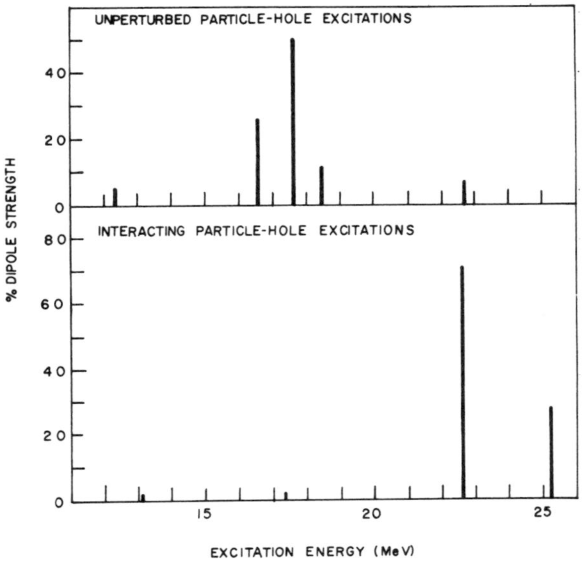

Figure 5
Comparison of Elliott and Flowers' results for the $J^{\pi}=1^{-}$, T=1 excited states of O^{16} with the unperturbed particle-hole excitations.

where we use the symbol Γ to denote both angular momentum, J and isospin T; λ distinguishes various states of the same Γ. The expansion coefficients and excitation energies are determined by evaluating the Hamiltonian matrix $<(ph^{-1})\Gamma|H|(p'h'^{-1})\Gamma>$ and diagonalizing it. The secular equation is of the form

$$\sum_{p'h'} <(ph^{-1})\Gamma|H|(p'h'^{-1})\Gamma> Y_{p'h'}(\lambda\Gamma) = (E_o+\omega_{\lambda\Gamma}) Y_{ph}(\lambda\Gamma). \quad (52)$$

The particle-hole matrix elements can be expressed in terms of two-particle matrix elements:

$$<(ph^{-1})\Gamma|(H-E_o)|(p'h'^{-1})\Gamma = \delta_{pp'}\delta_{hh'}(\varepsilon_p-\varepsilon_h) - \sum_{\Gamma_1}\hat{\Gamma}_1^2 W(phh'p';\Gamma\Gamma_1) <(h'p)\Gamma_1|V|(hp')\Gamma_1> \quad (53)$$

where W is a Racah coefficient and

$$\hat{\Gamma} \equiv (2J+1)^{1/2}(2T+1)^{1/2}. \quad (54)$$

Transition matrix elements for a general one-body operator W^{Γ} of spherical tensor rank Γ, for example an electromagnetic multipole operator, are given by

$$\langle\lambda\Gamma || W^{\Gamma} || o\rangle = \sum_{ph} Y^{*}_{ph}(\lambda\Gamma) \langle(ph^{-1})\Gamma \; || W^{\Gamma} || \; o\rangle \tag{55}$$

$$= \sum_{ph} Y^{*}_{ph}(\lambda\Gamma) \quad \langle p || W^{\Gamma} || h\rangle$$

where $\langle p \| W^{\Gamma} \| h\rangle$ is a single-particle reduced matrix element.

TDA calculations of the type described above have been reported by many authors. Some of the best known are those of Elliott and Flowers for the odd-parity states of O^{16}. Their results for the $J^{\pi}=1^{-}$, T=1 states of the giant dipole resonance are shown in the lower half of fig. 5. The main feature to be noted is that the particle-hole interaction effects a concentration of the dipole strength into a few states which are pushed upwards in energy. This is characteristic of T=1 excitations, for which the particle-hole interaction is predominantly repulsive. For T=0 excitations, the particle-hole interaction is predominantly attractive and the states are pushed down in energy (cf. fig.8).

In making comparison with experiment,[9] account must be taken of the particle emission broadening of each line in the spectrum. This is done phenomenologically in fig. 6 where it is seen that the TDA describes the gross structure of the gamma absorption cross section rather well.

The above formulation of the TDA is in the conventional Schrodinger approach of the shell model. We did this to make

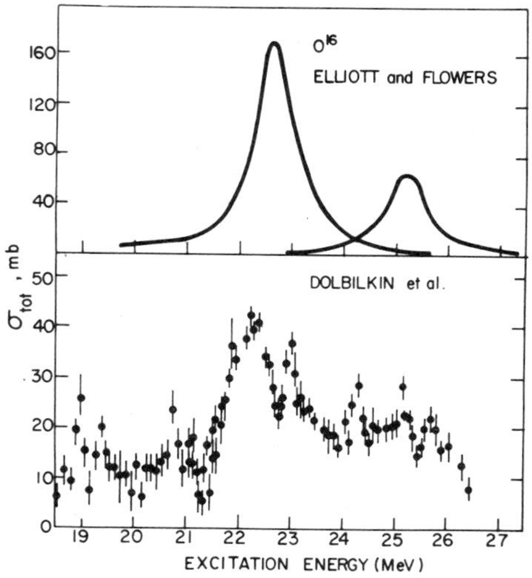

Figure 6
Comparison of the broadened Elliott and Flowers' results with total gamma-ray absorption measurements.[9]

an instructive comparison with the equations-of-motion approach. The equations-of-motion are rather excessively sophisticated for this particularly simple approximation; but they have the very significant advantage that they can much more readily be extended to higher orders of approximation, to include ground state correlation effects.

In equations-of-motion language, eq.(51) is expressed

$$|\lambda\Gamma\rangle = O^{\dagger}_{\lambda\Gamma} \, |o\rangle \tag{56}$$

with

$$O^{\dagger}_{\lambda\Gamma} = \sum_{ph} Y_{ph} \, (\lambda\Gamma) \; A^{\dagger}_{ph}(\Gamma), \tag{57}$$

when $A^{\dagger}_{ph}(\Gamma)$ is the angular momentum coupled particle-hole operator

$$A^{\dagger}_{ph}(\Gamma) = \sum_{m_p(m_h)} (j_p j_h m_p m_h|JM) \; (\tfrac{1}{2}\,\tfrac{1}{2}\;\tau_{op}\;\tau_{oh}|TM_T) \; a^{\dagger}_p \, a_{\bar{h}} \tag{58}$$

and $a_{\bar{h}}$ is shorthand for

$$a_{\bar{h}} \equiv (-1)^{j_h+m_h+\frac{1}{2}+\tau_{oh}} \qquad a_{j_h-m_h,\frac{1}{2}-\tau_{oh}} \tag{59}$$

In French's notation[10]

$$a^{\dagger}_p \equiv A^p$$

$$a_{\bar{h}} \equiv B^h \tag{60}$$

and

$$A^{\dagger}_{ph}(\Gamma) \equiv (A^p \times B^h)^{\Gamma}. \tag{61}$$

Inserting (57) into the equations of motion

$$\langle o| \; [O_{\kappa\Gamma}, H, O^{\dagger}_{\lambda\Gamma}]|o\rangle = \omega_{\lambda\Gamma} \langle o| \; [O_{\kappa\Gamma}, O^{\dagger}_{\lambda\Gamma}]|o\rangle \text{ all } \kappa, \tag{62}$$

gives the secular equation

$$\sum_{p'h'} \langle o|[A_{ph}(\Gamma), H, A^{\dagger}_{p'h'}(\Gamma)]|o\rangle \; Y_{p'h'}(\lambda\Gamma)$$

$$= \omega_{\lambda\Gamma} \sum_{p'h'} \langle o|[A_{ph}(\Gamma), A^{\dagger}_{p'h'}(\Gamma)]|o\rangle \; Y_{p'h'}(\lambda\Gamma). \tag{63}$$

Now part of the TDA is to assume that the ground state $|o\rangle$ is the particle-hole vacuum; i.e. the IPM ground state. With the special relationship that then exists between particle-hole creation operators and the particle-hole vacuum, the matrix on the right of (63) becomes the unit matrix, i.e.

$$\begin{aligned} \langle o|[A_{ph}(\Gamma),\ A^{\dagger}_{p'h'}(\Gamma)]|o\rangle &= \langle (ph^{-1})\Gamma|(p'h'^{-1})\Gamma\rangle \\ &= \delta_{pp'}\ \delta_{hh'} \end{aligned} \tag{64}$$

and the matrix on the left reduces to eq. (53). Thus the TDA secular equations of the equations-of-motion approach are identical to those of the shell model.

To go to higher orders of approximation, it is clear that 2 particle - 2 hole excitations can be included in the equations-of-motion excitation operator just as in the shell-model configuration space. But, in addition, it is straightforward to consider the simple particle-hole excitations of complicated non-IPM ground states. This is usually exceedingly difficult in the shell model.

Random Phase Approximation (RPA)

In the TDA it is explicitly assumed that the ground state is the IPM particle-hole vacuum state. In reality, of course, there are ground-state correlations, which show up as excited particle-hole components in the ground state. Excited particle-hole states can therefore be generated by exciting a nucleon from below the fermi surface (i.e. creating a particle-hole pair) or by de-exciting a nucleon from above to below (i.e. destroying a particle-hole pair), as illustrated in fig. 7.

For this reason, excitation operators in the RPA are expanded

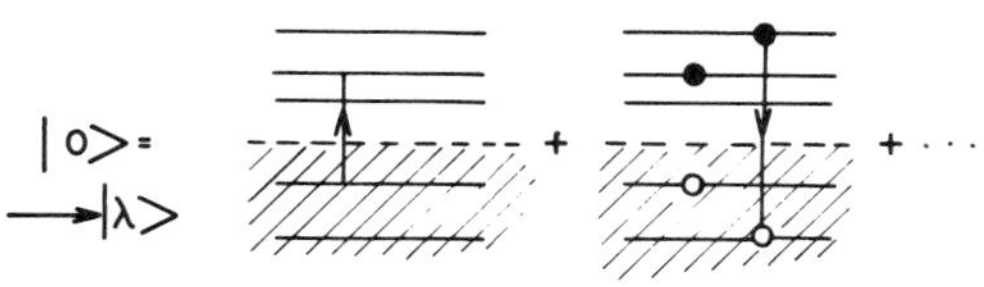

Figure 7
Illustration of two possible ways to generate excited particle-hole states of a closed-shell nucleus with a correlated ground state.

$$O^{\dagger}_{\lambda\Gamma} = \sum_{ph} \{Y_{ph}(\lambda\Gamma)\, A^{\dagger}_{ph}(\Gamma) - Z_{ph}(\lambda\Gamma)\, A_{ph}(\bar{\Gamma})\} \tag{65}$$

where

$$A_{ph}(\bar{\Gamma}) \equiv (-1)^{J+M+T+M_T} A_{ph}(J,-M,T,-M_T).^{\dagger} \tag{66}$$

In French's notation

$$A_{ph}(\bar{\Gamma}) \equiv (-1)^{\Gamma+p-h} (A^h \times B^p)^{\Gamma} . \tag{67}$$

To evaluate the equations of motion, some approximation must be assumed for the ground state. In the RPA the simplest possible ground state is used, namely the uncorrelated IPM state. This is blatantly inconsistent; for whereas ground state correlations are implicitly assumed in admitting the backward going excitation processes, they are now denied by the use of the IPM ground state. However, this inconsistency is not as bad as it appears at first sight. Recall that the double commutations were included in the equations of motion expressly for the purpose of making them as insensitive as possible to errors in the ground state. We shall see later just how sensitive the equations are to the neglect of correlations in this way.

Inserting the expansion (65) and the IPM ground state into the equations of motion one obtains the secular matrix equation:

$$\begin{pmatrix} A & B \\ B^* & A^* \end{pmatrix} \begin{pmatrix} Y \\ Z \end{pmatrix} = \omega \begin{pmatrix} I & 0 \\ 0 & -I \end{pmatrix} \begin{pmatrix} Y \\ Z \end{pmatrix} \tag{68}$$

where the sub-matrices A and B are defined

$$A^{\Gamma}_{php'h'} = \langle o | [A_{ph}(\Gamma),\ H,\ A^{\dagger}_{p'h'}(\Gamma)\ ]\ | o \rangle$$

$$= \langle (ph^{-1})\Gamma | (H-E_0) | (p'h'^{-1})\Gamma \rangle \qquad \text{(Hermitian)} \tag{69}$$

$$B^{\Gamma}_{php'h'} = -\langle o | [A_{ph}(\Gamma),\ H,\ A_{p'h'}(\bar{\Gamma})] | o \rangle$$

$$= \langle (ph^{-1})\Gamma;\ (p'h'^{-1})\bar{\Gamma}\ |\ H\ | o \rangle \qquad \text{(Symmetric)} \tag{70}$$

† Note that if the particle-hole creation operator $A^{\dagger}_{ph}(\Gamma)$ transforms under rotations like the spherical harmonic Y_{JM}, the destruction operator $A_{ph}(\Gamma)$ transforms like $Y^*_{JM} = (-1)^M Y_{J-M}$. For this reason we work with $A_{ph}(\bar{\Gamma})$ which transforms like $(-1)^{J+M} Y^*_{J-M} = (-1)^J Y_{JM}$ and is therefore a spherical tensor of rank Γ.

Clearly the A-matrix is identical to the Hamiltonian matrix of the TDA; cf. eqs. (63) and (53). The B-matrix, expressed in terms of two-particle matrix elements, is given by

$$B^{\Gamma}_{php'h'} = (1+\delta_{pp'})^{1/2}(1+\delta_{hh'})^{1/2} \sum_{\Gamma_1} \hat{\Gamma}_1^2 W(p\Gamma_1\Gamma h';p'h) \times <(p'p)\Gamma_1| V |(h'h)\Gamma_1>. \tag{71}$$

The sub-matrix I on the right hand side of (68) is the unit matrix and is the metric

$$<o|[A_{ph}(\Gamma), A^{\dagger}_{p'h'}(\Gamma)]|o> = \delta_{pp'}\delta_{hh'} . \tag{72}$$

The negative unit matrix is the metric for particle-hole destruction operators:

$$<o|[A^{\dagger}_{ph}(\bar{\Gamma}), A_{p'h'}(\bar{\Gamma})]|o> = -\delta_{pp'}\delta_{hh'} . \tag{73}$$

Thus we find that, with the introduction of particle-hole destruction operators, the metric matrix is no longer positive definite. With a general ground state and a general expansion of the excitation operators, the metric matrix may not even be diagonal. However it is always hermitian and can be diagonalized by a unitary transformation[4] and then followed by a scale transformation, cf. eq.(24), to reduce the metric matrix to the form

$$\begin{pmatrix} I & 0 \\ 0 & -I \end{pmatrix}$$

However, the dimensions of the positive and negative unit matrices differ if there is an unequal number of elementary creation and destruction operators - this situation may occur in the open-shell RPA equations for a nucleus with a non-scalar ground state, which is discussed later.

To solve the RPA equations it is convenient to express them in the form

$$\begin{pmatrix} A & B \\ -B^* & -A^* \end{pmatrix} \begin{pmatrix} Y \\ Z \end{pmatrix} = \omega \begin{pmatrix} Y \\ Z \end{pmatrix} \tag{74}$$

It is then a standard eigenvalue equation but for a non-hermitian matrix. Routines exist for the diagonalization of such matrices. Alternatively, since the matrices are invariably real, one can exploit the special symmetries of the RPA matrix in order to reduce the problem to eigenvalue equations for real symmetric matrices of half the dimensions.[11]

The properties of the RPA matrix equation (74) and of its

solutions are well known.[3] We therefore merely summarize some of the more important properties:

(i) RPA solutions occur in pairs. For every solution $\begin{pmatrix} Y \\ Z \end{pmatrix}$ with energy ω there is another solution $\begin{pmatrix} Z^* \\ Y^* \end{pmatrix}$ with energy $-\omega$. †If ω is positive the first solution corresponds to an excitation operator

$$O^\dagger = \sum \{Y_{ph} A^\dagger_{ph} - Z_{ph} A_{ph}\}$$

and its partner is its adjoint the de-excitation operator

$$-O = \sum \{Z^*_{ph} A^\dagger_{ph} - Y^*_{ph} A_{ph}\}$$

(ii) RPA solutions may be normalized to obey the orthogonality relations

$$(Y^\dagger(\kappa),\ Z^\dagger(\kappa))\begin{pmatrix} I & 0 \\ 0 & -I \end{pmatrix}\begin{pmatrix} Y(\lambda) \\ Z(\lambda) \end{pmatrix} = (Y^\dagger(\kappa),\ Z^\dagger(\kappa))\begin{pmatrix} Y(\lambda) \\ -Z(\lambda) \end{pmatrix}$$

$$= \frac{\omega_\lambda}{|\omega_\lambda|}\ \delta_{\kappa\lambda} \qquad (75)$$

This normalization corresponds to

$$\begin{aligned} \langle o|\ [O_\kappa,\ O^\dagger_\lambda]\,|o\rangle &= \delta_{\kappa\lambda} \qquad \omega_\lambda > o \\ \langle o|\ [O^\dagger_\kappa,\ O_\lambda]\,|o\rangle &= -\delta_{\kappa\lambda} \qquad \omega_\lambda < o\ . \end{aligned} \qquad (76)$$

(iii) Ground state transition matrix elements, for a hermitian one-body spherical tensor operator W^Γ of rank Γ, are given by

$$\langle\lambda\Gamma||\,W^\Gamma\,||\,o\rangle = \sum_{ph} \{Y^*_{ph}(\lambda\Gamma)\langle p||\,W^\Gamma\,||\,h\rangle + (-1)^\Gamma Z^*_{ph}(\lambda\Gamma)\langle p||\,W^\Gamma\,||\,h\rangle^*\}$$

or, if all the elements are real, by

$$\langle\lambda\Gamma||W^\Gamma||o\rangle = \sum_{ph} \{Y_{ph}(\lambda\Gamma) + (-1)^\Gamma Z_{ph}(\lambda\Gamma)\}\ \langle p||W^\Gamma||h\rangle\ . \qquad (77)$$

Some results of RPA calculations for the $J^\pi = 3^-$, $T = 0$ excitations of ^{16}O are given in fig. 8 in comparison with the corresponding TDA calculations. It is seen that the solutions are almost identical, except for the lowlying 3^- excitation which is highly collective. For this excitation the RPA backward going amplitudes are very large and effect a big difference between the TDA and RPA predictions for the electromagnetic transition matrix elements.

†In this brief review we ignore the possibility of zero or complex solutions for ω. This is discussed in detail in ref. 3.

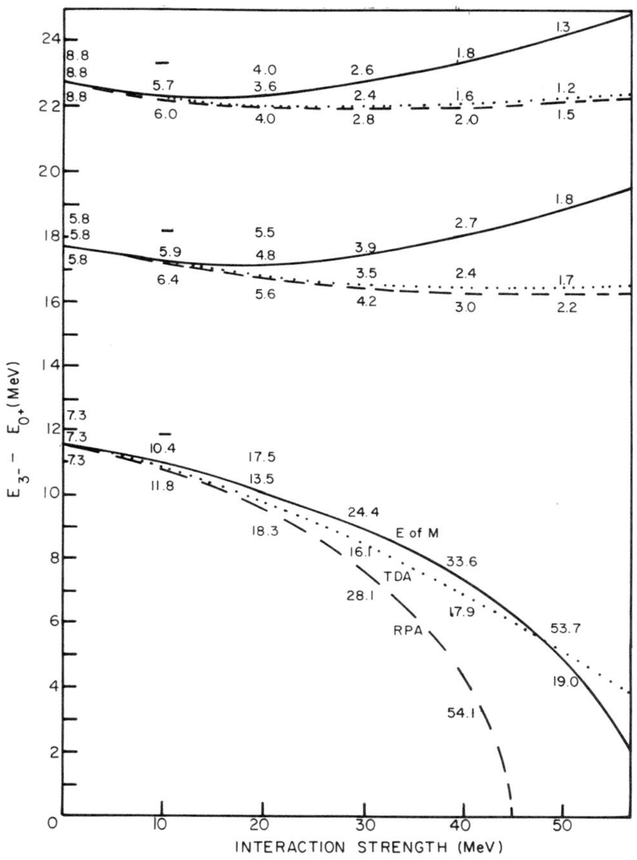

Figure 8
Comparison of TDA, RPA and more exact equations-of-motion (E of M) calculations of the $J^{\pi} = 3^-$, T = 0 excitations of O^{16}, as a function of interaction strength. Reduced transition probabilities are shown in oscillator units as numbers on the curves. Reading from top to bottom they refer to the E of M, TDA and RPA predictions, respectively. For details of the calculation see ref. 12.

More Exact Particle-Hole Calculations

There are two possibilities to improve on the RPA: one can solve the equations of motion with a better ground state or one can enlarge the space of the excitation operators. Results of an improvement of the first kind are shown in fig. 8. The O^{16} ground state for this calculation was obtained by a shell-model diagonalization in the zero plus all two particle-hole configurations up through the 2s-1d shells, after elimination of the spurious centre-of-mass states. It is seen that the specific inclusion of ground state correlations in this way has a repulsive effect (i.e. excitation energies are increased) Whereas the inclusion of the backward going amplitudes (cf. RPA vs. TDA) decrease the excitation energies.

If one were in addition to enlarge the excitation operator space, to include for example, two particle-hole excitation processes, this must lower the energies of the lowlying states, as one can readily show by a centroid argument.[3] However this argument is applicable only if one is using HF single particle energies. For one of the major effects of including two particle-hole excitations is to renormalize the effective single-particle energies (cf. fig.3). If one is using experimental, rather than HF single-particle energies, which include these and all other kinds of renormalizations to all orders, the remaining effect of coupling to two particle-hole excitations may be repulsive.

The Open-Shell RPA[5]

In the simple shell model description of a closed-shell nucleus, there is a single closed-shell configuration which gives the ground state. For an open-shell nucleus, there is a large number of valence-shell configurations which must be diagonalized to give not only the ground state, but also the lowlying positive parity states.

Thus the characters of lowlying positive parity excitations differ substantially between closed and open-shell nuclei. For example, all even nuclei, other than closed-shell nuclei, invariably exhibit a first excited state of $J^{\pi}=2^{+}$. This state may be of vibrational character, particularly in the superconducting nuclei, or it may be of rotational character. For closed-shell nuclei, without a valence space, the positive parity excitations involve particle-hole excitations and tend to lie at somewhat higher excitation energies.

Negative parity excitations, on the other hand, are generated in both cases by transferring a nucleon across a major shell boundary. Thus one is not surprised that the properties of negative parity excitations are substantially the same in open- and closed-shell nuclei. In particular, all nuclei exhibit highly collective lowlying octupole vibrations and a giant dipole resonance at higher energies.

Now the TDA and RPA were traditionally formulated in terms of a non-degenerate particle-hole vacuum ground state and were therefore not applicable to open-shell nuclei. In the above equations-of-motion formalism, however, the extension is almost trivial.

In this section, we consider only even N=Z nuclei with ground states of $J^{\pi}=0^{+}$, T=0. The extension to nuclei with non-scalar ground states is considered in the following section.

A schematic expansion of the ground state of an open-shell nucleus is illustrated in fig.9. It is seen to be analogous to the expansion, fig.7, for the ground state of a closed-shell nucleus. To obtain negative parity excitations, one can excite a nucleon from the valence to the 'empty' shell or from the 'full' to the valence shell. But in the presence of ground-state correlations it is clear from fig. 9 that negative parity particle-hole excitations can also be generated by de-exciting particle-hole pairs already present in the ground state. Thus in the open-shell RPA, we consider excitation operators of the form

$$O^{\dagger}(\lambda\Gamma) = \sum_{\alpha>\beta} \frac{Y_{\alpha\beta}(\lambda\Gamma)A^{\dagger}_{\alpha\beta}(\Gamma) - Z_{\alpha\beta}(\lambda\Gamma)\, A_{\alpha\beta}(\bar{\Gamma})}{\langle o|\,[A_{\alpha\beta}(\Gamma),\, A^{\dagger}_{\alpha\beta}(\Gamma)]\,\,|o\rangle^{1/2}} \qquad (78)$$

where the ordering of single-particle indices $\alpha>\beta$ is according to increasing single-particle energy (for negative parity excitation α and β are in different but adjacent shells). The denominator is included for normalization so that the equations reduce to standard RPA form.

As in the closed-shell TDA and RPA, one solves the equations of motion with an approximate ground state given by the first term in the expansion of fig. 9. But unlike the closed-shell case, this is not a single configuration. It must be obtained by a preliminary shell-model diagonalization within the valence space. This in itself may be a difficult calculation for which it will sometimes be necessary to make approximations. However, it is fundamentally a much simpler calculation than a shell-model calculation for the negative parity states which necessarily involve three major shell configurations.

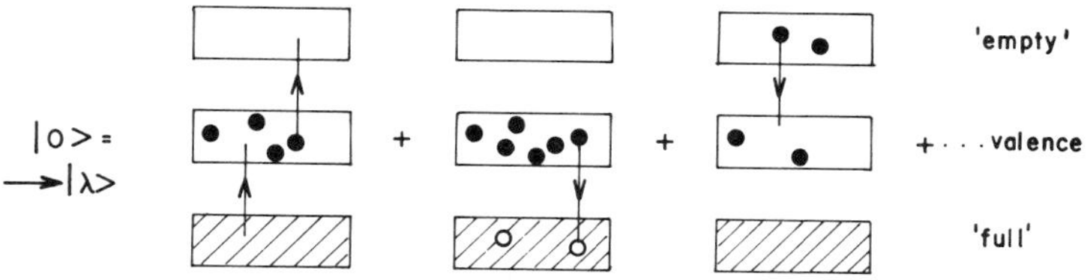

Figure 9
Illustration of possible ways to generate excited particle-hole states of an open-shell nucleus with a correlated ground state.

Insertion of the expansion (78) into the equations of motion (13), leads to the secular matrix equation

$$\begin{pmatrix} A & B \\ B^* & A^* \end{pmatrix}\begin{pmatrix} Y \\ Z \end{pmatrix} = \omega\begin{pmatrix} Y \\ -Z \end{pmatrix} \tag{79}$$

which is of standard RPA form. The sub-matrix A is again hermitian and is given by

$$A^{\Gamma}_{\alpha\beta\gamma\delta} = \frac{<o|[A_{\alpha\beta}(\Gamma),\ H,\ A^{\dagger}_{\gamma\delta}(\Gamma)]|o>}{<o|[A_{\alpha\beta}(\Gamma),A^{\dagger}_{\alpha\beta}(\Gamma)]|o>^{1/2}<o|[A_{\gamma\delta}(\Gamma),A^{\dagger}_{\gamma\delta}(\Gamma)]|o>^{1/2}} \tag{80}$$

The sub-matrix B is again symmetric and is given by

$$B^{\Gamma}_{\alpha\beta\gamma\delta} = \frac{-<o|[A_{\alpha\beta}(\Gamma),\ H,\ A_{\gamma\delta}(\bar{\Gamma})]|o>}{<o|[A_{\alpha\beta}(\Gamma),A^{\dagger}_{\alpha\beta}(\Gamma)]|o>^{1/2}<o|[A_{\gamma\delta}(\Gamma),A^{\dagger}_{\gamma\delta}(\Gamma)]|o>^{1/2}} \tag{81}$$

In deriving the open-shell RPA equation (79), from the equation of motion, we assume that the matrix elements $<o|[A_{\alpha\beta}(\Gamma), A^{\dagger}_{\gamma\delta}(\Gamma)]|o>$ are diagonal. This follows provided there are no two single-particle states of the same angular momentum and parity within the valence space. These matrix elements are then given by

$$<o|[A_{\alpha\beta}(\Gamma),\ A^{\dagger}_{\gamma\delta}(\Gamma)]|o> = \delta_{\alpha\gamma}\ \delta_{\beta\delta}\ (n_{\beta} - n_{\alpha}), \tag{82}$$

where n_{α} is the fractional occupancy of the single-particle state α in the shell-model ground state. With the shell ordering $\alpha>\beta$, the expression (82) is real and positive and hence has a real square root.

Unlike the closed-shell RPA, the sub-matrices A and B no longer reduce to simple expressions in terms of two-particle matrix elements of the interaction. To evaluate them it is necessary therefore to actually perform the double commutations. This is conveniently done using the second quantized tensorial techniques of French.[10]

In French's notation (cf. eq. (60)), where

$$\begin{aligned} A^{\dagger}_{\alpha\beta}(\Gamma) &\equiv (A^{\alpha} \times B^{\beta})^{\Gamma} \\ A_{\alpha\beta}(\bar{\Gamma}) &\equiv (-1)^{\Gamma+\alpha-\beta}(A^{\beta} \times B^{\alpha})^{\Gamma} \end{aligned} \tag{83}$$

the numerators in eqs. (80) and (81) become

$$\langle o|\,[A_{\alpha\beta}(\Gamma),\ H,\ A^{\dagger}_{\gamma\delta}(\Gamma)]\,|o\rangle$$
$$= (-1)^{\Gamma+\alpha-\beta}\,\hat{\Gamma}^{-1}\langle o|[(A^{\beta}xB^{\alpha})^{\Gamma},H,\ (A^{\gamma}xB^{\delta})^{\Gamma}]^{o}|o\rangle$$

$$\langle o|\,[A_{\alpha\beta}(\Gamma),\ H,A_{\gamma\delta}(\bar{\Gamma})]\,|o\rangle$$
$$= (-1)^{\alpha-\beta+\gamma-\delta}\,\hat{\Gamma}^{-1}\ \langle o|\,[(A^{\beta}xB^{\alpha})^{\Gamma},H,(A^{\delta}xB^{\gamma})^{\Gamma}]^{o}|o\rangle\ , \tag{84}$$

where we exploit the fact that the ground state has J=T=0, so that only the J=T=0 component of the double commutator has non-vanishing expectation. Each double commutator appearing on the right of (84) is characterized by a set of zero-, one-, and two-body matrix elements, just as a scalar Hamiltonian. Given these matrix elements one can therefore use conventional shell-model methods to calculate their ground state expectations.

From eq. (14) we have that

$$2[\ (A^{\alpha}xB^{\beta}),H,(A^{a}xB^{b})\ ]^{o} = [\ (A^{\alpha}xB^{\beta})^{\Gamma},[H,(A^{a}xB^{b})^{\Gamma}]\ ]^{o} + [(A^{a}xB^{b})^{\Gamma},[H,(A^{\alpha}xB^{\beta})^{\Gamma}]\ ]^{o} \tag{85}$$

The expansion of these double commutators is straightforward. Their two-body matrix elements,

$$V^{\Omega}_{rstu} \equiv \langle (rs)\Omega|\,[(A^{\alpha}xB^{\beta})^{\Gamma},[H,(A^{a}xB^{b})^{\Gamma}]]^{o}\,|(tu)\Omega\rangle\ , \tag{86}$$

for example, are given by

$$\begin{aligned}
V^{\Omega}_{rstu} = &-2\delta_{b\alpha}\,\delta^{\text{a.m.}}_{a\beta}\,\delta_{u\beta}(-1)^{\alpha-\beta+\Gamma}\,\frac{\hat{\Gamma}}{\hat{\beta}^{2}}\,\frac{\zeta_{t\beta}}{\zeta_{ta}}\,W^{\Omega}_{rsta}\\
&-2\delta_{a\beta}\,\delta^{\text{a.m.}}_{b\alpha}\,\delta_{r\alpha}(-1)^{\alpha-\beta+\Gamma}\,\frac{\hat{\Gamma}}{\hat{\alpha}^{2}}\,\frac{\zeta_{s\alpha}}{\zeta_{sb}}\,W^{\Omega}_{bstu}\\
&+2\delta_{tb}\,\delta_{u\beta}(-1)^{\alpha-\beta+\Gamma}\hat{\Gamma}W(\alpha a\beta b;\,\Omega\Gamma)\,\frac{\zeta_{b\beta}}{\zeta_{a\alpha}}\,W^{\Omega}_{rs\alpha a}\\
&+2\delta_{r\alpha}\,\delta_{sa}(-1)^{\alpha-\beta+\Gamma}\hat{\Gamma}W(ab\alpha\beta;\,\Gamma\Omega)\,\frac{\zeta_{a\alpha}}{\zeta_{b\beta}}\,W^{\Omega}_{b\beta tu}\\
&+4\delta_{r\alpha}\,\delta_{ub}(-1)^{\alpha-\beta+\Gamma}\hat{\Gamma}\sum_{\varepsilon}\hat{\varepsilon}^{2}W(\varepsilon t\Gamma b;\,a\Omega)W(\Gamma\beta\Omega s;\,\alpha\varepsilon)\,\frac{\zeta_{s\alpha}\zeta_{tb}}{\zeta_{s\beta}\zeta_{ta}}\,W^{\varepsilon}_{\beta sta}\\
&+4\delta_{ra}\,\delta_{u\beta}\,\hat{\Gamma}\sum_{\varepsilon}(-1)^{t-\alpha+\varepsilon}\hat{\varepsilon}^{2}W(as\Gamma\varepsilon;\,\Omega b)W(\varepsilon\alpha\Omega\beta;\,t\Gamma)\,\frac{\zeta_{sa}\zeta_{t\beta}}{\zeta_{sb}\zeta_{t\alpha}}\,W^{\varepsilon}_{bst\alpha},
\end{aligned} \tag{87}$$

where

$$\zeta_{rs} = (1 + \delta_{rs})^{-1/2} \tag{88}$$

$$\delta^{a.m.}_{ab} = 1 \quad \text{if } j_a = j_b \tag{89}$$
$$\quad\;\; = 0 \quad \text{if } j_a \neq j_b$$

and W^{Ω}_{abcd} is a matrix element of the two-body interaction

$$W^{\Omega}_{abcd} = \langle(ab)\Omega \mid V \mid (cd)\Omega\rangle. \tag{90}$$

Ground state transition matrix elements for a hermitian one-body spherical tensor operator W^{Γ} are obtained from eq.(18). One obtains

$$\langle\lambda\Gamma||W^{\Gamma}||o\rangle = \sum_{\alpha>\beta} \{Y^*_{\alpha\beta}(\lambda\Gamma)\langle\alpha||W^{\Gamma}||\beta\rangle+(-1)^{\Gamma}Z^*_{\alpha\beta}(\lambda\Gamma)\langle\alpha||W^{\Gamma}||\beta\rangle^*\}$$
$$\times\ (n_{\beta}-n_{\alpha})^{1/2}$$

or, if all the elements are real, by

$$\langle\lambda\Gamma||W^{\Gamma}||o\rangle = \sum_{\alpha>\beta} \{Y_{\alpha\beta}(\lambda\Gamma)+(-1)^{\Gamma}Z_{\alpha\beta}(\lambda\Gamma)\}\langle\alpha||W^{\Gamma}||\beta\rangle(n_{\beta}-n_{\alpha})^{1/2} \tag{91}$$

Some applications of the open-shell TDA and RPA models to the $J^{\pi}=1^-$, T=1 excitations of C^{12} are shown respectively in figs. 10 and 11.

C^{12} has frequently been regarded as a closed $(1p_{3/2})^8$ sub-shell nucleus, in order that it be accessible to standard (closed-shell) TDA and RPA calculations. Results of closed-shell TDA and RPA calculations, for a central Rosenfeld interaction, are shown in the upper parts of figs. 10 and 11, respectively. In reality, the ground state of C^{12} is nowhere near closed sub-shell. We therefore used an intermediate coupling shell model in the whole $1p_{3/2}$ and $1p_{1/2}$ valence space and performed open-shell TDA and RPA calculations. These results are marked E of M in figs. 10 and 11.

In comparing the closed- and open-shell calculations, it should be noted that exactly the same Hamiltonian was used for each. The difference is entirely due to the intermediate coupling ground state used in the latter calculations.

It may be observed that the main peak of the giant resonance is split into two in the open-shell calculations. For a closed-shell ground state, the $1p_{3/2}$ state is a particle state and the $1p_{1/2}$ state is a hole state. For an open-shell ground state each state behaves partly as a particle and partly as a hole state. Thus there are almost double the number of particle-hole

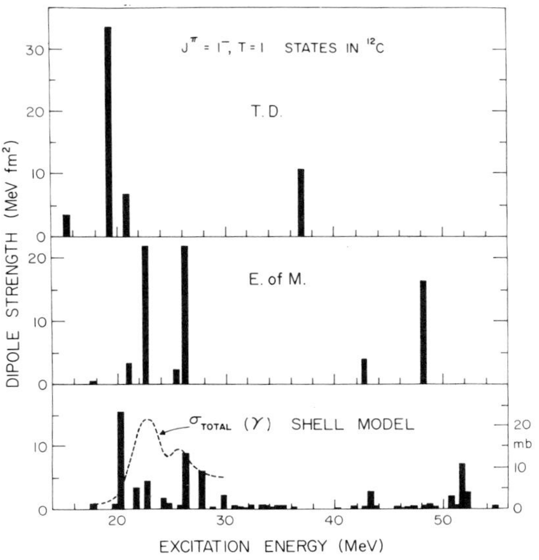

Figure 10
The $J^{\pi}=1^{-}$, T=1 giant dipole excitations of C^{12} in the closed-shell TDA (labelled T.D.), the open-shell TDA (labelled E.of M.) and in a more detailed shell model calculation. The γ-absorption cross section shown is experimental[13].

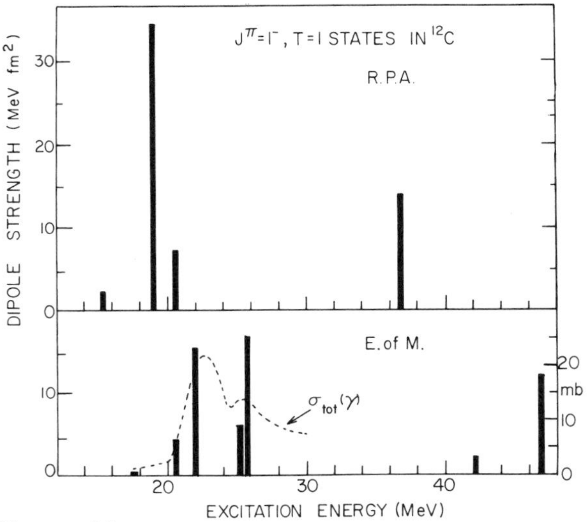

Figure 11
The $J^{\pi}=1^{-}$, T=1 giant dipole excitation of C^{12} in the closed-shell RPA (labelled R.P.A.) and in the open-shell RPA (labelled E.of M.)

excitations possible and there is a splitting of the dipole strength into two main peaks. The energy separation of the peaks is determined largely by the spin-orbit splitting of the 1p-shells.

Comparison of the TDA and RPA calculations, shows that they predict approximately the same structure for the giant resonance. However, the inclusion of the RPA backward going amplitudes effects an overall reduction in strength by about 20%.

To investigate the validity of the 1-particle-hole approximation, we also performed shell-model calculations within the complete $(1s)^{-1}(1p)^9$ and $(1p)^7(2s1d)^1$ configuration space. This space contains a total of 141 $J^\pi=1^-$, T=1 states including the seven 1-particle-hole states of the open-shell TDA. Such a shell-model calculation is possible for C^{12} but would involve far too many states for a heavier open-shell nucleus. The results are shown in fig. 10. The dipole strength, which in the TDA is vested in the few 1-particle-hole states, becomes distributed over a large number of more complicated states. Nevertheless the gross structure of the open-shell TDA remains clearly recognizable. The open-shell TDA may therefore be regarded as a success and one is encouraged to employ it, or better still the open-shell RPA, to other nuclei.

In addition to the giant dipole excitations, we have also calculated all the other negative parity excitations of C^{12}. The application of these results to the calculation of electron scattering form factors is given (by T.W. Donnelly) in the appendix to this series. The results are even more satisfying.

In concluding this section, I wish to mention the results of some open-shell RPA calculations that we have done[14] for the $J^\pi=1^-$, T=1 states of Ne^{20} and Si^{28}.

Apart from Ne^{20}, one must now face up to the fact that, it is computationally difficult to perform exact shell-model calculations in the complete 2s-1d valence space, even for the ground state. We have therefore chosen to work with PHF (projected Hartree-Fock) ground states, with variation after projection according to the prescription of ref.[15]. For Ne^{20}, however, it is possible to calculate the full shell-model ground state and make the comparison. This we have done and fig. 12 shows the results of open-shell RPA calculations performed with the shell-model and PHF ground states respectively. It will be seen that the results do not differ significantly. Fig. 13 shows the results of open-shell RPA calculations for ^{28}Si. We intend to calculate all the negative parity excitations of ^{28}Si and their electron scattering form factors.

EQUATIONS OF MOTION FOR NON-SCALAR NUCLEI[16]

In the equations of motion that we have considered so far, it has been implicitly assumed that J=T=0 for the nuclear ground state. As a result the excitation operators have well-defined spin and isospin. In this concluding section, the equations are generalized to nuclei of any spin and isospin using methods which have much in common with the multipole sum rule techniques of French.[10]

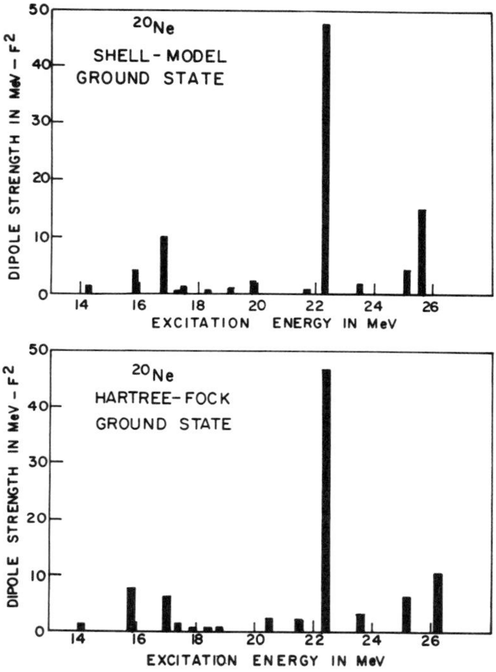

Figure 12
Open-shell RPA calculations for the $J^{\pi}=1^{-}$, T=1 excitations of Ne^{20} with a shell-model ground state and with a PHF ground state.

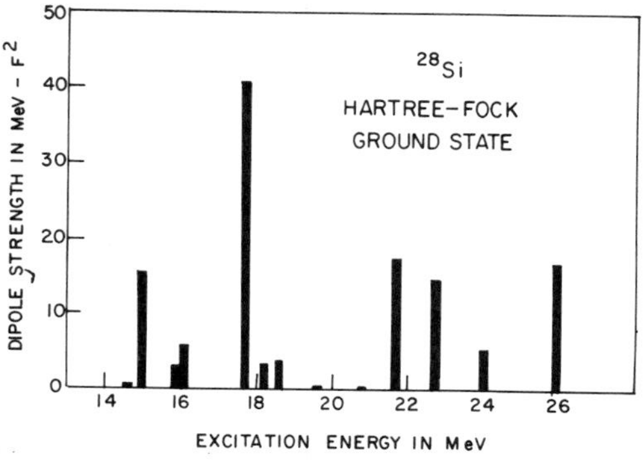

Figure 13
Open-shell RPA calculations for the $J^{\pi}=1^{-}$, T=1 excitations of Si^{28} with a PHF ground state.

Let $|\Delta>$ be the nuclear ground state and $|x\Lambda>$ an excited state, where Δ and Λ denote the spin and isospin and x distinguishes excited states of the same Λ. In parallel with eq.(3), we now define excitation operators $O^{\dagger}_{x\lambda}$ of tensorial rank λ such that

$$(O^{\dagger}_{x\lambda}|\Delta>)^{\Lambda} = |x\Lambda> \tag{92}$$

$$(O_{x\lambda}|\Delta>) = 0 . \tag{93}$$

In (92) the usual tensorial coupling is shown explicitly whereas (93) is required to vanish for all possible couplings.

Formally one can show that, not only do operators exist which satisfy eqs.(92) and (93) but there is a continuum of such operators; for example, the set of tensor operators

$$O^{\dagger}_{x\lambda_i} = N_i(|x\Lambda><\Delta|)^{\lambda_i} + \sum_{\mu,\nu\neq\Delta} C_{\mu\nu}(|\mu><\nu|)^{\lambda_i} \tag{94}$$

for arbitrary $C_{\mu\nu}$ and suitable normalization N_i.

Furthermore, unless $\Delta=\Lambda=0$, any λ_i which satisfies the triangle relations

$$|\Delta-\Lambda| \leq \lambda_i \leq \Delta + \Lambda, \tag{95}$$

is permissable. Thus the most general excitation operators which satisfy (92) and (93) are of mixed tensor rank:

$$Q^{\dagger}_{x\Lambda} = \sum_i O^{\dagger}_{x\lambda_i} . \tag{96}$$

While it may be preferable in an infinite operator space to work with excitation operators of definite tensor rank, in practice one is restricted to finite workable dimensions. It is therefore advantageous to adopt the most general expansion (96) in the hope of finding satisfactory solutions within the model space.

From the definition of the excitation operators we can write

$$[H,(Q^{\dagger}_{x\Lambda}]|\Delta>)^{\Lambda} \equiv H(Q^{\dagger}_{x\Lambda}|\Delta>)^{\Lambda}-(Q^{\dagger}_{x\Lambda}H|\Delta>)^{\Lambda} = \omega_{x\Lambda}(Q^{\dagger}_{x\Lambda}|\Delta>)^{\Lambda} \tag{97}$$

and hence

$$(<\Delta|Q_{y\Lambda})^{\Lambda}\ [H,(Q^{\dagger}_{x\Lambda}]|\Delta>)^{\Lambda} = \omega_{x\Lambda}(<\Delta|Q_{y\Lambda})^{\Lambda}(Q^{\dagger}_{x\Lambda}|\Delta>)^{\Lambda}$$

or explicitly

$$\sum_{ij} (<\Delta|O_{y\lambda_i})^{\Lambda}[H,(O^{\dagger}_{x\lambda_j}]|\Delta>)^{\Lambda} = \omega_{x\Lambda} \sum_{ij} (<\Delta|O_{y\lambda_i})^{\Lambda}(O^{\dagger}_{x\lambda_j}|\Delta>)^{\Lambda} \tag{98}$$

Making a recoupling, we obtain

$$\sum_{i,j,\Gamma} (-1)^{\Gamma} \hat{\Gamma} W(\lambda_j \lambda_i \Delta\Delta;\Gamma\Lambda) \langle\Delta\| (O_{y\lambda_i} [H,O^{\dagger}_{x\lambda_j}])^{\Gamma} \|\Delta\rangle$$

$$= \omega_{x\Lambda} \sum_{ij\Gamma} (-1)^{\Gamma} \hat{\Gamma} W(\lambda_j \lambda_i \Delta\Delta;\Gamma\Lambda) \langle\Delta\| (O_{y\lambda_i} O^{\dagger}_{x\lambda_j})^{\Gamma} \| \Delta\rangle. \qquad (99)$$

Now from (93) we also have

$$\langle\Delta\| ([H,O^{\dagger}_{x\lambda_j}] O_{y\lambda_i})^{\Gamma} \| \Delta\rangle = 0 = \langle\Delta\| (O^{\dagger}_{x\lambda_j} O_{y\lambda_i})^{\Gamma} \| \Delta\rangle. (100)$$

Combining (99) and (100), we obtain

$$\sum_{ij\Gamma} (-1)^{\Gamma} \hat{\Gamma} W(\lambda_j \lambda_i \Delta\Delta;\Gamma\Lambda) \langle\Delta\| [O_{y\lambda_i}, [H,O^{\dagger}_{x\lambda_j}]]^{\Gamma} \| \Delta\rangle$$

$$= \omega_{x\Lambda} \sum_{ij\Gamma} (-1)^{\Gamma} \hat{\Gamma} W(\lambda_j \lambda_i \Delta\Delta;\Gamma\Lambda) \langle\Delta\| [O_{y\lambda_i}, O^{\dagger}_{x\lambda_j}]^{\Gamma} \| \Delta\rangle \qquad (101)$$

By a similar argument it can also be shown that

$$\sum_{ij\Gamma} (-1)^{\Gamma} \hat{\Gamma} W(\lambda_j \lambda_i \Delta\Delta;\Gamma\Lambda) \langle\Delta\| [[O_{y\lambda_i}, H], O^{\dagger}_{x\lambda_j}]^{\Gamma} \| \Delta\rangle$$

$$= \omega_{x\Lambda} \sum_{ij\Gamma} (-1)^{\Gamma} \hat{\Gamma} W(\lambda_j \lambda_i \Delta\Delta;\Gamma\Lambda) \langle\Delta\| [O_{y\lambda_i}, O^{\dagger}_{x\lambda_j}]^{\Gamma} \|\Delta\rangle . \qquad (102)$$

We therefore take the average of (101) and (102) to obtain the more symmetric equations of motion

$$\sum_{ij\Gamma} (-1)^{\Gamma} \hat{\Gamma} W(\lambda_j \lambda_i \Delta\Delta;\Gamma\Lambda) \langle\Delta\| [O_{y\lambda_i}, H, O^{\dagger}_{x\lambda_j}]^{\Gamma} \|\Delta\rangle$$

$$= \omega_{x\Lambda} \sum_{ij\Gamma} (-1)^{\Gamma} \hat{\Gamma} W(\lambda_j \lambda_i \Delta\Delta;\Gamma\Lambda) \langle\Delta\| [O_{y\lambda_i}, O^{\dagger}_{x\lambda_j}]^{\Gamma} \|\Delta\rangle , \qquad (103)$$

where the symmetrized double commutator is defined as in eq. (14).

In the special case of a scalar ground state, i.e. $\Delta=0$, these equations reduce to the equations of motion, (13), used previously.

The philosophy behind the derivation of eq.(103) is exactly as in the scalar equations. The equations of motion are exact but in deriving them we exploit formally exact properties to shape them into a form suitable for practical approximate calculations.

Excitation operators of mixed rank, eq.(96), are included even though the equations are simpler, and still exact, with operators of definite rank. The gain is the possibility of finding operators which are simpler in other respects. For example, in some situations satisfactory excitation operators of mixed tensor rank may be found in a model space of one-body operators, whereas it may be necessary to include many-body operators to obtain equally good results for excitation operators of fixed tensor rank.

In practice, an approximation has to be made for the ground state $|\Delta>$. Now while we may have some confidence in the accuracy of an approximate model ground state to predict one-body and possibly two-body densities, we would have little faith in its reliability for many-particle correlations. For this reason we have again included the maximum number of commutations in order that the quantities between the double bars in eq. (103) should be of minimum particle rank.

For the exact ground state, eqs.(101), (102) and (103) are completely equivalent. They cease to be equivalent, however, for an approximate model ground. What is more eqs. (101) and (102) lead to secular equations which may not be symmetric. The symmetric eq. (103) is therefore adopted which leads, as we now show, to eigenvalue equations of certain diagonalizability and to orthogonal eigenvectors, whatever the approximations used.

Assume that we have an approximate ground state $|\Delta>$ and a set of boson-like operators $\{\eta^\dagger_\alpha(\lambda_i)\}$; this set may or may not include the adjoint operators $\eta_\alpha(\lambda_i)$. We then expand

$$Q^\dagger_{x\Lambda} = \sum_{\alpha i} X_{\alpha i}\, \eta^\dagger_\alpha(\lambda_i) \tag{104}$$

which inserted into the equations of motion (103) gives the matrix equation

$$AX = \omega RX. \tag{105}$$

The matrices A and R are both hermitian and are given by

$$\begin{aligned} A_{\alpha i,\beta j} &= \sum_\Gamma (-1)^\Gamma \hat{\Gamma}\, W(\lambda_j\lambda_i\Delta\Delta;\Gamma\Lambda)\, <\Delta||\, [\eta_\alpha(\lambda_i),\, H,\, \eta^\dagger_\beta(\lambda_j)]^\Gamma \,||\Delta> \\ &= A^*_{\beta j,\alpha i} \end{aligned} \tag{106}$$

$$\begin{aligned} R_{\alpha i,\beta j} &= \sum_\Gamma (-1)^\Gamma \hat{\Gamma}\, W(\lambda_j\lambda_i\Delta\Delta;\Gamma\Lambda)\, <\Delta||\, [\eta_\alpha(\lambda_i),\, \eta^\dagger_\beta(\lambda_j)]^\Gamma \,||\Delta> \\ &= R^*_{\beta j,\alpha i}\,. \end{aligned} \tag{107}$$

This hermiticity is guaranteed for the equations of motion (103) whatever approximations are made for $|\Delta>$. This is not necessarily the case for the equations (101) or (102) however.

The matrix R is the metric matrix. Since it is hermitian it can be diagonalized by a unitary transformation, as discussed previously (see discussion surrounding eq. (23)). The transformed basis operators then necessarily form an orthogonal set. Frequently the original set will have been overcomplete, especially when we allow components of different tensor rank. If this were the case it would show up by the occurence of vanishing eigenvalues of the metric matrix. This means that the corresponding operators have vanishing norms and are therefore the spurious operators of the overcomplete space. Eliminating the spurious operators and renormalizing the non-spurious, one can in general reduce the equations of motion to the form

$$\begin{pmatrix} A & B \\ B^{\dagger} & C \end{pmatrix}\begin{pmatrix} Y \\ Z \end{pmatrix} = \omega \begin{pmatrix} I & 0 \\ 0 & -I \end{pmatrix}\begin{pmatrix} Y \\ Z \end{pmatrix} \tag{108}$$

which can also be written in standard eigenvalue equation form:

$$\begin{pmatrix} A & B \\ -B^{\dagger} & -C \end{pmatrix}\begin{pmatrix} Y \\ Z \end{pmatrix} = \omega \begin{pmatrix} Y \\ Z \end{pmatrix} \tag{109}$$

This equation has less symmetry than, for example, the scalar RPA equations. As a consequence its solutions will not in general occur in pairs with energies $\omega = \pm|\omega|$. The negative energy solutions are then no longer the adjoints of the positive energy solutions and must be regarded as non-physical. This loss of symmetry is directly attributable to the lack of symmetry between the eqs.(92) and (93) which define the excitation and de-excitation operators.

The above matrix equation is diagonalizable and leads to solutions with the orthonormality relations

$$(Y^{\dagger}(x),\ Z^{\dagger}(x)\ )\begin{pmatrix} Y(y) \\ -Z(y) \end{pmatrix} = \delta_{xy}, \quad \omega(x) > 0 \tag{110}$$

In concluding this section we derive expressions for the ground state transition matrix elements of a general one-body operator $W(\Omega)$ of tensor rank Ω. Such an operator can be expressed in second quantized form

$$W(\Omega) = \sum_{\alpha\beta} \hat{\Omega}^{-1} \langle\alpha\| W(\Omega) \| \beta\rangle\, A^{\dagger}_{\alpha\beta}(\Omega). \tag{111}$$

The ground state transition matrix elements for $W(\Omega)$ can be written

$$\langle x\Lambda|(W(\Omega)|\Delta\rangle)^{\Lambda} = \sum_{i} (\langle\Delta|O_{x\lambda_i})^{\Lambda}(W(\Omega)|\Delta\rangle)^{\Lambda}$$

which can be recoupled and combined with eq. (93) to give

$$\langle x\Lambda\| W(\Omega) \| \Delta\rangle = (-1)^{\Omega}\hat{\Omega}^{-1}\sum_{i\alpha\beta\Gamma} (-1)^{\Gamma}\hat{\Gamma}\hat{\Lambda}\, \langle\alpha\|\ W(\Omega) \| \beta\rangle W(\Omega\lambda_i\Delta\Delta;\Gamma\Lambda)$$
$$\times\ \langle\Delta\| [O_{\lambda_i}, A^{\dagger}_{\alpha\beta}(\Omega)]^{\Gamma} \| \Delta\rangle. \tag{112}$$

Thus in solving the equations of motion it is useful to calculate the transition amplitudes

$$G^{\Omega}_{\alpha\beta}(x\Lambda) = (-1)^{\Omega}\,\hat{\Omega}^{-1}\sum_{i\Gamma}(-1)^{\Gamma}\hat{\Gamma}\hat{\Lambda}\,W(\Omega\lambda_i\Delta\Delta;\Gamma\Lambda)\langle\Delta\|\,[O_{\lambda_i},A^{\dagger}_{\alpha\beta}(\Omega)]^{\Gamma}\,\|\Delta\rangle. \tag{113}$$

Transition matrix elements are then readily obtained for any operator $W(\Omega)$ from

$$\langle x\Lambda\|\,W(\Omega)\,\|\,\Delta\rangle = \sum_{\alpha\beta} G^{\Omega}_{\alpha\beta}(x\Lambda)\ \langle\alpha\|W(\Omega)\|\,\beta\rangle \tag{114}$$

The Open-Shell RPA For A Non-Scalar Nucleus

The application of the above equations of motion to generalize the open-shell RPA to nuclei with non-scalar ground states is straightforward. The excitation operators are expanded in terms of particle-hole creation and destruction operators as for scalar nuclei (cf. eq. (78)):

$$Q^{\dagger}_{x\Lambda} = \sum_{\lambda}\sum_{\alpha>\beta}\{Y_{\alpha\beta}(x\lambda)\ A^{\dagger}_{\alpha\beta}(\lambda) - Z_{\alpha\beta}(x\lambda)\ A_{\alpha\beta}(\bar{\lambda})\}. \tag{115}$$

The main difference is that there is now a sum over particle-hole operators of different tensor rank λ. Since the open-shell RPA is applied to negative parity excitations the single-particle indices α and β are restricted to different but adjacent major oscillator shells.

The expansion (115), inserted in the equations of motion (103), immediately yields the open-shell RPA equations in the same way as for scalar nuclei. However, there are a few significant differences:

The main calculational difference is that we must now evaluate the ground state expectations of tensor two-body operators, viz.

$$\langle\Delta\|[(A^{\alpha}xB^{\beta})^{\lambda},\ H,\ (A^{a}x\ B^{b})^{\lambda'}]^{\Gamma}\|\,\Delta\rangle, \tag{116}$$

whereas for a scalar ground state only $\Gamma=0$ occurred.

Another difference is that the set of particle-hole operators $A^{\dagger}_{\alpha\beta}(\lambda)$, for different λ, may be non-orthogonal and overcomplete. A transformation to linearly independent combinations is achieved by diagonalization of the metric matrix, as described above.

After diagonalization of the metric matrix, elimination of the spurious operators and renormalization of the remaining non-spurious operators, the equations reduce to the general form of eqs.(108) or (109). These equations have less symmetry than the scalar RPA equations.

The above general open-shell RPA theory has been developed and applied by C. Ngo-Trong[16] to the negative parity excitations of the even Nickel isotopes. The ground states of these nuclei still have J=0 but since N>Z (with the exception of Ni^{56}) they have T≠0. The interest was therefore to investigate the isospin splitting of the negative parity states in these nuclei. We present results here for the $J^{\pi}=1^{-}$ states of the giant dipole resonance.

For a nucleus with ground state isospin $T_0 \neq 0$ the giant dipole resonance is expected to split into isospin T_0 and T_0+1 components. This splitting has been investigated by other authors[17] in simpler situations e.g. for closed-shell nuclei.

The ground state for our calculation was calculated for the valence neutrons in the $(2p_{3/2}, 2p_{1/2}, 1f_{5/2})$ shells, using the Oak Ridge Rochester shell-model code[18] and the two-body matrix elements of Auerbach[19]. The interaction used for the particle-hole calculation was a central Yukawa potential with a Gillet-Sanderson[20] exchange mixture.

The results are shown for the $Ni^{58,60,62,64}$ isotopes in figs. 14-17, respectively.

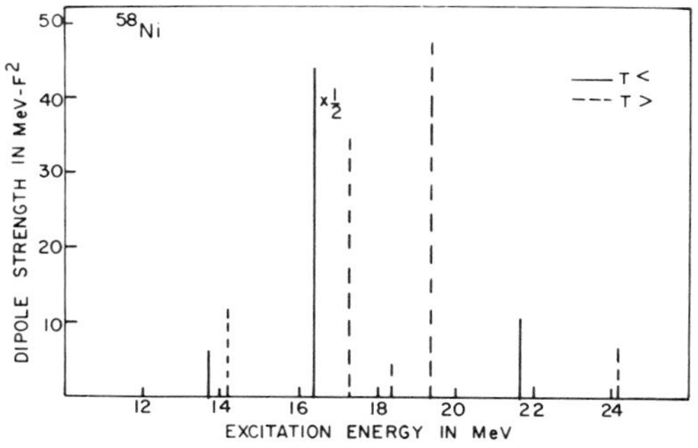

Figure 14
Calculated dipole strength in the region of the giant dipole resonance in Ni^{58}.

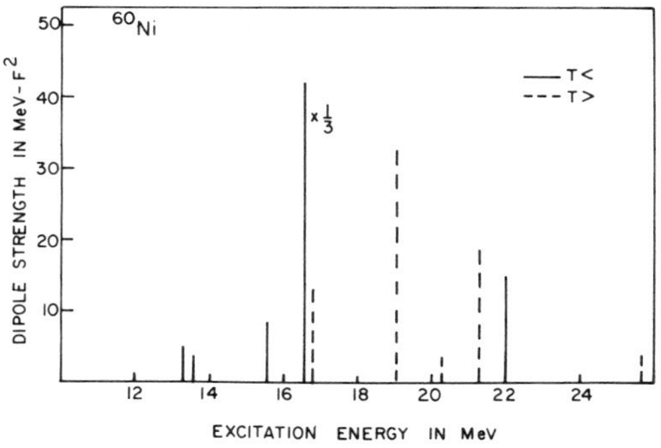

Figure 15
Calculated dipole strength in the region of the giant dipole resonance in Ni^{60}.

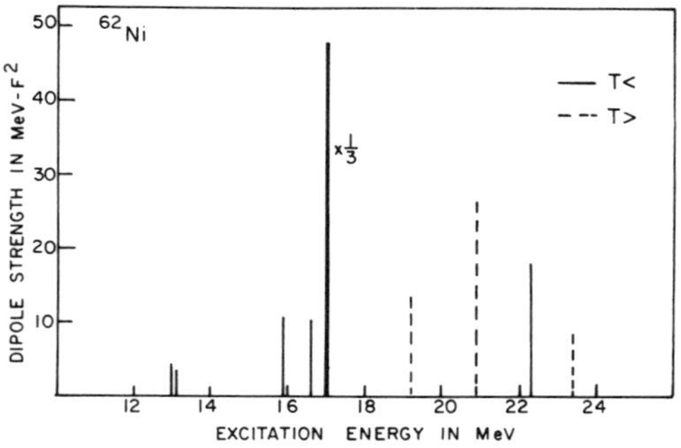

Figure 16
Calculated dipole strength in the region of the giant dipole resonance in Ni^{62}.

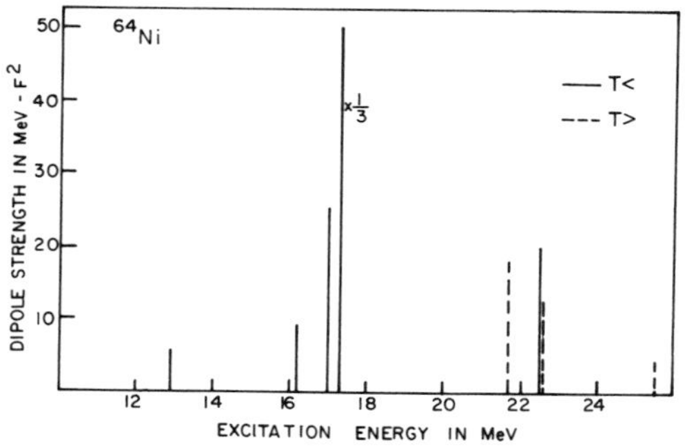

Figure 17
Calculated dipole strength in the region of the giant dipole resonance in Ni^{64}.

In a $T_0=0$ nucleus, like Ni^{56}, all the dipole strength is in the $T_> = 1$ states; there is no non-spurious strength in the $T_<=0$ states. From our results, it is seen that the separation of the $T_<$ and $T_>$ strength distributions increases with T_0 and that the strength shifts from the $T_>$ into the $T_<$ states.

The isospin splitting is approximately in agreement with the formula[17]

$$\Delta E = U(T_0+1)/T_0, \tag{117}$$

if the symmetry energy

$$U = 38(N-Z)/A \text{ MeV}. \tag{118}$$

is taken.

Unfortunately there are some inconsistencies in the available data[21] concerning the isospin structure of the giant resonance in Nickel isotopes. Nevertheless our results are in excellent agreement with the position, widths and gross structure of the photo-neutron and photo-proton cross sections available.

This example, we consider to be another major success of the equations-of-motion approach, which as far as we know is the only formalism which lends itself to such a detailed microscopic description of these phenomena. For although the giant dipole and other negative parity excitations are simple in terms of the excitation process, they would be exceedingly complicated, for something like the Nickel isotopes, in terms of stationery state wave functions.

APPENDIX: APPLICATION TO INELASTIC ELECTRON SCATTERING

T.W. DONNELLY, Department of Physics, University of Toronto

This appendix includes a brief summary of the main features which make electroexcitation a particularly useful means of studying the nucleus. It provides a way of obtaining the charge, convection current and magnetization current densities insofar as any nuclear model which describes these current densities also contains a description of inelastic electron scattering. We are interested here in using the equations-of-motion formalism as developed by D.J. Rowe in the main text to yield such current densities and to study the electron scattering form factors within this framework. Results are presented for C^{12} with comparisons between the Tamm-Dancoff approximation (TDA) and the open-shell random phase approximation (OSRPA) and between these theoretical results and experimental data. This is a particularly advantageous case to consider since extensive electron scattering studies in the TDA have been presented in the past[22] and recently D.J. Rowe and S.S.M. Wong[5] (with whom this work is a collaboration) have used the OSRPA for C^{12}.

We begin the inelastic electron scattering study after the equations of motion (eq.(13)) have been solved for the Y and Z coefficients involved in the expansion of the operators which produce the excited states from the ground states (eq.(78)). The reduced matrix elements of any hermitian one-body operator W_{JT} are then simply expressed in terms of single-particle reduced matrix elements by eq.(91), reproduced here for completeness:

$$\langle\lambda JT \vdots\vdots W_{JT} \vdots\vdots 0\rangle$$

$$= \sum_{\alpha>\beta} \{\mathbf{Y}^*_{\alpha\beta}(\lambda JT) + (-1)^{\Gamma} Z^*_{\alpha\beta}(\lambda JT)\}\langle\alpha \vdots\vdots W_{JT} \vdots\vdots \beta\rangle$$

$$\times \sqrt{n_\beta - n_\alpha} \qquad \text{(A1)}$$

where the notation is the same as the main text except for the symbols $\vdots\vdots$ which are used here to emphasize that the matrix elements are reduced in both angular momentum and isospin spaces.

Indeed as outlined below, in electroexcitation we are interested in just this type of operator and so need only the single-particle reduced matrix elements $\langle\alpha\vdots\vdots W_{JT}\vdots\vdots\beta\rangle$ to complete the picture. We now summarize some of the salient features of inelastic electron scattering.[23]

We consider a process where an electron with initial four-momentum† $k_\mu=(\underset{\sim}{k},iE)$ is scattered to a state with final four-momentum $k'_\mu=(\underset{\sim}{k}',iE)$ from a nucleus which has initial and final four-momenta p_μ and p'_μ respectively. Momentum conservation implies that the four-momentum q_μ transferred by the single virtual photon exchanged when we study this process in first Born approximation satisfies $q_\mu = k_\mu - k'_\mu = p'_\mu - p_\mu$. We define the three-momentum transfer $\underset{\sim}{q}$ and energy transfer ω by $q_\mu = (\underset{\sim}{q}, i\omega)$ with $q = |\underset{\sim}{q}|$. Now in first Born approximation the cross section for detecting the final electron at given values of q,ω and Θ can be written

$$\frac{d\sigma}{d\Omega} = \frac{4\pi\ \sigma_M\ (E,\Theta)}{1+\frac{2E}{M_T}\sin^2\Theta/2}\ \left(\tfrac{1}{2} + \tan^2\Theta/2\right)\ F^2(q)\ . \qquad \text{(A2)}$$

Here

$$\sigma_M = \left(\frac{e^2\cos\Theta/2}{2E\sin^2\Theta/2}\right)^2$$

is the Mott cross section describing the scattering of the electron from a point charge and M_T is the mass of the nuclear target which occurs in the recoil factor in the denominator. $F^2(q)$ is the form factor, that is the function which contains all the information on the structural details of the nuclear four-current density. We have made the approximations $k^2 \simeq E^2$ and $q^2 \gg \omega^2$, which are valid for high energy electrons and for large momentum transfers to excited states in the giant resonance region. The form factor can in turn be decomposed into transverse and longitudinal parts, each having no angle dependence:

$$F^2(q) = F_T^2(q) + F_L^2(q)/\left(\tfrac{1}{2}+\tan^2\Theta/2\right) \qquad \text{(A3)}$$

and the transverse form factor itself can be split into electric and magnetic form factors:

$$F_T^2(q) = F_E^2(q) + F_M^2(q)\ . \qquad \text{(A4)}$$

†We use $\hbar = c = 1$.

If we restrict ourselves to the electroexcitation of states $J^{\pi}T$ from scalar ground states $0^{+}0$ we obtain for these three form factors:

$$F_L^2(q)\,\Big|_{J^{\pi}T} = \frac{1}{2T+1}\,|\langle J^{\pi}T\vdots\vdots\,\hat{L}^{C}_{J,T}(q)\vdots\vdots\,0^{+}0\rangle|^2 \tag{A5}$$
$$J=0,1,2,\ldots;\ T=0,1;\ \pi=(-)^J$$

$$F_E^2(q)\,\Big|_{J^{\pi}T} = \frac{1}{2T+1}\,|\langle J^{\pi}T\vdots\vdots\,\hat{T}^{E}_{J,T}(q)\vdots\vdots\,0^{+}0\rangle|^2 \tag{A6}$$
$$J=1,2,\ldots;\ T=0,1;\ \pi=(-)^J$$

$$F_M^2(q)\,\Big|_{J^{\pi}T} = \frac{1}{2T+1}\,|\langle J^{\pi}T\vdots\vdots\,\hat{T}^{M}_{J,T}(q)\vdots\vdots\,0^{+}0\rangle|^2 \tag{A7}$$
$$J=1,2,\ldots;\ T=0,1;\ \pi=(-)^{J+1}$$

that is, proportional to the squares of reduced matrix elements of one-body operators $\hat{L}^C$, $\hat{T}^E$ and $\hat{T}^M$. These operators are simply multipole projections of expressions involving the charge, convection current and magnetization current densities:

$$\hat{L}^{C}_{JM,T}(q) = \int d\underset{\sim}{x}\; M^M_J(\underset{\sim}{x})\,\hat{\rho}^{(T)}(\underset{\sim}{x}) \tag{A8}$$

$$\hat{T}^{E}_{JM,T}(q) = \frac{1}{q}\int d\underset{\sim}{x}\,[(\nabla\times \underset{\sim}{M}^M_{JJ}(\underset{\sim}{x}))\cdot\hat{\underset{\sim}{j}}^{(T)}(\underset{\sim}{x}) + q^2\,\underset{\sim}{M}^M_{JJ}(\underset{\sim}{x})\cdot\hat{\underset{\sim}{\mu}}^{(T)}(\underset{\sim}{x})] \tag{A9}$$

$$i\hat{T}^{M}_{JM,T}(q) = \int d\underset{\sim}{x}\,[\underset{\sim}{M}^M_{JJ}(\underset{\sim}{x})\cdot\hat{\underset{\sim}{j}}^{(T)}(\underset{\sim}{x}) + (\nabla\times \underset{\sim}{M}^M_{JJ}(\underset{\sim}{x})\cdot\hat{\underset{\sim}{\mu}}^{(T)}(\underset{\sim}{x})] \tag{A10}$$

where

$$M^M_J(\underset{\sim}{x}) = j_J(qx)\;Y^M_J(\Omega_x)$$
$$\underset{\sim}{M}^M_{JL}(\underset{\sim}{x}) = j_L(qx)\;\underset{\sim}{Y}^M_{JL1}(\Omega_x)\ .$$

At this point it is necessary to insert the nuclear physics, that is to specify more explicitly the four-current density. This involves adopting a model: for example, we may consider oscillating charged fluids and study the Goldhaber-Teller model or the oscillating liquid drop model. Or, as we shall do in the present work which is basically shell model in nature, we use the single-particle description of the four-current density:

$$\hat{\rho}^{(T)}(\underset{\sim}{x}) = \tfrac{1}{2}\,e_i^{(T)}\,\delta(\underset{\sim}{r}_i-\underset{\sim}{x}) \tag{A11}$$

$$\hat{\underset{\sim}{j}}^{(T)}(\underset{\sim}{x}) = \tfrac{1}{2}[e_i^{(T)}\,\delta(\underset{\sim}{r}_i-\underset{\sim}{x})\frac{1}{iM}\,\nabla]_{sym} \tag{A12}$$

$$\hat{\underset{\sim}{\mu}}^{(T)}(\underset{\sim}{x}) = \frac{1}{2}\,\frac{\mu_i^{(T)}}{2M}\,\delta(\underset{\sim}{r}_i-\underset{\sim}{x})\;\underset{\sim}{\sigma}(i) \tag{A13}$$

which places the nucleons of mass M at the positions $\underset{\sim}{r}_i$ with isoscalar and isovector charges and magnetic moments given by

$$e_i^{(0)} = 1, \; e_i^{(1)} = \tau_3(i)$$
$$\mu_i^{(0)} = \mu_p + \mu_n = \mu^S \;, \; \mu_i^{(1)} = (\mu_p - \mu_n)\tau_3(i) = \mu^V\tau_3(i). \tag{A14}$$

We now have a complete framework for studying inelastic electron scattering in the equations-of-motion formalism. We begin by showing which states are of particular interest in such a study. The first point is to note that at large angles the transverse form factor $F_T{}^2(q)$ is dominant over the longitudinal form factor $F_L{}^2(q)$ because of the $\tan^2 \theta/2$ factor (see eq. A3). Let us examine the behaviour of typical reduced matrix elements of the convection and magnetization current densities. Using harmonic oscillator wave functions these matrix elements have the following dependence on $y=(\frac{1}{2}bq)^2$, where b is the oscillator parameter:

$$\text{convection current:} \quad \sim y^{k/2}e^{-y}p_C(y) \tag{A15}$$

$$\text{magnetization current:} \quad \sim y^{k/2}e^{-y}p_M(y) \tag{A16}$$

where k = J-1 for electric and J for magnetic multipoles. The $p_C(y)$ and $p_M(y)$ are simple polynomials in y and appear typically (e.g. the 1d5/2 - 1p3/2 matrix elements) as

$$p_C(y) = A + By$$

$$p_M(y) = \qquad Cy + Dy^2$$

namely with matrix elements of the magnetization current being one order higher in y than the corresponding matrix elements of the convection current. So if the momentum transfer q is reasonably large (>200 MeV), y is large and the magnetization piece dominates over the convection piece. Furthermore since the magnetization current (eq. A13) involves the magnetic moments given in eq. A14, with

$$\left(\frac{\mu^V}{\mu^S}\right)^2 \sim 28$$

we find that T=1 states are far more strongly excited than T=0 states. Another important feature of inelastic electron scattering can be seen by looking at eqs. A15 and A16, where we see that if y is large (as it is for medium to high values of q) then the state with the highest multipolarity J is most strongly excited. This means as we see below that the 4^- operator plays a vital role in describing the giant resonance region at medium to high values of momentum transfer. Another example[24,25] is the 6^- particle-hole state in Si^{28} excited as an M6.

Let us proceed to results for C^{12} using the equations-of-motion formalism. We may take the ground state to be a closed 1p3/2-shell and set the Z-coefficients in eq.(78) to zero, that is, have no backward-going amplitudes. This is the TDA (eqs. (51)(52)) and we consider the following pure 1p-1h states:

$(1p1/2)(1p3/2)^{-1}$		1^+	2^+		
$(2s1/2)(1p3/2)^{-1}$		1^-	2^-		
$(1d3/2)(1p3/2)^{-1}$	0^-	1^-	2^-	3^-	
$(1d5/2)(1p3/2)^{-1}$		1^-	2^-	3^-	4^-
$(1p1/2)(1s1/2)^{-1}$	0^-	1^-			

In first Born approximation we cannot excite the 0^- states so these are disregarded. The other states of given $J^{\pi}T$ are configuration-mixed using some residual interaction (here we use the Serber-Yukawa and the Rosenfeld-Yukawa interactions[22,5]) yielding the forward-going amplitudes (the Y-coefficients in eq. (51)). With the formalism outlined here we can then study the electroexcitation of these states, but before presenting results we mention the other approximation considered, namely the open-shell random phase approximation (OSRPA). In this the ground state is taken from an intermediate-coupling shell model calculation with all the 1p nucleons active. Moreover, the excitation operators which build the excited states from this ground state are allowed to have forward and backward going amplitudes (eq.(78)). These two approximations have been discussed in some detail in the main text and our primary aim in the remainder of this appendix is to show where the results of inelastic electron scattering calculations are sensitive to the nature of the approximations made.

In fig. A1 we show the form factor at $\Theta=135°$ and q=166 MeV as a function of the excitation energy ω. The data are recent results[26] and have been presented here for clarity as a smooth continuous curve through the experimental points. In the lower part of the figure are the TDA results for T=1 states obtained using a Serber-Yukawa residual interaction with single-particle and single-hole energies taken from C^{13} and C^{11}. The positions of the main peaks are in fairly good agreement with the data, namely the 1^- peak at 18.1 MeV, the 2^- giant quadrupole at 19.4 MeV and the 1^- giant dipole resonance region between 22 and 26 MeV. Of course the states in this latter region will be broadened by the continuum (neutron threshold is 18.7 MeV) and so should not appear as the sharp spikes plotted here. In the upper part of the figure we present results using the shell model for the even parity states (SHMDL) and the OSRPA for the odd parity states built on the intermediate-coupled shell model ground state.

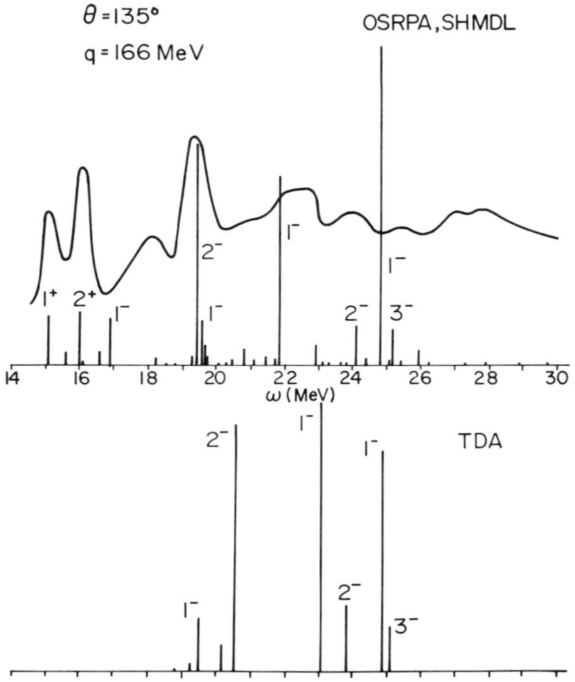

Figure A1
The inelastic electron scattering form factors for C^{12} at $\Theta=135^{\circ}$ and q=166 MeV are shown as functions of excitation energy ω. TDA refers to the Tamm-Dancoff approximation, OSRPA to the open-shell random phase approximation and SHMDL to the shell model calculation of the even parity states. In TDA only the T=1 odd parity states are shown; in OSRPA plus SHMDL both T=0 and T=1 for both even and odd parity states are shown. The data are from ref. 26 represented as a smooth curve for clarity.

The resulting agreement with experiment is now excellent insofar as the energies and overall gross structure are concerned. In fact the largest disagreement is the 1^- state at 18.1 MeV which is obtained about 1 MeV too low; all other states are within a few hundred keV of their experimental positions. In this comparison the relative magnitudes of the calculated results (TDA versus OSRPA and SHMDL) are as shown, but the normalization to the data is arbitrary. The question of the absolute amplitude will be discussed below, but we wish only to note here that there is an overall reduction in strength in going from closed- to open-shell approximations. Figures A2, A3 and A4 show more results at different values of momentum transfer.

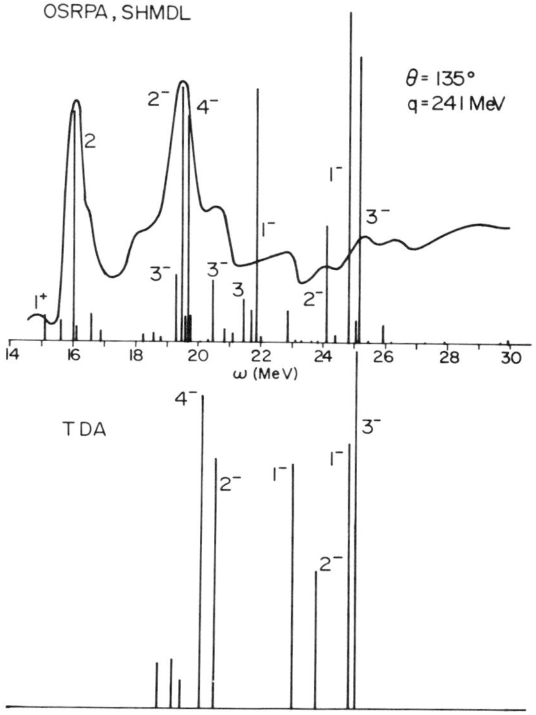

Figure A2
The inelastic electron scattering form factors for C^{12} at $\Theta=135^{o}$ and q=241 MeV are shown as functions of excitation energy ω. TDA refers to the Tamm-Dancoff approximation, OSRPA to the open-shell random phase approximation and SHMDL to the shell model calculation of the even parity states. In TDA only the T=1 odd parity states are shown; in OSRPA plus SHMDL both T=0 and T=1 for both even and odd parity states are shown. The data are from ref. 26 represented as a smooth curve for clarity.

The importance of the points mentioned earlier can be seen here: (1) working at large Θ most of the strength is in the transverse form factor, (2) working at high q the magnetization current is dominant and we are seeing mainly T=1 states (in fact in the OSRPA plus SHMDL calculation both T=0 and T=1 states are included although the former only yield the small

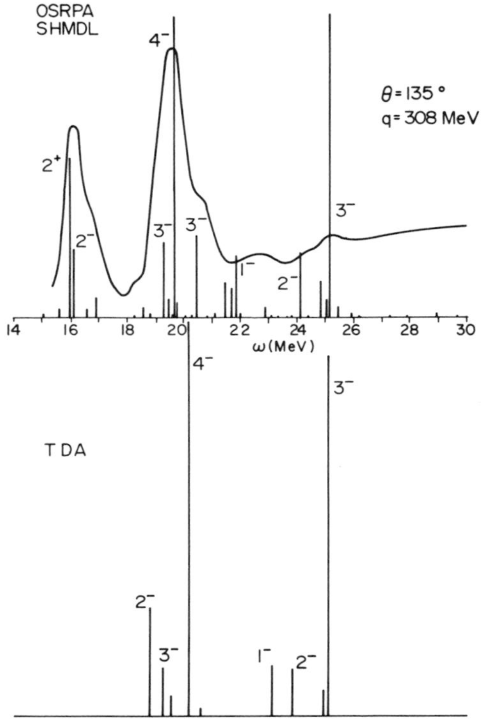

Figure A3
The inelastic electron scattering form factors for C^{12} at $\Theta=135^{o}$ and q=308 MeV are shown as functions of excitation energy ω. TDA refers to the Tamm-Dancoff approximation, OSRPA to the open-shell random phase approximation and SHMDL to the shell model calculation of the even parity states. In TDA only the T=1 odd parity states are shown; in OSRPA plus SHMDL both T=0 and T=1 for both even and odd parity states are shown. The data are from ref. 26 represented as a smooth curve for clarity.

spikes seen below the much larger T=1 excitation spectrum) and (3) the states of high angular momentum such as the 4^- state become very prominent at large values of q. In the remaining figures we show the form factors as functions of momentum transfer at $\Theta=135°$ and at fixed values of excitation

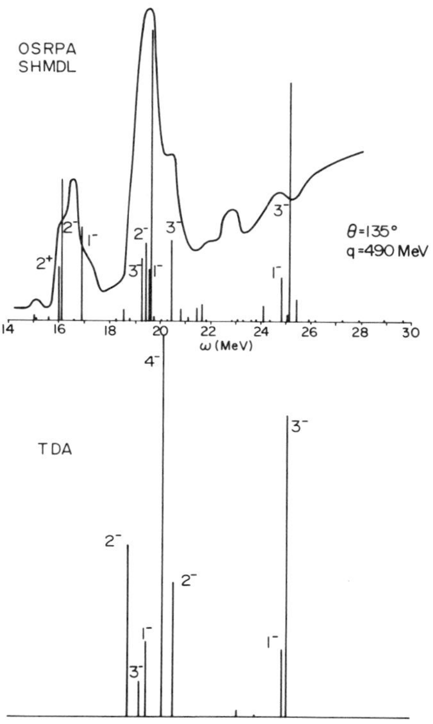

Figure A4
The inelastic electron scattering form factors for C^{12} at $\Theta=135^{\circ}$ and q=490 MeV are shown as functions of excitation energy ω. TDA refers to the Tamm-Dancoff approximation, OSRPA to the open-shell random phase approximation and SHMDL to the shell model calculation of the even parity states. In TDA only the T=1 odd parity states are shown; in OSRPA plus SHMDL both T=0 and T=1 for both even and odd parity states are shown. The data are from ref. 27 represented as a smooth curve for clarity.

energy. In figs. A5 and A6 the importance of using the intermediate-coupled open shell description of C^{12} for the even parity 1^{+} and 2^{+} states is clear: in TDA the form factors for these states are four times larger than experiment, whereas the SHMDL results are shown directly as

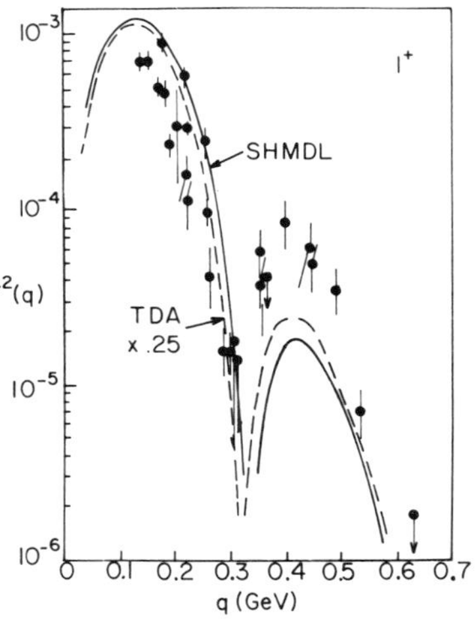

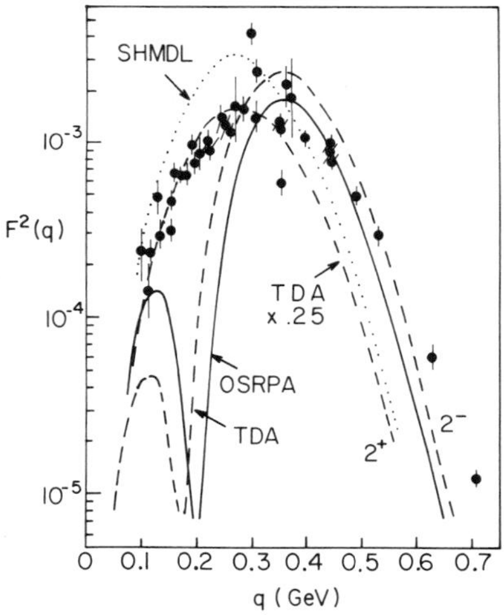

Figure A5 (left)
The form factor at $\Theta=135°$ for the 1^+ state near 15 MeV shown as a function of momentum transfer q. The dashed curve is the TDA result reduced by a factor of four, while the solid curve is the SHMDL result with no scaling factor. The data are from several sources (see ref. 22 for a more detailed presentation of this information).

Figure A6 (right)
As for fig. A5 but for the 16 MeV complex containing the 2^+ and 2^- states in that region. The TDA results for the 2^+ have been reduced by a factor of four, whereas the SHMDL results have not been scaled.

calculated and agree quite well with experiment. The situation is similar for the odd parity states shown in figs. A6, A7 and A8 where the TDA results are high (for instance about two times too large for the 2^- giant quadrupole state) whereas the OSRPA results are generally reduced in amplitude and are in closer agreement with experiment.

The 4^- M4 form factor is a particularly good case to examine since there is only one state in either TDA or OSRPA to be compared. Here the reduction is approximately 25%, whereas a factor of two is apparently needed. The situation is less clear for the 2^- states in fig. A8, since in OSRPA there are two states to be compared with the one in TDA; however the reduction exists there as well and as with most of the odd parity states is not as much as required to bring the results into quantitative agreement with experiment. The qualitative agreement is very good though in that the q-dependence of the form factors is well-reproduced. The most sensitive case studied was the 1^- state shown in fig. A7 where the Serber-Yukawa (SY) and Rosenfeld-Yukawa (RY) results are quite different at low values of q.

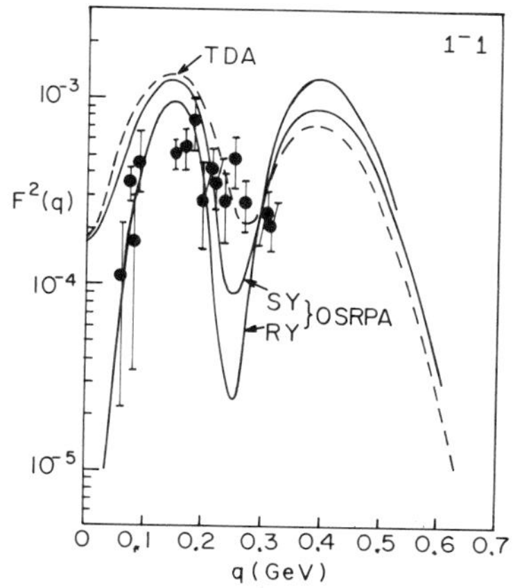

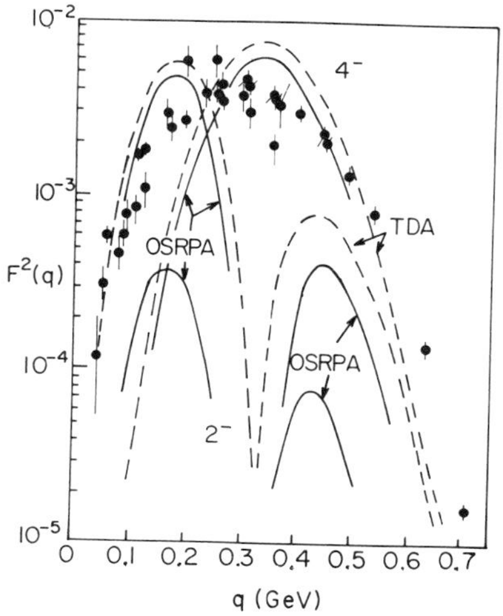

Figure A7 (left)
As for fig. A5, but for the 1^- state near 18 MeV. The two OSRPA curves are for calculations using different residual interactions: Serber-Yukawa (SY) and Rosenfeld-Yukawa (RY).

Figure A8 (right)
As for fig. A5, but for the 19 MeV complex containing the giant quadrupole 2^- and the 4^- strength in that region. In TDA there is only one 2^- near 19 MeV, while in the OSRPA there are two (shown as the two solid curves, the larger being for the 2^- at 19.4 MeV and the smaller for the 2^- at 20.8 MeV).

The wealth of information available in such studies of inelastic electron scattering should be apparent. One can obtain far more than simply the energies and spins and parities of the states concerned; indeed there is much information about the nucleus in the four-current density obtained with a given nuclear model. The present models for example show the effect of the open-shell ground state and of building the excited states onto this with both forward- and backward-going amplitudes. We see that the magnitudes of the form factors are reduced in going from TDA to OSRPA for these T=1 odd parity states, but that still more reduction is required for reproduce the magnitudes as measured experimentally. One possible line of attack, strongly suggested by the equations-of-motion formalism, is to include 2ph operators in the expansion of the excitation operators. Another is to study other approximations for the ground state, for example deformed Hartree-Fock or perhaps states of a given SU(3) representation. In fact the equations-of-motion formalism can be generalized to include non-scalar ground states[16] (see the main text) and at present a study of electroexcitation in the Ni region is in progress.

REFERENCES

1. Anderson, P.W. Phys. Rev. 112, 1900 (1958); Valatin, J.G. Phys. Rev. 122, 1012 (1961).
2. Sawada, K. Phys. Rev. 106, 372 (1957); Baranger, M. Phys. Rev. 120, 957 (1960); Sawicki, J. Nucl. Phys. 23, 285 (1961).
3. Rowe, D.J. Rev. Mod. Phys. 40, 153 (1968); Nucl. Phys. A107, 199 (1968); Nuclear Collective Motion - Models and Theory (Methuen, London, 1970).
4. Rowe, D.J. J. Math. Phys. 10, 1774 (1969).
5. Rowe, D.J. and Wong, S.S.M. Nucl. Phys. A153, 561 (1970).
6. Thouless, D.J. The Quantum Mechanics of Many-Body Systems (Academic Press, N.Y., 1961).
7. See A.M. Lane, Nuclear Theory (Benjamin, N.Y., 1964) and references quoted therein for a review of linearized equations-of-motion methods.
8. Elliott, J.P. and Flowers, B.H. Proc. Roy. Soc. 242A, 57 (1957).
9. Dolbilkin, B.S., Korin, V.I., Lazareva, L.E. and Nikolaev, F.A. JETP Letters 1, 148 (1965).
10. French, J.B, in Many-Body Description of Nuclear Structure and Reactions, ed. C. Bloch (Adacemic Press, N.Y., 1966).
11. Ullah, N. and Rowe, D.J. Nucl. Phys. A163, 257 (1971).
12. Rowe, D.J., Ullah, N., Wong, S.S.M., Parikh, J.C. and Castel, B. Phys. Rev. C3, 73 (1971).
13. Bezić, N., Brajnik, D., Jamnik, D. and Kernel, G. Nucl. Phys. A128, 426 (1969).
14. Wong, S.S.M., Rowe, D.J. and Parikh, J.C., to be published.
15. Castel, B. and Parikh, J.C. Phys. Rev. C1, 990 (1970).
16. Ngo-Trong, C. and Rowe, D.J., to be published.
17. Fallieros, S., Goulard, B. and Venter, R.H. Phys. Letts. 19, 398 (1965); Goulard, B., Hughes, T.A. and Fallieros, S. Phys. Rev. 176, 1345 (1968).
18. French, J.B., Halbert, E.C., McGrory, J.B. and Wong, S.S.M. in Advances in Nuclear Physics, Vol.III, eds. M. Baranger and E. Vogt (Plenun Press, N.Y., 1969).
19. Auerbach, N. Phys. Rev. 163, 1203 (1967).
20. Gillet, V. and Sanderson, E.A. Nucl. Phys. A91, 292 (1967).
21. Min, K. and White, T.A. Phys. Rev. Letts. 21, 1200 (1968); Goryachev, B.I., Ishkanov, B.S., Kapitonov, I.N., Piskarev, I.M., Shevchenko, V.G. and Shevchenko, O.P. Sov. J. Nucl. Phys. 11, 141 (1970); Ishkanov, B.S., Kapitonov, I.M., Piskarev, I.M., Shevchenko, V.G. and Shevchenko, O.P. Sov. J. Nucl. Phys. 11, 272 (1970); Diener, E.M., Amann, J.F. and Paul, P. Phys. Rev. C3, 2303 (1971).
22. Donnelly, T.W. Phys. Rev. C1, 833 (1970).
23. de Forest, T., Jr. and Walecka, J.D. Advan. Phys. 15, 1 (1966).
24. Donnelly, T.W., Walecka, J.D., Walker, G.E. and Sick, I. Phys. Lett. 32B, 545 (1970).
25. Donnelly, T.W. and Walker, G.E. Ann. of Phys. (N.Y.) 60, 209 (1970).

26. Yamaguchi, A., Terasawa, T., Nakahara, K. and Torizuka, Y. Phys. Rev. C3, 1750 (1971).
27. Donnelly, T.W., Walecka, J.D., Sick, I. and Hughes, E.B. Phys. Rev. Lett. 21, 1196 (1968).

NUCLEAR SYMMETRIES AND DISTRIBUTIONS*

J. B. FRENCH
Department of Physics and Astronomy
University of Rochester

1. INTRODUCTION

The purpose of these lectures is to discuss a kind of spectroscopy[1-3] which is well adapted to studying some general aspects of nuclear structure and to searching for general simplicities in complicated systems. Conventional spectroscopy to the contrary is usually concerned with highly detailed questions about the construction of matrices, the properties of individual states and transitions between pairs of them, and other things of that sort. There is a danger that when one proceeds exclusively in that way he may lose sight, among the detail, of some of the general features. That simplicities actually show up in many-particle spectra is suggested by Fig. 1 which shows the exact cumulative density or distribution function resulting from a shell-model calculation as compared with the corresponding function for a Gaussian density. The agreement is striking. Further examples are given later.

It will be recognized immediately that our example, and most of the later ones as well, involves a comparison of two calculations, not of a calculation with data. "Complete" theoretical spectra however do not lend themselves to such comparison; this is because one has made, in any such calculation, a restriction of the single-particle space, the higher-lying states in the multiparticle spectra being then inadequately treated. Nevertheless the simplicities indicated above do have their consequences in real spectra, as we shall see. In fact it will become clear also that our spectroscopy will allow us to study many details of low-lying states and not only those general features which are our main concern.

*Supported in part by the U.S. Atomic Energy Commission.

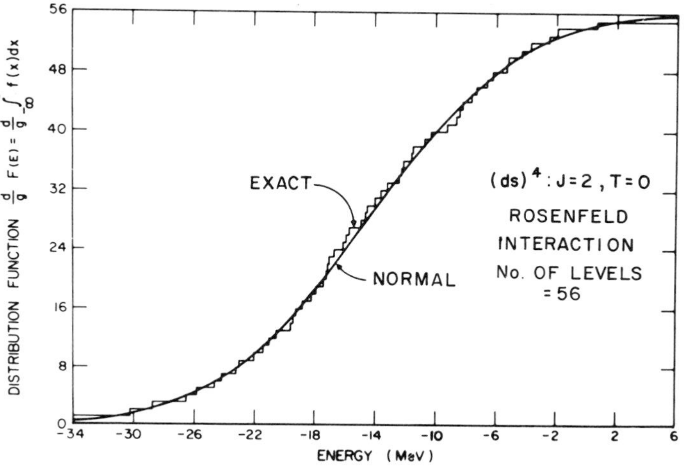

Figure 1
Exact distribution function and its Gaussian approximation for $(ds)^4$ with J = 2, T = 0.

2. FERMION SPACES

Since the formal operations of spectroscopy are concerned with vector spaces and transformations of them, we start by considering some elementary properties of the spaces ("particle spaces") with which spectroscopy deals, and of the operators which function in these spaces. To begin with, there is in general an underlying finite single-particle space, for example the 24 states of the (ds) shell when our interest is with low-lying states of Mg^{24}. The multi-particle spaces are powers of the 1-particle one, the m-particle space being the m'th power; by this we mean that there exists a basis for $\underset{\sim}{m}$ (as we shall write the m-particle space) in which the vectors are products of m single-particle states. Of course we really need the antisymmetrized power in order to look after the Pauli principle; if we think about things in a number representation (second quantization) with creation operators A_i, one for each single-particle state, then the ordinary products of the A_i correspond to the antisymmetrized powers of the single-particle states, this of course being the very great advantage of this way of describing things.

We may write the m-particle space as $\underset{\sim}{m}=(\underset{\sim}{m}-\underset{\sim}{1})\times\underset{\sim}{1}$ or $\underset{\sim}{m}=(\underset{\sim}{m}-\underset{\sim}{2})\times\underset{\sim}{2}$ and these familiar decompositions form the basis for most detailed calculations (by fractional parentage coefficients). If we focus too much on these decompositions rather than on $m=\underset{\sim}{1}\times\underset{\sim}{1}\times\underset{\sim}{1}\times\ldots\times\underset{\sim}{1}$ (m times) we in fact lose sight of a great deal of the general structure and the simplicities of the many-particle system. We shall see this as we go along. At the opposite extreme we could ignore completely the direct-power or particle structure of the spaces and then almost the only thing to be done is along the line of random matrix calculations in which one makes broad statistical assumptions about the system or its Hamiltonian. Since, with our m-fold decomposition, we shall also make use of statistical notions and terms, it is important to understand that, in the sense described above, our procedures will be quite the opposite of random matrix ones. In everything which we do the "particle" or "direct-power" structure of the spaces will be a dominant feature, and will indeed be responsible for the simplicities which we are seeking. In another sense however we shall aim for an intermediate position; random matrix theories are largely uninteresting because they take no account of particle structure; conventional spectroscopy usually takes account of far too much of it, focussing too much on the details. We want to have the structure but without the detail.

When we speak of an operator in any given space we imply that, when acting on any vector in that space, it gives back another one, having then no matrix elements which lead outside the space. Thus when we speak of a Rosenfeld interaction in $(ds)^4$ we are really referring to the <u>restriction</u> to the (ds) shell of an operator which is more broadly defined. Of course many different operators may have the same restriction to a given space; in the p shell for example all 2-body central Hamiltonians belong in a family defined by only two parameters (if two such H's have these parameters the same, their p-shell restrictions are identical).

If we proceed according to the usual way of dealing with vector spaces we would regard an operator in $(ds)^4$ as distinct from one in $(ds)^5$ because the spaces are different. But it would be unwise to maintain this point of view because the direct-power structure imposes a relationship between the spaces which is vitally important to us in our search for simplicities, and which is of course responsible for the fact that nuclei come in families (if we properly restrict the excitation energy); thus we have the (ds) nuclei, $f_{7/2}$ nuclei, and so forth.

One often speaks of an operator as being "small"

compared with another, but unless we have in mind some measure of size this remark is without meaning. The appropriate way when dealing with operators in a given space is to assign to each one a real number, a norm, (to P we assign $||P||$, or $||P||_{\alpha}$ if we wish to specify the space α). The norm must satisfy the following conditions[4]:

1) $||P|| \geqslant 0$; $||P|| = 0$ only for the zero operator

2) $||PQ|| \leqslant ||P|| \times ||Q||$

3) $||P + Q|| \leqslant ||P|| + ||Q||$. Equality holds only if $Q = \lambda P$.

4) $||kP|| = |k| \times ||P||$ where k is a constant.

(1)

We must of course have some prescription for calculating the norm of a given operator. The norm which will be significant for us is the Euclidean norm, defined by

$$\{||P||_{\alpha}\}^2 = \langle\langle P^+P\rangle\rangle^{\alpha} = d(\alpha)\ \langle P^+P\rangle^{\alpha} \qquad (2)$$

where $\langle Q\rangle^{\alpha}$ denotes the expectation value of Q averaged over an orthonormal basis for α (we shall often say "averaged over the space α", a rough usage but appropriate because $\langle Q\rangle^{\alpha}$ is invariant under a unitary transformation of the basis; the Euclidean norm is then a property of the operator and space but not of the representation); $d(\alpha)$ is the dimensionality of α, and of course then $\langle\langle P\rangle\rangle^{\alpha}$ is the trace. We see then that $||P||_{\alpha}$ is, to within a dimensionality factor, the R.M.S. matrix element of P in the space α. If γ_i are the eigenvalues of P in α then obviously $||P||^2_{\alpha} \sim \sum_i |\gamma_i|^2$ so that the first condition for a norm is satisfied. The last one is trivial; the reader should convince himself about the other two.

The norm of course also defines an orthogonality of the operators, P and Q being orthogonal in α if $\langle P^+Q\rangle^{\alpha} = 0$. The Euclidean norm is important to us because we shall be able to calculate it in a wide variety of cases and because it defines the widths or the spread in the eigenvalues of the operator.

For wavefunctions we have of course the usual unitary norm, which we are in fact using when we speak of the intensity with which a certain basis state is found in an eigenfunction, or when we say e.g. that a state has a 10% 2-particle-2-hole admixture. An important feature of the two norms is that they are compatible in that

$$||P\Psi|| \leqslant ||P|| \times ||\Psi|| \tag{3}$$

where of course the vector norm is taken for $P\Psi$ and Ψ and the operator norm for $||P||$. A little thought will show that this condition and those which define the norm are essential if the norm is to represent properly the magnitude of the operator. Since the average $\langle P^+P\rangle^{\underset{\sim}{\alpha}}$ is also a measure of P in $\underset{\sim}{\alpha}$ we may at times refer to this also as a norm, thereby departing slightly from the formal definition.

If we think about the eigenvalues of P, plotted on a line, we see that $||P||^2$ is the second moment taken about the origin. It is convenient to renormalize the distribution to unity and to deal with the <u>second central moment</u> or <u>variance</u>, defined as

$$\sigma^2(\underset{\sim}{\alpha}) = d^{-1}(\underset{\sim}{\alpha})||P-\mathcal{P}||^2 = \langle(P^+-\mathcal{P})(P-\mathcal{P})\rangle^{\underset{\sim}{\alpha}} \tag{4}$$

the second moment about the eigenvalue centroid $\mathcal{P}$. If the eigenvalues are distributed in a Gaussian fashion, the half-width at half maximum of the distribution is 1.18σ; we shall speak of σ as the <u>width</u> (though "half-width" would be better). We shall often arrange that $\mathcal{P} = 0$ and then, the norm being simply the width, we see that the spread in eigenvalues is a natural measure of the size of an operator. Similarly the squares of the widths (the variances) simply add when we add together two orthogonal operators.

Consider now in particular the Hamiltonian operator H which acts as before in the space $\underset{\sim}{\alpha}$. It follows of course that $\underset{\sim}{\alpha}$ is an <u>eigenspace</u> of H, meaning that the eigenfunctions, Ψ_i, of H supply a basis for $\underset{\sim}{\alpha}$. If we consider any state (not necessarily an eigenstate of H) in the space $\underset{\sim}{\alpha}$ we can ask ourselves about the intensity with which this state appears in the eigenfunctions Ψ_i. Considering this as a function of the eigenenergy E_i, we have a distribution which we can examine in terms of its moments. If $\phi_\alpha = \sum_i B_{\alpha i}\Psi_i$ we have for the centroid energy $\mathcal{E}(\alpha)$ and the variance $\sigma^2(\alpha)$ that

$$\mathcal{E}(\alpha) = \sum_i |B_{\alpha i}|^2 E_i = \langle\phi_\alpha \; H \; \phi_\alpha\rangle$$

$$\sigma^2(\alpha) = \sum_i |B_{\alpha i}|^2 (E_i - \mathcal{E}(\alpha))^2 = \langle\phi_\alpha (H - \mathcal{E}(\alpha))^2 \phi_\alpha\rangle \tag{5}$$

and similarly for the higher moments.

If instead we consider the intensity for a set of states ϕ_α forming an orthonormal basis for $\underset{\sim}{\alpha}$ we find easily that the p'th moment is $M_p(\underset{\sim}{\alpha}) = \langle H^p \rangle^{\underset{\sim}{\alpha}}$. This gives for the centroid, expressed in terms of the individual-state centroids, the obvious result that

$$E(\underset{\sim}{\alpha}) = d^{-1}(\underset{\sim}{\alpha}) \sum_{\alpha \in \underset{\sim}{\alpha}} E(\alpha) \quad . \tag{6}$$

For the variance we find, after a little algebra, that

$$\sigma^2(\underset{\sim}{\alpha}) = d^{-1}(\underset{\sim}{\alpha}) \sum_{\alpha \in \underset{\sim}{\alpha}} \{\sigma^2(\alpha) + [E(\alpha) - E(\underset{\sim}{\alpha})]^2\} \tag{7}$$

which has a more complicated structure. The first term contributes to the width because the individual states have widths (giving then a "spreading" width); the second term contributes when not all of the states are centered at the same energy (giving a "displacement" width). The separation is of course representation-dependent, for the first term vanishes in an eigenfunction representation; in this case the $E(\alpha)$ become eigenvalues and we have returned to our original notion of a width.

All this becomes of consequence when, as will be vitally important to us, we consider not an eigenspace $\underset{\sim}{\alpha}$ but one of its subspaces $\underset{\sim}{\beta}$. If we have the (disjoint) decomposition $\underset{\sim}{\alpha} = \Sigma \underset{\sim}{\beta}$ then the normalizing trace for the operator H decomposes similarly

$$\langle\langle H^2 \rangle\rangle^{\underset{\sim}{\alpha}} = \sum_{\underset{\sim}{\beta}} \langle\langle H^2 \rangle\rangle^{\underset{\sim}{\beta}} \tag{8}$$

and the equations above for the moments are unchanged in form. However, in general a representation for a given $\underset{\sim}{\beta}$ such that the first term in (7) should vanish will in general not exist. The exception of course is when the $\underset{\sim}{\beta}$'s are also eigenspaces; in that case the $\underset{\sim}{\beta}$ traces also define norms and the square of the $\underset{\sim}{\alpha}$-norm reduces to a sum of squares of the separate subspace norms; an example, for H, would be if $\underset{\sim}{\alpha} \equiv$ all m-particle states and $\underset{\sim}{\beta}_i$ all those with angular momentum J_i. When however $\underset{\sim}{\beta}$ is not an eigenspace then $\langle H^2 \rangle^{\underset{\sim}{\beta}}$ does not give a norm for H; nor in fact does it give one for the restriction of H to $\underset{\sim}{\beta}$, this because intermediate states not in $\underset{\sim}{\beta}$ must come into the evaluation. It nonetheless does define a quite proper width for $\underset{\sim}{\beta}$, one which has a very significant decomposition into partial widths corresponding to the intermediate states. Thus if we introduce projection operators $P(\underset{\sim}{\beta}_j)$ we are able to write

$$\sigma^2(\underset{\sim}{\alpha}) = \sum_i \sigma^2(\underset{\sim}{\beta}_i) = \sum_{i,j} \sigma^2(\underset{\sim}{\beta}_i;\underset{\sim}{\beta}_j)$$

$$\sigma^2(\underset{\sim}{\beta}_i;\underset{\sim}{\beta}_j) = \langle H\, P(\beta_j) H \rangle^{\underset{\sim}{\beta}_i} \quad . \tag{9}$$

The variance $\sigma^2(\beta_i;\beta_i)$ defines the internal width, the others with $j \neq i$ collectively define the external width of β_i. To the extent that the external width is different from zero we have mixing and although we could attempt to define a representation which comes closest to diagonalizing H in β_i, and even though β_i might contain some eigenfunctions of H, we could not diagonalize it exactly. The same kind of concept is of course employed in forming the complex well in optical-model scattering and in many other cases.

3. THE LIMITING SPECTRUM FOR NON-INTERACTING PARTICLES

Instead of further studying at this time the properties of widths let us pay some attention to the complete distribution, asking then about the nature of the many-particle spectrum itself. This is obviously an exceedingly complicated question, and so we begin by considering the case of non-interacting particles distributed over a set of single-particle states, in which case the Hamiltonian is a (0 + 1)-body operator. In one sense this is a trivial problem because we can produce the multiparticle energies by simply combining the single-particle energies; but if the number of single-particle states, N, and of particles, m, are large, the number of multiparticle states, $\binom{N}{m}$, may be enormous and we then need better ways of proceeding. In fact the search for such ways has long dominated the conventional theories of nuclear level densities.

Choosing a single-particle representation which diagonalizes H we have

$$H = \sum \varepsilon_i n_i \quad \text{with} \sum \varepsilon_i = 0 \tag{10}$$

where we have chosen a suitable zero for the energy and have moreover taken H to be traceless (physically this could mean that the true Hamiltonian is $H + n\lambda$ where we have introduced a "chemical potential" which is fixed by the position of the Fermi surface as a function of particle number; formally the new traceless H is a one-body operator which will be seen later to be unitarily irreducible, a property which more generally will be important to us.

Let us form the m-particle system by combining $(\underset{\sim}{m}-\underset{\sim}{1})$ and $\underset{\sim}{1}$. If we agree that N is very large compared with m, we may at first ignore the Pauli-principle "blocking effects" and then, using a continuous representation, we have

$$\rho_m(E) = \int \rho_{m-1}(E')\, \rho_1(E - E')dE' \quad . \tag{11}$$

Since the average of H^p is given by $\langle H^p\rangle^{\underset{\sim}{m}} = \int \rho_m(E)E^p dE$ we find easily, on making a change of variables, $W = E-E'$, and a binomial expansion, that

$$\langle H^p\rangle^{\underset{\sim}{m}} = \int \rho_{m-1}(E')\rho_1(W) \sum_{r=0}^{p} \binom{p}{r}(E')^r W^{p-r}$$

$$= \sum_{r=0}^{p} \binom{p}{r}\langle H^r\rangle^{\underset{\sim}{m-1}} \cdot \langle H^{p-r}\rangle^{\underset{\sim}{1}} \quad . \tag{12}$$

We find then that $\langle H^2\rangle^{\underset{\sim}{m}} = \langle H^2\rangle^{\underset{\sim}{m-1}} + \langle H^2\rangle^{\underset{\sim}{1}}$ which we rewrite as $\sigma^2(\underset{\sim}{m}) = \sigma^2(\underset{\sim}{m-1}) + \sigma^2(\underset{\sim}{1})$, and which implies that the variances add linearly for the various particles; thus

$$\sigma^2(\underset{\sim}{m}) = m\sigma^2(\underset{\sim}{1}) \quad . \tag{13}$$

The same additivity is found for the third central moment (the powers of H define central moments because $\langle H\rangle^{\underset{\sim}{1}} = 0$) and we are tempted to conclude that it might obtain for the p'th central moment in general. But this is not correct, since for $p = 4$ we find additivity instead for the more complicated quantity

$$K_4(\underset{\sim}{m}) = \langle H^4\rangle^{\underset{\sim}{m}} - 3(\langle H^2\rangle^{\underset{\sim}{m}})^2 \quad . \tag{14}$$

We can understand the whole thing by going back to the $\rho_m(E)$, which is expressed as a convolution, a complicated kind of multiplication. But the Fourier transforms are easily seen to be multiplicative under convolution; thus

$$g_m(t) \equiv \int e^{itE}\rho_m(E)\,dE = \int dE \int dE'\rho_{m-1}(E')\rho_1(E-E')e^{itE'} \cdot e^{it(E-E')}$$

$$= g_{m-1}(t) \cdot g_1(t) \tag{15}$$

and then

$$g_m(t) = \{g_1(t)\}^m$$

$$\ln g_m(t) = m \ln g_1(t) \quad . \tag{16}$$

Thus the logarithm of the characteristic function (as the Fourier transform of the density is called) is additive under convolutions. If moreover we expand $\ln g(t)$ in powers of t as

$$\ln g(t) = \sum_p K_p \frac{(it)^p}{p!} \quad , \quad g(t) = \exp\{\sum_p K_p \frac{(it)^p}{p!}\} \tag{17}$$

(a Maclaurin expansion) we see that the individual quantities K_p, the cumulants, are similarly additive when we combine two independent systems (remember that the Pauli principle blocking will modify the additivity as we shall see in detail later, because the distributions are only independent when these effects are ignored). Taking successive derivatives of g(t) evaluated at t=0 we find the cumulants expressed in terms of the moments; K_o is uninteresting, involving only the dimensionality (which in probability theory is usually taken unity), K_1 is the centroid $\mathcal{E}$, K_p for $p \geqslant 2$ is expressible in terms of the central moments μ_ℓ of order up to p, and is "p-linear" in the sense that multiplying H by a constant ω multiplies K_p by $(\omega)^p$. The lowest few cumulants are:

$$\begin{aligned} K_1 &= \mu_1 = \quad ; \quad K_2 = \mu_2 = \sigma^2 \\ K_3 &= \mu_3 \quad ; \quad K_4 = \mu_4 - 3\mu_2^2 \\ K_5 &= \mu_5 - 10\mu_3\mu_2 \\ K_6 &= \mu_6 - 15\mu_4\mu_2 - 10\mu_3^2 + 30\mu_2^3 \quad . \end{aligned} \tag{18}$$

We have grown accustomed to the fact that the variances, which are quadratic quantities, add linearly when we combine independent systems; but now we see that there exist quantities of arbitrary order, namely the cumulants, for which the same linear law applies.

For a continuous Gaussian distribution centered about the origin, $g(t) = d \exp(-\frac{1}{2}\sigma^2 t^2)$ so that all K_p with $p > 2$ vanish; and this conversely defines a Gaussian distribution. The precise shape of a more general distribution is of course fixed by the K_p but, since these are not invariant under a scale change of H, we should for discussing the shape introduce the "shape parameters" $K_p/(K_2)^{p/2} = K_p/(\sigma)^p$ which are invariant and appropriate because of the norm property of σ. But now we notice that, since the K_p vary linearly with m, the higher shape quantities decrease as $m^{1-\frac{1}{2}p}$ as the particle number increases, the distribution then approaching Gaussian. We have really given here a derivation of the most elementary central limit theorem which says that under repeated convolution a function is converted into a Gaussian, this of course being a much more general result than the simple one that a convolution of two Gaussians is itself Gaussian, but one nonetheless whose genesis is pretty much the same.

The most important subspace decomposition of $\underset{\sim}{m}$ will be into configurations defined by partitioning the single-particle states into ℓ subsets and assigning to each of these a definite number of particles; thus

$N=\Sigma N_i$ and $m=\Sigma m_i$, an example being the partitioning of (ds) into $d_{5/2}+d_{3/2}+s_{1/2}$ which generates 45 12-particle configurations, $(5/2)^{12}, (5/2)^{11}(3/2), (5/2)^{11}(1/2), (5/2)^{10}(3/2)^2, (5/2)^{10}(3/2)(1/2)$ and so forth, each of which is described by an ℓ-dimensional "vector" which we write as $\vec{\underset{\sim}{m}}$. For non-interacting particles we get by a simple extension of the above argument a similar additivity, $K_p(\vec{\underset{\sim}{m}})=\Sigma m_i K_p(\underset{\sim}{0}\underset{\sim}{0}..\underset{\sim}{1}_i...\underset{\sim}{0})$ and a corresponding approach to normality for large particle number. A good example[5] is shown in Fig. 2.

Similarly the additivity of M_J, the z-component of angular momentum, implies that we may expect a Gaussian energy dependence for states of fixed M_J, which leads in turn to a similar dependence for fixed J. The extension to distributions in which other Lie-group symmetries are fixed should then follow from the fact that the corresponding Lie algebras contain "additive" operators like J_z; actually little formal study has been yet made of these more general cases.

Figure 2 has shown how well the Gaussian form represents the spectrum in a typical case of non-interacting particles, and Fig. 1 for interacting particles with fixed (J,T); they indicate incidentally that the

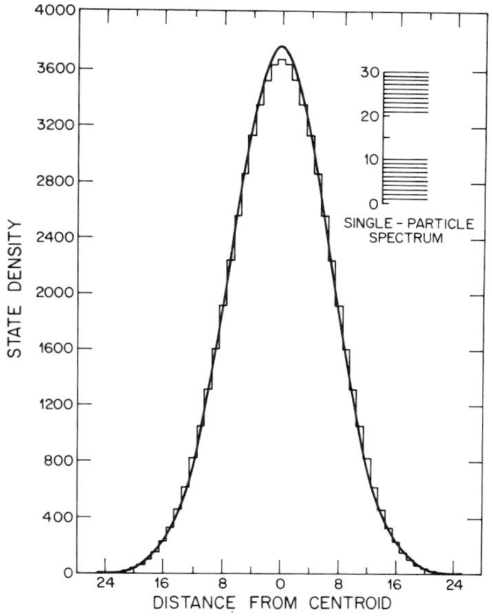

Figure 2
The exact 5 particle-5 hole state density and its Gaussian approximation for 10 non-interacting particles with the single-particle spectrum shown in the inset.

Pauli effects on the shape are quite small. However the extension to interacting particles of the simple arguments above is not obvious or elementary since with interactions the energies are no longer additive and the convolution result for the density no longer applicable; we turn therefore to a more detailed method of study via direct calculation of the moments of the distribution, dealing first however with the distribution over all the states of m-particles, just as we have done above for the non-interacting case.

4. SCALAR MOMENTS AND POLYNOMIAL PROPAGATORS

We wish to construct distributions which take account of the Pauli-principle effects and the residual interactions. We assume that the departures from normality are not too great and that they specifically do not involve rapid fluctuations in the density versus energy* (which would require high-order cumulants for their description). The experience is that this assumption is excellent, (see Figs. 1,2); we consider its formal basis later. We can now proceed by evaluating the low-order cumulants, the technical problem involved here being that of evaluating traces of powers or products of operators over the spaces of interest, $\underset{\sim}{\alpha}$, since, as we have seen, the moments, and hence the cumulants, are defined by such traces. We consider first the case $\underset{\sim}{\alpha}=\underset{\sim}{m}$ (all m-particle states); this distribution may be called <u>scalar</u> because of its behavior with respect to the U(N) group of transformations in the single-particle space, as we see later. The method of evaluation is to express the m-particle cumulants as linear combinations of cumulants for simpler systems; we may regard this process as one of <u>propagating</u> cumulant information throughout the various subspaces of the complete space. In doing this we encounter <u>polynomials</u> in the number operator n (in more complex cases Casimir operators or more general functions of Lie-group generators are encountered) which in fact represent the density operators of the system. Hence we see the origin of the title of this section.

We are interested in traces of complicated operators. Consider first a k-body operator; in a configuration-space representation this is a symmetrical sum of terms involving k particles, e.g. we have

*We imply by this that the density smoothed over ΔE containing several levels is smooth. On a finer scale of course the spectrum is discrete and the density a sum of delta functions.

$\sum_{i<j} V_{ij}$ for k = 2; in the number representation such an operator is expressible as a linear combination of "unit operators" $\psi_\alpha(k)\psi_\beta^\dagger(k)$ where $\psi_\alpha(k)$ is a k-particle state operator. In order to evaluate its m-particle trace we introduce a simple product representation for the states, these then being pictured e.g. as [•| |•|•| |•| |] which shows, for N = 8, particles in states #1,3,4,6. No more than k particles can be transferred by any k-body unit operator. Obviously only the diagonal operators have a trace, and all of these, since they differ only by numbering, have the same trace. A particular one of them gives unity on states with particles in slots 1...k and zero on all others. The totality of such states is the number of ways in which the other (m-k) particles can distribute over the other (N-k) states. Then easily for this operator, and thus for a general k-body operator,

$$\langle\langle O(k)\rangle\rangle^{\underset{\sim}{m}} = \binom{N-k}{N-m}\langle\langle O(k)\rangle\rangle^{\underset{\sim}{k}};\ \langle O(k)\rangle^{\underset{\sim}{m}} = \binom{m}{k}\langle O(k)\rangle^{\underset{\sim}{k}} \qquad (19)$$

where the second form follows on dividing out the dimensionality factors. As always << >> is the trace and <> the average. The coefficients are binomial coefficients.

If it were an elementary operation to write in normal* form the operator H^p whose average gives the p'th moment, we could just use the result above and our problem would be finished; but this is not so and accordingly we use an indirect approach which avoids the necessity of conversion to normal form and introduces new ways of thinking about the information contained in the distributions.

In a product of operators of given rank the maximum particle rank is (less than or equal to) the sum of the individual ranks. But also, since $\binom{m}{k}$ is a polynomial in m of order k, it follows that $\langle O\rangle^{\underset{\sim}{m}}$ is a polynomial of order $\underline{u}$ where $\underline{u}$ is the maximum rank of O. As such it can be expressed in terms of its own values at any convenient (u+1)=dimensional set of m-values, say $m=t_i$ with i=0,1,...k. The explicit solution to this elementary problem is given in terms of Lagrange polynomials, as

$$\langle O\rangle^{\underset{\sim}{m}} = \sum_{j=0} \prod_{\substack{i=0\\(i\neq j)}} \frac{m-t_i}{t_j-t_i}\langle O\rangle^{\underset{\sim}{t_j}} \ . \qquad (20)$$

*i.e.as a sum of k-body operators of various (particle-rank)k. The word "normal" here is not related to "Gaussian".

For example if we have u=2 and take t_i=0,2,4, we find

$$\langle O\rangle^{\underset{\sim}{m}} = \frac{1}{8}(m-2)(m-4)\langle O\rangle^{\underset{\sim}{0}} - \frac{1}{4}m(m-4)\langle O\rangle^{\underset{\sim}{2}} + \frac{1}{8}m(m-2)\langle O\rangle^{\underset{\sim}{4}} \quad (21)$$

which may be verified by substituting m=0,2,4. The form (20) applies not only to averages but also to products of averages, and thus to cumulants, as long as we take u to be the sum of the individual u's; this follows from the fact that the product of two polynomials is itself a polynomial, the orders being additive.

As a trivial application we have, for the width of the scalar distribution with a 1-body operator, that

$$\sigma^2(\underset{\sim}{m}) = m\frac{(N-m)}{(N-1)}\,\sigma^2(\underset{\sim}{1}) \Rightarrow m\sigma^2(\underset{\sim}{1}) \text{ if } m \ll N \,. \quad (22)$$

a form which follows at sight by taking t_i = 0,N and 1, the first two values not contributing since the subspaces contain only a single state and are therefore dispersion-free. Note that the (N-m)/(N-1) factor gives the Pauli correction to the elementary central-limit-theorem value derived earlier. But besides that we can clearly proceed now to evaluate low-order cumulants even in the presence of interactions, something which we could not do before.

We shall be interested in three choices of the set t_i. (1) $t_i=0,1,\ldots u$; (2) $t_i=0,1,2,\ldots u_1$, $N,N-1,\ldots N-u_2$ with $u=u_1+u_2+1$; (3) $t_i=0,1,\ldots u_1$, $N,N-1,\ldots N-u_1$ with $u_1=[\frac{1}{2}u]\equiv\frac{1}{2}u$ or $\frac{1}{2}(u-1)$ according as u is even or odd. In this case we require also that $\langle O\rangle^{\underset{\sim}{t}}=(-1)^u\langle O\rangle^{\underset{\sim}{N}-\underset{\sim}{t}}$. The significance of the last condition, that the operator should have a definite symmetry under hole-particle transformations will be seen shortly; note that if u is even we are actually writing the polynomial as if it were of order (u+1), but there is no harm done since one of order u is a special case of a polynomial of higher order.

In each of these sets the t_i's come in "runs" of successive integers and then, not surprisingly, we get binomial coefficient forms from the Lagrange polynomials. For case (2), which except for the symmetry condition in (3) includes both of the others, we find for the cumulants

$$K_p(\underset{\sim}{m}) = \binom{m}{u_1+1}\binom{N-m}{u_2+1}\Big\{(u_1+1)\sum_{t=0}^{u_1}\binom{N-t}{u_2+1}^{-1}\frac{(-1)^{t-u_1}}{(m-t)}\binom{u_1}{t}K_p(\underset{\sim}{t})$$
$$+(u_2+1)\sum_{t=0}^{u_2}\binom{N-t}{u_1+1}^{-1}\frac{(-1)^{t-u_2}}{(N-m-t)}\binom{u_2}{t}K_p(\underset{\sim}{N}-\underset{\sim}{t})\Big\} \quad (23)$$

which, it must be stressed, takes full account of the

interactions and the Pauli-principle effects. A major advantage of this form over that which will result from the first set ($t_i=0,1,\ldots u$) is that the "input information", the set of $K_p(\underset{\sim}{t}_i)$, is much simpler. As an example we evaluate the width for a (0+1+2)-body H (which then includes two-body residual interactions). We have $u=4$ and taking $u_1=2, u_2=1$ we have

$$\sigma^2(\underset{\sim}{m})=\frac{m(m-1)(m-2)(N-m)(N-m-1)}{(N-1)(N-2)(N-3)}\left\{\frac{\sigma^2(\underset{\sim}{N-1})}{(N-m-1)}-\frac{(N-3)}{(m-1)}\sigma^2(\underset{\sim}{1})\right.$$

$$\left.+\frac{(N-1)}{2(m-2)}\,\sigma^2(\underset{\sim}{2})\right\}\ . \qquad (24)$$

This gives the width in terms of the one- and two-particle widths and the one-hole width, each of which can be written directly in terms of the Hamiltonian matrix elements. We thus have a quite practical form for the scalar width when there are two-body interactions, and since the centroid is even more easily given (take $u_1=1, u_2=0$ or simply $u_1=2$) we have the Gaussian approximation to the distribution.

The first t_i set gives for the p'th cumulant

$$K_p(\underset{\sim}{m})=(u+1)\binom{m}{u+1}\sum_{t=0}^{u}\frac{(-1)^{t-u}}{(m-t)}\binom{u}{t}K_p(\underset{\sim}{t}) \qquad (25)$$

in which, since N does not explicitly appear, the Pauli effects are implicit in the relationships between the cumulants on the "defining net" (i.e. on the line $t=0\ldots u$). To use this, as it stands for the variance with a (0+1+2)-body interaction we would need to calculate the variances for up to four particles.

The third t_i set, with the implied symmetry condition gives

$$K_p(\underset{\sim}{m})=(u_1+1)\binom{m}{u_1+1}\binom{N-m}{u_1+1}\sum_{t=0}^{u_1}\binom{N-m}{u_1+m}^{-1}\frac{(-1)^{t-u_1}(N-2x)}{(m-t)(N-m-t)}K_p(\underset{\sim}{t}) \qquad (26)$$

$$\xrightarrow[m\ \ll\ N]{\text{large } N}(u_1+1)\binom{m}{u_1+1}\sum_{t=0}^{u_1}\frac{(-1)^{t-u_1}}{(m-t)}\binom{u_1}{t}K_p(\underset{\sim}{t}) \qquad (27)$$

where $u_1=[\frac{1}{2}u]$ and $x=t$ if u is even, $x=m$ if u is odd.

We shall see shortly, by U(N) group considerations, that every operator effectively decomposes into operators which satisfy one of the $h\rightleftarrows p$ symmetries used above for (3); the meaning of the qualifying adjective will also become clear. Moreover, since only the polynomial and symmetry properties have been used, the forms given apply not only to powers but to more general products of operators with a definite $h\rightleftarrows p$ symmetry, and thus, after the decomposition is made, to the separate terms in the cumulants.

Observe that, since $\binom{a}{b} \sim a^b/b!$, the exact forms for $K_p(\underset{\sim}{m})$ have m^u as the strongest m-dependence, which is appropriate since $K_p(\underset{\sim}{m})$ is generated by a u-body operator. However in the large-N case, correct to terms in m/N compared with unity, we encounter only m^{u_1} where $u_1=[\frac{1}{2}u]$; indeed the approximate form for a given u is identical with the exact form for u_1. We see from this that, if we ignore certain Pauli blocking corrections, the effective particle rank of the cumulant generating operator reduces to half its exact value (this does not of course mean that the higher-rank parts vanish in the "low-density" limit ($m/N<<1$) but only that these parts are traceless); we shall shortly see why this comes about.

Neither the exact nor the limiting (large N) forms contain an explicit indication of an approach to normality for large particle number. Since our real interest is with the large-N systems let us consider the limiting form (27). Then for the p'th cumulant of a k-body operator we have u=pk and $u_1=\frac{1}{2}pk$ or $\frac{1}{2}(pk-1)$ according as pk is even or odd. The shape parameters, K_p/σ^p, then behave for large m as m^0 or $m^{-\frac{1}{2}}$ in the two cases. The odd moments for an odd-rank interaction do fall off (slowly) with increasing particle number but otherwise this effect, required for normality, is missing.

To see what goes on here, consider the one-body operator (k=1) for which we know that the approach to normality does obtain. We have, from (27) with p=4, that $K_4(\underset{\sim}{m}) \Rightarrow \frac{1}{2}m\{(4K_4(\underset{\sim}{1})-K_4(\underset{\sim}{2}))-m(2K_4(\underset{\sim}{1})-K_4(\underset{\sim}{2})\}$; but now we observe that the m^2 term cancels because of the condition $K_4(2)=2K_4(1)$. In other words the p=4 cumulant-generating operator is asymptotically a 1-body operator only, rather than 2-body, this coming about because of a cancellation not explicit in the propagation equation. For K_2 on the other hand no such cancellation occurs and we have simply $K_2(\underset{\sim}{m})=mK_2(\underset{\sim}{1})$. Thus the shape parameters decrease with m and we approach normality.

For the general case with interactions, normality requires then a similar reduction in the rank of the cumulant generating operators for even p>2 and no such reduction for p=2. Formally we rely on the expansion of the $(m-t)^{-1}$ kernel in (27) as

$$\frac{1}{m-t} = \frac{1}{m} + \frac{t}{m}\,\frac{1}{m-t} = \frac{1}{m} + \frac{t}{m^2} + \frac{t^2}{m^3(m-t)} = \cdots\cdots \qquad (28)$$

and the elementary result, for an r'th order polynomial $P^r(t)$, that

$$\sum_{t=0}^{u} (-1)^t \binom{u}{t} P^r(t) = 0 \text{ if } r < u \, . \tag{29}$$

But if then a cancellation occurs for the leading power in $K_p(\underset{\sim}{t})$ we get non-vanishing contributions only by taking higher powers of t in the kernel expansion. These bring with them higher inverse powers of m, corresponding really to the fact that, if the cumulant-generating function increases less rapidly than maximum with t on the defining net, then the m-dependence of the cumulants is correspondingly slower. In fact the entire argument is quite self-contained (and, as should be easily seen produces also the combinatorial result given).

The argument will be complete only when we convince ourselves that no cancellation can occur for the variance. That this is so follows from the facts that, for the square of a k-body operator, $u_1=k$ while for $t<k$ the operator vanishes. Thus the m-particle variance is directly proportional to the 2-particle variance. We have seen this before for the traceless 1-body operator; it is now seen to be true generally for any k-body operator with a definite $h\rightleftarrows p$ symmetry. Observe also that in (26,27) we may take t=k as the lower limit, rather than t=0.

Not all low-rank operators give rise to Gaussian spectra for large particle number; a simple exception is the square of a traceless one-body operator. Can we express in a more satisfactory way the necessary conditions which we have found? First we must understand the reduction from $u\to u_1$ which comes about for operators with a definite $(h\rightleftarrows p)$ symmetry when we make the "low density" ($m/N<<1$) approximation. Consider the variance for a traceless 1-body operator, writing separately the terms from the 1-body and 2-body parts of H^2.

$$H = \sum \varepsilon_i n_i \text{ with } \sum \varepsilon_i = 0 \quad \text{(traceless)}$$

$$H \xrightarrow{h\rightleftarrows p} \sum \varepsilon_i (N-n_i) = -\sum \varepsilon_i n_i = -H \quad (\therefore \text{antisymmetric}).$$

We have now

$$\sigma^2(\underset{\sim}{m}) = \frac{m(N-m)}{(N-1)}\, \sigma^2(\underset{\sim}{1}) = \left\{\binom{m}{1} - \frac{2}{(N-1)}\binom{m}{2}\right\}\sigma^2(1) \tag{30}$$

which we recognize, by the binomial forms, as the desired separation. We observe as above that the 2-body contribution is of relative order 1/N. In more detail

$$H^2 = \sum \varepsilon_i\varepsilon_j\; n_i n_j = \underbrace{\sum \varepsilon_i^2 n_i}_{\text{1-body}} + \underbrace{\sum_{i\neq j} \varepsilon_i\varepsilon_j n_i n_j}_{\text{2-body}}$$

$$\langle H^2\rangle^{\underset{\sim}{1}} = \frac{1}{N}\sum \varepsilon_i^2 \qquad \langle (H^2)_{\text{1-body}}\rangle^{\underset{\sim}{2}} = \frac{2}{N}\sum \varepsilon_i^2$$

$$\langle H^2\rangle^{\underset{\sim}{2}} = \frac{2}{N}\sum \varepsilon_i^2 + \frac{2}{N(N-1)}\sum_{i\neq j}\varepsilon_i\varepsilon_j$$

$$\sigma^2(\underset{\sim}{m}) = \langle H^2\rangle^{\underset{\sim}{m}} = m\;(H^2)_1\rangle^{\underset{\sim}{1}} + \binom{m}{2}\langle (H^2)_2\rangle^{\underset{\sim}{2}}$$

$$= \frac{m}{N}\sum \varepsilon_i^2 + \frac{m(m-1)}{N(N-1)}\sum_{i\neq j}\varepsilon_i\varepsilon_j\;.$$

But

$$\sum_{i\neq j}\varepsilon_i\varepsilon_j = \sum_{i=1}^{N}\varepsilon_i\sum_{\substack{j=1\\(\neq i)}}^{N}\varepsilon_j = \Big(\sum_{i=1}^{N}\varepsilon_i\Big)^2 - \sum_{i=1}^{N}\varepsilon_i^2 = -\sum \varepsilon_i^2$$

and this then gives us again the result (30). In this example, and more generally too, tracelessness and $h \rightleftarrows p$ symmetry are closely related; we shall see that they both stem from irreducible behavior with respect to the U(N) group. We see also that for traceless operators a self-correlation in the ε_i terms is necessary in order to produce a non-vanishing trace and it is this restriction which gives us the 1/N reduction.

It is easy to show, by the methods of the next section, that a k-body operator with a definite $h \rightleftarrows p$ symmetry is expressible in terms of products of 1-body traceless operators. Then the same self-correlation is required for non-vanishing trace in the low-density case, and we get thereby the general reduction from u to u_1. In fact this kind of consideration has been used for a long time, in the Ursell-Mayer expansion of classical statistical mechanics (discussed very elegantly by Brout and Carruthers[6]). The problem here is to evaluate the trace of $\exp(\beta V)$ and, since the procedure is to classify terms in a given power of V in the series expansion, the results are directly applicable.

For the square of a 2-body operator V, expressed in configuration space, we encounter terms $V_{ij}\ V_{k\ell}$, $V_{ij}V_{j\ell}$ and $(V_{ij})^2$ which in an obvious notation (different subscripts not being allowed equal) may be pictured as ⁞⁞ , V and 0 ; the separation also is one by particle rank the terms being respectively 4,3, and 2-body terms. However, by the argument above, the first two terms are traceless for large N, and only the 2-body term in V^2 remains. For $(V)^3$ there are eight kinds of terms, pictured respectively as , and  which are respectively operators of rank 6,5,4,4,3,4,2,3, the  term for example having the form $V_{ij}^2 V_{ik}$ which of course is also traceless. Taking cumulants, rather than moments makes for a further simplification since it is easy to see then that <u>disconnected</u> terms such as  encountered for H^4, which do contribute to the large-N moments, do not contribute to the cumulants. For K_3 then the only large-N contributions come from the last two diagrams pictures above; (respectively of ranks 2,3); for K_4 the contributing diagrams are , and  which are of ranks 2,3,4, the structures here being $(V_{ij})^4$, $(V_{ij})^2 V_{ik}V_{kj}$ and $V_{ij}V_{jk}V_{k\ell}V_{\ell i}$. The highest rank terms in each case are of order p described by the classical "ring" diagrams , etc. and it would be the dominance of these terms which would make impossible an approach to normality for large particle number. Since one would expect anyway that K_3 and K_4 should measure the principal deviation from normality (and since $K_3/\sigma^3 \to 0$ for large m) we would say that asymptotic normality should result in large-N systems when the fourth-order ring terms  are traceless (or of small trace). The implications of this result (which could also be expressed in terms of the more usual perturbation-theoretic diagrams) are being studied for various kinds of interactions by Dr. S. Y. Li. For configurations the same arguments apply almost without modification, but for more complex symmetries little or nothing has been done. However, rather than discuss further the source of normality we turn now to considerations needed for practical applications.

5. UNITARY-GROUP DECOMPOSITIONS: QUALITATIVE DISCUSSION

A three-body operator is usually regarded as a much more complicated thing than a 2-body operator (so much so in fact that spectroscopists are very reluctant to consider three-body interactions at all); but it is not always so. If F(2) is 2-body then G(3)=(n-2)F(2) is three-body; for, since (n-2) is (0+1), the product can only be (2+3), but, by the vanishing of (n-2), it vanishes in the two-particle space. However (n-2)F(2) has exactly the same complexity as F(2); it represents a three-body operator in which (assuming that F(2) itself cannot be similarly simplified) the "action" factors into a 2-body action and a simple counting. We see also that some three-body operators may be simpler than two-body ones, so that the particle rank is not necessarily a good guide to complexity.

In order to find such a guide, and at the same time to understand the connections encountered above between tracelessness and $h \rightleftarrows p$ symmetry, and to produce physically relevant decompositions of H and effective ways of evaluating moments for configurations, we consider briefly the unitary-group decomposition of operators[3]. Essentially all the formal problems become simpler when we do this.

Starting with a (number-conserving) k-body operator F(k) we can "promote" it as above to higher rank, the resultant (k+r)-body operator being $\binom{n-k}{r}F(k)$. This is true as long as the new operators do not vanish. It is clear that if we take r̲ too large we must get a vanishing; e.g. if (k+r)>N, since our spaces cannot admit more than N particles and hence there is no such thing as a (t>N)-body operator. Another example would be with the 1-body angular-momentum operator J_z in a set of spherical shells; clearly if r=N-1 we get a vanishing in this case.

Thus a k-body operator (or operator space) can be promoted in a natural way to q-body with q=(k+1), (k+2),...N. In many, but not all, cases the promoted operators behave (except for a number-dependent scaling) precisely as do their parent F(k); in these cases the promotion is invertible. If however we go too far we lose the operator or part of it, and in that case the operation is not invertible and the promoted operator is simpler than its parent.

Reversing the argument we ask whether we may represent our F(k) operator itself as a promotion of a lower-rank operator or as a sum of such promotions; in short can we write

$$F(k) = \sum_{\nu=0}^{k} \binom{n-\nu}{k-\nu} \mathcal{F}(\nu) \qquad (31)$$

where we could agree, without loss of generality that no part of $\mathcal{F}(\nu)$ can be further demoted? We are then representing the k-body operator as being produced by a set of operators $\mathcal{F}(\nu)$, each of which has been generated in a different (ν-particle) space. What would be the significance of the $\mathcal{F}(\nu)$?

The simplest group-theoretical consideration answers these questions for us. To begin with, all the m-particle states (or state operators) form an irreducible $\binom{N}{m}$-dimensional representation of the group U(N) of unitary transformations among the single-particle states. To see this, consider first the fact that the commutator of two 1-body operators is itself a 1-body operator (a product is (1+2)-body and the maximum rank goes down by unity on taking the commutator).

Being closed under commutation they are therefore the generators of some Lie group, which rather obviously is U(N); writing a basis for the one-body operators as $U_{ij}=A_iB_j$ we find easily that

$$[U_{ij}, U_{rs}] = \delta_{jr}U_{is} - \delta_{is}U_{rj} \tag{32}$$

which settles the matter. Since the m-particle states are converted into each other by any number-conserving operators, they supply a representation of U(N); since moreover (the proof of this is left to the reader) any two m-particle states can be connected by some U(N) generator, a one-body operator, it follows that the representation is irreducible. Similarly $\psi^+(m)$, the adjoint state or state operator, transforms irreducibly though not by the same representation (instead the pair of representations generated by $\psi(m)$ and $\psi^+(m)$ are contragradient to each other). Then the $\binom{N}{m}^2$-dimensional set of products $\psi^+_\alpha(m)\psi_\beta(m)$ generates a certain number of irreducible representations; since the vacuum is scalar it follows then that for F(k) to be supported by m (i.e. not to vanish identically in $\underset{\sim}{m}$) it must transform like at least one representation which is common to those generated by the density $\psi^+_\alpha(m)\psi_\beta(m)$.

It will turn out that, as we increase m from 0 to N/2, we generate one new representation at each step as well as all the older ones; as we go beyond m = N/2 we lose them one at a time, the more complex ones (those first generated in the larger spaces) disappearing first. It is clear then that we can label the representations by a single parameter ν which describes the simplest spaces which will support the operator (ν being the corresponding number of particles or holes for that space). The dimensionalities are then obviously $\binom{N}{\nu}^2 - \binom{N}{\nu-1}^2$.

The definition of $F(\nu)$, as an operator which survives in $\underset{\sim}{m} = \underset{\sim}{\nu}$ but which cannot be denoted further, now defines it as an operator with a given unitary symmetry transforming as D^ν, so that (31) gives exactly the U(N) group decomposition. Moreover $F(\nu)$ survives in $\underset{\sim}{m}$ with $m < \nu < N-m$ (i.e. which have not less than ν particles and ν holes) but vanishes for other m. The promotion of F(k) to rank s is then certainly invertible as long as $s \leqslant N-k$, and, beyond that, for $s \leqslant N-\nu_{max}$ where ν_{max} is the highest unitary symmetry of F(k). For $s > N-\nu_{max}$ the promotion is no longer invertible (which suggests to us a "sieve" process for the explicit separation of F(k) according to ν).

Observe also the important fact that, for operators

of fixed particle rank, those which have low unitary rank grow more rapidly in magnitude as the particle number increases, this because of the n-factor representing a particle counting or, more formally, multiple pairwise contractions of A_i and B_i operators. Thus not all 2-body operators which have the same norm in the two-particle space are the same size for large m. Besides that, since taking the trace over all m-particle states is a unitary-scalar operation, it follows that, in the variance or norm integral, no cross-terms appear with different ν so that the decomposition is an orthogonal one. Putting together the last two remarks (and realizing that $\nu = 0$ cannot contribute to σ^2) we see that there is an <u>a priori</u> favoring for operators to behave, in spaces with very many particles, like effective 1-body (i.e.,ν=1) operators and thus to give rise to Gaussian spectra. It is clear of course, that for $\nu>0$, $F^{\nu}(k)$ is traceless, but beyond that it is traceless to order ν, in the sense that it remains traceless even when multiplied by any $(k'<\nu)$-body operator; thus the unitary group classification also supplies a heirarchy of tracelessness.

The irreducible and fully contracted operator, $\mathcal{F}(\nu)$, has a remarkable symmetry under hole$\rightleftarrows$particle conjugation (by which we mean formally that every $A_i \rightleftarrows B_i$ and thus to within a phase $\psi_\alpha \rightleftarrows \psi_\alpha^+$). This converts a ν-particle operator into a ν-hole operator;normally then on carrying out the $[\psi,\psi^+]$ commutation (or anticommutation) we would pick up terms of lower particle rank; but since conjugation cannot change the symmetry ν (why not?) we see that these commutator terms should have unitary rank ν and particle rank $<\nu$ which is impossible. They then must vanish and so $\mathcal{F}(\nu)$ has a definite h$\rightleftarrows$p symmetry; putting in the phases, we find

$$\mathcal{F}(\nu) \xrightarrow{h \rightleftarrows p} (-1)^{\nu(2\nu-1)} \mathcal{F}^+(\nu) \qquad (33)$$

Conversely an operator with definite unitary and particle ranks is necessarily irreducible and fully contracted, and the two ranks are equal.

A little thought will show that the unitary decomposition of an operator is connected with representing the operator according to certain averages and "fluctuations" about these averages. If we identify the particles according to the single-particle states in which they are to be found, we see that the ν=(0+1) parts of an operator derive from averaging the effect on one of the particles (so that the matrix elements are proportional to (n-1); the effect on the particle not averaged over is then represented by a 1-body operator which would vanish if the original operator were irreducible with $\nu=2$. If we carry out a further averaging on the 1-body operator we separate thereby the $\nu=0$ and $\nu=1$ parts, the latter being then, in its turn, a kind of lower-order fluctuation. And similarly for higher-rank operators and also for operators which are not number conserving.

Going beyond the scalar distribution (over m) we are led naturally to introduce configurations, at which stage the unitary averagings become very significant indeed. For example the single-particle energy for a particle outside a closed shell derives in part from a true 1-body kinetic energy operator and in part from a 2-body operator which has been averaged over the shell as described above; in this case of course the averaging is over a subset of the single-particle states, this finer average being appropriate for configurations. It should be clear in fact that the relevant group is now a subgroup of U(N), in which we consider separate transformations in the ℓ single-particle subspaces defined by the partition $N=\Sigma\, N_i$. We may regard this group as a "direct-sum" subgroup and write it as $U(N_1)\dotplus(N_2)\dotplus\cdots\dotplus U(N_2)\equiv\Sigma\, U(N_i)$. Since the separate "orbits", the single-particle states in the set N_i, which need not have any relation to a spherical orbit, are disjoint, this group is as easy to handle as the original U(N), and we see immediately that, whereas the classification before produces a single ν, we now have $\nu_1,\nu_2,..\nu_\ell$ which we may consider together as an ℓ-dimensional vector $\vec{\nu}$.

However our earlier restriction to number-conserving operators is inadequate because such an operator does not in general behave in that way in the separate orbits; an ordinary pairing interaction for example destroys two particles in one orbit and makes two in another. In a single orbit we have a (k,μ) classification according to the structure $\{\psi_\alpha(k+\mu)\psi_\beta^\dagger(k-\mu)\}$ or $\{(A)^{k+\mu}(B)^{k-\mu}\}$ where $k\geqslant 0$, $|\mu|\leqslant k$, and $\mathbf{k}+|\mu|\leqslant N$. An operator of this type, $F(k,\mu)$, creates 2μ particles, or destroys -2μ; it acts between pairs of fixed-number spaces so that we expect the representations to be defined by two parameters. Since the product of two operators with different μ is traceless it follows that such operators have no representations in common, so that μ may be taken as one of the parameters.

For the other parameter we can be guided by our earlier procedure. μ defines the particle number difference for two spaces between which $F(k,\mu)$ operates and k, which we may continue to regard as the "particle rank", the average. As we carry out contractions and divide out the n-factors, μ remains unchanged while ν decreases by unity for each step, the value of ν for an operator which cannot be demoted then giving the smallest(average) particle number to be associated with the reduced operator. We expect then that we may label the representations as $D^{\nu}(\mu)$.

For example the set $\{A^2\}$, of which a characteristic member is A_iA_j, is not contractible and belongs to $D^1(1)$; more generally $\psi_\alpha(k)$ belongs to $D^{k/2}(k/2)$ and $\psi^{+}_{\alpha}(k)$ to $D^{k/2}(-k/2)$ (adjointing changes the sign of μ). $\{A^2B\}$ transforms as two representations, $D^{\frac{1}{2}}(\frac{1}{2})$ which is the result of a contraction under which $A^2B \to nA$, and $D^{3/2}(\frac{1}{2})$ representing the fluctuation. Then, as long as $k \leq \frac{1}{2}N$, $F(k,\mu)$ transforms as $D^{\nu}(\mu)$ with $\nu=\mu, \mu+1, \ldots k$ if $\mu>0$, and more generally, for either sign of μ, as $|\mu|, (|\mu|+1) \ldots k$. If $k>\frac{1}{2}N$ the upper limit is replaced by $(N-k)$. The dimensionality for $D^{\nu}(\mu)$ is seen from the dimensionalities of its defining pair of spaces, to be

$$d(\nu,\mu) = \binom{N}{\nu-\mu}\binom{N}{\nu+\mu} - \binom{N}{\nu-\mu-1}\binom{N}{\nu+\mu-1} =$$

$$= \frac{(N-2\nu+1)}{(N+1)} \binom{N+1}{\nu-\mu}\binom{N+1}{\nu+\mu} \ . \qquad (34)$$

The extension of (31) is

$$F(k,\mu) = \sum_{\nu} \binom{n-\nu-\mu}{k-\nu} \mathcal{F}(\nu,\mu) = \sum_{\nu} \mathcal{F}(\nu,\mu)\binom{n-\nu+\mu}{k-\nu} \qquad (35)$$

where $\mathcal{F}(\nu,\mu)$ is then the contracted part of $F(k,\mu)$ which transforms as $D^{\nu}(\mu)$. The behavior of $\mathcal{F}(\nu,\mu)$ under $h \rightleftarrows p$ conjugation is given by (33) just as for the $\mu=0$ case.

Consider now the forms and representations encountered in the ℓ-orbit decomposition of a k-body operator. We first have the orbital generalizations of particle rank, $(k,0) \to (\vec{k},\vec{\mu})$ which is itself a partial classification by symmetry, and after that the introduction of the irreducible representations $(\vec{k},\vec{\mu}) \to (\vec{\nu},\vec{\mu})$. A two-body operator, k=2, divides into seven classes according to orbital structure. These are as listed and diagrammed (a block denoting a single orbit so that for example [··][ ] → [ ][··] describes a transition induced by a standard pairing interaction). The form of the operator and of its matrix elements is also indicated; in the pairing case we have $A^2|B^2$ indicating a behavior like A^2 in one orbit and B^2 in a distinct one, the corresponding matrix elements having the structure $W_{\alpha\alpha\beta\beta}$ where α and β refer to orbits (not of course to single states). We have now:

(1) Single-orbit; $A^2B^2|$ $W_{\alpha\alpha\alpha\alpha}$; [··] → [··]

(2) Two-orbit Pairing; $A^2|B^2$; $W_{\alpha\alpha\beta\beta}$; 

(3) Two-orbit Multipole; $AB|AB$; $W_{\alpha\beta\alpha\beta}$; 

(4) Two-orbit Hole-particle; $A^2B|B, AB^2|A$; $W_{\alpha\alpha\alpha\beta}$;

(5) Three-orbit Pairing; $A^2|B|B, A|A|B^2$; $W_{\alpha\alpha\beta\gamma}$;

(6) Three-orbit Multipole; $AB|A|B$; $W_{\alpha\beta\alpha\gamma}$; 

(7) Four-orbit; $A|A|B|B$; $W_{\alpha\beta\gamma\delta}$; 

It should be clear that, for a given identification of the orbits diagrammed and a given vector $\vec{\mu}$ (not allowing for a sign change), these interactions give respectively 3,1,4,2,1,2,1 irreducible representations. Besides this we may have a multiplicity arising from the different assignments of orbits; for example there are $\binom{\ell}{2}$ different such assignments for the pairing interaction (2). We shall of course need also the decomposition for a 1-body operator, which involves only two orbital structures, $AB|$ and $A|B$, the first of which yields two representations for each orbit, while the second is irreducible. Note very carefully that these same representations are found also among those for the two-body operator and, since there is no $\vec{k}$-orthogonality, the contributions to widths are coherent.

6. UNITARY-GROUP DECOMPOSITIONS: SOME FORMAL RESULTS

We consider in a more formal way some of the results of §5 and show specifically how they lead to configuration widths and centroids. Further details are given in reference (3).

Unitary-group representations are usually discussed in terms of Young shapes. The space defined by a shape contains a vector which is symmetrized according to the rows and one which is antisymmetrized according to the columns. Hence it follows, since $A_i^2 = B_i^2 = 0$, that, with A_i or B_i operators alone, we can generate only 1-columned representations, while, quite generally, we are restricted to 1- and 2-columned representations. We have that $A_i \sim \square$, and B_i as its contragradient; writing the column structure, these are [1] and [N-1] respectively. Then, by the elementary combination rule, we have that $\psi(m) \sim [m]$, $\psi^+ \sim [N-m]$, both irreducible (as we saw before by a different argument), and

$\psi(k) \times \psi^{+}(k) \sim \Sigma\ [N-\nu,\nu]$ where the ν sum runs from 0 to k and each representation occurs once only. The D^{ν} representation encountered before for the number-conserving operator is then identified as $[N-\nu,\nu]$. In exactly the same way we find that

$$D^{\nu}(\mu) \equiv [N-\nu+\mu,\ \nu+\mu]. \tag{36}$$

For the multi-orbit case we introduce, just as before, $(\nu_i,\ \mu_i)$ for the i'th orbit.

Several ways are available for the explicit decomposition of an operator into irreducible parts. One procedure is via a unitary scalar demotion; as long as $k \leqslant N/2$ then a unit reduction in the rank, carried out via a scalar operation, annihilates the $\nu=k$ part of an operator $F(k,\mu)$; continuing this operation we are left finally with the lowest - ν part and then, comparing the results at the various stages, we can recapture the others. We are really using here the notion of a mathematical sieve. It should be seen easily that DF defined by

$$D_{\pm}\ F = \sum_i\ [A_i[B_i F]_{\mp}]_{\pm} \tag{37}$$

where D_+ is used for bosonlike operators (even total number of A, B operators) and D_- for fermionlike, is exactly the operation required; observe that any $\mathcal{F}(\nu,\mu)$ operator satisfies $D\ \mathcal{F}(\nu,\mu)=0$. It will be recognized immediately that, with F the Hamiltonian operator, the D operation is encountered in the sum-rule analysis of pickup and stripping reactions (in which case the summation is over a spherical orbit) and more generally, in spectral function sum rules. For simple operators such as H the necessary demotions are easily carried out; an elegant explicit form for the most general case, which produces all the $\mathcal{F}(\nu,\mu)$ for a given $F(k,\mu)$, has been given by Chang[3].

Instead of demoting the operator, we can promote it and use the sieve action generated by the fact that in $\underset{\sim}{m}$ with $m=N-s$ only a limited number of symmetries, those with $\nu \leqslant s$, can survive. Evaluating the matrix elements in the small spaces and reconstructing the operator thereby, we once again achieve a decomposition. The details are left as an exercise.

A less sophisticated way, but quite adequate for H relies on the $h\rightleftarrows p$ symmetry of the irreducible contracted parts of the operator. It involves recognizing the natural matrix element forms which enter when we make the partial contractions discussed in §5. We illustrate it by reducing a two-body operator defined in a single orbit. We have

$$H = -\tfrac{1}{4} W_{ijk\ell} A_i A_j B_k B_\ell \tag{38}$$

where $W_{ijk\ell}$ is the 2-body matrix element $\langle 1_i 1_j | H | 1_k 1_\ell \rangle$. If we contract over (j,ℓ) we encounter $W_{ijkj} = \varepsilon(i,k)$ say (summation always being understood), which of course defines a 1-body operator $\varepsilon(i,k) A_i B_k$. This operator may itself be further reduced by introducing the trace and the traceless one-body operator. Let

$$\varepsilon = N^{-1} \varepsilon(i,i)$$

$$\xi(i,j) = \varepsilon(i,j) - \delta_{ij}\varepsilon \text{ ; then } \xi(i,i) = 0 . \tag{39}$$

We have now, correct to multiplying constants, the $(\nu=0)$ part $\varepsilon\binom{n}{2}$ and the $(\nu=1)$ part $(n-1)\xi(i,j)A_i B_j$; on subtracting these from the original H we are left with the $(\nu=2)$ part, the constants being determined by satisfying the $h \rightleftarrows p$ symmetry condition (33) for each contracted fixed-ν part.

The result, for the two-body H, is

$$\begin{aligned} H = {} & \frac{\varepsilon}{(N-1)} \binom{n}{2} + \frac{(n-1)}{(N-2)} \xi(i,j) A_i B_j \\ & + \left\{ H - \frac{\varepsilon}{N-1} \binom{n}{2} - \frac{n-1}{N-2} \xi(i,j) A_i B_j \right\} \end{aligned} \tag{40}$$

which are respectively the $\nu=0,1,2$ parts of H. Since the two-body matrix elements of $\binom{n}{2}$ and $(n-1)\xi(i,j)A_i B_j$ are respectively given by

$$\begin{aligned} \langle t\ u | \binom{n}{2} | r\ s \rangle &= \delta_{tr}\delta_{us} - \delta_{ts}\delta_{ur} \\ \langle t\ u | (n-1)\xi(i,j)A_i B_j | rs \rangle &= \xi(t,r)\delta_{us} + \xi(u,s)\delta_{tr} \\ & \quad - \xi(t,s)\delta_{ur} - \xi(u,r)\delta_{ts} \end{aligned} \tag{41}$$

we see that the $(\nu=2)$ Hamiltonian is

$$-\tfrac{1}{4} W^{\nu=2}_{ijk\ell} A_i A_j B_k B_\ell \tag{42}$$

with

$$\begin{aligned} W^{\nu=2}_{ijk\ell} = {} & W_{ijk\ell} - (N-1)^{-1}\{\delta_{ik}\delta_{j\ell} - \delta_{i\ell}\delta_{jk}\}\varepsilon \\ & -(N-2)^{-1}\{\delta_{ik}\xi(j,\ell) + \delta_{j\ell}\xi(i,k) - \delta_{i\ell}\xi(j,k) - \delta_{jk}\xi(i,\ell)\}. \end{aligned} \tag{43}$$

The $(\nu=1)$ part is best written as a one-body operator. However to evaluate the width we should first add in the one-body H; dropping off the $\nu=0$ part (which cannot contribute) we are left with

$$H^{\nu=1}(\text{1-body}) = \sum \varepsilon_{ij} A_i B_j \text{ with } \varepsilon_{ij} = 0 \tag{44}$$

so that the total $\nu=1$ H is then

$$H^{\nu=1} = \sum\{\varepsilon_{ij} + \frac{n-1}{N-2}\,\xi(i,j)\}A_i B_j = \sum \varepsilon^{\nu=1}(i,j)A_i B_j \tag{45}$$

in which we notice the role of the induced number-dependent single-body energies.

The ν-orthogonality now enables us to write separate widths for the $(\nu=1,2)$ parts. Moreover we can use the symmetrical form (26) for the widths and, taking note of the fact that any $F(\nu)$ vanishes for $(t<\nu)$ particles, we see that, as discussed in §4, the fixed-ν width for m particles is simply a polynomial multiple of that for ν particles. Specifically we have, writing $u_1=\nu$, that

$$\sigma^2(\underset{\sim}{m}) = \sum_{kk'\nu=1}^{2} \binom{m-\nu}{k-\nu}\binom{m-\nu}{k'-\nu}(\nu+1)\binom{m}{\nu+1}\binom{N-m}{\nu+1}\binom{N-\nu}{\nu+1}^{-1} \cdot \tag{46}$$
$$\cdot \frac{N-2\nu}{(m-\nu)(N-m-\nu)} \langle H^{\nu}(k)H^{\nu}(k')\rangle^{\underset{\sim}{\nu}}.$$

The ν-particle averages are simply the average products of matrix elements (43,45) of the contracted H^{ν}'s. Observe that the 1-body and 2-body terms, and the cross-term, have different particle-number dependences.

All this manipulation may seem excessive since, as we have indicated in (24), the scalar-width problem is simple anyway. The real and quite overwhelming advantage is that, since configurations define independent group structures for each orbit, the same procedure solves the otherwise much more difficult problem of evaluating widths in the configuration case. Combining the orbit decomposition of §5 with the natural extension of the unitary scalar decomposition given here, we produce a polynomial form for the configuration moments. The results are given in reference (3) where it will be seen that there enter Hartree-Fock-like energies representing the average interactions of particles in one orbit with those in another, and other quantities of that kind.

There is one special but important case where things are so simple that the results can be written at sight, namely when our orbits are spherical and there is no radial degeneracy (i.e. no two orbits enter which differ only in radial quantum number). The special simplification here is that no interactions have a ν-1 character in any orbit; for, by spherical symmetry, no J-preserving interactions can split the single-particle or hole states for any one orbit; but if there is no such splitting there is no $\nu=1$ residue after the averaging which produces $\nu=0$. Since now $\nu=0$ does not contribute to the width we should subtract it out and discard it; we do this[3] by modifying the two-body matrix elements via a special case of (43), viz

$$W_{rstu} \to W_{rstu} - \delta_{rt}\delta_{su} d^{-1}(2:rs) \sum_{J,T} (2J+1)(2T+1) W^{J,T}_{rsrs} . \tag{47}$$

Here the subscripts are spherical orbit labels and, to make things simpler, we have agreed that the ordering is such that $r \le s, t \le u$ (in which case a $\delta_{ru}\delta_{st}$ term disappears). d is simply the two-particle dimensionality for particles in the orbits specified (which may be the same). Inspection (a very careful one) now yields for the configuration variance the form[3] (where $\vec{m}$ denotes the ℓ-dimensional vector)

$$\sigma^2(\underset{\sim}{\vec{m}}) = \sum_{\substack{r<s\\ t<u}} \frac{(N_r - m_r)(N_s - m_s - \delta_{rs}) m_t (m_u - \delta_{tu})}{(N_r - \delta_{rt} - \delta_{ru})(N_s - \delta_{st} - \delta_{su} - \delta_{rs}) N_t (N_u - \delta_{tu})} \cdot \sum_{J,T} (2J+1)(2T+1)\{W^{JT}_{rstu}\}^2 . \tag{48}$$

It is an important fact, which might be clear from the equation, that this form yields not only the configuration variance but also its decomposition into parts which represent the interaction with a definite other configuration. These partial variances, to which we return briefly later, are important in studying symmetries (in this case the extent to which configurations are admixed) and in giving measures for the importance, at various excitation energies, of various irreducible parts of the interaction.

To be able to specify isospin is important for many reasons. We achieve a finer decomposition than otherwise; we can study heavier nuclei where the low-lying states have isospins larger than the minimum possible values (which at low energies would be automatically produced by a distribution which ignores isospin); we can study isobaric mass differences and the goodness of isospin. If we use an isospin formalism everything is fully soluble but the forms encountered are more complex than above (because we have then a direct-product as well as direct-sum decomposition). A less elegant but easy way is to use a p-n formalism in which we speak separately of proton and neutron orbits; in this case the forms above still apply and we can recover isospin by the elementary subtraction procedure often used for this purpose. Without going further into the details let us agree now that we are able to calculate centroids and widths for configurations with or without a specification of isospin.

7. BINDING ENERGIES AND SPECTRA

The spectral distributions give us of course a theory of state densities which takes account of residual interactions and of the orbital structure. We come later to this most obvious application and turn first to some less obvious ones which are really in the domain of detailed spectroscopy.

We therefore need to discuss levels and not just densities. We can of course easily convert a discrete spectrum into a continuous one, by a "windowing" or "smoothing" operation; how can we do the reverse? One answer comes from an inspection of Fig. 1; the continuous distribution function F(E) might be regarded as a best fit to the exact staircase function if it passed through the mid-points of the vertical steps, i.e. if, at the eigenenergy E_j ($j=1,2.3,\ldots$), it takes on the value $(j-\frac{1}{2})$. But now, following Ratcliff[2], we turn the argument around and use this relationship to <u>define</u> the approximate eigenenergy as that value of E for which $F(E)=(j-\frac{1}{2})$, this then to be regarded as the best we can do with a given continuous distribution. There are in fact some variants on this; for one thing, if the system has symmetry degeneracies and our distribution includes representations with different dimensionalities (for example different J-values) then the vertical steps are not all equal. In this case we really need to know the ordering of the symmetries if we wish to deduce the spectrum from the distribution. In some cases we may in fact "know" this for the lowest levels (0^+-2^+ for most even-even nuclei, for example), and in some others we may make the assumption that the ordering is as given by experiment. Of course the procedure here, just as in the simpler case, gives also a value for the ground-state energy, a quantity of fundamental importance. We may often however choose to determine this from the distribution by first fixing a low-lying excited state and then using experiment to fix the ground-state energy from that. One often gets a dramatic improvement in the accuracy on doing this, not surprisingly because, in so doing, we move in from the extreme tail of the distribution where the accuracy will be best. This can be seen[2] from Fig. (3) which shows, for spaces of varying dimension, how far from the centroid certain specific values of the Gaussian distribution function are to be achieved. Experience has been that an individual distribution is untrustworthy if one goes say more than 3.5σ away from the centroid, but of course this restriction would not apply to the total distribution given as a sum of subspace distributions. Another variant, which

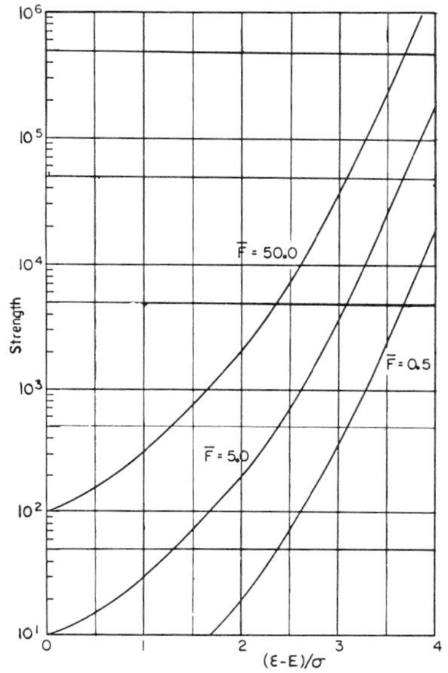

Figure 3
Specific values of the Gaussian distribution function, located with respect to the centroid E and given as function of the dimensionality.

we shall make use of in connection with level densities, fixes the reference energy not by the distribution function but by the level or state density.

Figure 4 compares the exact and Gaussian distribution spectra for the rather simple case already considered in Fig. 1; this in fact is relatively unfavorable since the particle number and dimensionality are both small. Still the general fit, reached with only two parameters, is rather good. The spectra however are quite different in appearance, the exact one showing density fluctuations while the distribution one displays only the slow semi-monotonic variation characteristic of the Gaussian. The natural question is to what extent "physics" as opposed to "noise" is contained in the fluctuations, being therefore lost when we filter them out as we do via the low-moment distributions.

This is not an easy question but we can learn something about it by considering the "pattern" of the deviations between the two spectra as a function of the energy. Consider the energy variation of

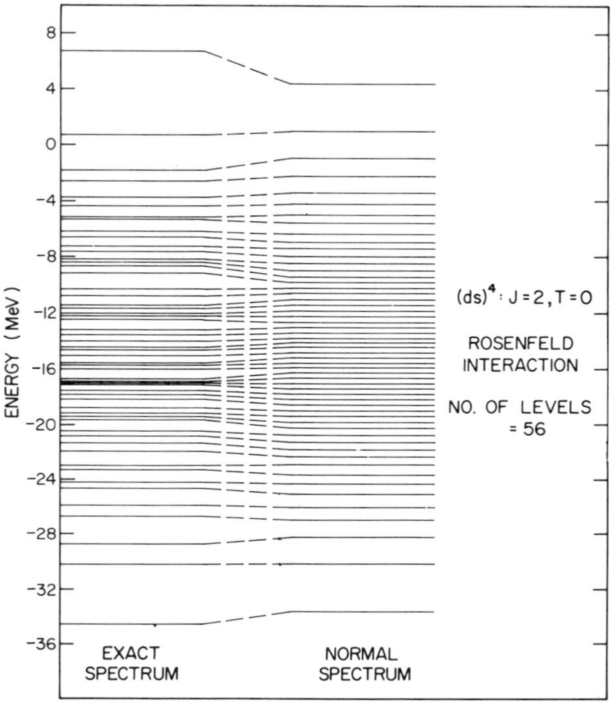

Figure 4
The J=2, T=0 spectrum for $(ds)^4$ and its normal approximation.

$\Delta_r = E_r(\text{exact}) - E_r(\text{distribution})$, where r is the level number. The level spacing supplies a natural scale with which to measure how rapidly this quantity varies. A rapid fluctuation in the sign of Δ_r with level number could be interpreted as "noise", the kind of thing which could only be slowly reduced by taking higher and higher moments; we are of course suspicious of high-moment phenomena since one cannot, except perhaps very near the ground state, expect models to reproduce them (their detailed nature is therefore beyond us). If on the other hand the Δ_r show a slowly-varying sign pattern, and in particular if this is characteristic, say, of third-or fourth-moment departures from Gaussian, then they correspond to a "secular" variation quite different from noise. Proceeding with low-moment corrections to Gaussian should then give major improvements in the spectra.

Instead of applying this test to the simple case of Fig. 4, we use the much more interesting $(ds)^{12}$ J=T=0 spectrum of Zuker[7], derived by diagonalizing the 839-dimensional matrix constructed with the realistic interaction of Kahana, Lee and Scott.[8] The cumulants are calculated from the eigenvalues which are believed accurately enough given to permit fourth-cumulant calculations. The pattern of deviations from Gaussian makes it clear that they are not describable as "noise". Levels #1-153 and 725-839 show negative deviations while the ones between #154-724 show positive, altogether with only a half dozen exceptions. Each of these regions spans about the same energy. Even with four moments we have a simple sign pattern, with five sign regions instead of three, so that we have not, even with four moments, reached the irreducible errors which would constitute fluctuations or noise.

On the other hand even the Gaussian approximation is not bad for the spectrum, even for locating the ground-state energy. For a scale here it might be borne in mind that the projected Hartree-Fock approximation predicts the ground-state of Si^{28} about 10 MeV too high. Table 1 shows that the fixed-(m,J,T) distributions do very well indeed for the ground state.

Table 1
Given are the R.M.S. differences between Zuker's exact J=T=0 $(ds)^{12}$ spectrum and the distribution spectra defined by two, three, and four moments. In computing the "modified" error the five highest and five lowest states are omitted. The dimensionality is 839, and the spectrum span is 88 MeV

No. of Moments	R.M.S. Error	Modified R.M.S. Error	Ground State Error
2	.81 MeV	.66 MeV	-2.0 MeV
3	.36	.15	+4.8
4	.24	.11	+1.5

Given also is the R.M.S. deviation between the exact and distribution spectra, averaged over the entire spectrum and also averaged over all the states except the five highest and lowest. The purpose is to show that the errors are largely concentrated in the extremes of the spectrum; measured in terms of the local spacing this however is not so, the 4-moment error being then approximately one spacing unit over the entire spectrum.

Results for the finer distribution, fixed-(configuration, J,T), are not available. The extremely crude fixed-m Gaussian distribution gives about a 10 MeV error if we use the second-excited state as the reference energy (assuming then that the 0^+-4^+ difference would be given correctly by the interaction. Fixing isospin also hardly helps at all, but when we use the spherical-configuration distribution without isospin the error reduces to 5 MeV and with isospin to 4 MeV.

Many further results both for the (ds) shell and the 3-orbit (f,p) shell (closed $f_{7/2}$) are given in reference (3), all these being derived by use of configuration-isospin distribution (for which incidentally the computing time required is negligible). We reproduce in Fig. 5 the Gaussian fixed-(m,T) and fixed-(configuration, T) low-lying spectra for ^{65}Cu compared with the shell-model calculations of S.S.M. Wong[9] which give a beautifully normal spectrum; even allowing for the fact that the T=7/2 total spectrum span in this case is about 18 MeV compared with the 100 MeV or so found for many particles in the (ds) shell, the agreement is remarkably good, and of course it improves with increasing excitation energy. Do not forget

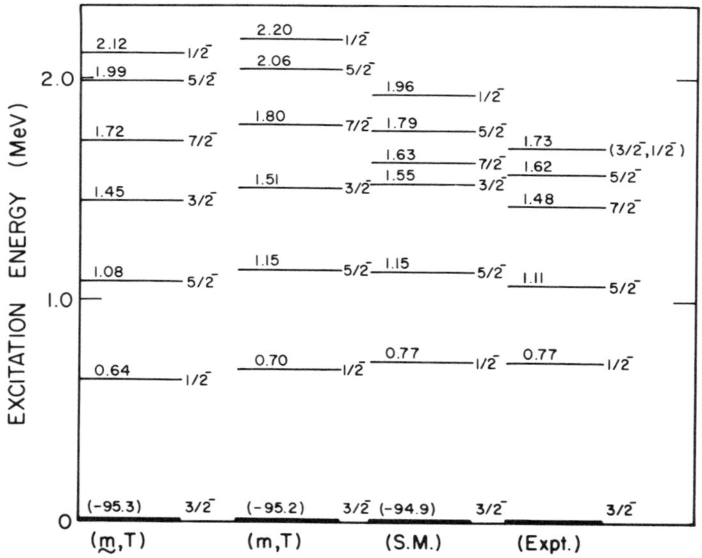

Figure 5
Spectra for ^{65}Cu. For ($\underset{\sim}{m}$,T) configurations and isospin are specified, for (m,T), only particle number and isospin.

however that the J-ordering is assumed here since fixed-J distributions are as yet technically difficult to produce. Perhaps then the most significant aspect is that the absolute energy is well given. The calculations here are with a Kuo-Brown[10] interaction with a small "secular" correction. Table 2 shows how well the absolute binding energies are reproduced throughout the entire 3-orbit (f,p) shell. Similar calculations have been made in the (ds) shell, with similar agreements, and are now being made in the 4-orbit (f,p) shell (in which also $d_{3/2}$-hole and $g_{9/2}$-particle

Table 2
Ground-state energies of (f,p) nuclei, as predicted by configuration-isospin distributions, by empirical binding energies, and by shell-model calculations. Energies are in MeV with respect to ^{56}Ni. Coulomb-corrections have been applied to empirical energies. Except where indicated a low-lying excited state has been used as a reference

	Distri-bution	B.E.	S.M.		Distri-bution	B.E.
^{60}Ni	-42.8	-42.8	-42.9	^{65}Ni	-84.8	-83.9
^{60}Cu	-45.9	-45.4	-	^{65}Zn	-102.5	-101.9
^{61}Ni	-50.9	-50.7	-51.1	^{65}Ga	-110.1[b]	-107.9
^{61}Cu	-57.2	-57.0	-56.9	^{65}Ge	-113.4[b]	-111.0
^{61}Zn	-60.3[a]	-60.5	-	^{66}Ni	-92.1	-92.8
^{62}Ni	-61.8	-61.3	-62.1	^{66}Cu	-102.2[a]	-101.4
^{62}Cu	-66.3	-65.9	-66.6	^{66}Zn	-113.3	-112.9
^{62}Zn	-73.2	-73.2	-	^{66}Ga	-119.3[a]	-116.9
^{63}Ni	-68.8	-68.1	-	^{66}Ge	-126.5[a]	-123.3
^{63}Cu	-77.5	-76.7	-77.4	^{67}Cu	-109.6[a]	-110.4
^{63}Zn	-83.7[a]	-82.3	-	^{67}Zn	-119.5	-119.8
^{63}Ga	-86.9[b]	-86.1	-	^{67}Ga	-129.5[a]	-127.9
^{64}Ni	-78.6	-77.8	-78.6	^{67}Ge	-135.6[b]	-133.0
^{64}Cu	-86.0[a]	-84.5	-	^{68}Cu	-116.7	-116.6
^{64}Zn	-94.8	-94.1	-	^{68}Zn	-129.7	-129.9
^{64}Ga	-98.9[a]	-96.2	-	^{68}Ga	-138.0[a]	-136.0
^{65}Cu	-95.3	-94.4	-94.9	^{68}Ge	-146.9	-144.9

[a]No "excited-state" correction, due to lack of data.
[b]Ground-state J-value is assumed to be 3/2.

excitations are permitted). The vector space dimensionalities encountered in the 3-orbit (f,p) cases range up to 35,000; we find in these no indication of any large systematic error depending on the dimensions, and in fact there are tests which indicate that the errors are small. When we extend to six shells we encounter, for Cu^{63}, dimensionalities of order 10^{12}; we are at present inclined to believe that even here the accuracy is acceptable (when we analyze things by configurations) but it is too early to be sure of that. There seems however a good possibility that we can determine the absolute energies, and certainly the level densities, in shell-model spaces of arbitrarily large size.

8. ORBIT AND STATE OCCUPANCIES

When we locate the ground-state by using a distribution which is a sum of partial distributions for specified symmetries we in principle learn, from the relative weights of the partial distributions, much also about the structure of the ground-state function; in particular we learn the intensities of the various contributing symmetries, e.g. the 45 jj-configuration intensities which contribute to the $(ds)^{12}$ states. In practice it would often be unwise to take too seriously the notion that the numbers derived from such approximate partial distributions refer to a single state; for experience indicates that in many cases there are, from state to state, more rapid fluctuations than can be dealt with by means of low-moment distributions. One way, which we shall use later, for interpreting the intensities is to regard them as average intensities taken over some finite energy domain, including a few states (Ratcliff[2]). Another way, especially appropriate for configuration intensities is to combine them to yield the orbit occupancies; obviously $\langle m_r \rangle = \sum m_r I(m_1 \ldots m_\ell)$ where the sum is over all configurations and the I's are relative intensities. In this case, since an averaging is already built in (the 45 jj configurations yield 3 orbit occupancies, or 6 if we consider protons and neutrons separately), we can indeed regard the results as applying to a single state.

The parameters so derived are the simplest parameters which enter into the description of the multiparticle density. As befits their fundamental significance they enter naturally into sum rules and, in the ground states of target nuclei, are measurable (in a spherical orbit representation) by stripping and pickup reactions. At higher excitation energies they are of equal interest, entering for example into the description of the internucleon cascade (Griffin[11]) which leads to compound nucleus formation, as well as in the γ-ray distributions which follow particle capture. A plot of the occupancies versus the single-particle energy yields the profile of the Fermi surface. Occupancies are also important parameters in the combinatorial theory of level density and in the B.C.S. treatment of pairing effects.

Table 3 gives ground state occupancies for the three-orbit (f,p)-shell calculations referred to above.

Table 3
Fractional occupancy (%) of the single-particle orbits for ground states of (f,p) nuclei, as predicted by the distribution method and detailed spectroscopic calculations

Nucleus	Distribution			Detailed Spectroscopy		
	$f_{5/2}$	$p_{3/2}$	$p_{1/2}$	$f_{5/2}$	$p_{3/2}$	$p_{1/2}$
^{60}Ni	2.0	40	13	5.5	38	8.3
^{61}Cu	3.5	48	18	6.0	47	13
^{61}Ni	6.2	44	18	10	48	17
^{62}Cu	4.8	55	26	5.2	55	24
^{62}Ni	8.3	49	27	9.3	46	30
^{63}Cu	8.9	58	32	8.7	59	32
^{64}Ni	17	50	48	22	48	40
^{65}Cu	19	61	45	20	60	43

Only those are given for which the detailed shell-model results are available. The agreement is very good. There has been so far no systematic comparison, for a wide range of nuclei, between theory and experiment for the ground-state occupancies, though this should now be quite feasible to make, since a simple theory is now available. Castel et al[12] have compared with experiment the (ds)-shell Hartree-Fock occupancies for both neutrons and protons with satisfactory agreement. The distribution theory occupancies have been similarly compared by Chang[13]; there is no formal difficulty here in producing separate neutron and proton parameters since, as explained above we really use a (p,n) formalism for the distributions. It is an interesting formal exercise to show how one would derive them in isospin language.

Turning now to the energy variation of the state

occupancies consider first the standard problem of the "Fermi profile" for non-interacting particles distributed over a set of single-particle states. The general results above apply immediately, but it is worthwhile to rederive the results for this especially simple case. We wish to plot, for a fixed multi-particle energy, E, the single-particle occupancy against the single-particle energy. We calculate this by partitioning the Fermi spectrum into two parts, $N \to [N-1_i, 1_i]$, which isolates the i'th single-particle state and defines for m particles two configurations $[m, 0_i]$ and $[m-1, 1_i]$. Then clearly we have

$$\langle n_i \rangle_E = \rho_{[m-1,1_i]}(E)/\rho_m(E) \qquad (49)$$

where of course $\rho_m(E)$ is the total density, the sum of the two fixed-configuration densities. This result is exact but in its application we shall use low-moment approximations (and could for accuracy further partition the (m-1) subspace). We are seeking $\langle n_i \rangle_E$ as a function of ε_i where $H = \Sigma \varepsilon_i n_i$.

More generally if we partition into a number of orbits, $N \to \Sigma N_r$, we have easily, for the centroid energies, that

$$E(\underset{\sim}{m}_1 \ldots \underset{\sim}{m}_\ell) = \sum m_r E(\underset{\sim}{1}_r) \ . \qquad (50)$$

For the variance we observe that uniformly moving all the states of an orbit can have no effect on the spread of the states of a single configuration. Thus the configuration variance cannot depend on the single-particle energies, and, when we write it in terms of the elementary variances, only single-orbit terms can appear. But then, using (30), we have

$$\sigma^2(\underset{\sim}{m}_1 \ldots \underset{\sim}{m}_\ell) = \sum \frac{m_r(N_r - m_r)}{(N_r - 1)} \sigma^2(\underset{\sim}{1}_r) \ . \qquad (51)$$

In terms of the single-particle averages

$$\langle\varepsilon\rangle_r = N_r^{-1} \sum_{i \in r} \varepsilon_i ; \langle\varepsilon^2\rangle_r = N_r^{-1} \sum_{i \in r} \varepsilon_i^2 \qquad (52)$$

we have then

$$E(\underset{\sim}{m}_1 \ldots \underset{\sim}{m}_\ell) = \sum m_r \langle\varepsilon\rangle_r$$
$$\sigma^2(\underset{\sim}{m}_1 \ldots \underset{\sim}{m}_\ell) = \sum \frac{m_r(N_r - m_r)}{(N_r - 1)} \{\langle\varepsilon^2\rangle_r - \langle\varepsilon\rangle_r^2\} \ . \qquad (53)$$

Similar results may be derived for the higher moments. It is worthwhile, as an exercise, to derive the results by the more general methods given above.

Applying them to the case in hand, we have

$$d(\underset{\sim}{m},\underset{\sim}{0}_i) = \binom{N-1}{m};\ E(\underset{\sim}{m},\underset{\sim}{0}_i) = m(N-1)^{-1}\sum{}'\varepsilon_i$$
$$\sigma^2(\underset{\sim}{m},\underset{\sim}{0}_i) = \frac{m(N-1-m)}{(N-1)(N-2)}\{\sum{}'\varepsilon_i^2 - (N-1)^{-1}(\sum{}'\varepsilon_i)^2\}$$
$$d(\underset{\sim}{m-1},\underset{\sim}{1}_i) = \binom{N-1}{m-1};$$
$$E(\underset{\sim}{m-1},\underset{\sim}{1}_i) = (m-1)(N-1)^{-1}\sum{}'\varepsilon_i + \varepsilon_i \rightarrow -(N-m)m^{-1}E(\underset{\sim}{m},\underset{\sim}{0}_i)$$
$$\sigma^2(\underset{\sim}{m-1},\underset{\sim}{1}_i) = \frac{(m-1)(N-m)}{m(N-1-m)}\sigma^2(\underset{\sim}{m},\underset{\sim}{0}_i) \qquad (54)$$

where d is the dimensionality, $\sum'$ is the sum over all single-particle states except #i, and the last form for $E(m-1,1_i)$ emerges when we take $\Sigma\varepsilon_i=0$ (as we may without loss of generality).

These equations yield Gaussian distributions for the two configuration distributions, the total density ρ_m of (49) then being given as the sum of these two (which is rather more accurate than simply using the one-orbit distribution). The overall accuracy will of course be better with the four-moment distributions. The occupancies for this case are given in Table 4, compared with the exact results, for 8 non-interacting particles in 16 single-particle states with unit spacing. The excitation spans 0-64 units; at the midpoint energy both the exact and distribution occupancies are precisely ½ for each state, and there are obvious hole-particle symmetries, $\langle n_i\rangle+\langle n_{(16-i)}\rangle=1$ for fixed E, and $\langle n_i\rangle\rightarrow 1-\langle n_i\rangle$ when $E\rightarrow 64-E$. The agreement is excellent.

Table 4
Single-state occupancies for 8 non-interacting particles in 16 states with unit spacing. The first line in each case gives the exact occupancies, and the second the four-moment approximations

State #	1	2	3	4	5	6	7	8
$\langle n_i\rangle$:E=5	1.00	1.00	1.00	0.86	0.86	0.71	0.71	0.57
	1.00	0.98	0.94	0.88	0.81	0.73	0.64	0.55
$\langle n_i\rangle$:E=15	0.87	0.83	0.78	0.72	0.68	0.62	0.58	0.53
	0.87	0.82	0.78	0.73	0.68	0.63	0.58	0.53
$\langle n_i\rangle$:E=25	0.67	0.63	0.61	0.59	0.56	0.55	0.53	0.51
	0.66	0.64	0.61	0.59	0.57	0.55	0.53	0.51

For non-interacting particles we have the simple test that

$$\sum_i \langle n_i \rangle_E \, \varepsilon_i = E \quad . \tag{55}$$

This is satisfied within 2 units for $10<E<54$ when the Gaussian distributions are used, and within 1 unit for $3<E<61$, and within 0.1 unit for $14<E<50$, when the four-moment distributions are used in the occupancy calculations.

Some studies have been made by Chang[13] for proton and neutron Fermi profiles in very large shell-model systems, and they display many interesting features. An account will be given later.

9. FIXED ENERGY AVERAGES

The general problem of evaluating, in any direct way, averages at fixed multiparticle energy is quite intractable, because the energy specification involves the multiparticle dynamics in an essential way. By using the centroid energies and widths as an "indicator" of the energy, we have in the last section been able to avoid the difficulties and have obtained simple results. In that case indeed things were particularly simple, because, in each of the representations used (the configurations), the operators in question (the number operators) behaved like multiples of unity. In general for Casimir operators of a group defining the representations we could do the same thing, the resulting averages then depending only on the assumption that the individual distributions are accurately enough given.

For more general operators one further <u>condition</u> must be met and one further <u>assumption</u> made. The condition is that the operator whose fixed-energy average we are seeking must not connect two representations, for, if it did, we would get serious perturbations where these overlap. This condition leads to no real difficulty; for example to measure the norm in configurations of a particular irreducible part of the Hamiltonian we need the configuration averages of operators $H^{\nu}(-\mu)\cdot H^{\nu}(\mu)$; this has intermediate states outside the beginning configuration but the left-hand operator returns us to that configuration. The point is that the configurations are eigenspaces for the quadratic operator. The same kind of consideration would apply for example if we were studying the emission of γ rays following particle capture; the γ intensities are given by operators, quadratic in the electromagnetic interaction, which can be decomposed as above.

Of course, if we are interested in more exotic symmetries, more work might be required for the decomposition, but the principle is the same.

The operator is then decomposed into operators for which the representation spaces are eigenspaces. Unlike the occupancy case they will however not in general give unit representations. This is where the necessary assumption comes in. If we assume that we may ignore the energy variation, over the width of a representation, of the operator expectation value, then we have the general result that

$$\langle O\rangle_E = \sum_{\underset{\sim}{\alpha}} \langle O\rangle^{\underset{\sim}{\alpha}} \rho_{\underset{\sim}{\alpha}}(E) / \sum_{\underset{\sim}{\alpha}} \rho_{\underset{\sim}{\alpha}}(E) \tag{56}$$

and it is this which we have used so far, and some results of which we report below. We remark in passing that corrections to this approximation can be made in terms of further averages involving O and H. The average $\langle OH\rangle^{\underset{\sim}{\alpha}}$ will for example tell us whether the states above the $\underset{\sim}{\alpha}$-centroid tend to have larger or smaller expectation values than those below; we could use the resulting number by assuming an energy-linear variation over $\underset{\sim}{\alpha}$. A very little of this has been done so far.

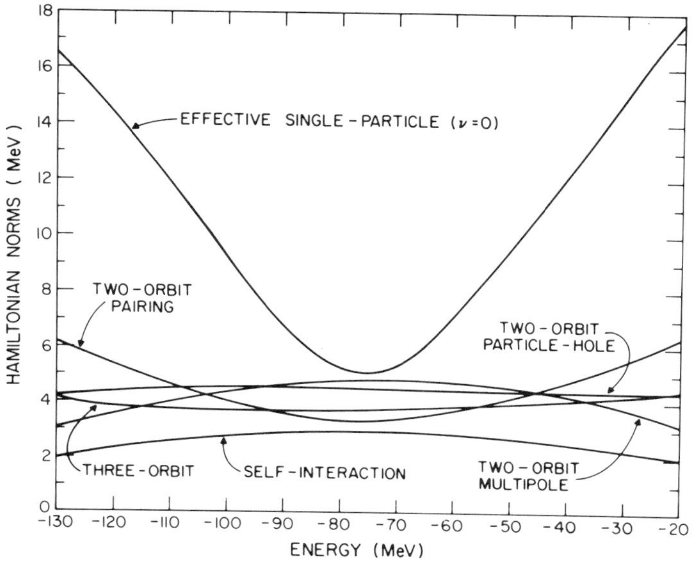

Figure 6
The variation with excitation energy of various orthogonal parts of a Rosenfeld interaction in $(ds)^{12}$ described in terms of three spherical orbits. The ground-state is at approximately -130 MeV.

Figure 6 shows,[3] for a Rosenfeld interaction in $(ds)^{12}$, the energy variation of the norms of the six parts of H, decomposed according to the structures given in §5. A further decomposition according to the orbits involved is of course feasible and in fact comes automatically from the calculations which give Fig.6. Remember that the energy indicated is the actual energy of the system taking account of the interactions. So far very little use has been made of these Hamiltonian decompositions but we envisage using them for comparing features of various interactions (including their renormalizations), in studies of what interactions dominate at various energies, and, more specifically, for evaluating R.M.S. matrix elements for use in studying the compound-nucleus cascade. Along the same line we do plan to study the general features of γ-ray emission at high excitations.

10. LEVEL DENSITIES

To a considerable extent there has been little contact between level density (theory and practice) and more detailed spectroscopy, in large part because the techniques and assumptions of the two domains are so different. This is unfortunate because measurements of level density and related quantities give us a great deal of information which we should know how to make better use of. It should be of major importance for example in the studies of effective interactions. The distributions which we have been constructing are themselves, on the other hand, partial densities, and so we have now a theory which can take account of arbitrary residual interactions and orbital structures and which, as we have seen, is applicable in all energy ranges and therefore capable of making contact with standard spectroscopy.

Since the subject is discussed elsewhere[3,14,15] (though still only in a preliminary way), we consider here only a few aspects of it. The first concerns the spin-cut-off factor σ_J which enters into the J-variation of the level density according to the standard form;

$$\rho(E;J) \simeq \rho(E)\,\frac{2J+1}{2\sigma_J^2}\,\exp -\{(J+\tfrac{1}{2})^2/2\sigma_J^2\}$$
$$\sigma_J^2(E) = \langle J_z^2\rangle_E\ . \qquad (57)$$

These come easily from the elementary central limit theorem. The problem is to evaluate σ_J as a function of the energy E and beyond that to verify the form given for $\rho(E;J)$. We do the first by evaluating the fixed-energy average, exactly as above, taking J_z^2 for the operator O. For the second we consider also the average of J_z^4 giving us then the "excess" correction to the distribution in J. Both of these operations are easy to carry out using the standard fixed-(configuration,isospin) distribution which of course yields also a reference energy from which to measure the excitation and the state density $\rho(E)$ which comes into (57).

As an example of how things turn out, we show in Fig. 7 a comparison of the distribution level densities for $(ds)^{12}$ with J=T=0 and the exact results of Zuker and Soyeur.[7]

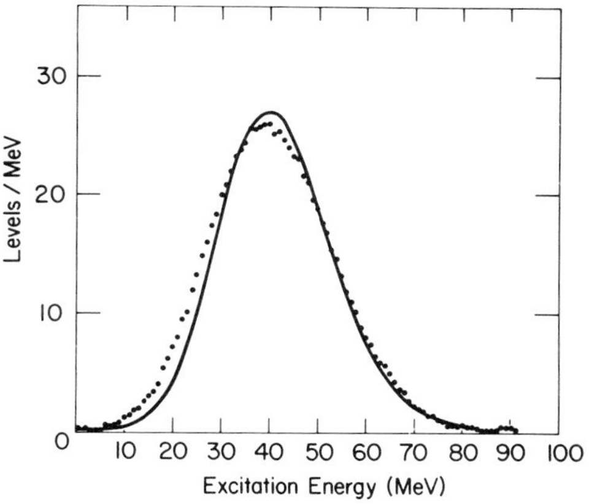

Figure 7
The (J=0,T=0) level density for $(ds)^{12}$ calculated from the configuration-isospin distribution decomposed according to J via the calculated spin-cut-off factor, compared with the exact results.

In this and in other cases the J_z^4 correction is small and so the conventional form is verified. Another example which shows the variation with energy and angular momentum for Cu^{63}, described via three spherical orbits, is given in reference (3). These examples, though complicated from the standpoint of spectra, are negligibly small from the standpoint of level densities. We have recently begun work on a serious study of level densities up to 20 MeV or so (Fig.7, though it extends to 80 MeV, gives only a partial density above, say, 10 MeV, one which is a very small contributor at 20 MeV), in which one finds spaces of dimensionality 10^{12} or larger. Present incomplete results, for Cu^{63} considered in terms of six orbits, are quite encouraging. Even in such large spaces the computing is not very onerous even though some thousands of configurations are encountered. In fact the observation by Chang[13] that the configuration widths show only a very small variation over the entire set of fixed-isospin configuration promises to make the whole thing quite simple, since we should be able to get along without any direct evaluation of the configuration widths at all.

Going in quite the opposite direction we might ask whether the fixed-J densities derived in this simple way are accurate enough to fix the low-lying spectra. We have no right to expect that treating the density as a double Gaussian (in E and J) is accurate enough for this purpose, but there is some evidence that this is meaningful for odd-even nuclei. It is not adequate however for even-even and probably not for odd-odd. With fourth-moment corrections things would improve, but to what extent is not yet clear.

When we consider things at high enough excitation we obviously shall find states of both parities because single-particle states of both parities will necessarily contribute. Moreover, contrary to what often happens at the very lowest energies, the multiparticle states of both parities must be interwoven, since each kind of single-particle state must contribute to each parity. In fact a little consideration of the dimensionalities involved will convince one that, at high enough excitations, the densities of both parities should be approximately equal. The question then, one which is significant in experiment, is how rapidly as we go up in energy does this equality set in in a given case. A separate formal question is whether, for a fixed set of single-particle states of both parities, the distributions of fixed parity might display a Gaussian form. We might expect this, from the fact that parity is a group symmetry and with groups we expect Gaussian.

The answer is "no" to the last question. Parity is determined by a group but it is not a U(N) Lie subgroup, and the earlier arguments for normality give us no reason to expect Gaussian behavior in this case; and in fact fixed parity does not lead to a Gaussian distribution. If, as an example, we go back to Fig.2, and imagine that the upper set of single particle states is of negative parity and the lower set positive, we see that the density of all states is quite unchanged; however k-(hole-particle) states have parity $(-1)^k$, and thus, since their centroids increase with energy, the (k=0-5) configurations produce a set of Gaussians alternating in parity as we go up in energy. There is a "stability" under addition of <u>all</u> the densities (examples are given in Fig. 4 of reference (1), and, for SU(4) representations in Fig. 9 ahead). But, if we take only every second one, we see a Gaussian with Gaussian holes; a plot is given in Fig. 1 of reference (14). For interacting particles the same argument applies, although, when we change from a set of identical-parity orbits to a mixed set, we have to put to zero those matrix elements which would violate parity. Thus the total density will not remain the same, but, since omitting some matrix elements does not change the particle-rank of the interaction, we would still encounter a Gaussian for that.

The practical problem of dealing with fixed parity causes us however no difficulty at all, for, as long as each orbit has a definite parity, so do the configurations and combining them appropriately solves the problem. An example from a preliminary study is given in Fig. 8, which shows, for partial densities

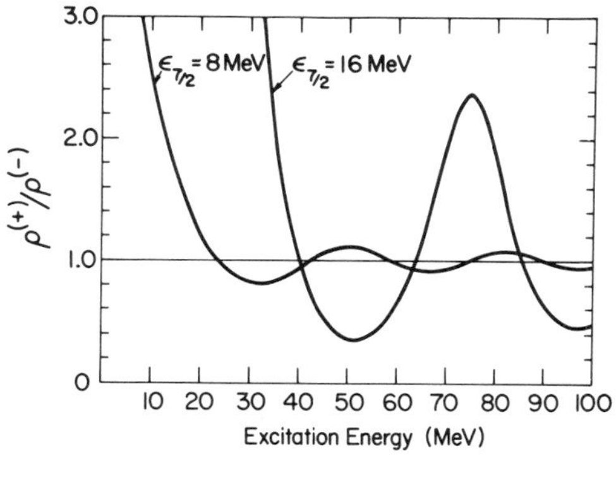

Figure 8
The parity ratio for the densities in $(ds+f_{7/2})^{12}$.The interaction is Brown-Kuo with two values for the $f_{7/2}$-$d_{5/2}$ single-particle difference. The number of states is 5.6×10^9.

in the complete $(ds+f_{7/2})^{12}$ space, the onset of the expected unit ratio for the densities. We plan shortly to investigate the parity ratio in much more detail.

11. MORE GENERAL SYMMETRIES

In dealing with the distribution over all states we learned (46) that the m-particle width for an irreducible operator of unitary rank ν is simply a multiple of the width for ν particles. If the operator is not irreducible, as for example in (24), we encounter a linear combination of widths. These facts, though elementary, are remarkable, for they tell us that certain information about the operator "propagates" throughout the line of fixed-m subspaces without being fragmented. With configurations similarly we see the same process, this time throughout the more general ℓ-dimensional "lattice" of representations. Other kinds of information not represented naturally by traces (which for example could only be expressed in that way by high-rank projection operators) do not propagate; it is perhaps equally clear that, for arbitrary subspaces, not even trace information will propagate undamaged.

We are led then to ask about the kinds of information and the kinds of lattices which will ensure some kind of propagation and about the rules for the propagation. We take for granted right away that the information must be in the form of traces over subsets, and that the latter must be defined by a group or set of groups. But beyond that we encounter a large number of extremely interesting questions, for example about the "geometry" of the lattice structure, most of which do not seem elementary and only a few of which are understood at all.

In the discussion of §5 we have stressed the very close relationship between the $\psi^+(m)\psi(m)$ operator representing the state density and the $\psi(k)\psi^+(k)$ k-body operator. For the purpose of evaluating fixed-m linear traces of a k-body operator $O(k)$, equation (19) shows that O may be represented as

$$O(k) \Rightarrow \langle O(k)\rangle^{\underset{\sim}{k}} \binom{n}{k} . \tag{58}$$

The $\binom{n}{k}$ operator is the only unitary-scalar k-body operator, and thus its appearance here is quite natural. But we have also, for the k-particle density operator, that

$$\rho(k) \equiv \sum_{\alpha} \psi_{\alpha}(k)\psi^+_{\alpha}(k) = \binom{n}{k} . \tag{59}$$

The middle form here defines a k-body operator which is easily seen to give unity on all k-particle states; similarly with $\binom{n}{k}$, and thus the equality. We may ask now by what inversion has the k-particle density operator entered when we dealt with a k-body operator in m-particle space; we would have expected the m-particle density.

The inversion is that of a hole⇄particle transformation. To a state operator $\psi_{\alpha}(m)$ there corresponds a hole-particle complement $\psi_{\alpha_c}(N-m)$, the product, taken in a prescribed order, defining the unique state $\psi(N)$; one of these is then a "template" for the other. If now $(\underset{\sim}{m},\underset{\sim}{\alpha})$ and $(\underset{\sim}{k},\underset{\sim}{\beta})$ are respectively m-particle and k-particle subsets, we make use of the standard hole⇄particle theorem that

$$\langle\psi_{\alpha}(m)O(k)\psi_{\alpha'}(m)\rangle = \langle\psi_{\alpha_c}(N-m)\tilde{O}(k)\psi_{\alpha'_c}(N-m)\rangle \qquad (60)$$

where $\tilde{O}$ is related to O by $A_i \rightleftarrows B_i$ for every single-particle state, this defining a separate kind of hole⇄particle involution. We can now turn the integral inside out and use the h⇄p theorem again. The result is that

$$\langle\langle\rho(\underset{\sim}{k},\underset{\sim}{\beta})\rangle\rangle^{\underset{\sim}{m},\underset{\sim}{\alpha}} = \langle\langle\rho(\underset{\sim}{N}-\underset{\sim}{m},\underset{\sim}{\alpha}_c)\rangle\rangle^{\underset{\sim}{N}-\underset{\sim}{k},\underset{\sim}{\beta}_c} = \langle\langle\tilde{\rho}(\underset{\sim}{N}-\underset{\sim}{m},\underset{\sim}{\alpha}_c)\rangle\rangle^{\underset{\sim}{k},\underset{\sim}{\beta}} . \qquad (61)$$

The traces then may be regarded as measures for the degree to which one space is contained in another, but note that, because of the inversion, which is contained in the other may be regarded as ambiguous.*

We now arrive easily at the result that

$$\langle\langle O(k)\rangle\rangle^{\underset{\sim}{m},\underset{\sim}{\alpha}} = \langle\langle\tilde{\rho}(\underset{\sim}{N}-\underset{\sim}{m},\underset{\sim}{\alpha}_c)O(k)\rangle\rangle^{\underset{\sim}{k}} \qquad (62)$$

is valid for an arbitrary subspace $\underset{\sim}{\alpha}$. Before asking how this is to be calculated we should first extend it to deal with operators of mixed particle rank ≤u, since, as discussed often above, these are what we need for the evaluation of moments. It is clear from (62) that for a general operator whose maximum particle rank is u we shall encounter a trace over all states with particle number ≤u; we label this set as $[\underset{\sim}{u}]$. The problem of finding the form in which the operator rank does not appear is quite analogous to the problem in electrostatics of calculating the potential due to a charged

*This has been noted by Augustus de Morgan in another connection. "If the Northern Hemisphere were land and all the Southern Hemisphere water, ought we to call the Northern Hemisphere an island or the Southern Hemisphere a lake?"

surface in terms of the potential on the surface rather than the charge density. We need in other words a surface Green's function for the defining space $t \leq u$. The result, valid for arbitrary $\underset{\sim}{\alpha}$ and for a general operator of maximum rank $\leq u$ is

$$\langle\langle O \rangle\rangle^{\underset{\sim}{m},\underset{\sim}{\alpha}} = \langle\langle \tilde{\rho}(\underset{\sim}{N}-\underset{\sim}{m},\underset{\sim}{\alpha}_c)\binom{u-m}{u-n} O \rangle\rangle^{[\underset{\sim}{u}]} \tag{63}$$

where the binomial operator may be written also as $(-1)^{n-u}\binom{m-n-1}{u-n}$. To verify this form we show that it is valid for an operator of arbitrary rank $p \leq u$, by using (62) along with the unitary scalar promotion of O, from p to (p+s), given by (31). Then

$$\langle\langle \tilde{\rho} O(p) \rangle\rangle^{\underset{\sim}{p+s}} = \binom{m-p}{s} \langle\langle O(p) \rangle\rangle^{\underset{\sim}{m},\underset{\sim}{\alpha}}; \quad 0 \leq s \leq m-p \tag{64}$$

and thus

$$\sum_{s=0}^{u-p} \langle\langle \tilde{\rho} O(p) \binom{u-m}{u-n} \rangle\rangle^{\underset{\sim}{p+s}} = \langle\langle O(p) \rangle\rangle^{\underset{\sim}{m},\underset{\sim}{\alpha}} \sum_{s=0}^{u-p} \binom{u-m}{u-p-s}\binom{m-p}{s}. \tag{65}$$

The combinatorial sum encountered here equals unity by an application of the standard Vandemonde convolution theorem[16] and thus the verification of (63) is complete. It is worthwhile to rederive some of the earlier results for averages over all m-particle states by using (63) which, we stress again, is completely general. This is left for the reader.

Suppose that the subspace $(\underset{\sim}{m},\underset{\sim}{\alpha})$ is a representation space for a group which may itself be a subgroup of a U(N) subgroup or belong to a more complicated structure of groups. We would then naturally decompose $[\underset{\sim}{u}]$ into a set of subspaces defined in the same way

$$[\underset{\sim}{u}] = \sum_{k=0}^{u} \sum_{\underset{\sim}{\beta}} (\underset{\sim}{k},\underset{\sim}{\beta})$$

and then the right hand side of (63) becomes a sum of $(\underset{\sim}{k},\underset{\sim}{\beta})$ traces. This will define a propagation of information in the sense discussed above, provided that the $\tilde{\rho}$ operator behaves as a multiple of unity in each of the $(\underset{\sim}{k},\underset{\sim}{\beta})$ representations. In other words things are particularly simple when the density operator for one subspace is a constant in each of the others (in which case we might think of the representations as not being "oriented" with respect to each other).

This condition obtains for the direct-sum subgroups which define the partitions, and also for the direct-product subgroups U(N/r) × U(r) which, for r = 2,4 respectively, define isospin and spin-isospin SU(4) symmetry. In these cases the density operators are polynomials in the various Casimir invariants of the system (and of the number operator which plays a similar role for U(N)); an elementary test of the condition that the density operators should be scalars is that, when we write the most general polynomial in these operators which has particle rank ≤k, it should contain a number of terms not less than the number of representations, p(k) say, which exist for (t≤k) particles. The conclusive test is that p(k) of these polynomial terms should be linearly independent on the defining (t≤k)-particle lattice, so that we can in fact construct a complete set of polynomials, each of which gives zero on all but one of the defining representations. Whether or not this procedure works depends of course on the lattice "geometry". When it does the construction is easy and completely solves the propagation problem; then for example we can express the m-particle widths, for the conventional H, in terms of the (1-4)-particle widths.

Explicit constructions for some cases (including identical-particle symplectic symmetry) are given in reference (1), and combinatorial aspects of the general case in which the density operators are scalar in the first paper of (2). An explicit application has been made[17] to SU(4) in the (ds) shell. Something about normality in this case is indicated by Fig. 9

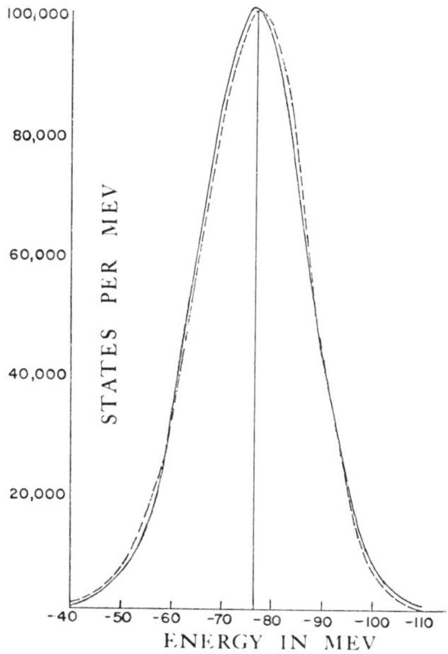

Figure 9
The sum of the Gaussian approximations to the eighteen SU(4) distributions in $(ds)^{12}$ compared with the scalar Gaussian approximation.

which shows, for $(ds)^{12}$, the distribution which results from combining the eighteen fixed-SU(4) distributions, assumed to be Gaussian. The sum is seen to be very close to a true Gaussian, an example of the "additive stability" which we have referred to earlier. In reference (17), estimates of the symmetry admixings in the ground state have been made by considering the contributions to the density near the ground state which arise from the various representations. It has been concluded that realistic interactions give in most cases strong admixings of a few representations.

If one goes further and considers, for example, SU(3) as a subgroup of U(N/4), or isospin as a U(2) subgroup of U(4), the simple behavior is not found. Quite recently some insight has been gained into the structures and forms which are encountered in these cases , and it appears now that some of them can be handled. In one subclass of these more complicated cases the density operators are expressible in terms of group generators though not via Casimir operators, this indicating a relative orientation of the representations. The natural classification of such density operators is then according to their tensorial rank, and we thus find ourselves constructing the tensorial generalizations of the usual operators. Analogous to (58) which expresses the (unitary-scalar) trace-equivalent Hamiltonian, non-scalar equivalences are encountered and these lead in fact to methods for decomposing the width according to the intermediate-state representations which contribute to it. One is dealing here with partial widths, the extension to more general symmetries of the concept encountered earlier for configurations. Since the partial width measures the R.M.S. matrix element connecting two representations, it is obvious that it enters naturally* into measures for the goodness of symmetries; this very large subject we hope to consider seriously before too long. It is hoped also to give then a more complete account of formal aspects of the symmetry problem.

*In the SU(4) case, referred to, we draw conclusions about the goodness of the symmetry without knowledge of the partial widths because we are willing to believe, in that case, that the few lowest-lying states do not belong predominantly to different symmetries. Then if various symmetries are encountered near the ground state they must be admixed.

12. FINAL REMARKS

The things we have discussed fall into three categories which we list now with a comment or two about each.

Fermion-Space Algebra: Things which should be further studied include; a classification of the various group structures which might be relevant; the "orientation" of the representations and the construction of the density operators; the goodness of symmetries; the notion of information* carried in a vector subspace and propagated throughout a lattice of them;

Normality: An elementary account of the source of normality in the simplest case has been given above; the extension to configurations is more or less immediate. The method used however does not give any real understanding of the deep connections between symmetries and statistics, nor does it apply directly to more general groups. The source of normality and the propagation of information are of course intimately connected; note especially the loss of information which is represented by the approach to normality.

Spectroscopy in Huge Spaces**: We have high confidence that our accuracy is good in spaces whose dimensionalities are only a few thousand, but beyond that we see no reason why, with proper partitioning, we should not have adequate accuracy in indefinitely large spaces and the indications here are favorable. Besides level densities and related problems (occupancies, spin and isospin cut-off factors, parity ratios) natural applications are to the study of compound-nucleus formation, of γ cascades following particle capture, of methods of comparing different interactions and the effects of renormalization, and of the energies at which various exotic excitations (multi-particle-hole for example) are first encountered. A report by F. S. Chang[13] giving the formalism and the computing programs required for some of these will shortly be made available.

ACKNOWLEDGMENTS
Almost all of the calculations referred to above, and many other things as well, are due to F. S. Chang. For enlightening discussions and in some cases for permission to borrow from unpublished work, I am

*An interesting discussion of this, in a very general context, is given in reference (18).

**We borrow this title from K. F. Ratcliff.

indebted to F. S. Chang, J. R. Huizenga, D. S. Koltun, P. E. Mugambi, J. C. Parikh, V. N. Potbhare, K. F. Ratcliff, T. H. Thio, F. C. Williams, S. S. M. Wong and A. Zuker. Some things in this account have drawn on lectures given at Panjab University, Chandigarh, India, at the invitation of Prof. H. S. Hans and Dr. R. K. Bansal, and under the auspices of the University, the Indian Council on Science and Technology, and NSF/AID.

REFERENCES

1. J. B. French, in "Nuclear Structure" (A. Hossain, et al, Eds.) North-Holland, Amsterdam (1967).
2. J. B. French, and K. F. Ratcliff, Phys. Rev. C3, 94 (1971); K. F. Ratcliff, Phys. Rev. C3, 117 (1971).
3. F. S. Chang, J. B. French and T. H. Thio, Annals of Physics 66, 137 (1971).
4. For an elementary account see J. H. Wilkinson, "The Algebraic Eigenvalue Problem" Oxford (1965).
5. F. C. Williams, private communication.
6. R. Brout and P. Carruthers, "Lectures on the Many-Electron Problem" Interscience, New York (1963).
7. A. Zuker, private communication, and M. Soyeur and A. Zuker, to be published.
8. S. Kahana, H.C.Lee and C.K. Scott, Phys. Rev. 185, 1378 (1969).
9. S.S.M.Wong, Nucl. Phys. A159, 235 (1970).
10. T.T.S.Kuo and G.E.Brown, Nucl.Phys. A114, 241 (1968), and private communication from T.T.S.Kuo.
11. J. Griffin, Phys. Rev. Letters 17, 478 (1966); C.K. Cline and M. Blann, Nuclear Physics A172, 225(1971).
12. B.Castel, et al, Nucl. Phys. A157, 137 (1970).
13. F.S.Chang, private communication.
14. J.B.French and F.S.Chang, "Proceedings of the Albany Conference on Statistical Properties of Nuclei" (1971, to be published).
15. P. E. Mugambi, Ph. D. Thesis, University of Rochester (1970).
16. J. Riordan, "Combinatorial Identities" Wiley, New York (1968).
17. J.B.French and J.C.Parikh, Physics Letters 35B, 1 (1971); J.C.Parikh, "Lectures on Group Symmetries in Nuclear Structure" Report UR-875-350 (Rochester, 1971), and to be published.
18. A.Akchurin, in "Philosophical Problems of Elementary Particle Physics" Progress Publishers, Moscow(1968).

FOLDED DIAGRAMS*

T.T.S. Kuo
Department of Physics
State University of New York at Stony Brook
Stony Brook, New York

I. INTRODUCTION

I would like to make a report on the subject of folded diagrams and its connection with the model-space eigenvalue problem. My talk will be based mainly on two recent papers of S.Y. Lee, K.F. Ratcliff and myself. I shall refer to these two papers hereafter as KLR(I)[1] and KLR(II)[2] respectively.

In nuclear structure theory, we often solve the nuclear eigenvalue problem in a truncated shell-model space. And as we are all familiar with, the results so obtained very often show extremely good agreements with experiments. These good agreements may not be entirely accidental, and hence we would all like to know why this type of shell-model calculation works so well. Let us look at this problem from the very beginning. Assuming that the nuclear many-body system can be described by the non-relativistic Schroedinger equation with a two-body nucleon-nucleon interaction potential, we should then solve the Schroedinger equation in principle in the entire Hilbert space. In conventional shell-model calculations, the Schroedinger equation is, however, solved in a much smaller Hilbert space which is usually referred to as the model space. Thus to compare these two approaches, our first step is to transform the nuclear Hamiltonian appropriate for the entire Hilbert space to a modified Hamiltonian which is appropriate for the model space. We shall refer to this modified Hamiltonian as the model-space effective Hamiltonian.

A main purpose of the folded diagrams is to facilitate the construction of the model-space effective Hamiltonian H_{eff}. Generally speaking, there are two types of H_{eff}. One is dependent on the energy eigenvalue, and the other is independent of the energy eigenvalue. The energy dependent H_{eff} has been derived by Bloch and Horowitz[3] and Feshbach.[4] It can also be obtained conveniently by using the Green's function method.[5] Because of its dependence on the energy eigenvalue, this type of H_{eff} may not be very convenient in application. For example, we need to use different H_{eff} for different energy levels. Furthermore, the conventional

*Work supported by the U.S. Atomic Energy Commission.

shell-model effective Hamiltonian is energy independent and it has been very successful in reproducing experimental data. For these two reasons, it seems to be highly desirable if we can construct a model-space effective Hamiltonian which is independent of the energy eigenvalue.

The derivation of an energy independent H_{eff} by the inclusion of the folded diagrams has been investigated before by Brandow[6], Baranger and Johnson,[7] Morita[8] and Oberlechner et al.[9]. I suppose that you may have had the same experience. After spending many hardworking hours or days in reading other people's work, one is often inspired to find a different method or formalism of his own. This was exactly what happened to us. Maybe it is due to my own prejudice; I do seem to like our folded-diagram formalism better than the others. The present formalism appears to have some desirable features. You may or may not agree with me. In any case, I will appreciate hearing your comments.

Because of the limitation of time, I shall not compare our formalism with the others mentioned above. It is highly recommended to read the references quoted above in order to compare the various formalisms. I should say that it is the works of Baranger and Johnson[7] and Brandow[6] which have motivated and influenced us the most.

II. MODEL-SPACE EIGENVALUE PROBLEM

We shall define briefly the model-space eigenvalue problem in this section. In nuclear structure calculations, it is common practice to solve the nuclear Schroedinger equation in a restricted Hilbert space which is usually referred to as the model space. For example, the model space for a shell-model calculation of the nucleus ^{18}O is usually taken as the oscillator 0d-1s shell, as shown in Figure 1. The projection operator which projects onto the model space may be written as

$$P = \sum_{i \in D} |\phi_i\rangle\langle\phi_i| \tag{1}$$

where D denotes the model space. ϕ_i is the single particle wave function defined by

$$H_0\phi_i = \varepsilon_i\phi_i \tag{2}$$

where H_0 is the unperturbed Hamiltonian which is related to the total Hamiltonian H by

$$\begin{aligned} H &= H_0 + H_1 \\ H_0 &= T + U \\ H_1 &= V - U \end{aligned} \tag{3}$$

where V is the two-nucleon interaction potential and U is a one-body potential which we can choose at will. The projection operator Q, the orthogonal complement of P, is then defined as

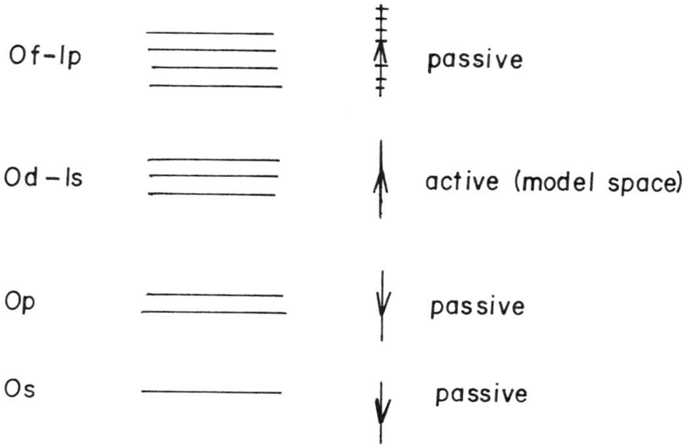

Figure 1

$$Q = 1 - P$$
$$PQ = QP = 0. \tag{4}$$

The Schroedinger equation in the entire Hilbert space is

$$H\Psi_\alpha = E_\alpha \Psi_\alpha. \tag{5}$$

Our purpose is to construct the model-space effective Hamiltonian H_{eff} for the model-space Schroedinger equation

$$H_{eff} P\Psi_\alpha = E_\alpha P\Psi_\alpha \tag{6}$$

where we note that E_α and Ψ_α are the same as those in Eq. (5) and P has been given by Eq.(1). Clearly H_{eff} is not identical to H. Feshbach[4] has shown that

$$H_{eff} = PHP + PHQ \frac{1}{E_\alpha - QHQ} QHP \tag{7}$$
$$= H_{eff}(E_\alpha).$$

This provides a formal relation between H and H_{eff}. We would like to point out that the above H_{eff} is dependent on the energy eigenvalue E_α and, in addition, H_{eff} in its present form is rather difficult to calculate. The model-space effective Hamiltonian derived earlier by Bloch and Horowitz[3] is of the same general structure as shown by Eq. (7). H_{eff} can also be derived by the Green's function method.[5] An advantage of using the Green's function method is that the removal of the unlinked diagrams can be visualized clearly. The H_{eff} derived from the Green's function method is also dependent on the energy eigenvalue. The solution of the model-space Shroedinger equation (Eq.(6)) will be made more complicated when H_{eff} itself is a function of E_α.

A merit of the folded diagrams is the removal of the E_α dependence of H_{eff}. As we shall show later, we have in fact found that by including the folded diagrams many features which are familiar in the conventional shell-model calculations emerge naturally. For example, for the calculation of the nucleus ^{18}F Eq. (6) can be reduced to a form where the eigenvalue is just

$$E_\alpha(^{18}F) + E_o(^{16}O) - E_o(^{17}O) - E_o(^{17}F)$$

where E_o denotes the ground state energy. The above quantity is just the "valence-interaction" energy which can be deduced easily from experimental data.

The main purpose of this paper is thus to derive H_{eff} and $P\Psi_\alpha$ of Eq. (6) using the folded-diagram formalism. Let us first define the folded diagrams.

III. FOLDED DIAGRAMS

The folded diagrams are needed for the factorization of the time development operator U(t,t') when it operates on a model-space basis vector. In the interaction representation, this operator can be written[10] as

$$U(t,t') = 1 = \sum_{n=1}^{\infty} I_n \qquad (8)$$

with

$$I_n = (-i)^n \int_{t'}^{t} dt_1 \int_{t'}^{t_1} dt_2 \ldots \int_{t'}^{t_{n-1}} dt_n \, H_1(t_2)H_1(t_2)\ldots H_1(t_n). \qquad (9)$$

Here we note that we use the Goldstone convention for the relative time orderings of the times $t_1, t_2 \ldots$ and t_n. Namely we use the time sequence $t_1 > t_2 > t_3 \ldots > t_n$, but in the evaluation of the matrix element of U(t,t') we include, of course, all possible contractions.

Let us now operate on a model-space basis vector $|0_j\rangle$ with the time development operator $U(0,-\infty)$. As an example, the vector $|0_j\rangle$ may represent a state composed of two valence particles and a closed ^{16}O core, namely

$$|0_j\rangle = [a_1^+ a_2^+]_j |0\rangle \qquad (10)$$

where $|0\rangle$ represents the unperturbed ^{16}O core state; the subscripts 1 and 2 denote the single-particle states (above the closed shells) and j represents the other quantum numbers necessary in specifying the state. By Eq. (8), (9) and (10), it is clearly seen that the quantity $U(0,-\infty)|0_j\rangle$ can be expressed as a collection of the time-ordered Goldstone diagrams. A typical such diagram is shown by the diagram (a) of Figure 2.

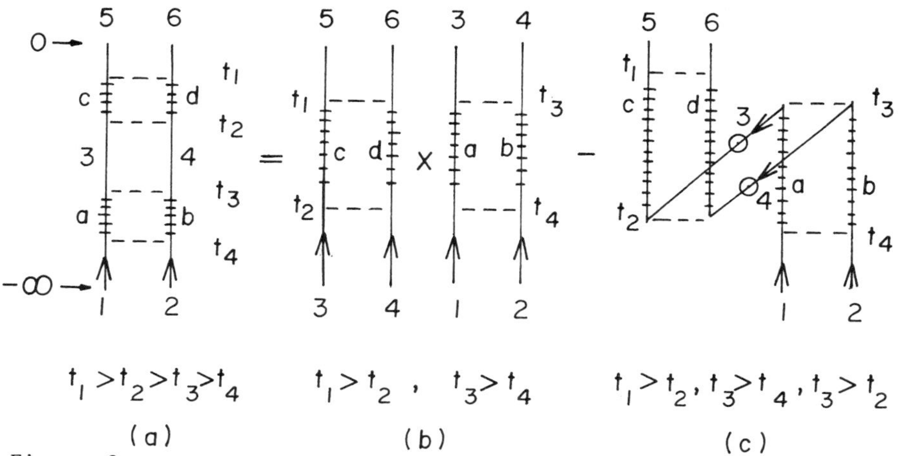

Figure 2

The purpose of Figure 2 is to illustrate the definition of the folded diagram. As shown, the time constraint for diagram (a) is $t_1>t_2>t_3>t_4$. We want to factorize (a) into the product of two diagrams as shown by (b) where there is no time constraint between (t_1,t_2) and (t_3,t_4). Clearly (b) is not equal to (a); (b) has introduced some time-incorrect contributions which must be corrected for. The diagram (c) is to correct for the time-incorrect contributions, so that we will have the diagrammatic identity

$$(a) = (b) - (c).$$

To see this we write down in the following contributions from the three diagrams:

$$(a) = (-i)^4 \int_{-\infty}^{0} dt_1 \int_{-\infty}^{t_1} dt_2 \int_{-\infty}^{t_2} dt_3 \int_{-\infty}^{t_3} dt_4 \exp[-it_1(\varepsilon_c+\varepsilon_d-\varepsilon_5-\varepsilon_6)$$

$$-it_2(\varepsilon_3+\varepsilon_4-\varepsilon_c-\varepsilon_d)-it_3(\varepsilon_a+\varepsilon_b-\varepsilon_3-\varepsilon_4)$$

$$-it_4(\varepsilon_1+\varepsilon_2-\varepsilon_a-\varepsilon_b)] \tag{11}$$

$$(b) = (-i)^2 \int_{-\infty}^{0} dt_1 \int_{-\infty}^{t_1} dt_2 \exp[-it_1(\varepsilon_c+\varepsilon_d-\varepsilon_5-\varepsilon_6)-it_2(\varepsilon_3+\varepsilon_4-\varepsilon_c-\varepsilon_d)]$$

$$\times (-i)^2 \int_{-\infty}^{0} dt_3 \int_{-\infty}^{t_3} dt_4 \exp[-it_3(\varepsilon_a+\varepsilon_b-\varepsilon_3-\varepsilon_4)-it_4(\varepsilon_1+\varepsilon_2-\varepsilon_a-\varepsilon_b)]$$

$$\tag{12}$$

$$(c) = (-i)^4 \int_{\infty}^{0} dt_1 \int_{\infty}^{t_1} dt_2 \int_{t_2}^{0} dt_3 \int_{\infty}^{t_3} dt_4 \exp[-it_1(\varepsilon_c+\varepsilon_d-\varepsilon_5-\varepsilon_6)$$
$$-it_2(\varepsilon_3+\varepsilon_4-\varepsilon_c-\varepsilon_d)-it_3(\varepsilon_a+\varepsilon_b-\varepsilon_3-\varepsilon_4)$$
$$-it_4(\varepsilon_1+\varepsilon_2-\varepsilon_a-\varepsilon_b)]. \qquad (13)$$

We note that in the above equations we have left out the factor

$$\langle 56|V|cd\rangle\langle cd|V|34\rangle\langle 34|V|ab\rangle\langle ab|V|12\rangle$$

which is common to all of them. We would like to point out that Eqs.(11), (12) and (13) have identical integrands; they differ only in the limits of the time integration. We shall refer to diagram (c) as the folded diagram whose value is clearly defined by Eq.(13). It is clearly seen that (a) = (b) - (c). The lines 3 and 4 in diagram (c) of Figure 2 have the appearance of hole lines, but they are in fact folded particle lines. Thus to distinguish them from hole lines we draw small circles on them as shown by the figure.

In Figure 2 we have used the notations defined in Figure 1, namely we draw a railed particle line to represent a passive particle line while the active particle line is represented by the ordinary (undressed) particle line. To facilitate further discussions, let us from now on use the following notations:

 $\equiv$ a propagator which is composed entirely of active lines. This will be referred to as an active propagator. (14)

 $\equiv$ a propagator which contains at least one passive line. This will be referred to as a passive propagator. (15)

As an illustration, Figure 2 becomes under the present notation

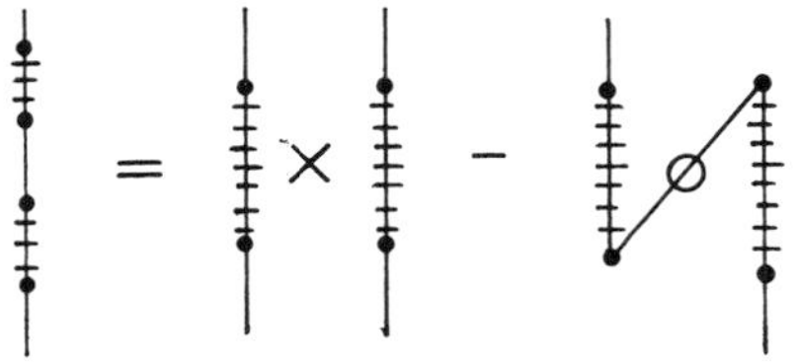

where the circular black dot represents an interaction vertex.

We can now generalize the folding operation illustrated in Figure 2 to obtain the following diagramatic equation:

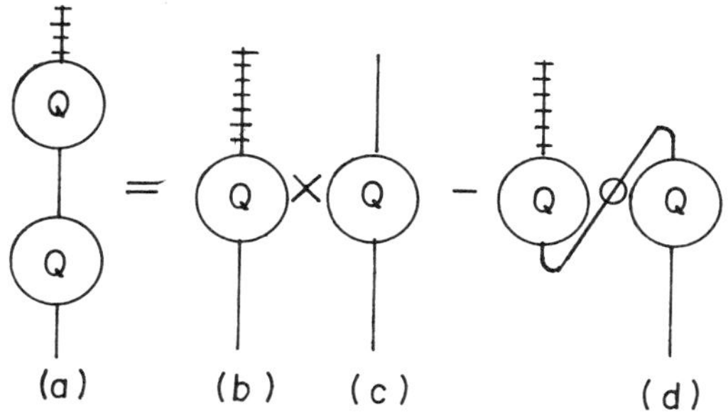

(16)

where the symbol Q represents a collection of any number of irreducible diagrams which contain at least one H_1 vertex and every one of which must be linked to at least one external active line. Here by irreducible we mean that the propagator between any two successive vertices must contain at least one passive line. In addition, we require that the incoming lines (i.e. those lines coming in from below the Q-box) to each Q-box must be composed entirely of active lines. We may refer to the Q-box with the outgoing lines composed also entirely of active lines as the closed Q-box, while that with the outgoing lines containing at least one passive line as the open Q-box. It is clearly seen that both types of Q-boxes are present in Eq. (16).

In figure 3 we show some typical diagrams which belong to the Q-box with two external active particle lines.
A point of interest is the "removal" of divergence by the folding operation. Consider the example shown by Figure 2. If $(\varepsilon_3+\varepsilon_4)$ is equal to $(\varepsilon_1+\varepsilon_2)$, then the diagrams (a) and (b) are both divergent since by Eq. (11) and (12) each diagram contains a factor

$$\frac{1}{(\varepsilon_1+\varepsilon_2) - (\varepsilon_3+\varepsilon_4)} .$$

But the diagram (c) of Figure 2 is not divergent as can be verified by carrying out the integral given in Eq. (13). Thus Figure 2 shows that the folded diagram (c) which is equal to the difference of two divergent terms is a finite quantity.

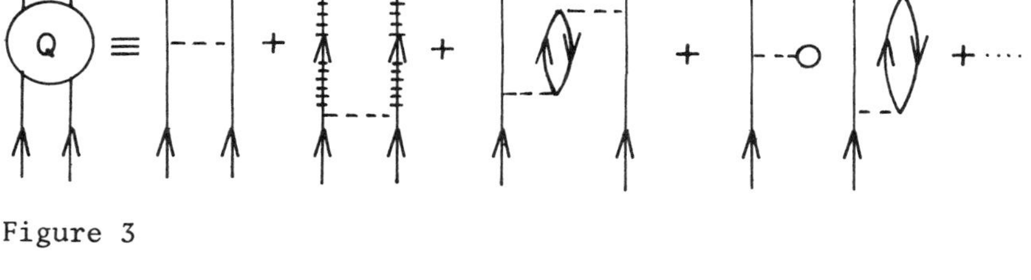

Figure 3

Since we shall use a complete degenerate model space in which all single particle energies are equivalent to each other, the expansion of $U(0,-\infty)|0j>$ will contain many diverging terms. As we will see in the next section, we will "fold" only those propagators which are entirely composed of active lines, so that we can remove all the divergent terms from the effective Hamiltonian.

Let us now compare the folding operation described here with that of Baranger and Johnson.[7] Consider again the example shown in Figure 2. Diagram (a) of the figure consists of 2 irreducible parts with time boundaries t_4 to t_3 and t_2 to t_1 respectively. The factorization of (a) into (b) of the figure extends in fact t_2 to $-\infty$ and t_3 to 0. This introduces time-incorrect contributions and therefore we must include the folded diagram to correct them. Thus in our approach the folded diagram arises from extending each of the irreducible parts of finite time duration in a chain of diagrams to that of infinite time. In the approach of Baranger and Johnson,[7] each of the finite-time irreducible parts in a chain of diagrams is shrunk to that of zero-time by bending the incoming and outgoing lines for each irreducible part to a common time. This of course also introduces time-incorrect contributions, and to correct them we need to include the folded diagrams. We see however that the origins of the folded diagrams of these two approaches are different.

As an illustration, we give in Figure 4 an example which compares the folding operation of Baranger and Johnson and that used in the present paper.

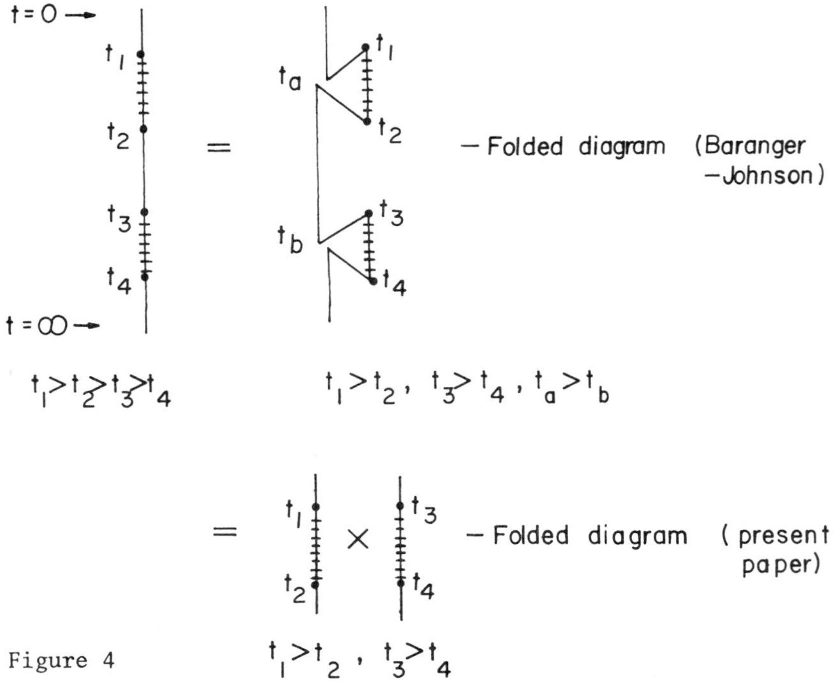

Figure 4

At the beginning of this section, we mentioned that the folded diagrams are needed for the factorization of the time development operator when it operates on a model space basis vector. Let us now discuss this in the following section.

IV. FACTORIZATION OF $U(0,-\infty)|0j\rangle$

In this section we shall discuss how to factorize

$$U(0,-\infty)|0j\rangle$$

where $|0j\rangle$ represents a model-space basis vector, such as the state vector given by Eq. (10). Among the various diagrams contained in $U(0,-\infty)|0j\rangle$, many are unlinked by which we mean that these diagrams contain one or more parts which are not linked to any external active lines. For example, the diagrams (a) and (b) shown in Figure 5 are both unlinked, because they both contain parts which are not linked to any external active (or valence) lines.

As discussed in KLR(I)[1], these unlinked diagrams can be factored into two parts; that which is linked to the external active lines and that which is not linked to the external active lines. This factorization is made possible by adding up all diagrams of identical structures except for the time orderings; this procedure is usually referred to as the generalized time ordering (g.t.o.). There are two groups of such "core" diagrams which can be factorized out. Let us denote them respectively as C_0 and Ψ_c^Q which are defined diagrammatically in Figures 6 and 7.

In these figures, $|0\rangle$ represents the unperturbed core state such as the nuclear model core state for ^{16}O. The quantity C_0 is just a number, but the symbol Ψ_c^Q represents a collection of wave functions which are created before time t=0 and are propagating at time t=0. These wave functions will be referred to as the wave functions "open to the future". The superscript Q is to denote this behavior. We can now write

$$U(0,-\infty)|0j\rangle = C_0 \times |\Psi_c^Q\rangle \times \{\text{diagrams which are linked to at least one external active line}\}. \quad (17)$$

The diagrams which are linked to at least one external active line can be grouped in terms of the Q-boxes defined in the previous section, thus we have the diagrammatic equation shown in Figure 8. The indices n, i, k, ℓ ... in Figure 8 are to be summed over all allowed values. The factorization of $U(0,-\infty)|0j\rangle$ can now be accomplished by folding all the active propagators in the lower parenthesis of Figure 8, and thereby extracting a factor from the lower parenthesis which is equivalent to the upper parenthesis of the same figure. To see this, let us apply the folding operation illustrated in Eq. (16) to the diagrams in the lower parenthesis of Figure 8. Thus we have the result shown in Figure 9.

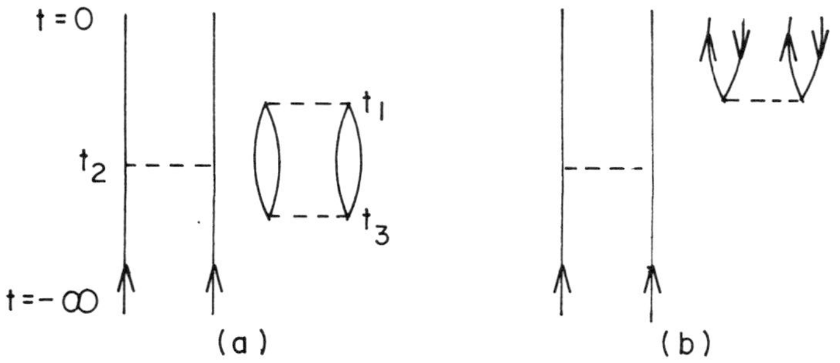

Figure 5

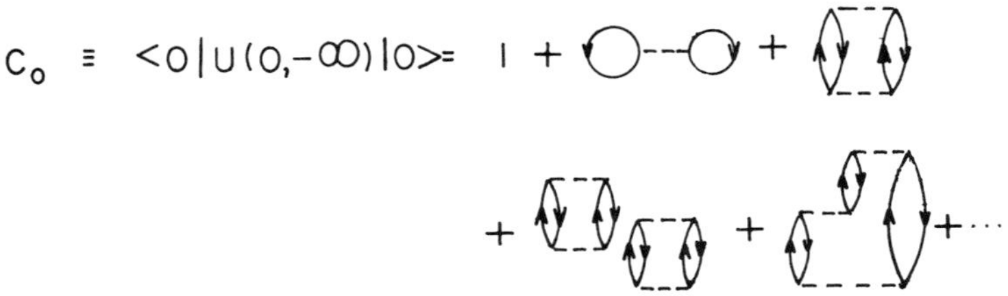

Figure 6

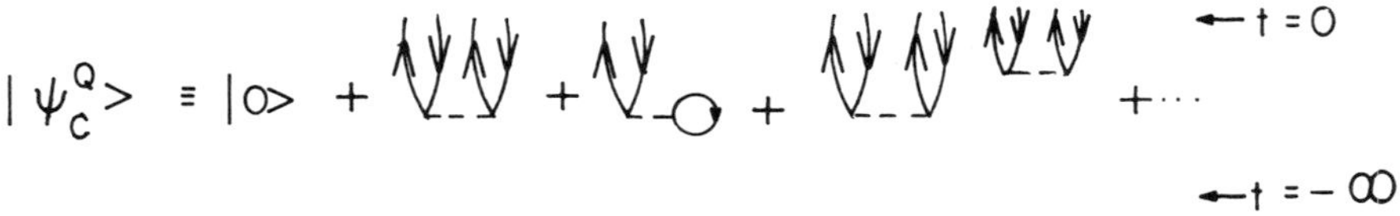

Figure 7

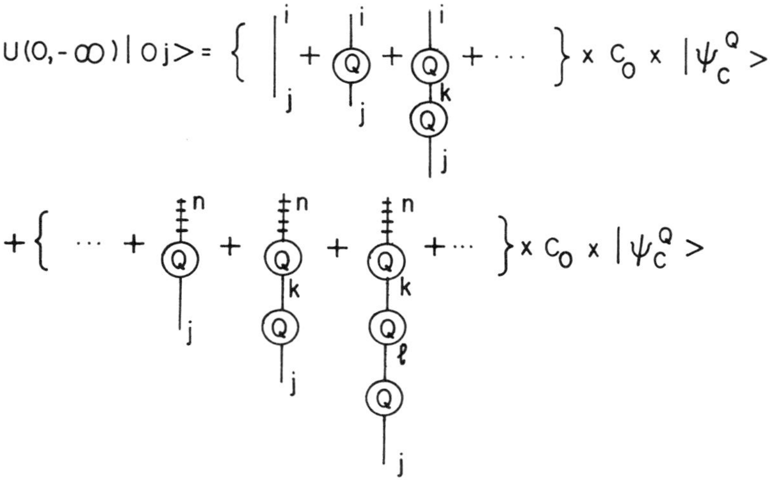

Figure 8

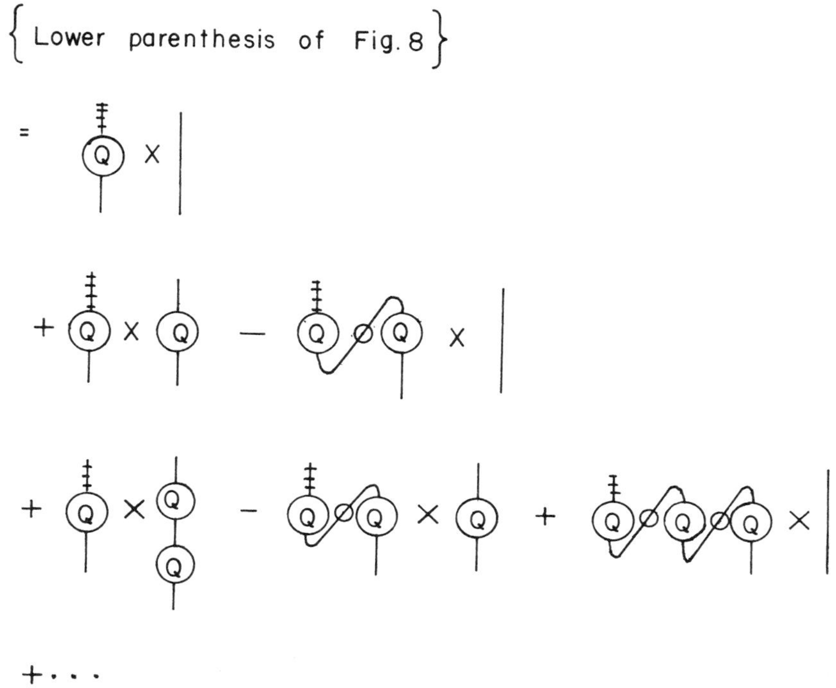

Figure 9

In Figure 9 the bare straight line at the end of each row is just the number 1. By collecting terms columnwise the quantity $U(0,-\infty)|0j\rangle$ can now be written readily as

$$U(0,-\infty)|0j\rangle = \sum_{i} U_Q(0,-\infty)|0i\rangle\langle 0i|U(0,-\infty)|0j\rangle \tag{18}$$

where

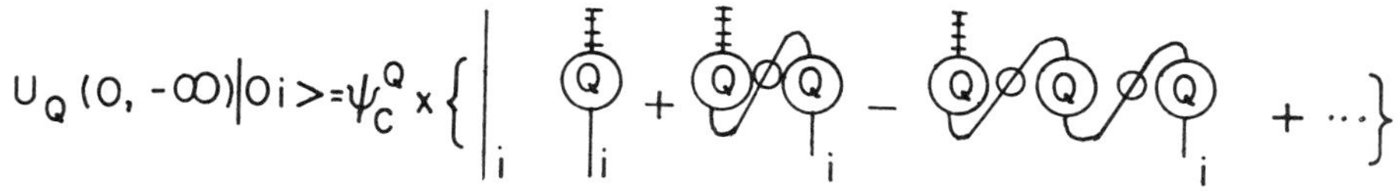

(19)

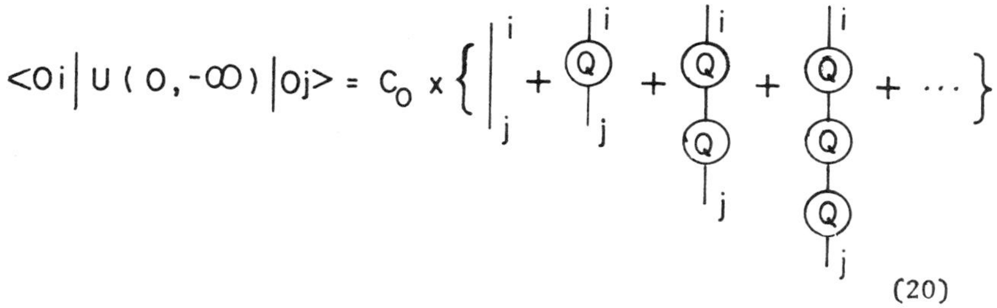

(20)

The above three equations will be referred to as the decomposition theorem. As shown, the quantity $U(0,-\infty)|0j>$ can be factorized into two parts: $U_Q(0,-\infty)|0i>$ and $<0i|U(0,-\infty)|0j>$. The former represents all the wave function diagrams, while the latter is just a c-number. When using a degenerate model space both $U(0,-\infty)|0j>$ and $<0i|U(0,-\infty)|0j>$ contain divergent terms, but $U_Q(0,-\infty)|0i>$ does <u>not</u> contain any divergent terms, as we discussed earlier. Recall that the quantities $\Psi_c{}^Q$ and C_0 appearing in the above equations have been defined in Figure 6 and 7.

V. SECULAR EQUATION

We shall first establish a one-to-one correspondence between Ψ_α, the true eigenfunction of H, and the model-space parent state ϕ_α which is written as

$$\phi_\alpha = \sum_{i=1}^{d} C_i^\alpha |0i>, \qquad \alpha = 1,2 \ldots d \tag{21}$$

where d is the dimensionality of the model space and $|0i>$ represents a model-space basis vector. As we will discuss later, there is no unique way to choose the coefficients C_i^α's. We shall choose them to satisfy

$$\sum_{i=1}^{d} C_i^{\alpha *} C_i^\beta = \delta_{\alpha\beta}. \tag{21-1}$$

As shown in KLR(I), we have

$$|\tilde{\Psi}_\alpha> = \frac{U(0,-\infty)|\phi_\alpha>}{<\phi_\alpha|U(0,-\infty)|\phi_\alpha>} = \frac{|\Psi_\alpha>}{<\phi_\alpha|\Psi_\alpha>}, \qquad \alpha = 1 \ldots d \tag{22}$$

provided that

$$<\phi_\alpha|\Psi_\alpha> \neq 0 \tag{23}$$

$$<\phi_\alpha|\Psi_1> = <\phi_\alpha|\Psi_2> = \ldots = <\phi_\alpha|\Psi_{\alpha-1}> = 0 \tag{24}$$

and $P\Psi_1$, $P\Psi_2$ ... $P\Psi_d$ are linearly independent. Here P is the projection operator defined in Eq. (1). Let us now outline the proof of the above results. The time development operator is given by

$$U(t,t') = e^{-iH(t-t')}. \tag{25}$$

We allow[11] t' to have a small imaginary part as it approaches $-\infty$. Then we have

$$\frac{U(0,-\infty)|\phi_1>}{<\phi_1|U(0,-\infty)|\phi_1>} = \lim_{\varepsilon\to 0^+} \lim_{t'\to-\infty} \frac{e^{-iH[0-t'(1-i\varepsilon)]}|\phi_1>}{<\phi_1|e^{-iH[0-t'(1-i\varepsilon)]}|\phi_1>} \tag{26}$$

where ϕ_1 is a model-space parent state. By inserting a complete set of eigenstates of H in front of $|\phi_1>$ for both the denominator and numerator, the above equation leads to

$$\frac{U(0,-\infty)|\phi_1>}{<\phi_1|U(0,-\infty)|\phi_1>} = \lim_{\varepsilon\to 0^+}\lim_{t'\to-\infty} \frac{\sum_\alpha e^{iE_\alpha t'}e^{E_\alpha \varepsilon t'}|\Psi_\alpha><\Psi_\alpha|\phi_1>}{\sum_\beta e^{iE_\beta t'}e^{E_\beta \varepsilon t'}<\phi_1|\Psi_\beta><\Psi_\beta|\phi_1>} .$$

Now if Ψ_1 is the lowest eigenstate of H and $<\Psi_1|\phi_1> \neq 0$, the real damping factor in the above equation will suppress all the other terms except the term containing Ψ_1 and E_1, namely

$$\frac{U(0,-\infty)|\phi_1>}{<\phi_1|U(0,-\infty)|\phi_1>} = \frac{|\Psi_1>}{<\phi_1|\Psi_1>} \equiv |\tilde{\Psi}_1> . \tag{27}$$

Note that by the lowest eigenstate we mean the eigenstate with its eigenvalue being the most negative (or the least positive if all the eigenvalues are positive). Note also that ε is chosen such that $t'\varepsilon \to -\infty$ when $t' \to -\infty$ and $t'\varepsilon \to 0$ when t' is finite. Thus starting from ϕ_1 we can construct $\tilde{\Psi}_1$ which is proportional to Ψ_1 as shown by Eq. (27), and Ψ_1 is the lowest eigenstate of H with $<\phi_1|\Psi_1> \neq 0$.

To continue this procedure, we construct another parent state ϕ_2 such that

$$<\phi_2|\Psi_1> = 0. \tag{28}$$

Then by repeating the above procedure, we readily obtain

$$\frac{U(0,-\infty)|\phi_2>}{<\phi_2|U(0,-\infty)|\phi_2>} = \frac{|\Psi_2>}{<\phi_2|\Psi_2>} \equiv |\tilde{\Psi}_2> \tag{29}$$

provided that

$$<\phi_2|\Psi_2> \neq 0. \tag{30}$$

In order to satisfy Eq. (28) and (30) simultaneously we must have $P\Psi_1$ and $P\Psi_2$ linearly independent. By continuing this procedure, the general results given by Eq. (22), (23) and (24) can be easily established. The parent states, ϕ_α, can be chosen to satisfy Eq. (21-1) when $P\Psi_1, \ldots P\Psi_d$ are linearly independent.

It should be emphasised that the d eigenstates so constructed are generally not the lowest d eigenstates of H; they are instead the d eigenstates of H which have non-zero overlap with the chosen model space.

Since $\tilde{\Psi}_\alpha$ is proportional to Ψ_α, we have clearly

$$H\tilde{\Psi}_\alpha = E_\alpha \tilde{\Psi}_\alpha . \tag{31}$$

Then by Eqs.(22) and (21) we can rewrite the above equation as

$$H\frac{\sum_i U(0,-\infty)|0i\rangle C_i^\alpha}{\langle\phi_\alpha|U(0,-\infty)|\phi_\alpha\rangle} = E\frac{\sum_i U(0,-\infty)|0i\rangle C_i^\alpha}{\langle\phi_\alpha|U(0,-\infty)|\phi_\alpha\rangle} \tag{32}$$

Here it is not appropriate to remove the denominators $\langle\phi_\alpha|U(0,-\infty)|\phi_\alpha\rangle$ from both sides of the above equation, because neither $U(0,-\infty)|\phi_\alpha\rangle$ nor $\langle\phi_\alpha|U(0,-\infty)|\phi_\alpha\rangle$ is finite by itself. Only the ratio of them is a well defined quantity. By applying the decomposition theorem (Eq. (18)) to the above equation, we have

$$H\,\frac{\sum_{ij} U_Q(0,-\infty)|0j\rangle\langle 0j|U(0,-\infty)|0i\rangle C_i^\alpha}{\langle\phi_\alpha|U(0,-\infty)|\phi_\alpha\rangle}$$

$$= E_\alpha\,\frac{\sum_{ij} U_Q(0,-\infty)|0j\rangle\langle 0j|U(0,-\infty)|0i\rangle C_i^\alpha}{\langle\phi_\alpha|U(0,-\infty)|\phi_\alpha\rangle}. \tag{33}$$

After defining

$$b_j^\alpha = \frac{\sum_i \langle 0j|U(0,-\infty)|0i\rangle C_i^\alpha}{\langle\phi_\alpha|U(0,-\infty)|\phi_\alpha\rangle}, \tag{34}$$

Eq. (33) becomes

$$H\sum_j U_Q(0,-\infty)|0j\rangle b_j^\alpha = E_\alpha \sum_j U_Q(0,-\infty)|0j\rangle b_j^\alpha. \tag{35}$$

By Eq.(19) and Figure 7, it is readily seen that

$$\langle 0k|U_Q(0,-\infty)|0j\rangle = \delta_{kj}\ . \tag{36}$$

Thus by multiplying Eq.(35) by $\langle 0k|$, it becomes

$$\sum_j \langle 0k|HU_Q(0,-\infty)|0j\rangle b_j^\alpha = E_\alpha b_k^\alpha \tag{37}$$

which is a model-space secular equation of the form shown by Eq.(6) if b_j^α is the projection of the true eigenfunction onto the model space. This is indeed so, since from Eqs. (21), (22), (18) and (34) we have

$$|\tilde{\Psi}_\alpha\rangle = \sum_j U_Q(0,-\infty)|0j\rangle b_j^\alpha\ . \tag{38}$$

Then by Eq.(36), we have

$$\langle 0k|\tilde{\Psi}_\alpha\rangle = b_k^\alpha\ . \tag{39}$$

Thus Eq. (37) is indeed of the form of Eq.(6), and hence the model-space effective Hamiltonian is given by

$$H_{eff} = HU_Q(0,-\infty). \tag{40}$$

Before we analyze the structure of Eq. (37) and H_{eff} shown above, we wish to point out that since $HU_Q(0,-\infty)$ is not hermitian the eigenfunctions of Eq.(37) are generally not orthogonal to each other for non-degenerate eigenvalues. Furthermore, the b_k's should not be normalized to one since they are just proportional to the projections of Ψ_α. It is Ψ_α which should be normalized to one. Thus we have generally

$$\sum_k b_k^{\alpha *} b_k^{\beta} \neq \delta_{\alpha\beta} \quad . \tag{41}$$

Once the b_k^α's are obtained, we can formally solve for the C_k^α's from Eq. (34), recalling that ϕ_α is given by Eq. (21). In practice, however, we hardly need to know the C_k^α coefficients. Thus we don't need to solve for them from Eq. (34). It is clearly very complicated to solve for the coefficients C_k^α from Eq. (34).

VI. EFFECTIVE INTERACTION

As shown by Eq. (37), we have derived the model-space secular equation with the effective Hamiltonian H_{eff} given by $HU_Q(0,-\infty)$. We would like to know whether or not this H_{eff} is independent of E_α. This can be answered by explicitly evaluating some matrix elements of H_{eff}. We shall do this in section (6.1). In section (6.2), we shall discuss the separation of the core energy from Eq. (37). In section (6.3) we shall discuss the removal of the stretchable diagrams from $H_1{}^{eff}$. In section (6.4), we shall discuss the separation of the one-body energies from the model-space secular equation. Finally, in section (6.5), we shall discuss a method for summation of the Q-box series, of which the effective interaction is composed.

6.1 Matrix Elements of H_{eff}

We consider the matrix element

$$\langle 0k|HU_Q(0,-\infty)|0j\rangle.$$

Recall that $H = H_0 + H_1$. Since $|0k\rangle$ is an eigenstate of H_0, we have by Eq. (36)

$$\langle 0k|H_0U_Q(0,-\infty)|0j\rangle = \delta_{kj}\langle 0k|H_0|0k\rangle \tag{42}$$

where the last matrix element is just the unperturbed energy for the state $|0k\rangle$. To discuss the matrix element of $H_1U_Q(0,-\infty)$, let us consider a case with two valence particle lines, such as the calculation for ^{18}O. As shown by Eq.(19), the various terms at $t = 0$ generated by operating $U_Q(0,-\infty)$ on $|0j\rangle$ are all passive states except the free propagator $|0j\rangle$. Since the passive states are orthogonal to $|0k\rangle$, a large number of the terms contained in $U_Q(0,-\infty)|0j\rangle$ will give zero contribution to

$$\langle 0k|H_1U_Q(0,-\infty)|0j\rangle.$$

Some non-vanishing matrix elements are shown in Figure 10 where each dotted-line vertex represents a H_1(= V-U) interaction. The t = 0 arrow is used to indicate the last H_1 interaction at t = 0. Note that all the interactions contained in $U_Q(0,-\infty)$ are at times earlier than t = 0, and we only integrate over the time variables contained in $U_Q(0,-\infty)$ as can be seen from Eqs. (8) and (9). Thus the contributions from (i) and (ii) of Figure 10 are

$$(i) = \langle cd|H_1|ab\rangle \qquad (43)$$

$$(ii) = \sum_{p_1,p_2,h_1,h_2} (-1)^1 \frac{\langle ch_2|H_1|p_1p_2\rangle\langle h_1f|H_1|h_1a\rangle\langle p_1p_2|H_1|fh_2\rangle}{(\varepsilon_a+\varepsilon_b-\varepsilon_{p_1}-\varepsilon_{p_2}+\varepsilon_{h_2})\ (\varepsilon_b-\varepsilon_{p_1}-\varepsilon_{p_2}+\varepsilon_{h_2}+\varepsilon_f)} . \qquad (44)$$

The above result can be easily obtained by carrying out the time integrals similar to those shown in Eqs. (11), (12) and (13), remembering that the time t = 0 is not to be integrated over. We note that in diagram (ii), f is a folded particle line. It should be emphasized that, because of the nature of $U_Q(0,-\infty)|0j\rangle$ as shown by Eq. (19), the propagators between any two successive interactions must contain at least one passive line. The passive line in (ii) is h_2, noting that the folded active line f is still counted as active. The sign $(-1)^1$ in Eq. (44) is because diagram (ii) is a <u>once</u> folded diagram, i.e. it arises from the third term in the parenthesis of Eq. (19) and this term is preceded by a minus sign. In general, there is a sign factor of $(-1)^n$ associated with a diagram folded n-times. When the diagrams are expressed in terms of the Q boxes which contain no folded lines within each Q box, the number of times of folding can be easily determined. But there may be some confusions in determining this number when diagrams are not expressed in terms of the Q-boxes. We shall discuss how to avoid this difficulty in section (6.5).

As shown by Eq. (44), the matrix elements of H_{eff} do not depend on E_α; they depend on the single particle energies, ε, defined in Eq. (22). Thus our H_{eff} is independent of the energy eigenvalue E_α.

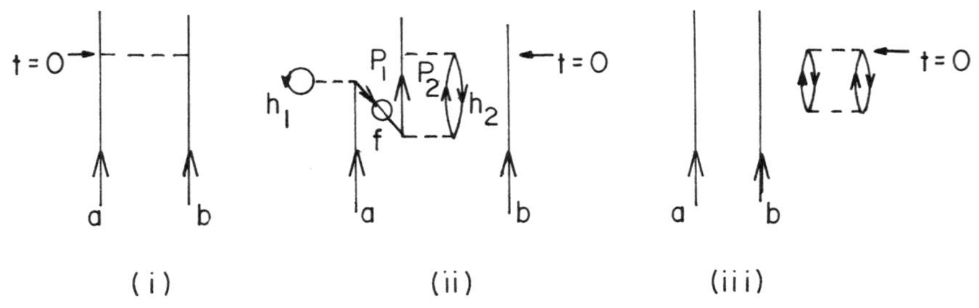

Figure 10

The diagram (iii) of Figure 10 belongs to the diagonal matrix element $<0k|H_1 U_Q(0,-\infty)|0k>$ where k represents the valence particles a and b. This diagram arises from attaching the last H_1 interaction at t = 0 entirely to $\Psi_c{}^Q$ of Eq. (19). We shall see in the next section that this type of diagram can be separated out as a whole to form the core energy which is, for example, just the true ground state energy of the ^{16}O nucleus.

6.2 Separation of Core Energy

Although all of the diagrams contained in a Q box are linked to at least one external active line, the quantity $U_Q(0,-\infty)|0i>$ as shown by Eq. (19) does contain parts which are not linked to any external active line. This is because of the presence of the $\Psi_c{}^Q$ factor in Eq. (19). Some such unlinked diagrams contained in $U_Q(0,-\infty)|0i>$ are shown in Figure 11. For the matrix element $<0k|H_1 U_Q(0,-\infty)|0j>$, diagram (a) of Figure 11 can give rise to the two types of diagrams seen in Figure 12.

Diagram (c) arises from attaching the H_1 interaction at t = 0 entirely to $\Psi_c{}^Q$, while diagram (d) arises from using the H_1 interaction to link up the free propagator and $\Psi_c{}^Q$. Diagram (b) of Figure 11 will give zero contribution to $<0k|H_1 U_Q(0,-\infty)|0j>$. We shall now show that the sum of all the diagrams arising from attaching H_1 to $\Psi_c{}^Q$ will give rise to the core energy.

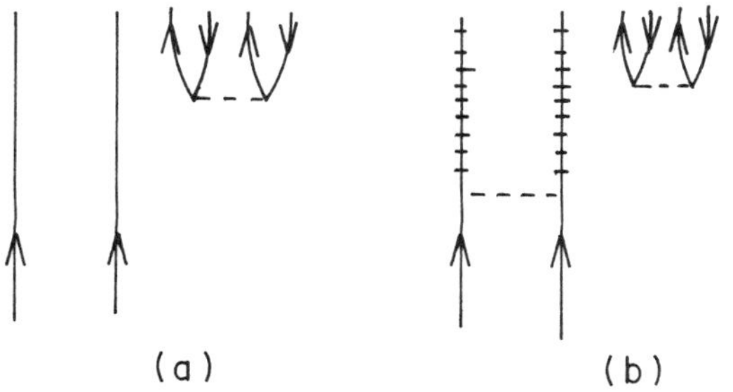

Figure 11

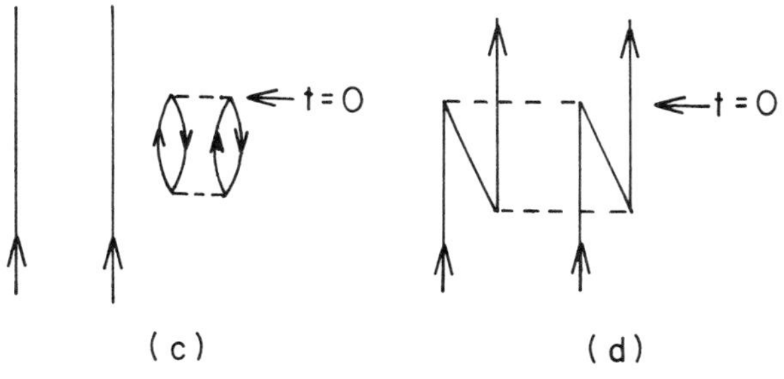

Figure 12

Similar to the way we proved Eqs. (22), (23) and (24), we can easily show that

$$|\Psi_c^Q\rangle = \frac{U(0,-\infty)|0\rangle}{\langle 0|U(0,-\infty)|0\rangle} = \frac{|\Psi_c\rangle}{\langle 0|\Psi_c\rangle} \tag{45}$$

if $\langle 0|\Psi_c\rangle \neq 0$. Here $|0\rangle$ is the unperturbed ground state of the core and $|\Psi_c\rangle$ is the true ground state of the core. As we mentioned earlier, the core may refer to the ^{16}O nucleus, for example. Thus $|\Psi_c^Q\rangle$ is proportional to the true ground state and we have

$$H\Psi_c^Q = E_c\Psi_c^Q \tag{46}$$

where E_c is the true ground state energy of the core.

The diagrams contained in Ψ_c^Q are shown in Figure 7. To show that these diagrams are consistent with the definition of Ψ_c^Q given in Eq. (45) we need to employ a decomposition theorem for $U(0,-\infty)|0\rangle$, namely

$$U(0,-\infty)|0\rangle = U_Q(0,-\infty)|0\rangle\langle 0|U(0,-\infty)|0\rangle. \tag{47}$$

We shall not carry out the proof of it here, since it can be proved readily using the same procedure as used for Eq. (18). Thus from Figure 7, we have

$$\langle 0|\Psi_c^Q\rangle = 1. \tag{48}$$

It then follows from Eqs. (46) and (47) that

$$\langle 0|H_1|\Psi_c^Q\rangle = E_c - \langle 0|H_0|0\rangle. \tag{49}$$

The matrix element $\langle 0k|H_1U_Q(0,-\infty)|0j\rangle$ clearly contains a term

$$\delta_{kj}\langle 0|H_1|\Psi_c^Q\rangle,$$

and hence it follows from Eq. (49) that we can write Eq. (37) as

$$\langle k|H_0|k\rangle b_k^\alpha + \sum_j \langle 0k|[H_1U_Q(0,-\infty)]_L|0j\rangle b_j^\alpha = (E_\alpha - E_c)b_k^\alpha \tag{51}$$

where $\langle k|H_0|k\rangle$ represents merely the unperturbed energy of the valence particles and

$$\langle 0k|[H_1U_Q(0,-\infty)]_L|0j\rangle \equiv \langle 0k|H_1U_Q(0,-\infty)|0j\rangle - \langle 0|H_1|\Psi_c^Q\rangle\delta_{kj}\ . \tag{52}$$

Here the subscript L denotes that all the diagrams must be linked to at least one external active line. Defining

$$H_1^{eff} \equiv [H_1 U_Q(0,-\infty)]_L, \tag{53}$$

the diagrams contained in $\langle k|H_1^{eff}|j\rangle$ can be expressed diagrammatically as

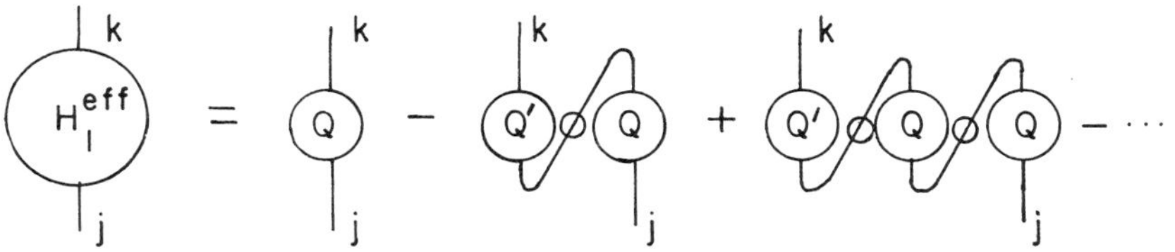

(54)

where we recall that the Q-box is composed of irreducible diagrams which are linked to at least one external active line. In addition, we note that both the incoming and outgoing lines of each Q box are now active. Note that there is no folded line within any Q box. The Q' box in Eq. (54) is different from the Q box in that the lowest order term in the former has two H_1 interactions while the lowest order term in the latter has one H_1 interaction. From Eqs. (19) and (52), it is seen that the relation between Q and Q' can be expressed symbolically as

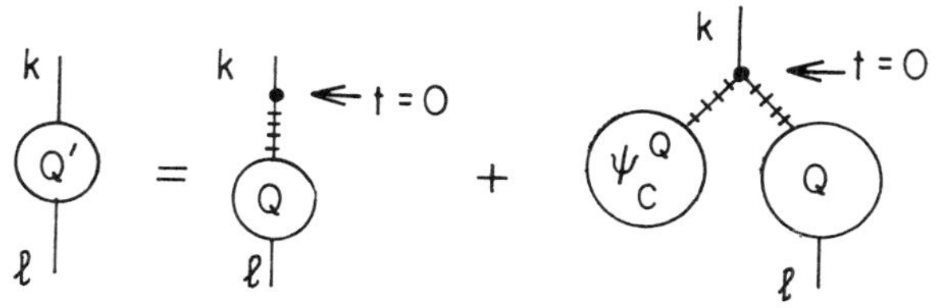

(55)

where the second term arises from linking up Ψ_C^Q, the open Q box and the final active propagator k with the last H_1 interaction at $t = 0$.

Thus we obtain a model space secular equation given by Eq. (51) where the eigenvalue is $E_\alpha - E_C$ rather than E_α; E_C is the core energy. For example, if we are calculating the energies of ^{18}O, the eigenvalue of Eq. (51) will be the energy of ^{18}O relative to the true ground state energy of ^{16}O.

6.3 Stretchable Diagrams

All of the diagrams contained in H_1^{eff} are irreducible and must be linked to at least one external active line. Some of these diagrams are, however, stretchable in the sense that they can be made reducible by stretching and/or shrinking any number of the free propagator lines within the time boundaries $-\infty$ and 0 and without bending any propagator line. By without bending we mean that a particle is not allowed to become a hole line as a result of the stretching and/or shrinking operation, and vice versa for the hole line. Consider now some examples shown in Figure 13.

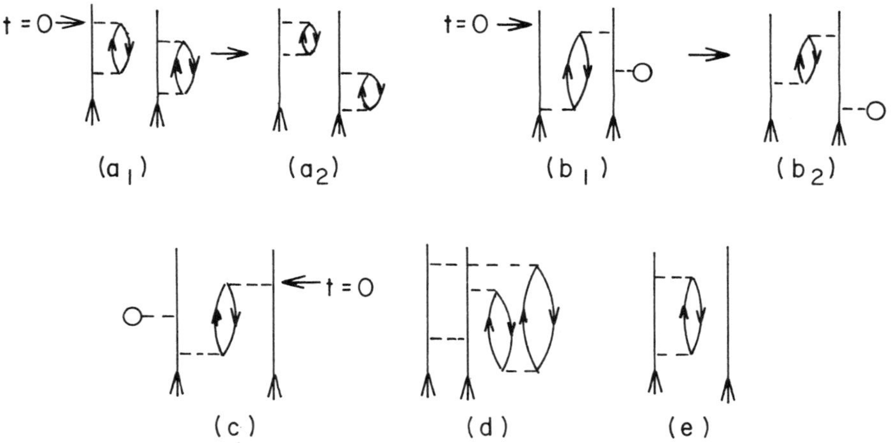

Figure 13

Here diagram a_1 is clearly a stretchable diagram, since it can be stretched into diagram a_2 which is reducible. Similarly, diagrams b_1 and d are strerchable. But diagram c is not stretchable, because to make it reducible we must stretch the bubble insertion past the t = 0 boundary which is not allowed. Diagram e is also not stretchable.

It can be shown that in the expansion of H^{eff} as shown by Eq. (54), all of the stretchable diagrams upon summation will cancel each other out. Thus, although in each of the Q, Q'-Q, Q'-Q-Q... terms there are stretchable diagrams, the net result will contain no stretchable diagrams when we sum up the whole series. Because of the limitation of time, I shall not prove this result here. A detailed discussion of the proof can be found in KLR(II).[2]

For simple cases, the above cancellation of stretchable diagrams can be easily visualized. Consider the case of diagram a_1 in Figure 13. In the expansion of H_1^{eff} shown by Eq. (54), the Q-box term contains diagram a_1. But the Q'-Q term contains also a term which is exactly equal to a_1 but with opposite sign since

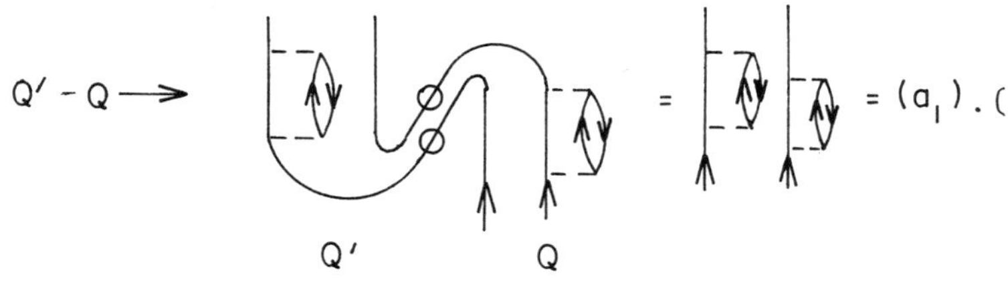

(56)

Thus a_1 is cancelled by the corresponding contribution from the Q'-Q term. In obtaining the equality (56), we have made use of the following property for free propagators

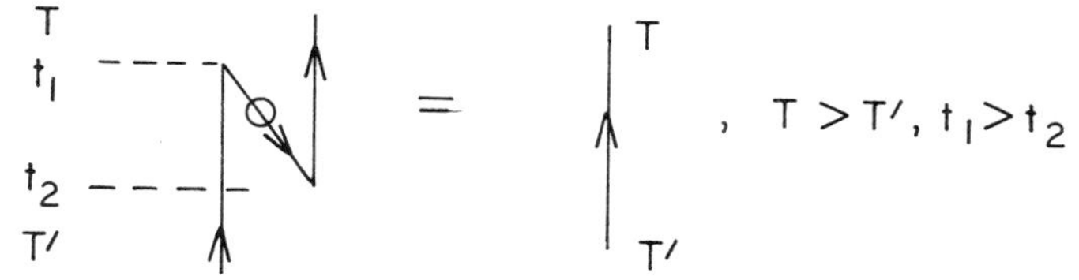

$$, \quad T > T', \ t_1 > t_2 \qquad (57)$$

This is simply because

$$e^{-i\varepsilon_k(T-t_1+t_1-t_2+t_2-T')} = e^{-i\varepsilon_k(T-T')} . \qquad (58)$$

. Let us now define the valence-disconnected diagram. If a diagram is composed of more than one disconnected part, each of which contains at least one H_1 interaction and is linked to at least one external active line, we call this diagram a valence disconnected diagram. As an example, diagram a_1 of Figure 13 is a valence-disconnected diagram. The valence-disconnected diagrams are obviously stretchable, and hence the expansion of H_1^{eff} does not contain any valence-disconnected diagrams. This result is very helpful in simplifying the model-space secular equation.

6.4 Separation of 1-Body Energies

As an example, let us consider the case of ^{18}F. For it, we take the model space as composed of a valence proton and neutron in the s-d shell. Then the effective interaction for it can be divided into the following four types of diagrams:

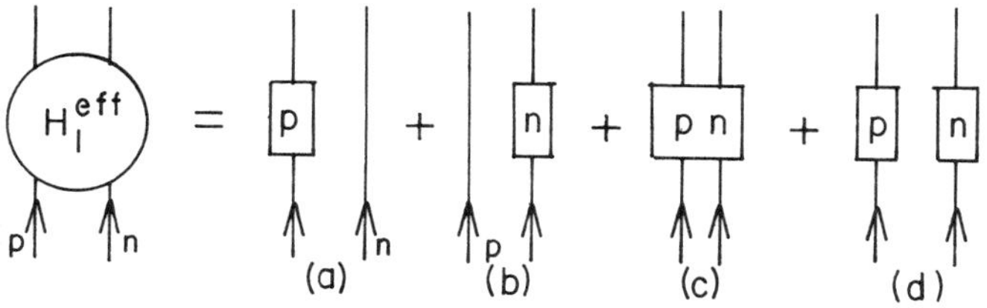

$$H_1^{eff} = \boxed{p} + \boxed{n} + \boxed{p\ n} + \boxed{p}\boxed{n} \qquad (59)$$

$$(a) \qquad (b) \qquad (c) \qquad (d)$$

Here (a) and (b) are of one-body nature, namely all interactions are attached to one active line, leaving the other free. (c) is the two-body part of H_1^{eff} where the two active lines are connected by at least one interaction. (d) is composed of 2 disconnected one-body parts and hence is stretchable. Thus (d) is identically zero, i.e. we don't need to calculate any diagrams of this category.

Suppose now we are solving Eq. (51) for ^{17}O employing a model space composed of one basis vector. Then we readily see

$$E_0(^{17}O) - E_c(^{16}O) = \varepsilon_n + n \qquad (60)$$

where the n-box is identical to (b) in (59). E_0 is the lowest eigenenergy of ^{17}O with non-vanishing overlap between the corresponding true eigenfunction and the model-space basis vector. E_c is the ground state energy of ^{16}O. Thus we can group terms (a) and (b) of (59) respectively with the corresponding unperturbed single particle energies to form the true energies of ^{17}F and ^{17}O.

Thus for ^{18}F, we can write Eq. (51) as

$$[(E_p-E_c) + (E_n-E_c)]b^{\alpha}_{pn} + \sum_{p'n'} <pn|H_1^{eff}(2)|p'n'>b^{\alpha}_{p'n'}$$

$$= (E_{\alpha}-E_c)b^{\alpha}_{pn} \quad (61)$$

where E_p and E_n are correspondingly the lowest eigenenergies of the ^{17}F and ^{17}O systems consistent with the chosen model-space basis vectors. E_{α} is the energy of the ^{18}F system and E_c is the ground state energy of ^{16}O. H_1^{eff} (2) refers to the two-body part of the effective interactions, namely term (c) of Eq. (59). Eq. (61) is a remarkable result. It provides a relation among the eigenenergies of ^{16}O, ^{17}O, ^{17}F and ^{18}F. The eigenvalue of it is indeed the conventional valence-interaction energy mentioned at the end of section II.

Eq. (61) is very similar to the conventional shell-model secular equation where the experimental "single-particle" energies of ^{17}F and ^{17}O are employed in the calculation of the spectrum of ^{18}F. There is, however, a rather significant difference. E_p and E_n in (61) are the true energies of the ^{17}O and ^{17}F systems, rather than the single-particle energies in the conventional sense. For example, E_n should be the lowest ^{17}O experimental energies of $j = 5/2^+$, $1/2^+$ and $3/2^+$ for a s-d shell calculation of ^{18}F. But in the conventional shell-model calculations, E_n is deduced from an average of the experimental spectrum over the single-particle spectroscopic factors. This averaging procedure does not seem to be necessary in the present formalism.

We can easily generalize Eq.(61) to the case with more than 2 valence particles. For example, when we have 3 valence particles, the secular equation is simply

$$[(E_1-E_c) + (E_2-E_c) + (E_3-E_c)]b^{\alpha}_{123} + \sum_{1'2'3'} <123|H_1^{eff}(2)$$

$$+ H_1^{eff}(3)|1'2'3'>b^{\alpha}_{1'2'3'} = (E_{\alpha}-E_c)b^{\alpha}_{123} \quad (62)$$

where $H_1^{eff}(2)$ is the 2-body effective interaction and $H_1^{eff}(3)$ is the 3-body effective interaction in which all 3 valence lines are connected with each other via interactions. For instance the diagram in Figure 14 is a diagram belonging to $H_1^{eff}(3)$. The meanings of the other notations in Eq. (62) are self-explanatory.

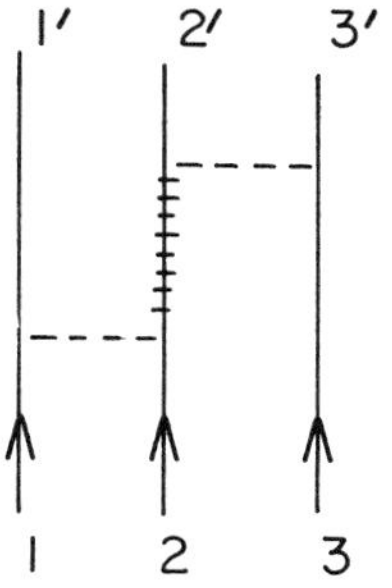

Figure 14

6.5 Summation of Folded Diagrams

As shown by Eq. (54), H_1^{eff} is given as a series of Q and Q' boxes. It will be very useful if there is a convenient method with which we can sum up the series so that the calculation of H_1^{eff} becomes simplified. I shall describe here the method proposed in KLR(II).[2]

We can write symbolically the diagrams contained in a Q box for H_1^{eff} as

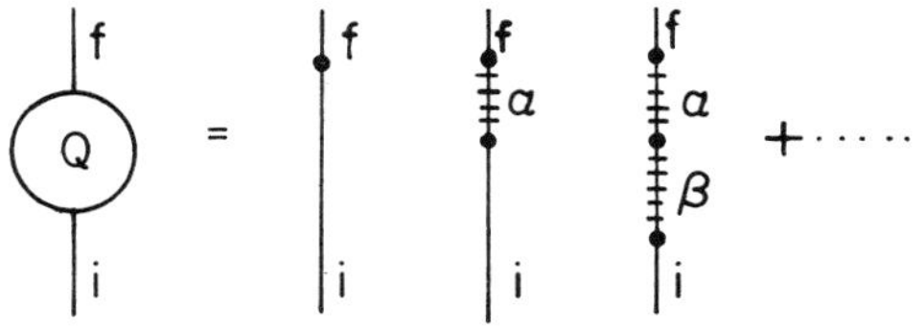

(63)

where α, β... denote the passive propagators and the circular dots represent the H_1 interaction. Here the last interaction is always at time t = 0. The symbols f and i are used to denote the final and initial active propagators. Algebraically we can write Eq. (63) as

$$<f|Q(\varepsilon_i)|i> = <f|H_1|i> + \sum_\alpha \frac{<f|H_1|\alpha><\alpha_1|H_1|i>}{\varepsilon_i-\varepsilon_\alpha}$$

$$+ \sum_{\alpha\beta} \frac{<f|H_1|\alpha><\alpha|H_1|\beta><\beta|H_1|i>}{(\varepsilon_i-\varepsilon_\alpha)(\varepsilon_i-\varepsilon_\beta)} + \ldots \qquad (64)$$

where ε_i is the sum of the single-particle energies of all the single-particle propagators contained in the propagator i, and similarly for ε_α, ε_β,... . Note that the starting energies in (64) are all ε_i, and thus we use the symbol $Q(\varepsilon_i)$ to specify this property. We remember that all of the diagrams in a Q box are linked to at least one external active line, and hence in (64) we retain only those terms which satisfy this property.

The diagrams contained in the Q'-Q term can be expressed as

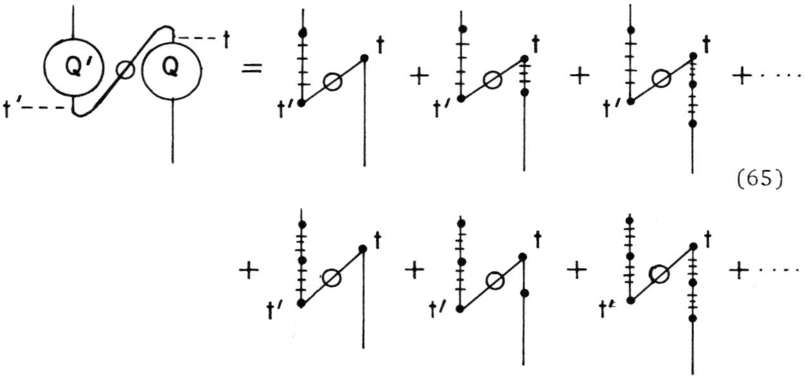

(65)

Note here all of these diagrams are just <u>once</u> folded. The time constraint is that t > t' where t' is the time for the earliest interaction in the Q' box while t is for the last interaction in the Q-box. Of course the last interaction in the Q' box is at time t = 0, and all other interactions occur prior to this time. Thus except for the first term in (65), each term in (65) represents more than one Goldstone time-ordered diagram which together satisfy the above mentioned time constraints. Although the evaluation of each single term in (65) is straightforward, most of the diagrams are off-energy-shell and this will cause difficulty in the summation of the series. For example, the following diagram is off-energy-shell:

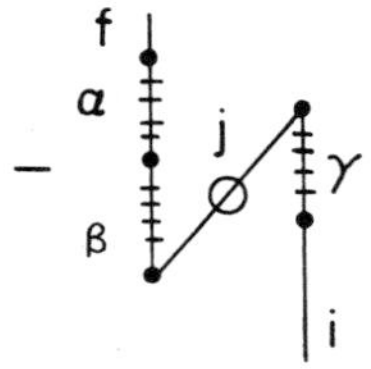

$$= - \frac{\langle f|H_1|\alpha\rangle\langle\alpha|H_1|\beta\rangle\langle\beta|H_1|j\rangle\langle j|H_1|\gamma\rangle\langle\gamma|H_1|i\rangle}{(\varepsilon_i-\varepsilon_\beta+\varepsilon_j)(\varepsilon_i-\varepsilon_\beta-\varepsilon_\gamma+\varepsilon_j)(\varepsilon_i-\varepsilon_\alpha-\varepsilon_\gamma+\varepsilon_j)(\varepsilon_i-\varepsilon_\alpha)} . \qquad (66)$$

By off-energy-shell we mean that the above energy denominator can not be factorized into two parts, one part depending only on f, α, β, j, and i and the other part depending only on j, γ and i. Such factorization is however possible when we sum up all of the once folded diagrams shown in (65).

Recall that the Q' and Q box differ only in that the former contains terms at least of second power in H_1 while the latter begins with terms of first power in H_1. Thus we can write

$$\langle f|Q'(\varepsilon_i)|a\rangle = \langle f|Q(\varepsilon_i)|a\rangle - \langle f|H_1|a\rangle . \qquad (67)$$

In the following I shall merely give the final result without proving it because, again, of the limitation of time. With Q and Q' given by (64) and (67), it can be shown[2] that all of the once-folded diagrams shown in (65) can be summed up to give

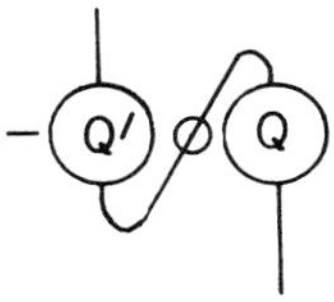

$$= \sum_{A} \langle f| \frac{Q'(\varepsilon_i)-Q'(\varepsilon_A)}{\varepsilon_i-\varepsilon_A} |A\rangle\langle A|Q(\varepsilon_i)|i\rangle \qquad (68)$$

where A represents the folded active propagator between Q' and Q. For the case of a degenerate model space we have $\varepsilon_i = \varepsilon_A$, and (68) becomes

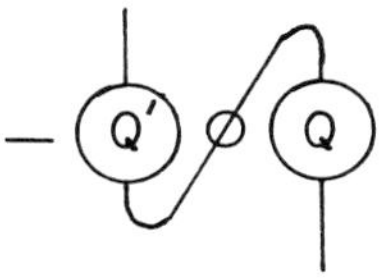

$$= \langle f| \frac{dQ(\varepsilon)}{d\varepsilon} \Big|_{\varepsilon_i} PQ(\varepsilon_i)|i\rangle. \qquad (69)$$

This is a very useful result. Suppose we decide to let Q be composed of the following diagrams, for the case of ^{18}F,

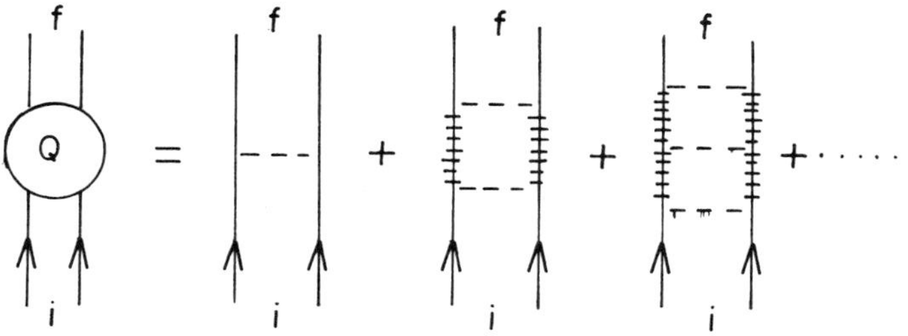

(70)

then (69) tells us that the sum of all of the once-folded diagrams is related to the energy derivative of (70) evaluated at ε_i. The energy derivative of Q can easily be obtained by numerical methods. The choice of Q in (70) is just the Brueckner G matrix. The operator P in (69) is just the model-space projection operator defined in Eq. (1).

Eq. (69) indicates that we can use the following scheme for calculation. We first decide what type of diagrams to keep in the Q box. A more appropriate choice for Q as compared with that shown in (70) is

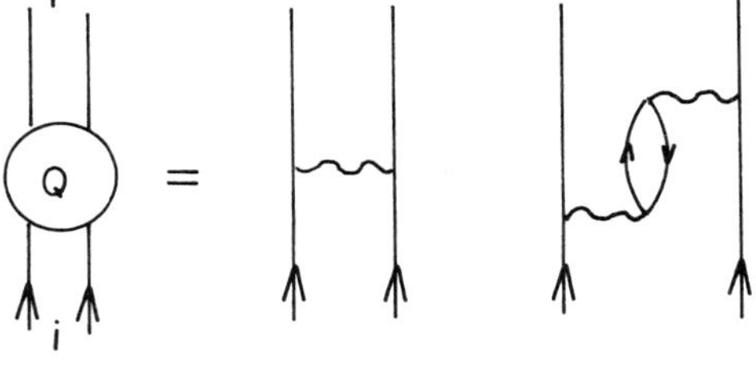

(71)

where each wavy line represents the Brueckner G matrix. We wish to point out that the expression for each individual folded diagram as shown by (66) does not allow us to perform partial summation to obtain the G matrix conveniently. This is due to the off-energy-shell inconvenience. But for the Q box shown above, such partial summation is straightforward. Then to obtain the contribution of all the once-folded diagrams, we need all the matrix elements of both Q and $(dQ/d\varepsilon)|\varepsilon_i$ within the model space.

We may expect that the twice-folded diagrams may depend on the second energy derivation of Q. It is indeed so. We have found that

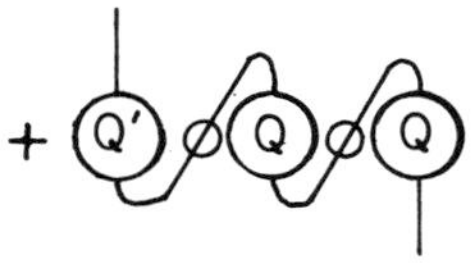

$$= \langle f|\frac{1}{2}\frac{d^2Q}{d\varepsilon^2}\,PQPQ + \frac{dQ}{d\varepsilon}\,P\,\frac{dQ}{d\varepsilon}\,PQ|i\rangle \qquad (72)$$

where, just as in (69), Q, $dQ/d\varepsilon$ and $d^2Q/d\varepsilon^2$ are all to be evaluated at energy ε_i. For thrice-folded diagrams, we found[2]

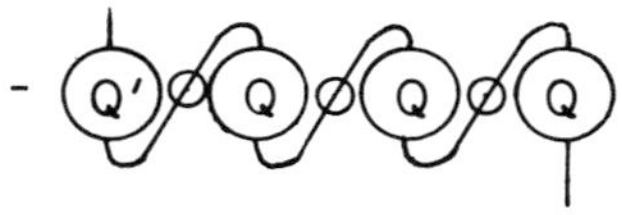

$$= \langle f|\frac{1}{6}\frac{d^3Q}{d\varepsilon^3}\,PQPQPQ + \frac{1}{2}\frac{d^2Q}{d\varepsilon^2}\,PQP\,\frac{dQ}{d\varepsilon}\,PQ + \frac{1}{2}\frac{dQ}{d\varepsilon}\,P\,\frac{d^2Q}{d\varepsilon^2}\,PQPQ + \frac{dQ}{d\varepsilon}\,P\,\frac{dQ}{d\varepsilon}\,P\,\frac{dQ}{d\varepsilon}\,PQ|i\rangle \qquad (73)$$

where again Q and its derivatives are to be evaluated at energy ε_i. There are some regularities in the above equations. For the Q'Q ... Q diagram with n folded active propagators, the sum of the powers of energy derivatives is n. For a particular Q box which has k folded propagators to its right, we associate with it a factor

$$\frac{1}{m!}\frac{d^mQ}{d\varepsilon^m} \text{ with } 0 \le m \le k.$$

For the Q' box, we associate with it a factor

$$\frac{1}{m!}\frac{d^mQ}{d\varepsilon^m} \text{ with } 1 \le m \le n.$$

Hence the effective interaction H_1^{eff} of Eq. (54) can now be expressed as terms composed of Q and its energy derivatives, as shown by Eqs. (69), (72) and (73). One nice feature is that this series expansion may converge fairly rapidly! For nuclear structure calculations using the Reid nucleon-nucleon potential, we found[5] that the variation of Q as a function of ε is very closely linear and the magnitude of $dQ/d\varepsilon$ is generally $\sim 10^{-2}$. Thus it is quite likely that we may ignore all diagrams with more than 2 folded propagators for the case mentioned above. One must of course perform some calculations to check how good is this truncation.

In general, we expect that the contribution from the folded diagrams will be of opposite sign to the one-Q box term in (54) and about 5-10% as large, for a typical nuclear structure calculation using a realistic nucleon-nucleon interaction.

A final remark I would like to make before I finish this section is about the removal of the stretchable diagrams contained in H_1^{eff}. As we discussed in section (6.3), the stretchable diagrams of H_1^{eff} will cancel with each other exactly if we sum up the whole series composed of Q, Q'-Q, Q'-Q-Q, etc. But now we perform the summation for Q'-Q, Q'-Q-Q ... one at a time. Thus each of the expressions (69), (72), and (73) contains stretchable diagrams. Thus when using these expressions for calculations, we should include the stretchable diagrams in evaluating $Q(\varepsilon_i)$ and its energy derivatives. It is interesting then to realize that although H_1^{eff} as a whole does not contain any stretchable diagrams, it seems to be more convenient to include them in calculations!

VII. CONCLUDING REMARKS

We have seen that by including the folded diagrams we have succeeded in deriving a model-space Hamiltonian which is very similar to the conventional shell-model Hamiltonian. Our model-space Hamiltonian is independent of the energy eigenvalue, and the core energy as well as the true 1-body energies can be separated out from the eigenvalue of the model-space secular equation. This does appear to be very nice. I feel that this can not be accomplished without the inclusion of the folded diagrams. For example, in the Green's function formalism the effective Hamiltonian will be dependent on the energy eigenvalue. In addition, we can not separate out the true one-body energies although the core energy can be separated out as well.

Another interesting point is that the folded diagram series for the expansion of the model-space effective Hamiltonian may converge very rapidly. As we recall, the folded diagram series can be summed up to terms involving the energy derivatives of the Q-box. For typical nuclear structure calculations with a reasonable choice for the diagrams to be included in the Q-box, we have found numerically that the variation of Q with energy is very nearly linear, implying that the higher order energy derivatives of the Q-box are very small. Thus the folded diagram series may indeed converge rapidly in powers of the energy derivatives of the Q-box. This feature is again very nice. An open question is, of course, then what diagrams are to be included in the Q-box? I don't know the answer. It would be very worthwhile to find a way to regroup the various diagrams which belong to the Q-box so that the convergence of the diagrammatic expansion of the Q-box may be enhanced. The first step, I believe, is to sum up all the ladder V interactions in the Q-box to form the Brueckner G matrix. Mr. S.Y. Lee of Stony Brook is currently investigating whether the Q-box expansion will show some convergent behavior or not when grouped in powers of G.

An important subject which I have not included here is about the effective operators. Since we are working in a truncated Hilbert space, any physical operator 0 which is appropriate for the entire Hilbert space must be modified in order to use it in the truncated Hilbert space, just as we have done for the energy Hamiltonian. In KLR(I) we have investigated the expectation value of an operator. Namely we derived a folded-diagram expansion for the effective operator 0_{eff} such that

$$<P\Psi_\alpha|0_{eff}|P\Psi_\alpha> = <\Psi_\alpha|0|\Psi_\alpha> \qquad (74)$$

where Ψ_α is the normalized true eigenfunction for the entire Hilbert space and P is the projection operator defined in Eq.(1). We have not quite succeeded in finding an expansion for $0'_{eff}$ such that

$$<P\Psi_\alpha|0'_{eff}|P\Psi_\beta> = <\Psi_\alpha|0|\Psi_\beta>, \ \alpha\neq\beta. \qquad (75)$$

Further investigations are in progress.

REFERENCES

1. T.T.S. Kuo, S.Y. Lee and K.F. Ratcliff, to be published in Nuclear Physics around December, 1971. This paper will be referred to as KLR(I).
2. K.F. Ratcliff, S.Y. Lee and T.T.S. Kuo, to be published in Nuclear Physics. This paper will be referred to as KLR(II).
3. C. Bloch and J. Horowitz, Nucl. Phys. 8 (1958) 91.
4. H. Feshbach, Ann. Phys. 19 (1962) 287.
5. T.T.S. Kuo, "Reaction Matrix Theory for Nuclear Structure" in the Proceedings of the Fourth International Symposium on Light-Medium Mass Nuclei (Oct. 1970), to be published by the University of Kansas Press.
6. B.H. Brandow, Rev. Mod. Phys. 39 (1967) 771; B.H. Brandow in "Lectures in Theoretical Physics", Vol. 11, Boulder 1968 (K.T. Mahanthappa, ed.), Gordon and Breach, New York, 1969.
7. M. Johnson and M. Baranger, Ann. Phys. 62 (1971) 172.
8. T. Morita, Prog. Theo. Phys. 29 (1963) 351.
9. G. Oberlechner et al., Nuovo Cimento 68B (1970) 23.
10. See, for example, P. Nozieres, The Theory of Interacting Fermi Systems, W.A. Benjamin, Inc., New York (1965).
11. D.J. Thouless, "The Quantum Mechanics of Many-Body Systems", Chap. IV, Academic Press, New York/London, 1961.

THE DROPLET MODEL*

WILLIAM D. MYERS
Lawrence Berkeley Laboratory
University of California

In the last few years there has been a surge of interest in what might be called the "two-part approach" to nuclear properties. The basis of this extremely productive approach is the division of various nuclear phenomena into a "macroscopic" and a "microscopic" part. I will attempt to show what motivated this approach and how it is related to the main stream of nuclear physics. However, my main emphasis will be on progress which has been made in understanding the macroscopic aspects of nuclei.

INTRODUCTION

The statistical theory of nuclei known colloquially as the "liquid drop model" is the best method available for predicting average nuclear properties. It is more broadly applicable and gives more accurate results than most people realize. In what follows I will cite some examples of the power and scope of the old liquid drop model theory. Then I will show how this approach fits into the hierarchy of methods applicable to the description of many-particle systems. Later, the Hill-Wheeler box will be used to illustrate how the total energy of a quantum system can be treated by statistical methods. Finally, I will describe a generalization of the liquid drop model that W. J. Swiatecki and I have developed and I will give some examples of applications of this new model.

*Work performed under the auspices of the U. S. Atomic Energy Commission.

LIQUID DROP MODEL

The major success of the liquid drop model is in the prediction of nuclear masses. Figure 1 shows how well mass decrements are predicted for nuclei throughout the periodic table.[1] Part of the reason for the excellent agreement between experiment and theory that can be seen in this figure is shown in Fig. 2. The quantity plotted here is that part of the binding energy per particle of real nuclei that is due to nuclear forces alone. Shell effects, the Coulomb energy and the influence of the neutron excess have all been removed. The liquid drop model predicts that when this quantity is plotted against $A^{-1/3}$ the points should lie on a straight line whose intercept is the volume energy coefficient and whose slope is the surface energy coefficient. The figure shows that the liquid drop model gives an excellent description down to A = 50 and is not bad even down to A = 10.

The next figure shows how well fission barriers are predicted by the liquid drop model.[2] Here the experimental fission barriers of a number of nuclei are plotted against the nuclear fissility x. The smooth curve represents the predicted values and the points lying near the line are experimentally measured or inferred from half-lives.

RELATIONSHIP OF MANY-BODY THEORIES

The name "liquid drop model" is unfortunate since it tends to conjure up in people's minds an image of some sort of semi-classical approach whose success may well be fortuitous. Examination of how statistical theories like the liquid drop model fit into the spectrum of possible approaches to many-body theory shows that this isn't the case.

In Fig. 4 we see first of all that an "independent particle approach" is applicable if the force is "weak". By weak we mean that the binding energy of the two-body system is small compared to the kinetic energies involved. The most fundamental method used in this approach is Brueckner theory. Calculations performed in this way deal explicitly with the problem of the hard core in the two-nucleon potential. If non-local or velocity-dependent forces are used, or forces with a soft core, then one can treat the problem using the Hartree-Fock methods which are familiar from atomic physics. Often one does not even attempt to fit the scattering data but instead uses "effective interactions" chosen for mathematical simplicity

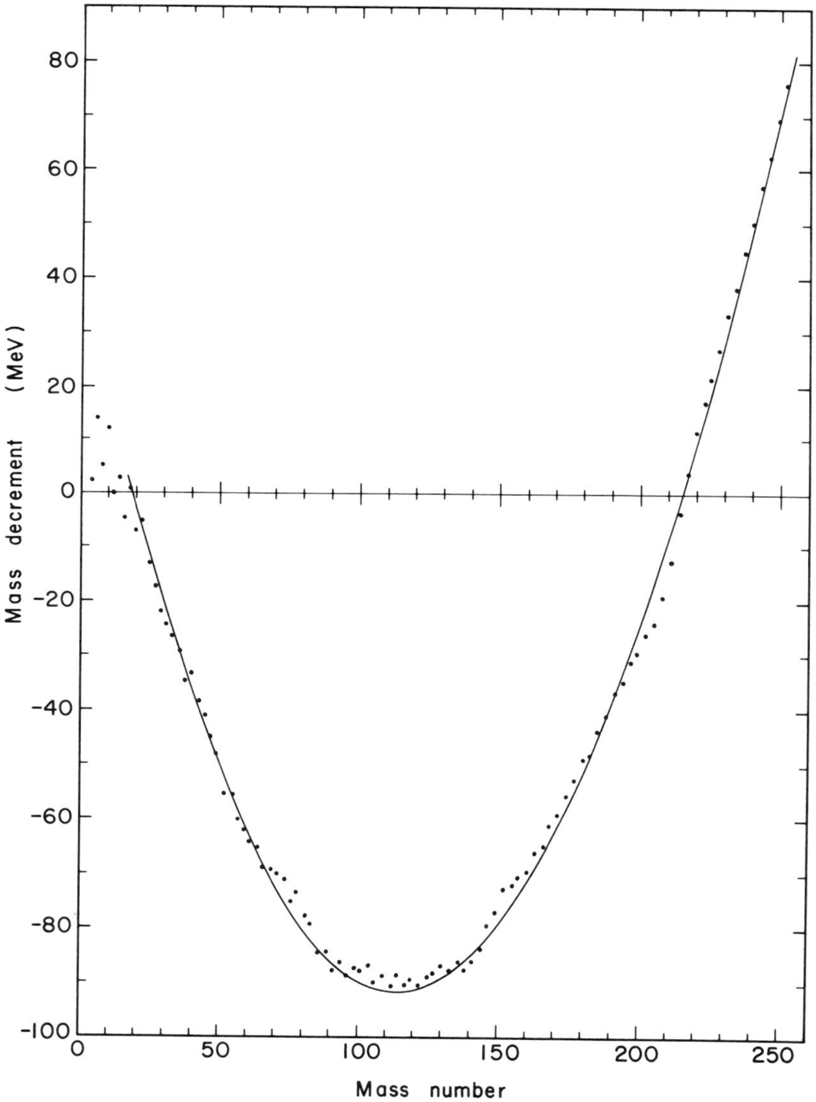

Figure 1
The mass decrements of 97 beta-stable nuclei are compared with the smooth curve corresponding to a liquid-drop mass formula. Note that the over-all trend of the decrements is reproduced throughout the periodic table, including the light nuclei. The scatter of the points is due to shell effects.

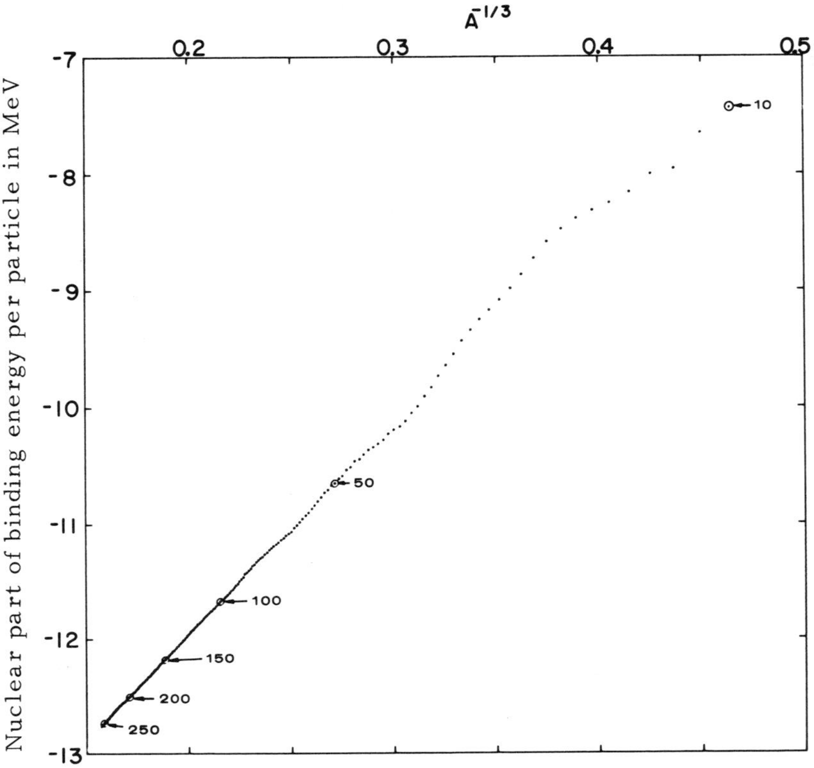

Figure 2
The Y-intercepts from Fig. 9 of Ref. 1 are plotted against $A^{-1/3}$, to exhibit the linear trend predicted by the liquid-drop formula with volume and surface terms but no curvature correction. The labels on the circled points are A values. Note the false zeros on both the ordinate and abscissa.

and containing parameters that are adjusted to reproduce the properties of finite nuclei. A further approximation can be made by neglecting exchange (i.e., doing a Hartree calculation) and changing the force so that agreement with experiment is maintained. Finally, if even the requirement of self-consistency is given up, we arrive at the shell model and Nilsson model.

If the number of particles is large, another stage of approximation is possible to what might be called the

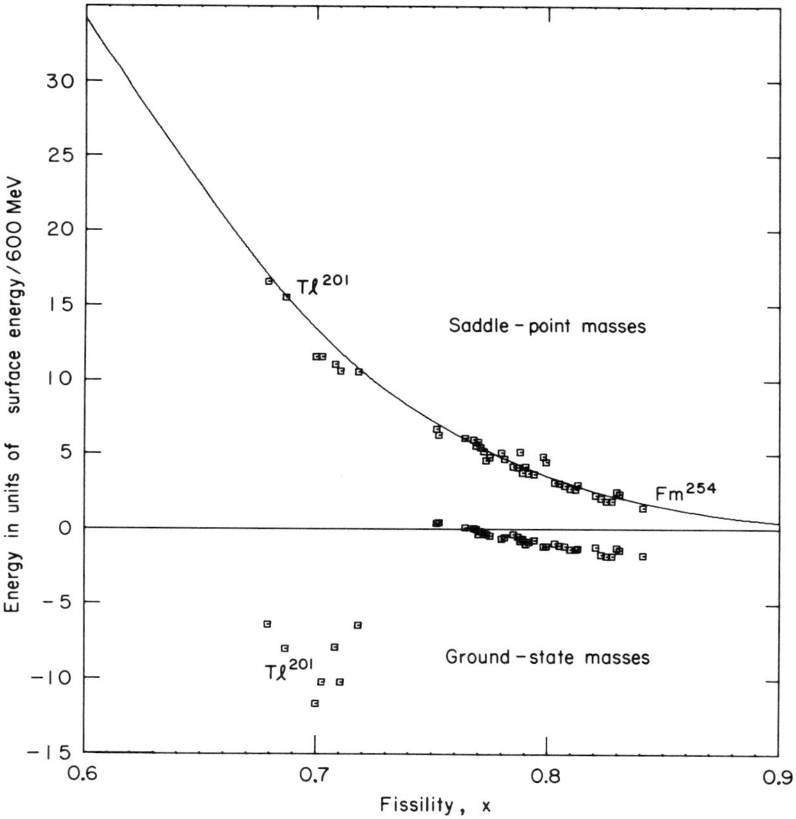

Figure 3
Comparison between calculated and experimental fission barriers.

"macroscopic approach". In this approach statistical methods are applied which ignore the discreteness of the particles. We find in our work that the Thomas-Fermi-Dirac method often employed in this connection is a useful approximation when $(\text{grad}\ \rho/\rho^{4/3}) \leqslant 10$. The fact that this criterion is met over most of the charge distribution of an atom explains why this method has been so useful there. Neglect of exchange effects gives the Thomas-Fermi method that we have used for the study of nuclei. Since the above criterion is also satisfied over the bulk region of a nucleus and in the surface

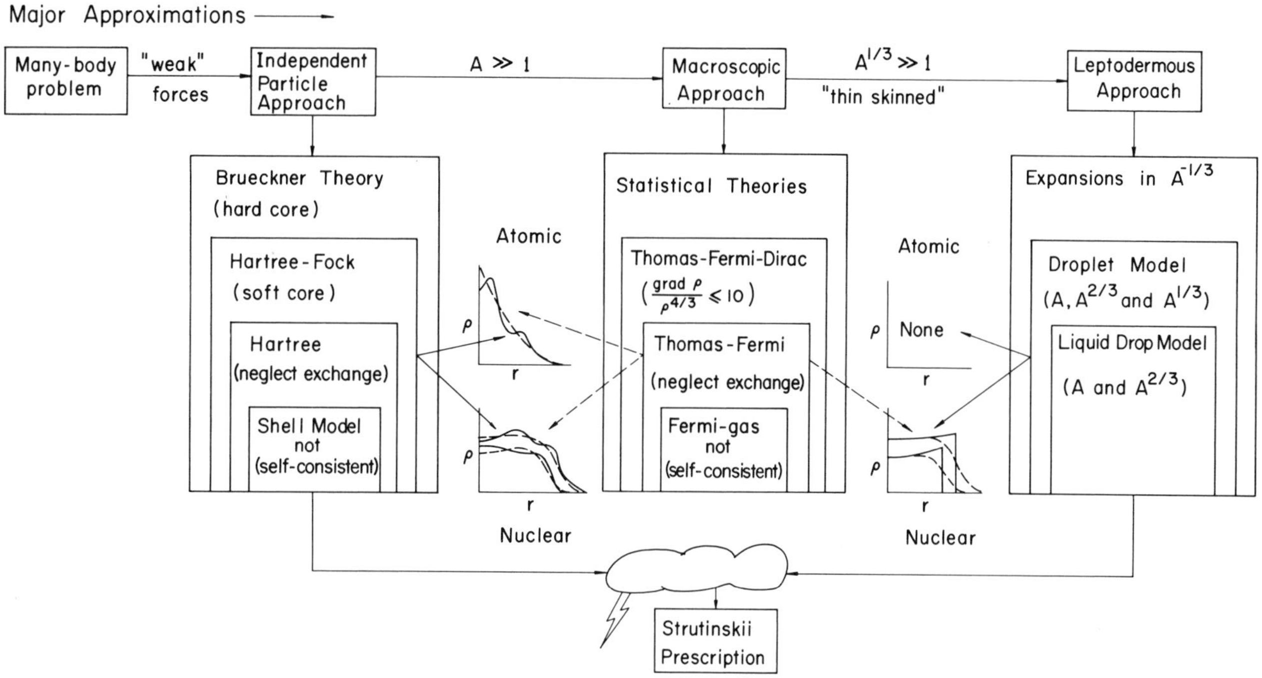

Figure 4
Relationships between Various Approaches to the Nuclear Many-Body Problem

out to the point where the density drops to 1/6 its central value, we expect this method to give a good representation of nuclear properties.[4] Much of the Thomas-Fermi work one sees in the current literature involves the additional assumption (which is not very well satisfied) that the range of the force be small compared with the diffuseness of the surface region.[4, 5] I've chosen to call these "gradient methods" since effects due to the finite range of the force are approximated by a term in the energy equation proportional to the gradient of the density.

Two small illustrations in this figure show schematically the sort of density distributions that Hartree-Fock theory predicts for atoms and nuclei. The dashed lines show how the Thomas-Fermi results appear. The density distributions obtained from the statistical approach are expected to be sufficiently accurate for the prediction of most average properties. They are inadequate only when one wishes to consider details that depend on the small deviations introduced by the single-particle structure of the system.

Another important level of approximation is possible if one is dealing with a saturating system (uniform density, energy $\propto A$) and if the surface diffuseness is small compared to the radius. For nuclei this means that $A^{-1/3} << 1$. This approach has been called the "leptodermous approach".[6] I will not have time to discuss the formal basis of the approach but I will describe its application to nuclei in the form of the droplet model.[3] The droplet model is a generalization of the liquid drop model but the choice of a name is a little unfortunate since it contributes to the misunderstanding many people have about the leptodermous approach. The mathematical procedures employed here have nothing whatever to do with assuming that the nucleus is a liquid drop. It is often thought that special assumptions about the nucleons being strongly correlated or about the flow of nuclear matter being irrotational are necessary to justify the droplet and liquid drop models. Such ideas are completely incorrect. Many authors have shown that all that is required for the droplet model to apply is that the system be leptodermous.[6-9]

Another reason the general applicability of the leptodermous approach is not commonly appreciated is that the liquid drop model seems to have no place in reaction theory. That this is true is due to the fact that until now one has used only light projectiles to which the model does not apply. We expect that the reaction theory for heavy-ion scattering will rely more heavily on the droplet model and the leptodermous approach.[10]

The solid line in the second nuclear density plot is a schematic representation that is appropriate when the leptodermous approach is valid. In such cases, where the surface thickness is small compared with the size of the system, only the location of the surface need be indicated. Of course no such idealization is possible for atoms since atoms are not leptodermous systems.

Since the droplet model only treats static properties, it is incorrect to describe it as a "classical approach". The word "classical" usually only refers to dynamical processes where $(\text{action}/\hbar) \gg 1$.

Recently there has been considerable interest in hybrid methods which attempt to combine the leptodermous approach with the shell model.[11-13] These methods, which have been so important in predicting islands of stability in the region of super-heavy nuclei and in the understanding of fission isomers, are based on the leptodermous approach combined with an estimate of shell effects obtained from the Nilsson model using the Strutinskii method.

The shell correction method that I just referred to has grown out of earlier attempts to treat the small deviations that remain when nuclear binding energies are calculated using the liquid drop model.[14] The liquid drop model gives a good account of binding energies that range in value from a few hundred MeV for light nuclei to a few thousand MeV for heavy nuclei, but small systematic deviations remain which can be as large as 10 MeV. Figure 5 shows these deviations plotted first against proton number then against neutron number. The magic numbers of 28, 50, 82, and 126 can be clearly seen. Even though these deviations are small, they can be very important in some special cases. One example is in the calculation of the mass difference between neighboring nuclei where the liquid drop model difference is small. Another situation where shell effects are important is in connection with the shape dependence of the binding energy. When the liquid drop model contribution to the energy is stationary, as it is at the spherical shape and at the fission saddle point shape, shell effects can also be important. Shell effects about the spherical shape lead to ground state deformations and those about the saddle point can lead to fission isomers. Figure 6 shows how important these effects can be for the deformation energy in the region of the heavy elements.[15, 16]

Even more interesting is the fact that shell effects at the barrier can also lead to asymmetric saddle point shapes and hence may provide the long-sought explanation of the asymmetric fission of the heavy elements.[15-19]

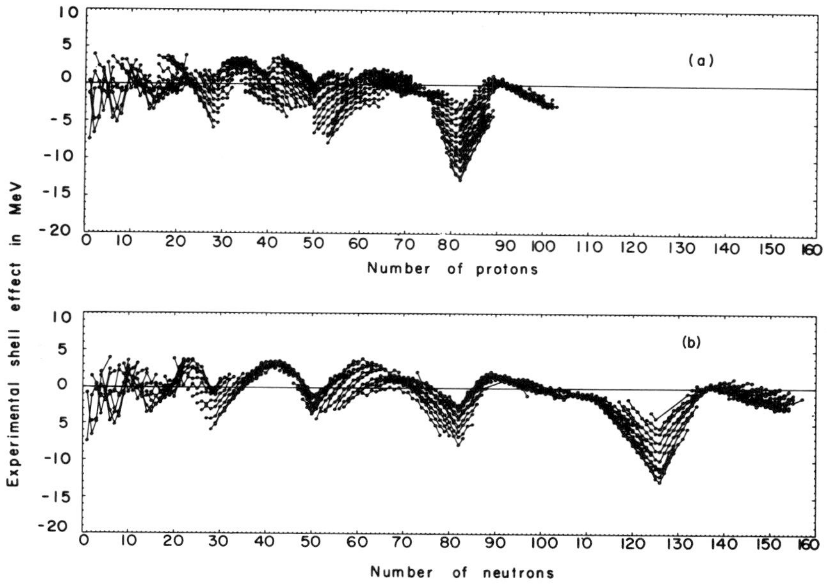

Figure 5
Shell effects in nuclear ground state masses.

HILL-WHEELER BOX

Even though the leptodermous approach is a type of many-body theory, there can be some question about when it is applicable. Most people would be willing to agree that a water drop (even a small one) is a leptodermous system, but they might be skeptical about a quantum system with a relatively small number of particles like a nucleus. In order to gain some insight into this question, let us consider the schematic model of an infinitely-deep square well potential in the shape of a cube.[6, 7] If the volume of the well is increased in proportion to the number of particles in the well, we have a model of a saturating system. Figure 7 illustrates how the possible single-particle states in this well can be represented as points of a cubic lattice in quantum number space. Examination of the expression given in this figure for the energies of the single-particle states shows that the isoenergy surfaces are spherical.

If all the states are occupied up to some Fermi-energy λ then the total number of particles N and the total energy

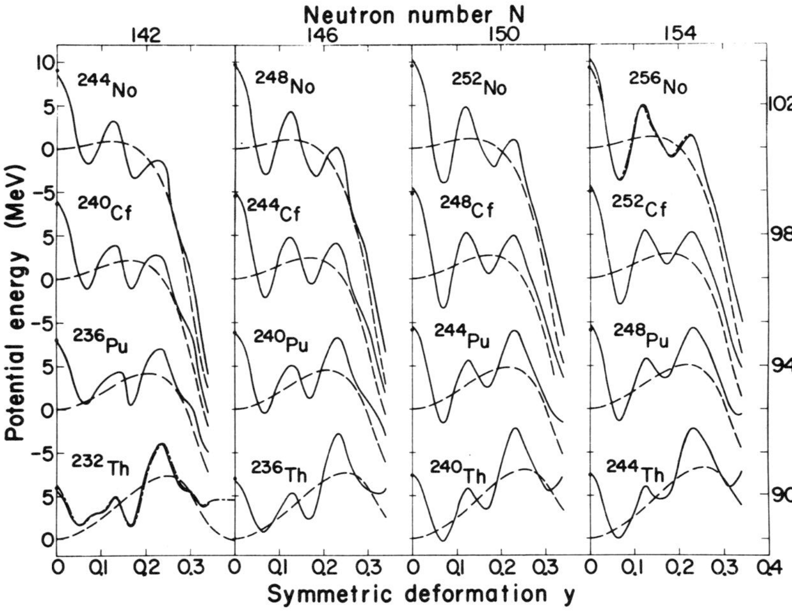

Figure 6
The dashed lines represent the calculated liquid drop model deformation energies of a number of actinide nuclei. The solid lines show how the deformation energy is changed when shell effects are added.

E are given approximately by the expressions

$$N = \frac{\lambda}{2}\int_0^{\lambda} n^2\, dn \quad , \qquad E = \frac{\lambda}{2}\int_0^{\lambda} E_{sp}\, n^2\, dn \quad . \tag{1}$$

Since the ground state of the system corresponds to particles filling the lowest energy levels, we can locate the Fermi surface λ by integrating out (in the octant of quantum number space that contains all possible states) from one isoenergy surface to another until we have the right number of particles. Then the total energy of this system can be calculated by integrating over the single-particle energies up to this same point. If λ is eliminated between these two expressions and appropriate corrections are made for end effects in the switch from sums to integrals, the following expression for E in terms of N is obtained:

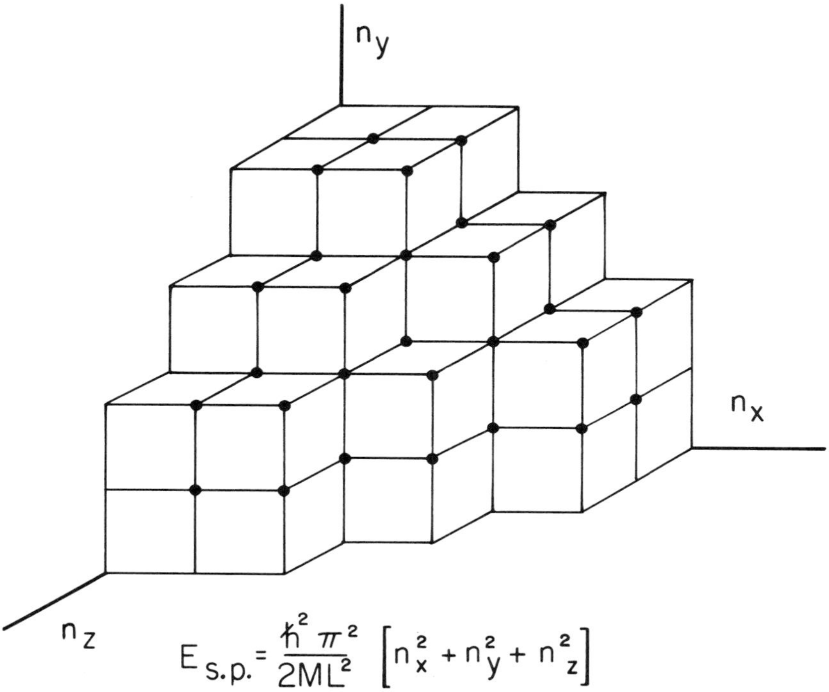

Figure 7
The lattice of possible single particle states in quantum number space for the energy levels in a cubical box.

$$E = T_f\left[\frac{3}{5}N + \frac{3(6\pi^2)^{2/3}}{16\pi}N^{2/3} + \frac{(9\pi-16)(6\pi)^{1/3}}{32\pi}N^{1/3}\right] . \qquad (2)$$

This is the sort of expression one expects for leptodermous systems. It is an expansion in powers of $N^{-1/3}$ that contains the leading terms proportional to N, $N^{2/3}$, and $N^{1/3}$.

For a box with 60 occupied levels (240 particles) at nuclear matter density the terms in Eq. (2) have the values

$$E = 4830 + 1845 + 225 + (14)\ \mathrm{MeV} \quad , \qquad (3)$$

where the (14) is the difference between the statistical value of E from Eq. (2) and the exact value obtained by summing over individual energy levels.

Figure 8, which is from Tsang's thesis,[6] shows how important it is to have all 3 terms in Eq. (2). Here the difference

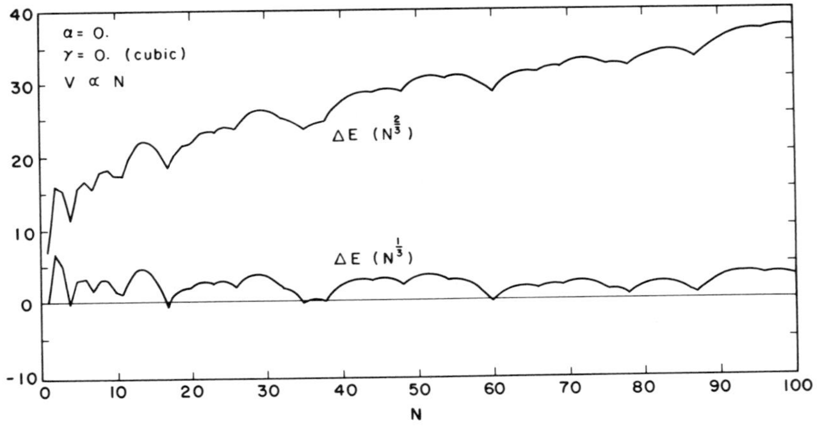

Figure 8
The difference between the actual and the calculated total energy of particles in a cubic box is plotted against the number of occupied levels. For the upper curve the calculated energy includes terms up to $N^{2/3}$ and for the lower curve terms through $N^{1/3}$ are included.

between a sum of individual particle energies and the smooth prediction of Eq. (2) is plotted against the number of occupied levels. The lower curve shows the shell effects that remain. The upper curve shows the large discrepancy that exists if one makes a liquid-drop-model-like assumption about the smooth part of the energy and does not retain the $N^{1/3}$ term in the smooth part of the total energy.

Figure 9, which is also due to Tsang, is a plot like the one we have just seen that goes all the way out to 1500 levels. For such a large system the energy consists of E = 181, 540 + 23, 966 + 1, 099 + (14), the shell correction being 7/1000 of a percent of the total energy.

At this point it is appropriate to ask ourselves why the leptodermous approach works so well. Swiatecki explained the reason in a couple of papers that are quite old now.[20, 21] It turns out that the bulk properties of the system are those of a free Fermi gas and the only deviations that occur are confined to the surface region in a thin layer whose thickness is of the order of the Fermi-wave length λ_f.

We see from this example that nuclei are probably quite nicely leptodermous and so we propose to make the fullest possible use of this fact; a decision which leads us directly to the droplet model.

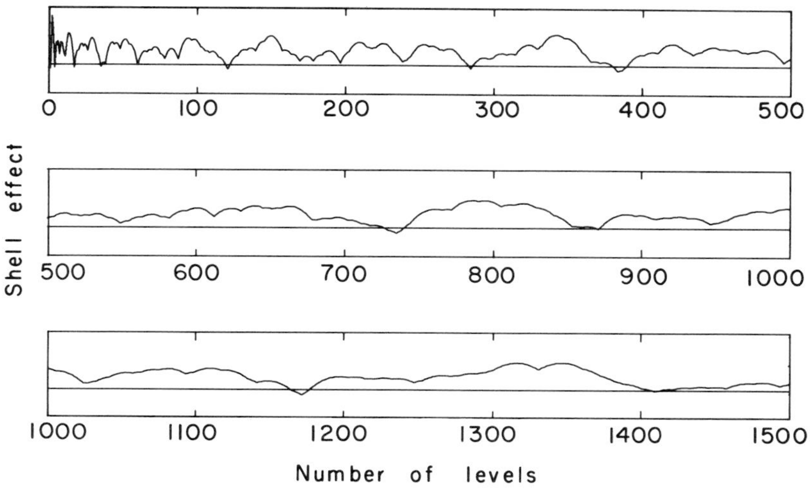

Figure 9
The difference between the actual and calculated total energy of particles in a cubic box is plotted against the number of occupied levels.

DROPLET MODEL

Our decision to generalize the liquid drop model was based on the belief that terms of order $A^{1/3}$ are important. We felt that the success of the old liquid drop model might be partly due to the fact that it is only possible to make comparisons over the relatively small range mass numbers provided by the known nuclei. Fitting to such a limited region may cause some of the coefficients in the mass formula to take on slightly incorrect values so as to compensate for the missing higher-order terms. With an improved theory we might expect to obtain more nearly correct values for these coefficients and some estimate of the values of the coefficients of the higher-order terms.

The droplet model is a generalization of the liquid drop model and it is based on the leptodermous approach which involves an expansion in powers of $A^{-1/3}$. Since nuclei are two-component systems (neutrons and protons), a second expansion parameter enters which is I^2, where $I = (N-Z)/A$. Figure 10 illustrates the nature of the refinements included in the droplet model. Retention of only the term in A constitutes the standard nuclear matter approach. Inclusion of the surface

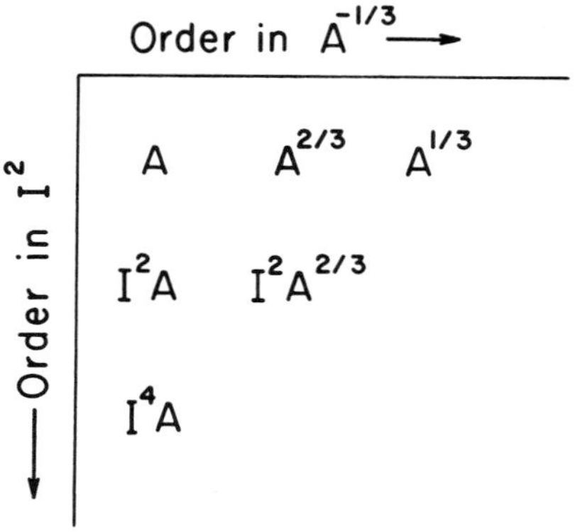

Figure 10

energy (proportional to $A^{2/3}$) and the volume symmetry energy (proportional to I^2A) corresponds to the liquid drop model. The droplet model is characterized by the inclusion of the next order of terms along the diagonal. It includes terms proportional to $A^{1/3}$, $I^2A^{2/3}$, I^4A. However, it is not enough to simply add terms like these to the mass formula with undetermined coefficients. For a broadly applicable model, we must fully understand the physical origin of the new terms that enter.

In order to include all effects that enter into these new terms, two degrees of freedom must be added to the liquid drop model.[3] (See also Refs. 22-25.) These new degrees of freedom are illustrated in Fig. 11. To begin with we lift the constraint that nuclear matter is incompressible. This type of variation is described by the quantity $\bar{\varepsilon}$, which is proportional to the deviation of the density from its nuclear matter value ρ_0. Next we remove the constraint that the neutron/proton ratio at each point must have the value N/Z. The "local asymmetry" is then given by $\bar{\delta}$ which need not be equal to I, and the possibility arises of the excess neutrons being forced to the surface to form a "neutron skin" of thickness t. The bars over ε, δ and ρ indicate that these symbols represent average values over the central region of the nucleus. We have also included in our work the possibility of spatial variations of ε and δ representing the redistribution of particles due to Coulomb repulsion.[3] However, in what follows this aspect of the problem will be omitted to simplify the discussion.

When the droplet model energy of a spherical nucleus is written as a function of N, Z, and $\bar{\varepsilon}$, $\bar{\delta}$ we find the expression

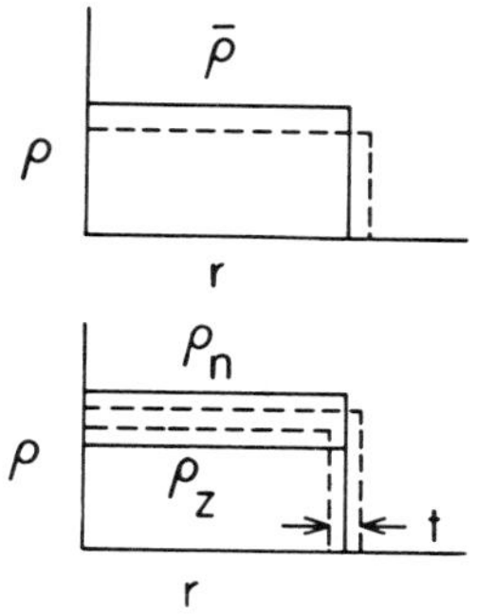

$$\bar{\epsilon} = -\tfrac{1}{3}\,(\bar{\rho}-\rho_0)/\rho_0$$

$$\bar{\delta} = (\bar{\rho}_n - \bar{\rho}_z)/\bar{\rho}$$

$$t = \tfrac{2}{3}\,(I-\bar{\delta})\,R$$

Figure 11
Schematic plots of nuclear densities against radial distance to illustrate the new degrees of freedom allowed in the droplet model.

$$\begin{aligned}E(N, Z;\bar{\varepsilon}, \bar{\delta}) = {} & [-a_1+J\bar{\delta}^2 + \tfrac{1}{2}(K\bar{\varepsilon}^2 - 2L\bar{\varepsilon}\bar{\delta}^2+M\bar{\delta}^4)]\,A \\ & + [a_2+Q(t/r_0)^2]\,(1+2\bar{\varepsilon})A^{2/3} \\ & + [a_3]\ A^{1/3} \\ & + c_1Z^2/\,[A^{1/3}(1+\bar{\varepsilon}) - \tfrac{1}{2}(t/r_0)]\ . \end{aligned} \tag{3}$$

Once this expression is available, the equilibrium values of $\bar{\varepsilon}$ and $\bar{\delta}$ can be obtained by minimizing the total energy. These values can then be substituted back into the expression to obtain the equilibrium energy as a function of N and Z alone.

If (for purposes of exposition) we choose to ignore Coulomb effects, the equilibrium energy expression takes the particularly simple form,

$$\begin{aligned}E = {} & -a_1A+a_2A^{2/3} + \left(a_3 - \frac{2a_2^2}{K}\right)A^{1/3} \\ & + JI^2A - \left(\frac{9}{4}\,\frac{J^2}{Q} - \frac{2a_2L}{K}\right)I^2A^{2/3} \\ & - \left(\frac{L^2}{2K} - \frac{M}{2}\right)\,I^4A \end{aligned} \tag{4}$$

Notice that the terms of this expression have been arranged so as to reproduce the pattern first presented in Fig. 10. None of the coefficients of the new terms we are considering are simple. Each consists of two contributions and the majority of these contributions arise not from single effects but from the competition between various higher-order terms. Approximate values for the various terms that enter are given below for the case of ^{208}Pb:

$$\begin{aligned} E = & -3261 + 652 + 42 \\ & + 260 - 148 \\ & + 10. \end{aligned} \tag{5}$$

APPLICATIONS

Prediction of nuclear ground state masses is one of the primary applications of the droplet model. While we do not expect much improvement over the already excellent fit of the liquid drop model to known nuclei, we do expect the droplet model to be superior in its prediction of the properties of nuclei far from stability. Some preliminary fits of the droplet model (or variations of it) have been performed,[26, 27] and Swiatecki and I are currently engaged in such a project. An essential feature of our work is the fitting of fission barriers as well as ground state masses. This is possible since the droplet model has been generalized to apply to arbitrarily-shaped nuclei and the various shape dependences and expected barrier properties have been worked out in some detail.[28] However, in the preliminary applications described below we have used only rough estimates of the parameters based on liquid drop model values and estimates from Thomas-Fermi calculations.[3]

One interesting application that is related to the prediction of ground state masses is the prediction of single-particle potential well parameters. An essential feature of most modern mass formulae[1, 14, 26, 29, 30] is the division of the binding energy into a smooth part, which is treated by statistical methods such as the liquid drop model, and "shell effects". The Strutinskii method referred to earlier is currently the most promising approach to the prediction of shell effects.[11] Without going into detail, let me merely remark that the Strutinskii method is based on the single-particle levels of the nucleus in question, and the neutron and proton single-particle potential well para-

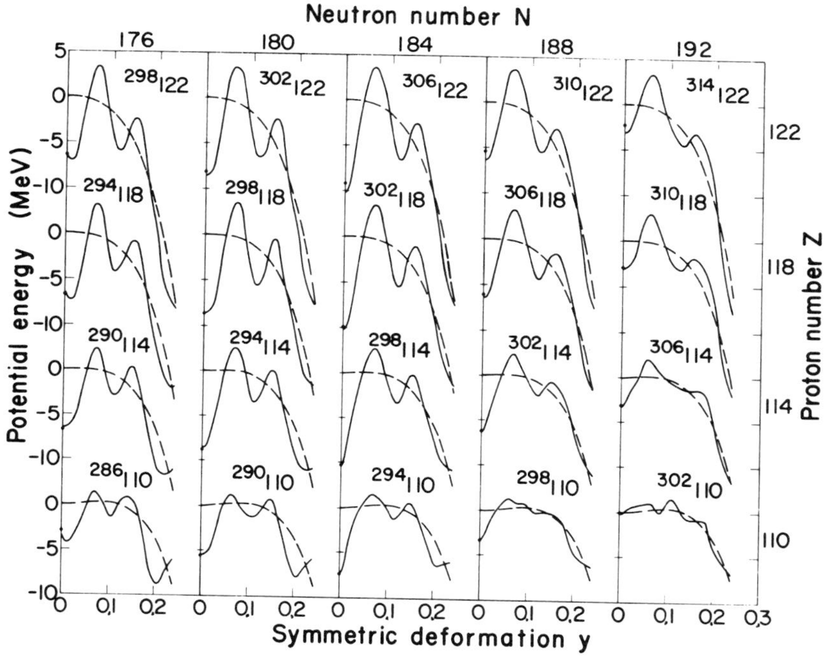

Figure 12
The dashed lines represent the calculated liquid drop model deformation energies of a number of superheavy nuclei. The solid lines show how the deformation energy is changed when shell effects are added.

meters must be known before these levels can be calculated. These parameters can be determined for some nuclei by fitting the observed excitation spectra and their variation with N and Z can be studied by considering nuclei throughout the periodic table. This approach is not sufficiently accurate for predictions for nuclei far from stability. Fortunately, a minor extension of the droplet model allows us to predict potential well parameters for any value of N and Z.[31] The fact that the predictions, which contain no free parameters, give good agreement for the known nuclei gives us confidence in their validity for use in calculating the single particle spectra and hence the deformation energies of super-heavy nuclei which are far from the usual region of stability. So it is possible to apply a unified approach to nuclear masses (and deformation energies) in which both the smooth part of the binding energy and the potential wells necessary for calculating shell corrections come from the droplet model. Figures 6 and 12 are examples of calculations that have

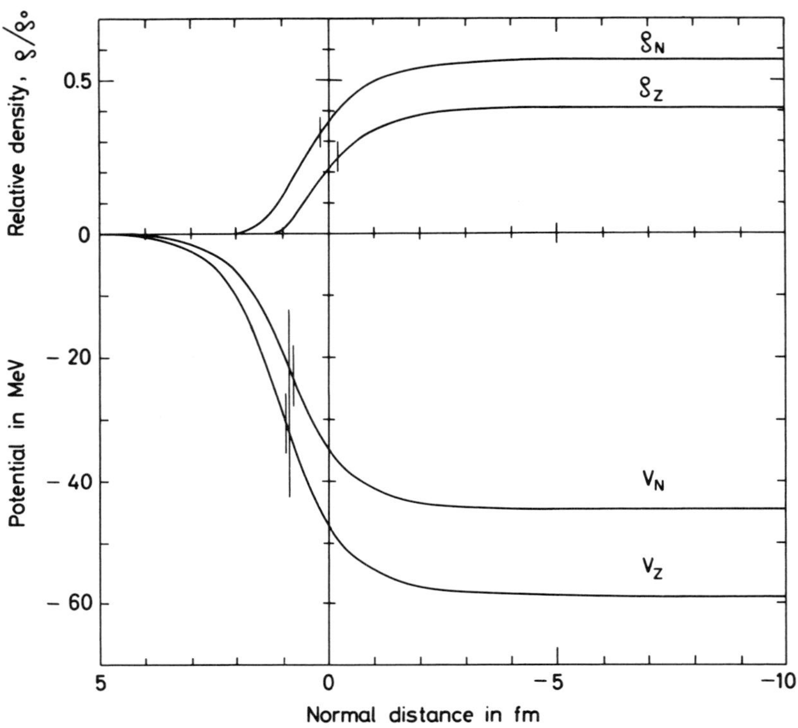

Figure 13
The Thomas-Fermi density distributions and single particle potential wells for a case of non-vanishing asymmetry $\bar{\delta}$. The curves are plotted relative to the effective location of the surface of the total density. In addition, the locations of the neutron and proton surfaces are indicated by small vertical bars. The separate locations of the neutron and proton potential surfaces are given by the smaller bars on either side of the long vertical bar that indicates their average position.

recently been performed using this method.[15, 16]

In order to establish the connection between the density distributions, which were already predicted by the droplet model, and the single-particle potential wells, a series of semi-infinite Thomas-Fermi calculations were carried out (similar to the one shown in Fig. 13) as a function of the asymptotic nuclear asymmetry $\bar{\delta}$. With the additional information provided by these calculations it becomes possible to predict potential well parameters with the droplet model. The agreement shown in Fig. 14 between the predicted values and the values determined by

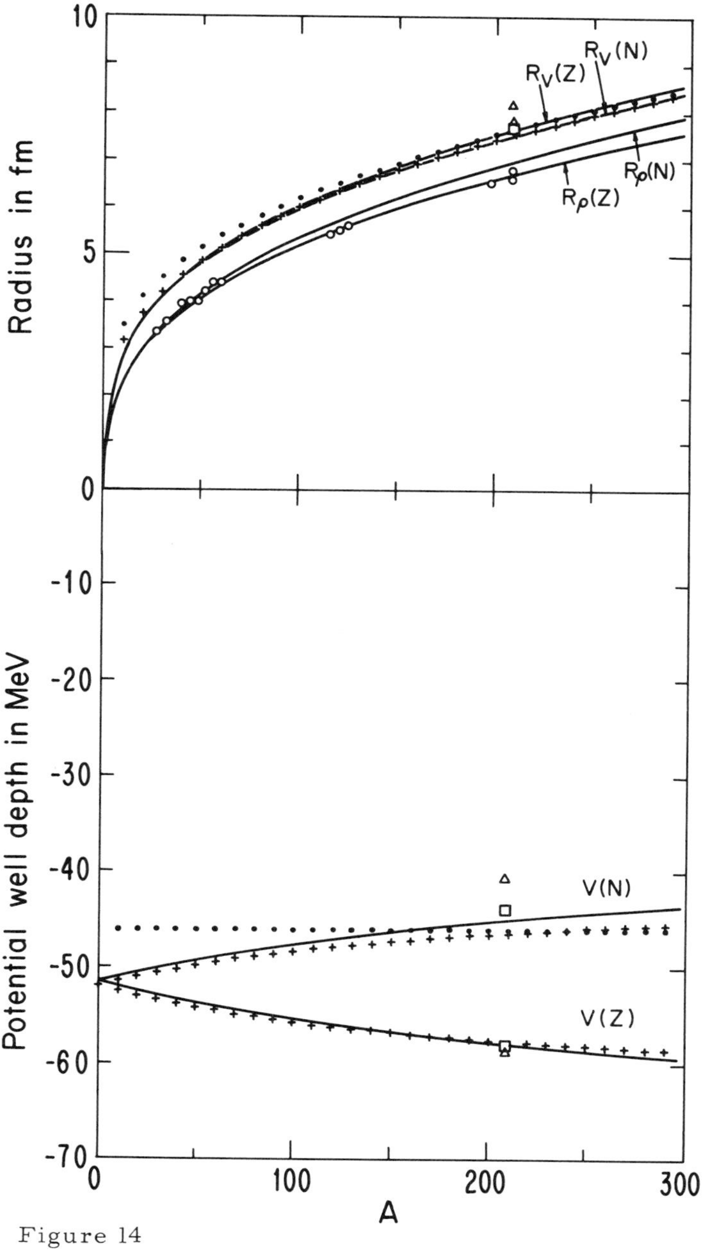

Figure 14
The predictions of the algebraic theory are plotted as solid lines against the mass number A for nuclei along beta stability. The various symbols represent experimental determinations of these same quantities.

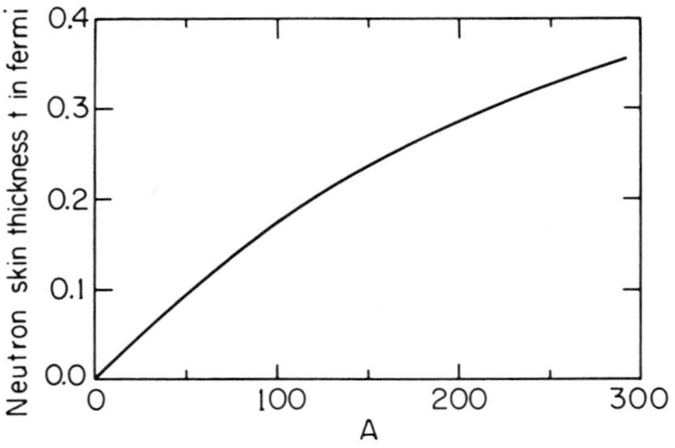

Figure 15
Neutron skin thickness t in fermi plotted as a function of the mass number A for nuclei along beta stability.

fitting to excitation spectra is good in spite of the fact that there are no free parameters.

Since the radii of both the neutron and proton density distributions are predicted by the droplet model, it is possible to investigate such things as the neutron skin thickness and the isotope-shift anomaly.[32] Figure 15 shows the predicted neutron skin thickness for nuclei along beta-stability throughout the periodic table. In a similar way, Fig. 16 shows the predicted isotone and isotope shifts. The agreement to be seen in this figure between these predictions and the experimental points would be more obvious if the experimental values had been corrected for the contribution to the isotope shift that is simply due to a change in deformation. One interesting aspect of this droplet model study of isotope shifts was the discovery that the previously rather widely-held belief that the anomaly was connected with finite nuclear compressibility is incorrect. Figure 17 shows that the anomaly would exist even for a completely incompressible nucleus and that smaller values of K even tend to reduce the size of the effect.

FINAL REMARKS

It is gratifying to see that the droplet model predictions agree reasonably well with experiment even though the various coefficients that enter have only been estimated in a rough way from previous liquid drop model and Thomas-Fermi work. Unlike

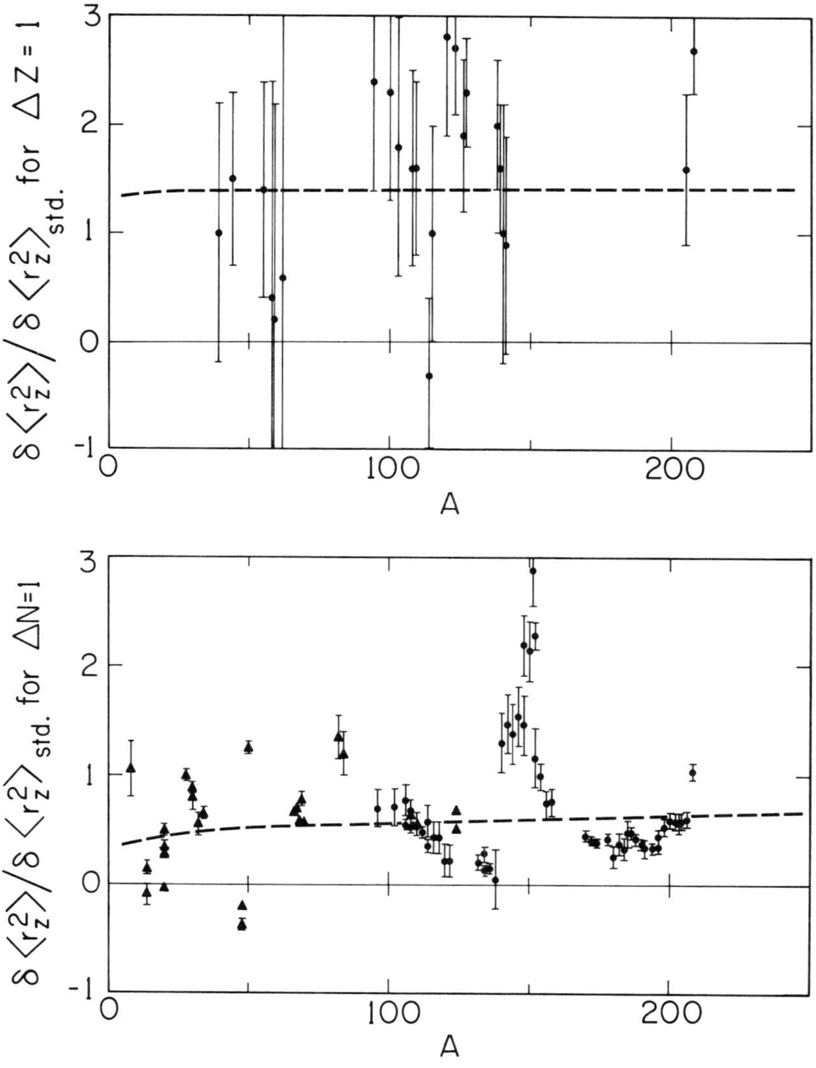

Figure 16
The ratio of the change in $\langle r_z^2 \rangle$ to the change expected from the liquid drop model is plotted against the mass number A. The upper part of the figure is for the case $\Delta Z = 1$ and the lower part is for the case $\Delta N = 1$. In both cases experimental points are shown as well as dashed lines representing the predictions of the droplet model for nuclei along beta stability.

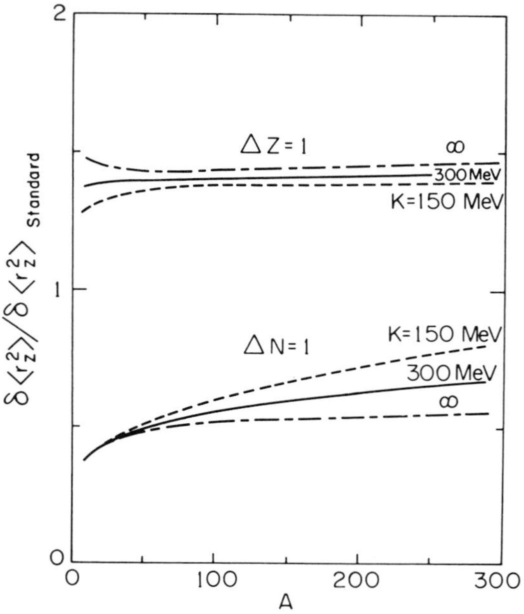

Figure 17
The ratio of the change in $\langle r_z^2 \rangle$ to the change expected from the liquid drop model is plotted against the mass number A for nuclei along beta stability. Curves for $\Delta Z = 1$ and $\Delta N = 1$ are shown for three different values of the compressibility coefficient K.

some of the earlier attempts at extending the liquid drop model by adding one specific effect or another as necessary to explain some particular result, the droplet model is a complete theory that carries the liquid drop model to one higher order in the small quantities $A^{-1/3}$ and I^2 in a consistent way. When one does not have a theory that is as precisely-defined as this one is, there is the hazard that the theory can always be brought into agreement with experiment by adding some terms or neglecting others. Fortunately the droplet model does not have this defect and confrontation with experiment seems possible on a broad front.

Before this confrontation can take place effectively, the free parameters of the theory must be fixed. A number of references to current work on this problem were given earlier. It will be interesting to see what values are finally agreed upon for such new coefficients as the nuclear compressibility and the curvature correction. It will also be interesting to see how much change there is in the old coefficients such as those for the surface and symmetry energies because of the improved functional form of the energy equation.

SUMMARY

In the discussion presented here we first began by considering some examples of the effectiveness of the liquid drop model and went on to discuss how this model and leptodermous approaches in general are related to other methods in many-body theory. Next the example of the Hill-Wheeler box was used to show that the leptodermous approach is applicable to quantum systems even with relatively small numbers of particles. The smooth part of the total energy was then separated from the shell effects and it was clear that an $A^{1/3}$ term must be included. After a brief introduction to the methods used to obtain the droplet model energy equation, a simplified version was used to display the most important new terms that enter. Finally, some applications of the theory were given and it was pointed out that comparison with a much wider range of phenomena is planned once the coefficients of the theory have been better determined.

REFERENCES

1. Myers, William D. and Swiatecki, Wladyslaw J., "Nuclear Masses and Deformations", Nucl. Phys. 81, 1 (1966).
2. Myers, W. D., and Swiatecki, W. J., "Anomalies in Nuclear Masses", Proceedings of the Lysekil Symposium, 1966, Arkiv. Fysik 36, 343 (1967).
3. Myers, William D., and Swiatecki, W. J., "Average Nuclear Properties", Ann. of Phys. (N. Y.) 55, 395 (1969).
4. Bethe, H. A., "Thomas-Fermi Theory of Nuclei", Phys. Rev. 167, 879 (1968).
5. Brueckner, K. A., Buchler, J. R., Jorna, S., and Lombard, R. J., "Statistical Theory of Nuclei", Phys. Rev. 171, 1188 (1968).
6. Tsang, Chin-Fu, (Ph. D. Thesis), "On the Microscopic and Macroscopic Aspects of Nuclear Structure with Applications to Superheavy Nuclei", Lawrence Radiation Laboratory Report UCRL-18899, 1969.

7. Hill, D. L., and Wheeler, J. A., "Nuclear Constitution and the Interpretation of Fission Phenomena", Phys. Rev. 89, 1102 (1953).
8. Hilf, E., and Süssmann, G., "Surface Tension of Nuclei According to the Fermi-gas Model", Phys. Letters 21, 654 (1966).
9. Strutinskii, V. M., and Tyapin, A. S., "Quasistatic Drop Model of the Nucleus as an Approximation to the Statistical Model", Zh. Eksperim. i Teor. Fiz. (U.S.S.R.) 45, 960 (1963) [Engl. Transl.: Soviet Phys.-JETP 18, 664 (1964)].
10. Swiatecki, W. J., "Prospects for Superheavy Nuclei", Proceedings of the International Conference on Nuclear Reactions Induced by Heavy Ions, Heidelberg, Germany, 15 July 1969.
11. Strutinsky, V. M., "Shell Effects in Nuclear Masses and Deformation Energies", Nucl. Phys. A95 , 420 (1967).
12. Nilsson, S. G., et al., "On the Nuclear Structure and Stability of Heavy and Superheavy Nuclei", Nucl. Phys. A131, 1 (1969).
13. Bosterli, M., Fiset, E. O., and Nix, J. R., "Single-Particle Calculations for Deformed Potentials Appropriate to Fission", Proceedings of the Second IAEA Symposium on Physics and Chemistry of Fission, Vienna, 1969.
14. For an extensive survey of the literature see: James Wing, "Systematic Comparison of Semiempirical Nuclidic Mass Equations", Proceedings of the Third International Conference on Atomic Masses, University of Manitoba, Winnipeg, 1967.
15. Nix, J. R., "Theory of Nuclear Fission and Superheavy Nuclei", Los Alamos Scientific Laboratory Report LA-DC-12488, 15 April 1971.
16. Bolsterli, M., Fiset, E. O., Nix, J. R., and Norton, J. L., article in preparation for submission to Phys. Rev. C.
17. Möller, P., and Nilsson, S. G., "The Fission Barrier and Odd-Multipole Shape Distortions", Phys. Letters 31B, 283 (1970).
18. Pauli, H. C., Ledergerber, T., and Brack, M., "Asymmetry in Nuclear Fission", Phys. Letters 34B, 264 (1971).
19. Pashkevitch, V. V., Dubna Preprint JINR-P4-5581 (1971).
20. Swiatecki, W. J., "The Nuclear Surface Energy", Proc. Phys. Soc. (London) A164, 226 (1951).
21. Swiatecki, W. J., "The Effect of a Potential Gradient on the Density of a Degenerate Fermi Gas", Proc. Phys. Soc. (London) A68, 285 (1955).
22. Swiatecki, W. J., "Nuclear Compressibility and Fission", Phys. Rev. 83, 178 (1951).

23. Swiatecki, W. J., "The Density Distribution Inside Nuclei and Nuclear Shell Structure", Proc. Phys. Soc. (London) A63, 1208 (1950).
24. Tyapin, A. S., "Compressibility and Polarizability of Nuclear Matter", Yadern. Fiz. 1, 581 (1965) [Engl. Trans.: Soviet J. Nucl. Phys. 1, 416 (1965)].
25. Bodmer, A. R., "Neutron and Proton Densities in Nuclei and the Semi-Empirical Mass Formula", Nucl. Phys. 17, 388 (1960).
26. Truran, J. W., Cameron, A. G. W., and Hilf, E., "Construction of Mass Formulas Designed to be Valid for Neutron-Rich Nuclei", Proceedings of the International Conference on the Properties of Nuclei Far From the Region of Beta-Stability, Leysin, Switzerland, August 1970.
27. Ludwig, S., et al., "Droplet Model Mass-Formula Fit", to be published in Nuclear Physics.
28. Hasse, Rainer W., "Studies in the Shape Dependence of the Droplet Model of Nuclei (Curvature and Compressibility Effects)", Lawrence Radiation Laboratory Report UCRL-19910, February 1971.
29. Zeldes, N., Grill, A., and Simievic, A., "Shell-Model Semi-Empirical Nuclear Masses (II)", Mat. Fys. Skr. Dan. Vid. Selsk. 3, no. 5 (1967).
30. Seeger, P. A., "A Mass Law With Model-Based Shape-Dependent Shell Terms", Proceedings of the International Conference on Properties of Nuclei Far From the Region of Beta-Stability, Leysin, Switzerland, August 1970.
31. Myers, William D., "Droplet Model Nuclear Density Distributions and Single-Particle Potential Wells", Nucl. Phys. A145, 387 (1970)
32. Myers, William D., "Droplet Model Isotope Shifts and the Neutron Skin", Phys. Letters 30B, 451 (1969).

BINARY CLUSTERING WITH REFERENCE TO FISSION

M. Harvey
Atomic Energy of Canada Limited
Chalk River Nuclear Laboratories
Chalk River, Ontario, Canada

In these talks I wish to discuss clustering in the broader context of the division of any A-particle system into two parts of mass A_1 and A_2 (= $A-A_1$). Thus the type of data that we are interested in can be found in the fission phenomenon as well as ion-ion collision.

Interest in fission has blossomed within the past 10 years with the discovery of isomeric fissioning states[1] and the possibility of discovery of a set (perhaps with one element?) of super-heavy nuclei far beyond the mass of the heaviest nuclei known today.[2-5] If one accepts the description of the fission process as a decay through a barrier then the isomeric fission data implies[3] states of shorter lifetimes which could arise from a barrier which has far more structure than that suggested by the liquid drop model. This structure can arise from the microscopic (granular) structure of the nucleus. Of course we have known from the very earliest data on fission that the microscopic structure plays a role because a nucleus like uranium prefers to fission spontaneously into fragments of unequal mass[6]. The particular mass division has always been assumed to arise from the shell structure of the parent nucleus and its fragments. The procedure followed in recent calculations[2,3,4,7] has been to add the shell structure to the fission barrier calculated from the liquid drop model using the Strutinsky technique.[3] Our aim in these lectures will be to develop a theory of fission which begins with a parent nucleus described in a shell model and ends with fragments also described in a shell model. We shall make contact with the liquid drop approach but in a different way to Strutinsky et al.[3]

By shell model I imply here that the many particle states of the system can be described in terms of the single particle states arising perhaps from some average self-consistent single particle potential. In fact I shall concentrate in these talks on a simple Slater determinantal state describing the A-particle system throughout the fission process.

Despite the voluminous literature on fission I would like to briefly introduce the basic approach to give the uninitiated a chance to put my talk into perspective.

A PERSPECTIVE

The gross features of the fission process are treated rather simply in the liquid drop model. Basically one assumes that, on fissioning, a nucleus (like uranium) divides into two fragments which then move apart under the repulsive Coulomb force. Only for small distances are the fragments under the attraction of the nuclear forces. One can guess that the potential energy between fragments then has the familiar form with respect to separation as given in Fig. 1.

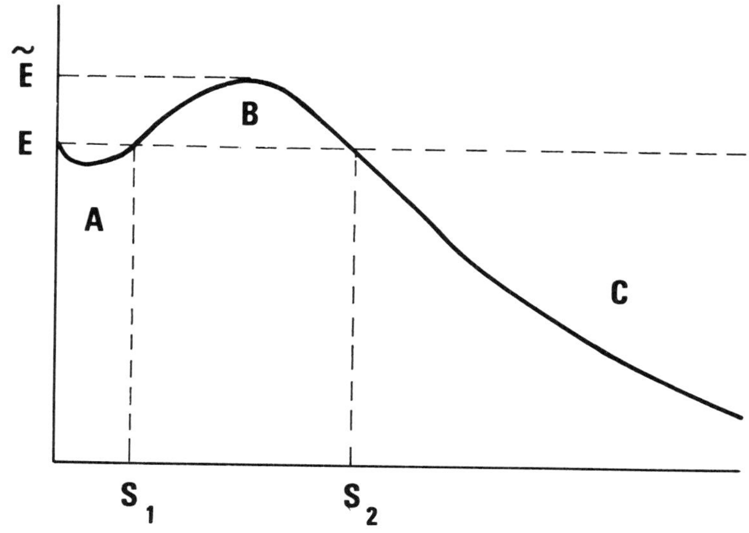

Figure 1
Schematic structure of the fission barrier for a heavy nucleus.

The problem of calculating lifetimes[4] is one of deducing the penetration P of the barrier (B) from the initial position A to the position of separated fragments C. The penetration factor is just the probability of penetrating the barrier per given assault. For n assaults per sec, the decay half-life is just given by

$$\tau = \frac{\ln 2}{nP} \text{ sec} \quad . \tag{1}$$

To get an estimate of P we can use the WKB method.[16] We assume the functional form W(S) of the potential curve as a function of separation S. We define separations S_1 and S_2 such that

$$W(S_1) = W(S_2) = E = \text{Total Energy.}$$

Now the W.K.B. method tells us

$$P = \exp\left\{-2 \int_{S_1}^{S_2} \sqrt{\frac{2\mu}{h^2}\,(W-E)}\; dS\right\} \tag{2}$$

For a parabolic dependence of W between S_1 and S_2

$$P = \exp\left\{-2\pi \;\; \frac{\tilde{E}-E}{\hbar\omega_f}\right\} \tag{3}$$

where

$$\omega_f = \sqrt{C/\mu}: \quad C = \text{curvature} = 2\tilde{E}/(S_2-S_1)^2$$

$$\mu = \text{effective mass}$$

$$\tilde{E}-E = \text{height of barrier.}$$

The number of assaults per second is $n = \omega/2\pi$. If we assume the vibrational energy of the parent nucleus is $\hbar\omega \approx 1$ MeV then

$$n = 1/(2\pi\hbar) = 2.4 \times 10^{20} \text{ sec}^{-1}$$

$$\tau = \frac{\ln 2\times 2\pi\times\hbar}{P} = \frac{10^{-20.54}}{P} \text{ sec} = \frac{10^{-28.04}}{P} \text{ yr} \tag{4}$$

If there is certainty of penetrating the barrier, P=1 and the lifetime is $10^{-20.54}$ sec. What is a realistic value for the penetration factor? By analysing the spontaneous decay rates of known nuclei[4] and equating to eq. 4 with eq. 3 one can estimate barrier heights and the fission frequency $\hbar\omega_f$ for the barrier. It turns out that $\tilde{E}-E \sim 6$ MeV and $\hbar\omega_f \sim 0.3$ MeV,

$$\text{i.e.} \quad P \approx 2.66 \times 10^{-55}$$

$$\text{or} \quad \tau \sim 10^{24} \text{ yr} \qquad (5)$$

If the barrier height is decreased to 5 MeV then

$$P = 3.3 \times 10^{-46}$$

$$\text{and} \quad \tau \sim 10^{17} \text{ yr}$$

Thus a change of ∿20% in the barrier height can make a difference of a multiplicative factor of 10^7 in the lifetime measured in years.

Thus one must calculate the barrier height to an accuracy better than ∿20% in order to predict log τ with errors less than ±7. The superheavy (eka-lead) nuclei are being predicted at the moment[3] with lifetimes of $\sim 10^7$ yr with models for the barrier height in which the accuracy is not much better than ∿20%. Whether the lifetime is long enough to find a superheavy nucleus in old ores[20] or whether it is so short (∿1 yr) to allow discovery only in the lab[7] or whether the barrier is practically negligible ($\tau \sim 10^{-21}$ sec) so that superheavy nuclei only exist as a figment of theorists' imaginations is still very much an open question. This question however is still fascinating enough to stimulate the development of better models.

Looked at a different way, the fission process involves the reorganisation of binding energy of an A-particle system from close interaction to separated fragments. This results (in uranium) in a transfer of about 200 MeV of internal energy of the A-particle system into kinetic energy and excitation energy of the fragments. Thus one is interested in getting barrier heights correct to better than 1/2% of this energy transfer - a formidable task when one considers the crudeness of present models.

It is because such small changes in barrier height can make such enormous changes in lifetime that the small fluctuations of the shape of the barrier due to shell effects can have such a large effect. Perhaps it is understandable that different groups can get such different estimates for the lifetimes of superheavy nuclei. Nilsson et al.[4] predict $\sim 10^7$ yr while Malik et al. and Greiner et al.[7] predict ~ 1 yr (i.e. a superheavy can be produced in the lab but will not be found in nature).

For those who like relevance in their work the study of the fission process should make them happy. The ~ 200 MeV that is released in fission is of course what forms the basis of the nuclear power industry; the significance of this number is really quite easy to demonstrate. In 1 c.c. of ^{235}U there are about 0.05×10^{24} nuclei (atomic spacings being of the order of an angstrom) and so the amount of energy released if all nuclei fission is $\sim 10^{25}$ MeV $\equiv 1.6 \times 10^{19}$ ergs. If this "burn-up" is achieved every second the power output is 1.6×10^{12} watts. This is really an enormous power output; the hydro-electric station at Niagara Falls[8] generates power at the rate of 3.6×10^9 watts arising from a flow[19] of 130,000 cu. ft. of water every second falling through a distance of more than 167 ft. Of course knowing the potential of nuclear fission and converting to practise are two different things. In the first place ^{235}U is not isolated naturally - in natural uranium (used in Canadian Reactors) ^{235}U only forms $\sim 0.7\%$ of the whole. Thus the fissioning of ^{234}U in 1cc/sec burnup of natural uranium only produces a power of $\sim 10^{10}$ watts. Subtract thermal efficiency of extraction etc. and one can estimate the equivalence of about 1 Niagara. It is a formidable engineering problem to be able to handle this much power in a reactor and so modern nuclear-power stations have a lower power output per reactor with a slower burnup - but the power output is still appreciable. In the power station now being constructed in Pickering, Ontario there are four reactors each of which will have a power capacity of 0.5×10^9 watts.[9]

I have introduced the application of the fission process to power reactors partly for amusement but also so that you can better appreciate the significance of the seemingly innocent number of ~ 200 MeV released in the fission of uranium. One of the main problems for the modern-day physicist is to be able to recognize when a seemingly innocent piece of experimental data has important implications in other spheres, e.g. industrial, astrophysical, medical etc. To have this ability we must not isolate the nucleus too much from its environment in our probing but be well aware of the significance of a nuclear property to the outside world.

Let me return now to being a theoretical physicist. One of the basic problems in computing the fission barrier is to be able to define the parameter describing the fission process. For large separation of fragments this parameter is naturally chosen to be the distance between fragments. For small separation however the fragments overlap and lose their identity - thus one cannot strictly speak of their centres of mass. This is unfortunate because it is at small distances that we hope to understand how the nucleus breaks up. One needs therefore to define a fissioning parameter that is valid for all distances but is equivalent to the separation of the centres of mass for large separation.

Nilsson and co-workers[4] have described the deformation of the nucleus at small separation in terms of parameters λ which are the orders of the spherical harmonics Y^{λ} describing the shape of the nucleus ($\lambda = 2,3,4$ powers only have been assumed). We have taken another approach which forms the heart of these talks.

All of our calculations to date have been for "fissioning" light nuclei since in developing a model one chooses the simplest calculations first.[2a]

DISTRIBUTION OF BINDING ENERGY IN GROUND STATE [10]

In this section I wish to review some background information. First we assume that the lowest states of a nucleus can be described in terms of a single Slater determinant χ. This determinant may represent the lowest state directly as in a spherical nucleus or perhaps just represent the intrinsic state from which the lowest states can be projected. This Slater determinantal state is assumed to possess all the gross properties of the nucleus. Thus since fission involves the reorganisation of the total binding energy, we should expect to find the description of the gross properties of fission (at least) in the Slater determinantal state.

Let us now consider the division of the orbits α_i of χ into two sets $S_1(\alpha_1 \cdots \alpha_{A_1})$ and $S_2(\alpha_{A_1+1} \cdots \alpha_A)$. We make the usual assumption that the total energy of the determinantal state χ is given by the expectation value of a two body Hamiltonian.

$$E = \langle\chi|H|\chi\rangle = \sum_{i<j} \langle\alpha_i\alpha_j|H|\alpha_i\alpha_j-\alpha_j\alpha_i\rangle$$

$$= \sum_{i,j\epsilon S_1} + \sum_{i,j\epsilon S_2} + \sum_{i\epsilon S_1} \sum_{j\epsilon S_2}$$

$$= E_{A_1} + E_{A_2} + B_{A_1}.$$

One can consider that E_{A_1} and $E_{A_2} = E_{A-A_1}$ represent the internal energy in (hidden) clusters of mass A_1 and $A_2 = A-A_1$. The B_{A_1} represents the energy of the bond between the clusters. It is well known however that the occupied orbitals in a Slater determinant are undefined to within a unitary transformation.

Proof

Let $$\beta_i = \sum_j U_{ij}\ \alpha_j$$

where U is a unitary transformation. Then[21]

$$|\beta_1\text{---}\beta_A| = \sum_{j_1\text{--}j_A} U_{1j_1}\text{---}U_{Aj_A}\ |\alpha_{j_1}\text{---}\alpha_{j_A}|$$

$$= \sum_{[j_i]} U_{1j_1}\text{---}U_{Aj_A}\ P_{[j_i]}\ |\alpha_1\text{---}\alpha_A| \qquad (7)$$

The summation in the last expression is over all permutations $[j_i]$ of the numbers 1--A; the summation can be restricted to the permutation of the numbers rather than all values j_i=1--A (i=1--A) from the property that the Slater determinant $|\alpha|$ is zero if two or more rows are equivalent. The operator $P_{[j_i]}$ means the permutation operator taking the standard numerically increasing order of the numbers 1--A into the particular ordering $[j_i]$. Since $|\alpha|$ is a determinant

$$P_{[j_i]}|\alpha_1\text{---}\alpha_A| = (-1)^{[j_i]}\ |\alpha_1\text{---}\alpha_A|$$

where $(-1)^{[j_i]} = \pm 1$ depending on whether $[j_i]$ is an even or odd permutation. The summation in eq. (7) does not involve the Slater determinant $|\alpha|$ and is in fact a way of writing det $U = e^{i\delta}$. Thus $|\beta| \equiv |\alpha|$ to within an overall phase.

Thus there is an infinite number of ways of defining the orbits in the sets S_1 and S_2 as described above and hence no unique way of describing fragments of a given mass within the nucleus unless we impose some other condition. Normally, in the shell model, the condition is imposed that the occupied orbitals be eigenfunctions of the assumed single particle potential; this is not necessarily the best representation however when considering the initial stages of fission. What is likely to be a better representation?

If the A-particle nucleus is going to binary fission adiabatically then the initial structure of its fragments is likely to be that for which the binding energy B_{A_1} associated with the bond between fragments is minimal - (after all a chain will break at its weakest link). This condition uniquely defines the structure of the sets of orbits S_1 and S_2. To find the "minimal" set of orbits[10] we write

$$B_{A_1} = <\chi_A|H(A)|\chi_A> - <\chi_{A_1}|H(A_1)|\chi_{A_1}> - <\chi_{A_2}|H(A_2)|\chi_{A_2}> \tag{8}$$

where the Hamiltonian is given an A-dependence for generality. The minimal condition leads to

$$\delta B_{A_1} = 0 = (\delta\chi_{A_1}|H(A_1)|\chi_{A_1}) + <\chi_{A_1}|H(A_1)|\delta\chi_{A_1}>$$
$$+ <\delta\chi_{A_2}|H(A_2)|\chi_{A_2}> + <\chi_{A_2}|H(A_2)|\delta\chi_{A_2}> \tag{9}$$

where variation is considered only in the space of the occupied orbitals and hence $\delta<\chi_A|H(A)|\chi_A> \equiv 0$. Following the common derivation of the Hartree Fock equations[11] we write

$$\delta\chi_{A_1} = \eta\, a_t^\dagger\, a_r\, \chi_{A_1} \tag{10a}$$

where orbit $r \in S_1$ and hence $t \in S_2$. Because the space of variation is closed we can immediately see that

$$\delta\chi_{A_2} = -\eta\, a_r^{\dagger}\, a_t\, \chi_{A_2}. \tag{10b}$$

For separate variation of a wave function and its complex conjugate the minimal condition reduces to

$$<\chi_{A_1}|a_r^{\dagger} a_t H(A_1)|\chi_{A_1}> - <\chi_{A_2}|a_t^{\dagger} a_r\, H(A_2)|\chi_{A_2}>=0. \tag{11}$$

Writing the Hamiltonian in a second quantised form[22]

$$H(A) = \frac{1}{4} \sum_{\alpha\beta\gamma\delta} h^{A}_{\alpha\beta\gamma\delta}\; a_{\alpha}^{\dagger} a_{\beta}^{\dagger} a_{\delta} a_{\gamma}, \tag{12}$$

with $\quad h^{A}_{\alpha\beta\gamma\delta} = <\alpha\beta|H(A)|\gamma\delta-\delta\gamma>,$

we can write the minimal condition as

$$\sum_{s\epsilon S_1} h^{A_1}_{tsrs} - \sum_{s\epsilon S_2} h^{A_2}_{tsrs} = 0 \tag{13a}$$

where $r\epsilon S_1$ and $t\epsilon S_2$.

The similarity with the standard Hartree-Fock method is apparent; in fact the first (or second) term in eq. (13a) when equated to zero is the condition that the binding energy of the set S_1 (or S_2) is an extremum (and hopefully a maximum). Of course the set of orbits that maximise the binding in set S_1 does not necessarily leave the structure of the remaining orbits of the set of A orbits such that they will maximise the binding in set S_2. In general both sets (clusters, fragments) will have to sacrifice internal binding energy in order that the binding between fragments is a minimum; hence the two terms in eq. 13a.

We have expressed the minimal condition in eq. (13a) in terms of the minimal set of orbits. It is usual to consider a standard representation $(\alpha,\beta,\gamma$ --) for the A-occupied orbits. Now expressing the minimal set in terms of the standard set we find

$$|r> = \sum_{\alpha} C_{r\alpha}|\alpha>$$

and eq. (13a) becomes

$$\sum_{\alpha\gamma} C_{\alpha t}^{*} C_{\gamma r} W_{\alpha\gamma} = 0 \tag{13b}$$

with

$$W_{\alpha\gamma} = \sum_{\beta\delta} \sum_{s\epsilon S_1} C_{\beta s}^{*} C_{\delta s} h_{\alpha\beta\gamma\delta}^{A_1} - \sum_{s\epsilon S_2} C_{\beta s}^{*} C_{\delta s} h_{\alpha\beta\gamma\delta}^{A_2}$$

For the minimal set of states the matrix W is diagonal - thus the problem reduces to finding the eigenfunctions $C_{\gamma r}$ of the matrix W. But since W is a function of the C we have a familiar self-consistency condition to be solved iteratively.

Before solving for this minimal set let us think a little about the kind of solution we should expect. If the nucleus is to fission into heavy fragments then nuclear matter to the left of the nucleus will move off to the left and matter to the right will move off to the right. Thus if the minimal set describes the initial stages of fission the orbits should be such that some have their centres of mass to the left of the centre of mass of the whole nucleus and some will have their centres of mass to the right. Note that since all orbits are occupied the centre of mass of the whole nucleus does stay at the origin. This intuitive physical picture of the fission process suggests an easier construction of the "minimal" set. We take the occupied orbitals and diagonalise the dipole operator z in the direction of fission. Since we are only diagonalising within the occupied set of orbitals then we are merely making a unitary transformation among them and the total Slater determinant is left undisturbed. By ordering the eigenfunctions of the dipole operator with respect to the eigenvalue we can clearly order our new set of single particle orbitals with respect to their centres of mass. As an example consider two like fermi particles occupying the zero- and one-quantum states of a one dimensional harmonic oscillator well. The functional form for these orbits are

$$|0\rangle = \sqrt{\frac{1}{\sqrt{\pi}\, b}} \exp(-z^2/2b^2) \tag{14a}$$

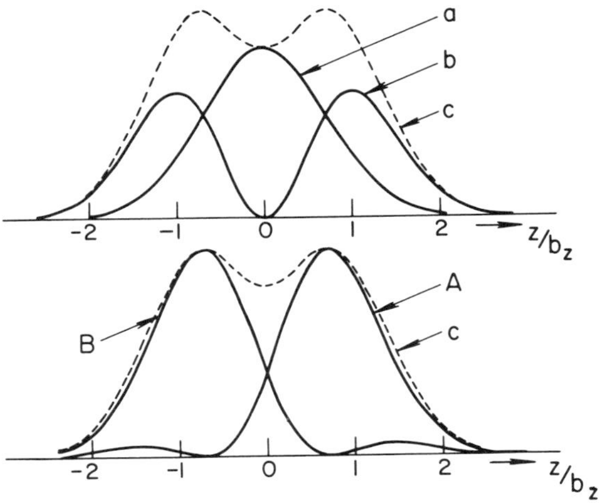

Figure 2
Dashed curve shows the density distribution of a two particle Slater determinant with the lowest zero and one quantum states of a one dimensional harmonic oscillator occupied. Solid curves at top show the zero and one quantum states (eq. 14) and those at bottom the "dipole" represented states of eq. 15.

$$|1\rangle = \sqrt{\frac{1}{\sqrt{\pi}\, b}}\ \sqrt{2}\ z/b\ \exp(-z^2/2b^2). \tag{14b}$$

Clearly if we diagonalize the operator z we shall generate orbits

$$\phi^{\pm} = \sqrt{1/2}\ \{|0\rangle \pm |1\rangle\}. \tag{15}$$

In Fig. 2 I show the density distribution of the determinant $\chi = \big||0\rangle\ |1\rangle\big| \equiv |\phi^{+}\phi^{-}|$. In the upper part of Fig. 2 the density is made up from the sum of the density distributions of the orbits $|0\rangle$ and $|1\rangle$ and in the lower part this same density is made up from the ϕ^{+} and ϕ^{-} components. In this way one can see that whereas the orbits $|0\rangle$ and $|1\rangle$ had their centres of mass at the origin, the orbits $\phi^{\pm}$ clearly have their centres of mass to the right and left respectively of the origin. This can be done for any determinantal state. In Fig. 3 I show the contour diagrams on the x-z plane for the density distributions of the axially symmetric

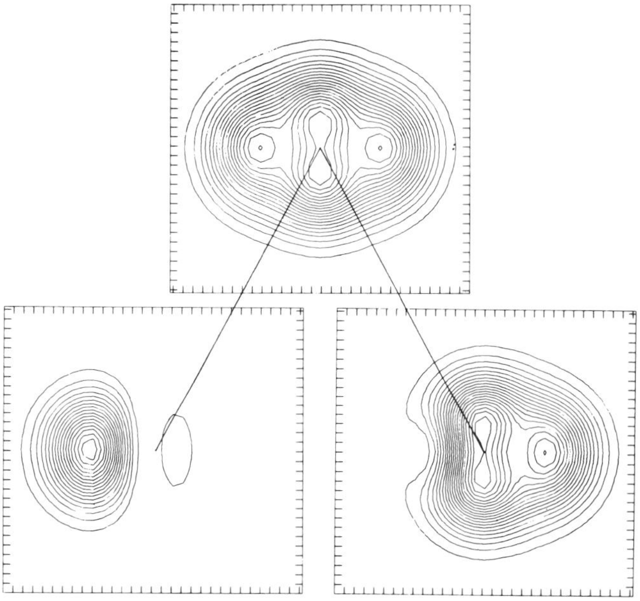

Figure 3
Top diagram shows the density distribution of the SU_3 intrinsic state for ^{20}Ne and how this is made up from the four particle (bottom left) and sixteen particle (bottom right) states of the dipole representation.

intrinsic state for ^{20}Ne in the spherical SU_3 shell model and also the density distribution for fragments of mass 16 and 4 described by the "dipole" representation. It is interesting to note that the four particle fragment looks a little like a spherical ball (α-particle?) whereas the ^{16}O cluster is clearly in a highly excited state. The whole ^{20}Ne structure can thus be viewed as ^{16}O holding an α-particle in its mouth.

I have now introduced three representations of the occupied orbitals, i.e. (a) the ordinary eigenfunctions of the single particle Hamiltonian; (b) the "minimal" set; and (c) the "dipole" set. Let us now examine how the binding energy between fragments of different masses differs in the three sets. As an example we consider the binding energy distribution in ^{16}O using a Hamiltonian of two body kinetic and potential energy operators

$$H = T_A + V \qquad (16)$$

$$T_A = \frac{1}{2mA} \sum_{i<j} (p_i - p_j)^2 = \frac{1}{2m} \sum_i p_i^2 - \frac{1}{2mA} (\sum_i p_i)^2 \quad (17)$$

$$V(r_{ij}) = (W + MP_{ij}^{r} + BP_{ij}^{\sigma} - HP_{ij}^{\tau})U(r_{ij})$$
$$U(r_{ij}) = -V_a \exp(-r_{ij}/a_1^2) + V_r \exp(-r_{ij}^2/a_2^2) \quad (18)$$

Here clearly T_A is the kinetic energy operator of the relative coordinates after taking out the centre of mass motion. The potential form chosen is that suggested by Volkov[12] with B=H=0. Parameters for the first three of Volkov's forces are given in Table 1.

Table 1

FORCE	W	M	$-V_a$	a_1	V_r	a_2
1	0.4	0.6	83.34	1.6	144.86	0.82
2	0.4	0.6	60.65	1.8	61.14	1.01
3	0.4	0.6	106.67	1.5	106.67	1.05

In Fig. 4 I show the structure for E_{A_1}, E_{A_2} and B_{A_1} for the "minimal" set of single particle orbits. For each orbit the filling was n^+, n^-, p^+, p^- where n^+ means neutron with spin up, etc. Notice the structure of the B_{A_1} curve from which one deduces that the "fragments" of mass four and eight are weaker bound than the fragments with neighbouring mass. A similar set of curves is shown in Fig. 5, which is for the dipole set of single particle orbits - differences occur only for the very light fragments (mass 1 or 2). Before showing the next figure I would like you to note first the apparent binding of fragments. The four particle system is apparently bound by ∿90 MeV for example. In Fig. 6 I show the same set of curves again for the normal single particle orbits with the filling s_0, p_0, p_{+1} p_{-1}. These curves correspond to those for the minimal set only for mass 1 and 2 but for heavier masses provide a fragmentation with binding between fragments that is larger than the minimal or the equivalent dipole. It is instructive to compare these curves in Figs. 4 - 5 with the ground state binding energies $\tilde{E}$ of nuclei of mass A_1 and A_2

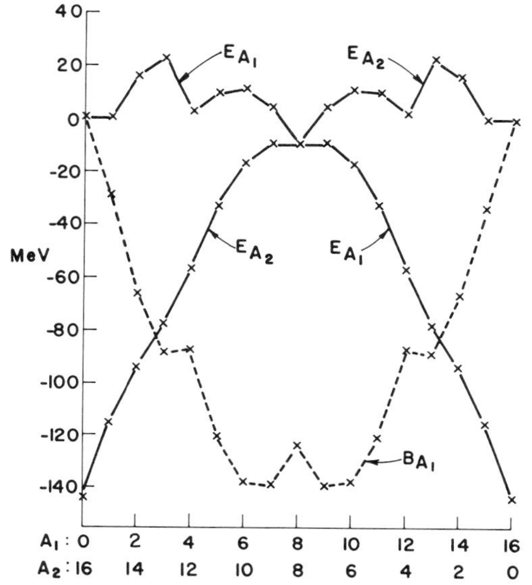

Figure 4
Energies of (hidden) clusters E_{A_1}, E_{A_2} and the binding energy B_{A_1} between the clusters for states in the "minimal" representation of the occupied orbits of ^{16}O.

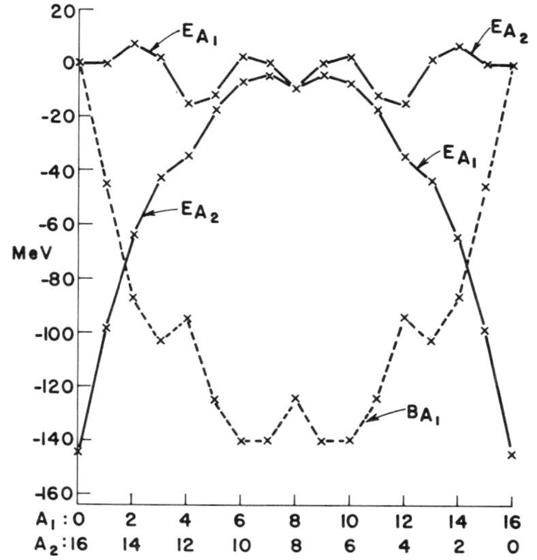

Figure 5
As in Fig. 4 but for the dipole representation.

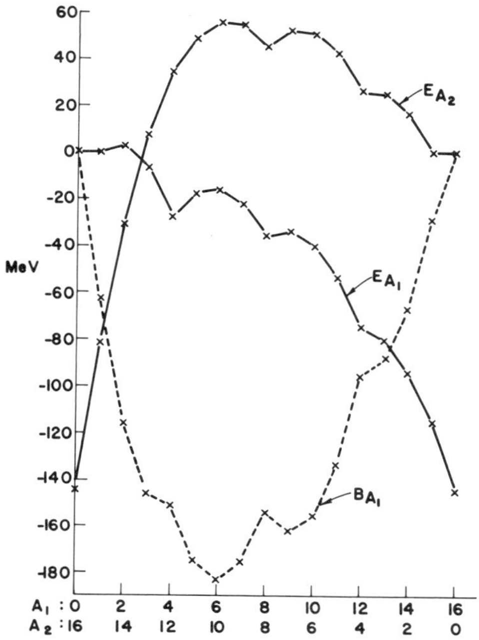

Figure 6
As in Fig. 4 but for the familiar shell model ordering of orbits.

and the Q-value $Q_{A_1} = \tilde{E}_{16}-\tilde{E}_{A_1}-\tilde{E}_{A_2}$ as shown[23] in Fig. 7. If fragments of masses A_1 and A_2 were in their ground states then B_{A_1} should correspond to the Q-value. Two points to notice: the shapes of the curves B_{A_1} in Figs. 4-5 and $B_{A_1}=Q_{A_1}$ in Fig. 7 are very similar but the Q-values[23] are much smaller. Typically for mass 4 the Q-value is 16 MeV to be compared with the 90 MeV for the "minimal" and "dipole" sets. Thus since pulling the ^{16}O nucleus apart takes only 16 MeV (without help from the Coulomb force[24]) there is clearly a redistribution of binding energy within the ^{16}O nucleus as it begins to separate, i.e. fragments sink into lower states with energy going into the bond between fragments to help fission. Our hypothesis however is that there are "weaknesses" in the nucleus before fission which predetermine how the nucleus will eventually break up. The rough analogy here is with the score in a piece of glass which will determine where the glass will break if tapped (i.e. energy applied). Of course

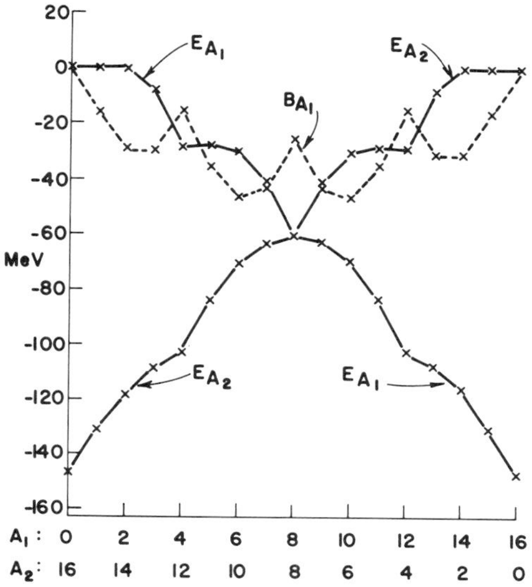

Figure 7
Binding energy of nuclei for $A_1(A_2)=0$ to 16 (E_{A_1} and E_{A_2}) and the difference $B_{A_1}=Q_{A_1} = E_{16}-E_{A_1}-E_{16-A_1}$. The observed binding energy has been increased by 0.7 MeV per pair of protons.

the analogy is not complete because the glass on either side of the score does not change its structure after breaking as must the nucleus.

With this background information I wish to discuss now a model in which we could pull the nucleus apart into fragments of any mass. At each stage of separation we could calculate energy systematics and hopefully be able to calculate the barrier for fission. I shall discuss this new approach in general but demonstrate it in a very simple example of the α-clustering cluster of ^8Be and fission of ^{16}O. In the ^8Be case I shall compare with the molecular orbital technique and show that it has several distinct advantages.

A DOUBLE SHELL MODEL FOR BINARY FISSION

In the previous discussion we have noted the non-uniqueness of the definition of fragments within a nucleus without some further condition. Whichever way we define a fragment the orbits are always represented in terms of the A-orbits of the original nucleus. In the shell model these are the lowest A-orbits of the single particle well. For separated fragments the structure of each fragment is represented in terms of the lowest A_1 (or A_2) orbits in the wells associated with each fragment; again this can be considered as a special case of the filling of the lowest A-orbits in each well. We have therefore proposed the following "double shell model" to describe the process of a nucleus undergoing adiabatic binary fission.[25] We suppose that at *each stage* of the separation the structure of each fragment of mass A_+ and A_- can be represented in terms of the lowest A $(= A_+ + A_-)$ orbits in two wells at distances Z_+ and Z_- from the origin.[26] The scheme is described in Fig. 8 for harmonic oscillator wells. Thus the structure of the fragment of mass A_+ is given in terms of the lowest A orbits in the + well and the fragment of mass A_- is given in terms of the lowest A orbits in the - well. The constraints in the system are that at each stage of the fission process the orbits of the + fragment must be orthogonal to those of the - fragment. Can this be achieved? The answer is yes. In fact given the structure of one of the fragments, that of the other is *uniquely* determined for finite separation. To demonstrate this we write the A_+ orbits ϕ_i^+ in terms of a basic set of lowest states $|j>^+$ in the + well;

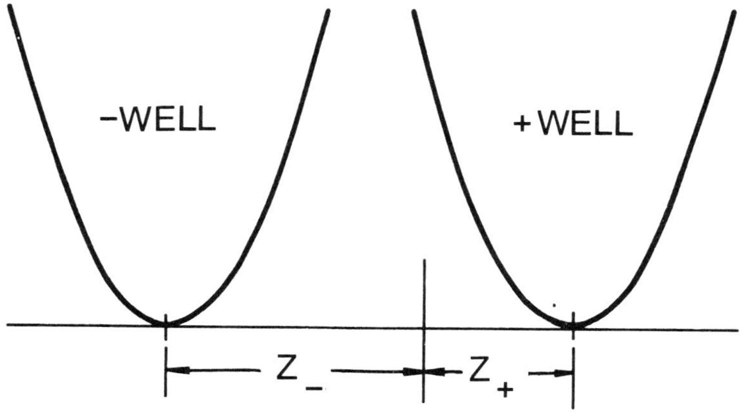

Figure 8
Schematic representation and definition of the double single particle wells.

$$\phi_i^{+} = \sum_{j=1}^{A} c_{ij} \,|j\rangle^{+} \qquad i = 1---A_+ \tag{19}$$

Similarly the A_- orbits ϕ_i^- can be written in terms of the basic set of lowest states $|j\rangle^-$ in the - well;

$$\phi_i^{-} = \sum_{j=1}^{A} d_{ij} \,|j\rangle^{-} \qquad i = 1---A_-. \tag{20}$$

The structure of the A particle system is given by the Slater determinant

$$\chi_Z = |\phi_1^{+}\phi_2^{+}---\phi_{A_+}^{+} \quad \phi_1^{-}---\phi_{A_-}^{-}|. \tag{21}$$

Here Z is a parameter defining the distance apart of the two wells; for example, to try to keep the centre of mass of the system centred at the origin we might choose $Z_+ = A_- Z/A$ and $Z_- = -A_+ Z/A$.

Since all the ϕ_i^- are to be occupied we are free to make any unitary transformation among them without changing the actual structure of the A_- fragment. We make such a transformation and seek the particular solution such that

$$d_{ij} = 0 \qquad \text{for} \qquad j \geq A_+ + i+1.$$

The number of non-zero coefficients d is therefore

$$AA_- - A_-(A_- -1)/2 = A_+A_- + A_-(A_- +1)/2$$

and is illustrated in Fig. 9.

The orthonormality constraints between the ϕ^- orbits lead to $A_-(A_-+1)/2$ conditions and the orthogonality conditions with the ϕ_+ orbits lead to a further A_-A_+ conditions. Since the number of conditions equals the number of constraints then the d-coefficients are uniquely determined once the c-coefficients have been chosen unless some of the conditions become identical or trivial. For example in determining the condition of orthogonality between orbits in the two wells we have to know the overlaps $\langle j^-|k^+\rangle$. For very large distance between the wells these go to zero as

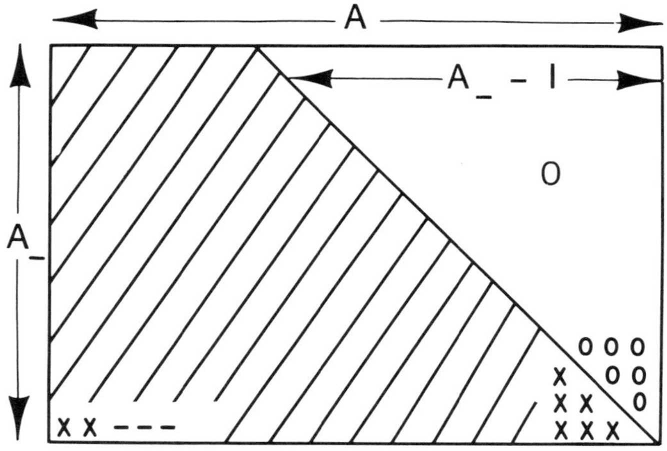

Figure 9
Distribution of non-zero d-coefficients

we shall show explicitly later - then there is no restriction on the structure of one fragment due to the other. For small separations however, which is what we are interested in, the overlaps of the orbits in the two wells are non-zero and the condition of orthogonality is sufficient to allow the calculation of the d-coefficients.

The number of degrees of freedom (α) is determined by the number of ways of choosing the c-coefficients and the shape and size parameters of the two wells. In determining the degrees of freedom in choosing the c-coefficients we note that we can first bring the orbits of the + fragment to a standard form by a unitary transformation among them without changing the overall structure. We can therefore choose the c_{ij} such that

$$c_{ij} = 0 \quad \text{for} \quad j \geq A^- + i+1.$$

The number of non-zero c-coefficients is therefore

$$AA_+ - A_+(A_+-1)/2 = A_+A_- + A_+(A_++1)/2.$$

The number of orthonormality conditions among the c-coefficients is $A_+(A_++1)/2$ leaving a final degree of freedom of A_+A_-. Note that this number is symmetrical in A_+ and A_- showing that there is the same number of degrees of freedom by choosing the ϕ_i^+ orbits and then finding the ϕ_i^- by orthogonality as there is by choosing the ϕ_i^- and finding the ϕ_i^+.

The size and shape parameters of each well together with the A_+A_- c-coefficients can be chosen so as to minimise some function of the energy at each stage of the separation. It is by no means easy to define exactly what function of the energy one should be minimising. Let me return to this however, after showing a simple example so that you can become more familiar with the model.

Suppose we have two like fermi particles in a one dimensional harmonic oscillator well. The lowest state of the system is

$$\chi = \big|\,|0\rangle\;|1\rangle\big|. \tag{22}$$

Now we suppose we wish to pull the two particles apart to a situation in which each particle is in the lowest state of wells at $\pm\infty$. We suppose therefore that at any finite separation the structure of the orbit in the + well at a distance +Z from the origin is given by

$$\phi^+ = \alpha\;|0\rangle^+ + \sqrt{1-\alpha^2}\;|1\rangle^+. \tag{23}$$

We could now find the orbit ϕ^- in the well at -Z from the origin such that $\langle\phi^-|\phi^+\rangle \equiv 0$. We would have $A_+A_- = 1\times 1 = 1$ degree of freedom, i.e. the parameter α. Actually we can simplify this particular case by invoking symmetry, i.e. the structure of the orbital in the well at a distance -Z from the origin should be similar to that in the + well at a distance +Z from the origin - thus

$$\phi^- = \alpha|0\rangle^- - \sqrt{1-\alpha^2}\;|1\rangle^- \tag{24}$$

Symmetry dictates that the size parameter of the - well is the same as that for the + well. The symmetry condition also actually determines the single parameter, i.e.

$$\langle\phi^-|\phi^+\rangle = \alpha^2\,\langle 0^-|0^+\rangle - (1-\alpha^2)\langle 1^-|1^+\rangle + \alpha\sqrt{1-\alpha^2}\,(\langle 0^-|1^+\rangle - \langle 1^-|0^+\rangle) = 0. \tag{25}$$

The overlap between one dimensional harmonic oscillator orbitals in wells at a distance Z_1 and Z_2 from the origin can easily be determined in general. Each function is given in terms of the number of quanta n - thus an orbital in a well centred at Z_1 with oscillator parameter b_1 is given by

$$|n1\rangle = \sqrt{\frac{1}{\sqrt{\pi}\, 2^n b_1}}\; H_n\left(\frac{z-Z_1}{b_1}\right) \exp\left(-\frac{(z-Z_1)^2}{2b_1^2}\right) \tag{26}$$

where H_n is a Hermite polynomial.

Using the recurrence relation for Hermite polynomials it can be shown that the following recurrence relations exist for the overlaps:

$$\begin{aligned}
\langle n1|m2\rangle &= \Bigg(-\sqrt{\frac{n-1}{n}}\,(b_1^2-b_2^2)\,\langle n-2\;1|m2\rangle \\
&\quad - b_1\,(Z_1-Z_2)\sqrt{\frac{2}{n}}\,\langle n-1\;1|m2\rangle \\
&\quad + \sqrt{\frac{m}{n}}\,2b_1b_2\,\langle n-1\;1|m-1\;2\rangle\Bigg)\Big/(b_1^2+b_2^2) \\
&= \Bigg(+\sqrt{\frac{m-1}{m}}\,(b_1^2-b_2^2)\,\langle n1|m-2\;2\rangle \\
&\quad + b_2(Z_1-Z_2)\sqrt{\frac{2}{m}}\,\langle m1|m-1\;2\rangle \\
&\quad + \sqrt{\frac{n}{m}}\,2\,b_1b_2\,\langle n-1\;1|m-1\;2\rangle\Bigg)\Big/(b_1^2+b_2^2)
\end{aligned} \tag{27}$$

It can be shown by direct integration that

$$\langle 01|02\rangle = \sqrt{\frac{2\,b_1b_2}{(b_1^2+b_2^2)}}\;\exp\left(-\frac{(Z_1-Z_2)^2}{2(b_1^2+b_2^2)}\right) \tag{28}$$

Using the recurrence relation in eq. 27 together with the special case in eq. 28 all other overlaps can be computed. In particular

$$<11|02> = -b_1(Z_1-Z_2)\ \sqrt{2}\ <01|02>/(b_1^{\,2}+b_2^{\,2})$$

$$<01|12> = -\ <11|02>\ b_2/b_1$$

$$<11|12> = 2b_1b_2\{(b_1^{\,2}+b_2^{\,2})-(Z_1-Z_2)^2\}$$
$$<01|02>/(b_1^{\,2}+b_2^{\,2})^2.$$

Note incidentally that the overlap between oscillator orbits in different wells decreases like $\exp\{-(Z_1-Z_2)^2\}$ for large distance (Z_1-Z_2) between the wells.

Using these results one can solve eq. 25 for the parameter α and find

$$\alpha = \frac{(1+\sqrt{2}\ Z/b)}{\sqrt{1+(1+\sqrt{2}\ Z/b)^2}} \tag{29}$$

We see that in the limit $Z\rightarrow 0$ the $\alpha \rightarrow \sqrt{\tfrac{1}{2}}$ and the functions $\phi^{\pm}$ become those eigenfunctions from the diagonalisation of the dipole operator between the occupied orbitals $|0>$ and $|1>$ in a single oscillator well at the origin as given in eq. 15 and displayed in Fig. 2. For $Z\rightarrow\infty$ $\alpha\rightarrow 1$ and the function $\phi^{\pm} \rightarrow |0>^{\pm}$, i.e. the lowest quantum states in oscillator wells at $\pm\infty$. In Fig. 10 we show the density distribution of the states $\phi^{\pm}$ as a function of Z. Thus our procedure of defining the structure of fragments during an intermediate stage in the fission process does, in this simple case, agree with our intuitive ideas of waves moving apart.

α-CLUSTER STRUCTURE OF ^{8}Be

The simple example given at the end of the previous section in fact is sufficient to define an α-clustering model for ^{8}Be. We suppose now that at each stage of the separation of two α-particles the single particle orbits of the two α's are given in three dimensional oscillator wells at distances of $\pm Z$ from the origin along the z-axis. The two orbits are assumed to have the functional form

$$\phi_{\pm} = |0>_x\ |0>_y\ \phi_z^{\ \pm} \tag{30}$$

i.e. zero quanta in the x and y directions of single particle oscillator wells centred at the origin and

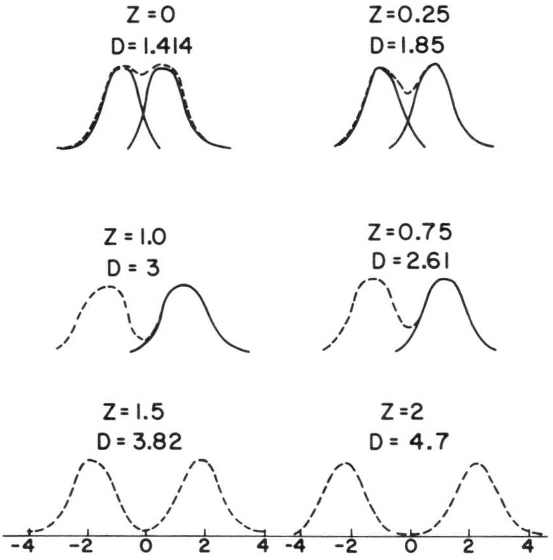

Figure 10
Separation of two single particle orbits in one-dimensional harmonic oscillator wells. The distance between the wells is 2Z and the distance between the peaks of the density distribution is D. For large Z, $D \approx 2Z$. Here b=1 for simplicity.

functions in the z-direction defined in eqs. 23 and 24 with eq. 29 which depend on the distance of the wells from the origin. The determinantal state $|\Phi_+{}^4\Phi_-{}^4|$ (where $\Phi^4 \equiv \Phi^{n+}\,\Phi^{n-}\,\Phi^{p+}\,\Phi^{p-}$) is then a cluster description for ^8Be with a cluster parameter Z. For Z=0 this intrinsic state is that for the spherical shell SU_3-model description of the lowest states in ^8Be with four particles in the 1s shell and four in the 1p-shell. For $Z=\infty$ the intrinsic state represents two α-particles with $1s^4$ structures at $\pm\infty$.

What is the relationship between the present cluster representation and that of the molecular orbital method[18] as discussed by Prof. Arima?

The molecular orbital cluster description uses single particle orbitals made up of the 1s states only in two oscillator wells separated from the origin, i.e. orbits

$$\psi_\pm = N^\pm\{|1s>^+ \pm |1s>^-\} \qquad (31)$$

The intrinsic state for ^8Be is thus $|\psi_+^4 \ \psi_-^4|$.

The orbits $|1s>^+$ and $|1s>^-$ are not orthogonal for finite separation of the oscillator wells, in fact the overlaps are given by eq. 28. The normalisation factors take the form

$$N^{\pm} = [2(1 \pm \exp(-Z^2/b^2))]^{-\frac{1}{2}}$$

For $Z\to\infty$, $N^{\pm} \to \sqrt{\frac{1}{2}}$ and the orbits $\psi^{\pm}$ tend between them to the filling of the 1s states in wells at $\pm\infty$. Note that a unitary transformation has to be performed such that the function of one well can be isolated from the other. For $Z\to 0$, $N^+\to\frac{1}{2}$ and $\psi_+ \to |1s>^0$, i.e. the 1s state of the well centred at the origin; also $N^-\to\infty$ with $\{|1s>^0 - |1s>^0\} \to 0$ such that in the limit of $Z\to 0$, $\psi^- \to |1p_0>^0$. Because of this limiting procedure great care has to be taken for $Z\to 0$, especially when using a computer. The orbits $\psi_{\pm}$ given above cannot be compared with the $\phi_{\pm}$ orbits of our new clustering technique directly since they differ at Z=0 and $Z=+\infty$; we can however construct from the $\psi_{\pm}$ orbits that have similar asymptotic values, i.e.

$$\tilde{\phi}_{\pm} = \sqrt{\tfrac{1}{2}}\,(\psi_+ \pm \psi_-) \tag{33}$$

Clearly for $Z\to 0$ $\tilde{\phi}_{\pm} \to \sqrt{\frac{1}{2}}\,\{|1s> \pm |1p>\}$ and for $Z\to\infty$ $\tilde{\phi}_{\pm} \to |1s>^{\pm}$. It is convenient to express the functions $\tilde{\phi}_{\pm}$ only in terms of orbits in the $\pm$ well respectively. Thus, by definition

$$\tilde{\phi}_+ = \sqrt{\tfrac{1}{2}}\,\{N^+(|1s>^+ + |1s>^-) + N^-(|1s>^+ - (1s)^-)\}$$

$$= \sqrt{\tfrac{1}{2}}\,(N^+ + N^-)|1s>^+ + \sqrt{\tfrac{1}{2}}\,(N^+ - N^-)|1s>^- \tag{34}$$

but

$$|1s>^- = \sum_n C_n \ |n>^+ \tag{35a}$$

where $|n> = |00n>$ is a representation of the oscillator function in Cartesian coordinates with zero quanta in the x and y directions and n quanta in the z. The expansion coefficient in eq. 35a can be deduced from eq. 27 with

$$C_n = <n+|0-> = (-)^n \frac{(\sqrt{2}\ Z)^n}{\sqrt{n!}} <0^+|0^-> \tag{35b}$$

and

$$\langle 0^+ | 0^- \rangle = \exp(-Z^2/b^2) \tag{35c}$$

Collecting terms in eq. 34 with eqs. 35 we find

$$\tilde{\phi}_+ = \tfrac{1}{2}\{(1+\exp{-Z^2/b^2})^{\frac{1}{2}} + (1-\exp{-Z^2/b^2})^{\frac{1}{2}}\} |0>^+$$
$$+ \tfrac{1}{2}\{(1+\exp{-Z^2/b^2})^{-\frac{1}{2}} - (1-\exp{-Z^2/b^2})^{-\frac{1}{2}}\}$$
$$\times \sum_{n\neq 0} (-)^n \frac{(\sqrt{2}Z)^n}{\sqrt{n!}} \exp - \frac{Z^2}{b^2} |n>^+ \tag{36}$$

The coefficients of the terms $|n>^+$ in eq. 36 are listed in Table 2 for various values of Z. It is seen that the percentage of the wave function outside the n=0 and 1 space is always less than ∿4.5%, i.e. the molecular orbital functions and those constructed using our new technique are almost equivalent.

Table 2

n \ Z	0	.1	.5	.75	1.0	1.6
0	.7071	.7552	.9020	.9544	.9823	.9993
1	.7071	.6522	.3790	.2195	.1048	.0068
2		-.0461	-.1340	-.1164	-.0741	-.0077
3		+.0022	+.0316	+.0412	+.0349	+.0058
4		-.0001	-.0056	-.0109	-.0123	-.0033
5			+.0008	+.0023	+.0035	+.0015
6			-.0001	-.0004	-.0008	-.0006
7					+.0002	+.0002
% (n≠0 or 1)	0	.4	4.3	4.1	2.4	.32

The determination of the structure of the determinantal function like $\chi_Z = |\phi_+{}^4\phi_-{}^4|_Z$ has concentrated on a description of the internal states of the "α-clusters" being correct in the limits $Z\to\infty$ - the relative function is incorrect however. To correct for the relative function between clusters we can use the generator coordinate method and construct a new intrinsic state

$$\overline{\chi} = \int f(Z)\, \chi_Z\, dZ.$$

Now one would like to choose the function f(Z) such that $\bar{\chi}$ approximates an eigenstate of the Hamiltonian - since $\bar{\chi}$ is in general not a state of definite angular momentum however the best it can be is the intrinsic state. The condition[18] that f(Z) must satisfy is that

$$\int \{(\chi_Z|H|\chi_{Z'}) - E(\chi_Z|\chi_{Z'})\} f(Z')dZ' = 0. \qquad (37)$$

If the integral in eq. 37 is replaced by a summation over discrete values of Z then eq. 37 reduces to an eigenvalue (E) problem with eigenfunction f(Z). This has been solved in collaboration with A. Jensen in the case of ^{8}Be with the Hamiltonian described in eqs. 16-18 for the intrinsic energy E; also the projected energies E_J of the states $P_J\bar{\chi}$ have been determined where P_J projects out states of definite angular momentum. The calculation is quite straightforward and the results (now in preprint form) agree with those of Horiuchi.[13]

GENERALISATION TO FISSION

Because of limited time I want to finish by discussing the generalisation of the ^{8}Be example to fission. Earlier in eq. 21 we introduced a Slater determinant of A orbitals which in the limit $Z\to\infty$ describe the internal structure of fragments of mass A_+ and A_-, i.e.

$$\chi_Z \equiv |\chi_Z^+ \chi_Z^-| = |\phi_1^+ \phi_2^+ --- \phi_{A+}^+ \phi_1^- --- \phi_{A-}^-| \qquad (38)$$

Now χ_Z does not necessarily have definite parity but this can be rectified by constructing

$$\tilde{\chi}_Z = N_p(1 \pm p)\chi_Z$$

where p is the parity operator with N_p a renormalisation coefficient and the ± is chosen depending on the parity of $\chi_{Z=0}$. The fact that χ_Z does not necessarily describe the correct relative motion between fragment means that we should also construct the generator coordinate function as in the ^{8}Be case.

$$\bar{\chi} = \int f(Z)\ \tilde{\chi}_Z. \qquad (40)$$

One should now solve the eq. 37 for the f(Z). Since $\bar{\chi}$ is to represent a decaying state into fission channel of mass A_+ and A_- then the solution of eq. 37 has to be divided into one in the interaction region which then matches onto a solution outside the interaction region in which the relative function of fragments A_+ and A_- is described by an outgoing Coulomb wave function[14].

Needless to say such a calculation has not been attempted. In the usual approach (translated into the present formalism) the whole fission process is reduced to a single particle with effective mass μ penetrating a potential $V_Z = \langle\chi_Z|\tilde{H}|\chi_Z\rangle$. Hopefully V_Z has the functional behaviour depicted by Fig. 1. It is usual to evaluate V_Z using the liquid drop model with the mass formula for which

$V_{Z=0}$ = binding energy of A particle system,

$V_{Z=\infty}$ = sums of binding energy of fragments.

In order to reproduce this liquid drop result we cannot use for $\tilde{H}$ the full Hamiltonian H. We know that the relative function between clusters is certainly wrong for large separation of clusters and would give a contribution to the relative kinetic energy operator which is certainly not included in $V_{Z=\infty}$. The exact relationship between $\tilde{H}$ and H is not clear to us. In order to reproduce the liquid drop model calculation for the potential V_Z we might consider a simple density dependent form

$$\langle\chi_Z|\tilde{H}|\chi_Z\rangle = a_1 \int (-1+a_2(\rho-\rho_o)^2 + a_4(\rho_N-\rho_p)^2) \qquad (41a)$$

$$(\rho+a_3(\frac{\partial\rho}{\partial r})^2)\,d\hat{r} + \text{Coulomb} \qquad (41a)$$

where the density function ρ is given by

$$\rho = \sum_{i=1}^{A^+} (\phi_i^+)^2 + \sum_{i=1}^{A^-} (\phi_i^-)^2 = \rho_+ + \rho_-$$

and $$\int \rho_+\, d\hat{r} = A_+, \quad \int \rho_-\, d\hat{r} = A_-. \qquad (41b)$$

There is nothing sacred about the form in eq. 41; it was merely chosen to reproduce the gross features of the mass formula with a saturation term at the density of nuclear matter ρ_0. From the Myers-Swiatecki mass formula[17] we can deduce

$$
\begin{array}{lll}
a_1 = +15.677 & \text{(volume)} & \\
a_2 = 21.585 & \text{compressibility} & \\
a_3 = -3.0 & \text{(surface)} & (42) \\
a_4 = 36.984 & & \\
\rho_0 = 0.22 & \text{density of nuclear matter.} &
\end{array}
$$

In some very recent calculations with Axel Jensen we have attempted to see what the effective potential V_Z would look like by pulling ^{16}O apart, i.e. a continuation of the work on the distribution of binding energies in ^{16}O.

In the case of ^{16}O there are four orbitals. In ^{12}C and α there are three and one orbital respectively. The number of degrees of freedom in choosing the structure of ^{12}C and α fragments from ^{16}O is therefore 3. We have reduced this to one by assuming that the structure of the ^{12}C orbitals is given by the eigenfunctions of the single particle Hamiltonian

$$\sqrt{(1-x^2)}\, H_o^+ + x(z-Z_+) \qquad (43)$$

where H_o^+ is the harmonic oscillator Hamiltonian of the plus well at distance $Z_+ = A_-Z/16$ from the origin. (The minus well is at distance $Z_- = -A_+Z/16$ from the origin). The choice of the dipole perturbation of the orbits is based on the work of distribution of binding energies. Thus by choosing x for a given Z the orbits of ^{12}C in the + well can be determined and those for the α-particle in the minus well found by orthogonality. The integral in eq. 41 is evaluated by numerical integration for a fixed oscillator parameter b = 1.6 and various x and Z parameters.

In Fig. 11 we show first the distribution of binding energy with ^{16}O. This is to be compared with Fig. 5 which was the same quantity with a Volkov potential and two body kinetic energy. The form in eq. 41 only reproduces the rough trends for the binding energies of ^4He, ^{12}C and ^{16}O (i.e. 19.3, 85.1, 119.6 MeV to be compared with experimental values of 28.3, 92.2, 127.6 MeV) perhaps it is not so surprising then that the distribution curve in Fig. 11 shows such a lack of structure - there is no hint of "weaknesses" in the system as was apparent with the Volkov force. Because of the poor fit to binding energies this calculation can at most be treated as a theoretical experiment to demonstrate the model.

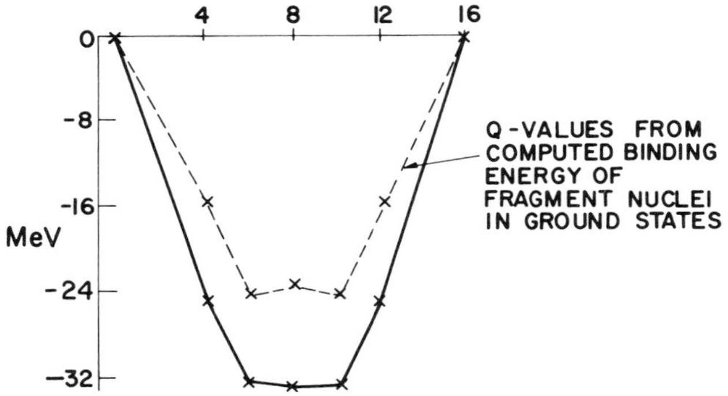

Figure 11
As in Fig. 5 but with the modified mass formula description of binding energies (see eq. 41a).

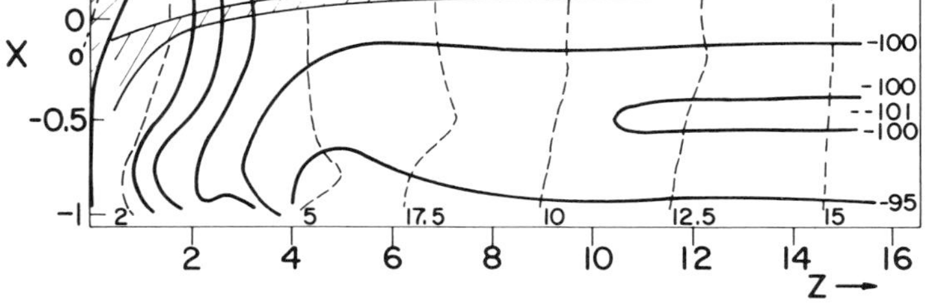

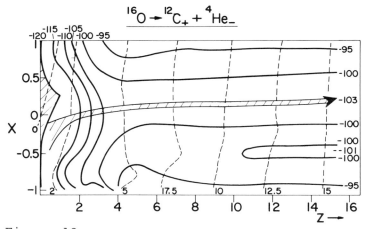

Figure 12
Solid curves show the energy contours (in MeV) for $^{16}O \rightarrow {}^{12}C + \alpha$ using the modified mass formula of eq. 41 with single particle functions defined for the parameter x of eq. 43 and distance Z between single particle wells. Dashed curves show the distance D (fm) between the centres of mass of the fragments.

In Fig. 12 is shown the energy contours on the (x,Z) plane for the breakup $^{16}O \rightarrow {}^{12}C + {}^{4}He$. In this figure x refers to the dipole diagonalisation parameter in the + well that contains ^{12}C; the structure of ^{4}He is determined by the orthogonality constraints of the model. The distance Z between the wells is not the distance D between the centres of mass of the fragments; this latter can be computed for each x,Z and the variation is shown by the dashed lines in Fig. 12. For very large Z then $D \rightarrow Z$. From this figure we can determine for each separation D the minimum energy and hence the path of "least resistance" to fission as shown by the hatched arrow. For small D (and Z) the exact position of the fission path is indeterminate (as it should be because of the indeterminacy of fragments in the ground state) but for larger D the position of the saddle point and the "chute" towards fission is well determined. In the fission path that is marked, both ^{12}C and ^{4}He end up in their ground states when $Z \rightarrow \infty$. We note that at $Z \approx 10.5$ and $x = -0.5$ another "chute" towards fission develops in which the ^{12}C and ^{4}He would end up in excited states. This "asymptotic" structure of the fragments is a fascinating subject and one which we have only just begun to study in our model. We can see however that even the "rate" at which one fragment approaches its ground state can determine the ultimate structure of the other fragment. It is tempting to relate the second "chute" with an isomeric fission mechanism.

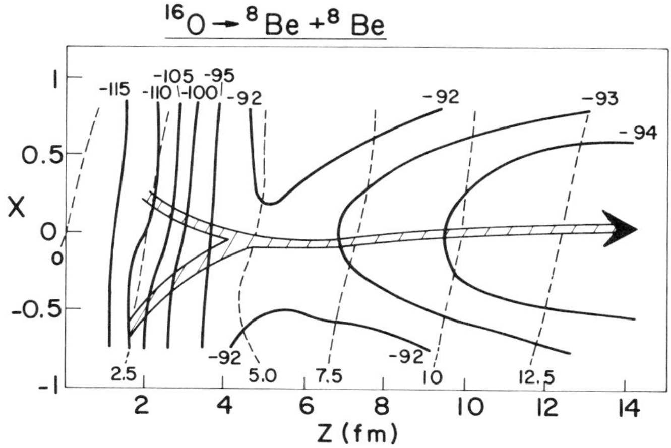

Figure 13
As in Fig. 12 for $^{16}O \rightarrow {}^{8}Be + {}^{8}Be$

In Fig. 13 we show the energy- and D-contours for the $^{16}O \to {}^{8}Be + {}^{8}Be$ fragmentation. Again the fission path is well defined except for very small Z or D. In this case no secondary fission path was evident.

In Fig. 14 we show the fission paths of least resistance for both the $^{16}O \to {}^{12}C + {}^{4}He$ and $^{16}O \to {}^{8}Be + {}^{8}Be$ breakup to present the results in the more familiar form of a fission barrier. Of course for such a light nucleus the asymptotic energies are greater than the ground state energies, i.e. the nucleus cannot spontaneously fission in its ground state. This feature does not change the principles of the model however. As is clear from Fig. 14 one could in principle compute the fission paths for each fragmentation and hence determine the fragmentation with the lowest barrier; in this way one would hope to explain the asymmetric mass division of fission fragments in a heavier nucleus.

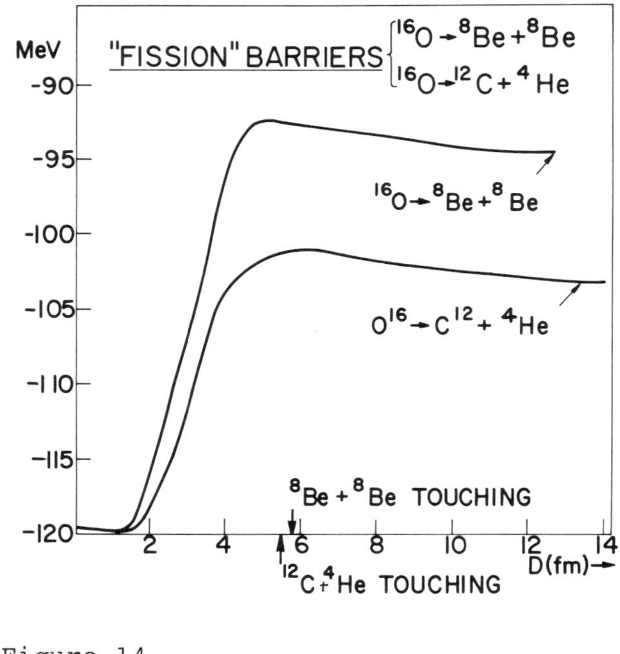

Figure 14
"Fission" barriers for $^{16}O \to {}^{12}C+\alpha$ and $^{8}Be + {}^{8}Be$ along the paths of least resistance plotted against the distance D between the centres of mass of the fragments.

CONCLUSIONS

Before applying the model to heavy or superheavy nuclei there are a number of points that have to be explored, e.g.

a) what is the best density dependent function to simulate the liquid drop results with use of the mass formulae? Should the compressibility term be proportional to $(\rho-\rho_0)^2$ (as we have assumed with Myers & Swiatecki[17]) or should we take a more exotic power dependence based on our knowledge of nuclear matter calculations? If one continues along this path one must also consider the dynamics, i.e. the calculation of the effective kinetic energy in the fission channel; only then can properties of fission be determined.

b) Is the dipole diagonalisation parameter x of one fragment sufficient freedom to describe the "condensation" of a fragment during the fission process? If so, which fragment (heavy or light) is best described by the dipole diagonalisation? In our $^{16}O \rightarrow {}^{12}C + {}^{4}He$ example we have done the calculation with the ^{4}He described by the dipole diagonalisation and ^{12}C determined by the orthogonality constraints and then we find a higher barrier. Does this mean that it is the heavier fragment that is best described by the dipole diagonisation?

c) Should one in fact construct the effective fission barrier or rather proceed directly with the generator coordinate solution? The advantage of the latter is that we could use the familiar nuclear Hamiltonian and one of the forms [15] for the potential energy that has been suggested in recent years. The model as it stands uses ordinary single particle harmonic oscillator orbitals and therefore use could be made of all the simplification techniques developed in recent years to calculate two body matrix elements. (Note that the model is not restricted to the harmonic oscillator wells - any type of single particle well could be used. The use of other types of wells however could mean that overlaps between the functions in the ± wells would be difficult to compute). The lifetimes would be computed through the use of a generalisation of reaction theory as for example suggested by Tobocman.[14] The disadvantage of this approach is that it seems to be a horrendous problem although it has been done in simple cases like ^{8}Be.

d) Is our cluster approach the best or should one generalise the molecular orbital method directly as is being investigated by Cusson, Dieperink & Kolb (private communication)?

Clearly at this stage we can ask more questions than we can answer. This however means we are in very exciting times with the prospects of the answers very shortly. Our own philosophy is that many of the answers can be learnt from a study of light nuclei where calculations are first of all easy and secondly can be analysed in great detail. It is in light nuclei that we think the connection between cluster models and fission models is to be found.

The collaborative assistance of Axel Jensen is gr tefully acknowledged. I would like to thank Professors A. Arima, H.J. Krappe, W.D. Myers, and F. Villars for providing considerable insight into the collective fission process.

REFERENCES AND NOTES

1a) Polyhanov et al. Sov. Phys. JETP 15(1962)1016.
1b) Perelygin et al. Sov. Phys. JETP 15(1962)1022.
1c) See also review talk by J.E. Lynn. AERE-R5891, Harwell (1968); Int. Symp. on Nuclear Structure (IAEA, Vienna 1968) p.463.
2a) Review by G.T. Seaborg, Ann. Rev. Nucl. Science 18 (1968)53.
2b) S.G. Nilsson, Conf. on Properties of Nuclear States, Montreal (1969) (Ed. M. Harvey et al., Les Presses de l'Universite de Montreal) p.149.
2c) G. Flerov, Ibid. p.175.
3) V.M. Strutinsky, Nucl. Phys. A95(1967)420.
4) Nilsson et al., Nucl. Phys. A131(1969)1.
5) Review by J.R. Nix, LA-DC 12488, Los Alamos 1971.
6) Milton & Fraser, Ann. Rev. Nucl. Sci. 16(1966)379.
7a) Malik et al., Bull. Amer. Phys. Soc. 16, no. 4, Washington 1971, paper BH5.
7b) J. Grumann, U. Mosel, B. Fink & W. Greiner, Z. Physik 228(1969)371.
8) Data on Niagara Falls and its Power Station is taken from Encyclopedia Britannica.
9) For a brief summary of the Canadian Nuclear Power Programme see the 1970-71 Annual Report of Atomic Energy of Canada Ltd. AECL-3913; Information Canada, Ottawa #CC1-1971.
10) The material of this section is taken from M. Harvey, Phys. Lett. 24B(1967)374 and M. Kohut & M. Harvey, Can. J. Phys. 46(1968)1491.
11) G.E. Brown, Unified Theory of Nuclear Models (John Wiley & Sons Inc., New York) 1964.
M. Baranger, Cargese Lectures in Theoretical Physics (Benjamin Inc., New York) 1963.

12) A. Volkov, Nucl. Phys. 74(1964)33.
13) H. Horiuchi, Prog. in Theor. Phys. 1971.
14) See for example W. Tobocman & M.A. Nagarajan, Phys. Rev. 138(1965)B1351.
15) For example, Heiner Meldner, Phys. Rev. 178(1969)B1815;
G. Saunier & J.M. Pearson, Phys. Rev. C1(1970)1353.
16) A. Messiah, "Quantum Mechanics" (North-Holland Pub. Co. Amsterdam 1964).
17) W.D. Myers & W.J. Swiatecki, Annals of Physics 55(1969)395.
18) D.M. Brink, Proc. of Int. School of Physics "Enrico Fermi" Course 36, Varenna 1965 (Ed. C. Bloch) Academic Press 1966.
19) Actually Niagara has a flow of 196,200 cu. ft./sec. of which 66,000 cu.ft./sec. flows over the actual falls for the tourists and the remainder is used for power.
20) It is interesting to note the talk of Professor Litherland, however, in which he discusses the possibility of observing the fission of light nuclei in excited states.
21) We use the notation

$$\chi = |\alpha| \equiv |\alpha_1 -- \alpha_A| \equiv a^\dagger_{\alpha_1} -- a^\dagger_{\alpha_A} |0>.$$

22) Here we consider the removal of the centre of mass kinetic energy. Thus the kinetic energy operator can be written $T = \frac{1}{2mA} \sum_{i<j} (\bar{p}_i - \bar{p}_j)^2$.
23) Actually the values shown in Fig. 4 are the experimental numbers with a Coulomb correction of $\sim$0.7 MeV per pair of protons. The curve shown therefore is an indication of the effect of nuclear forces alone.
24) It takes only $\sim$6 MeV if Coulomb effects are included.
25) The difference with the two centred shell model[7b] should be emphasised.
26) Actually fragments can be excited and the filled orbits may not therefore be the very lowest in the wells - we shall assume however that they never gain an excitation that requires the filling of orbits above the lowest A.

SHELL MODEL, QUARTETS AND CLUSTERS IN LIGHT NUCLEI*

AKITO ARIMA
Department of Physics
State University of New York at Stony Brook
Stony Brook, New York 11790

I. SHELL MODEL CALCULATIONS OF ^{20}Ne AND ^{22}Ne

The shell model Hamiltonian in the sd shell consists of three parts: the part corresponding to ^{16}O, $H_{^{16}O}$, the sum of the single particle Hamiltonians of the valence nucleons in the sd shell, H_S, and the residual interactions between the valence nucleons, V_{ij}. Thus,

$$H = H_{^{16}O} + H_S + \sum_{i>j} V_{ij} ,$$

where

$$H_S = \sum_{\substack{i \text{ in} \\ sd}} (T_i + U_i).$$

We can assume for V_{ij} either a phenomenological interaction or a more sophisticated interaction. For example, one may choose for V_{ij} a Gaussian central force, the parameters of which are fitted to reproduce the observed spectra of ^{18}O and ^{18}F as well as possible.[1] The Oak Ridge group, on the other hand, takes the Kuo-Brown interaction.[2]

Fig. 1 shows the level structure of ^{20}Ne.[1] The calculated energies of the ground state band members agree with the observations very well. There are, however, a number of 0^+ and 2^+ states between 6 MeV and 8 MeV excitation. Only one state of each spin can be explained by the normal shell model in which only the 1s and 0d orbits are taken into account. We will come back to these excited states. If we plot the excitation energies of the ground state band members versus I(I+1), where I stands for the spin, a beautiful rotational structure with interesting kinks can be observed. Kinks in a rotational structure were predicted to be in ^{22}Ne, too. Fig. 3 shows the excitation energies of the ground state band members in ^{22}Ne plotted versus I(I+1). This figure was drawn before the excitation energy of the 8^+ state was observed. Our predicted energy is 11.27 MeV.[1] The Chalk River group

*Work supported by the U.S. Atomic Energy Commission.

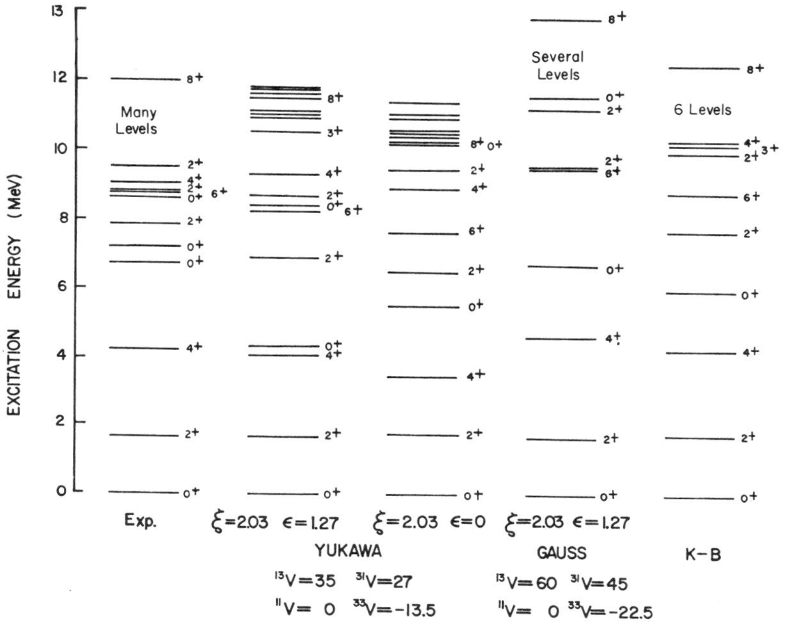

Figure 1
Energy levels of ^{20}Ne

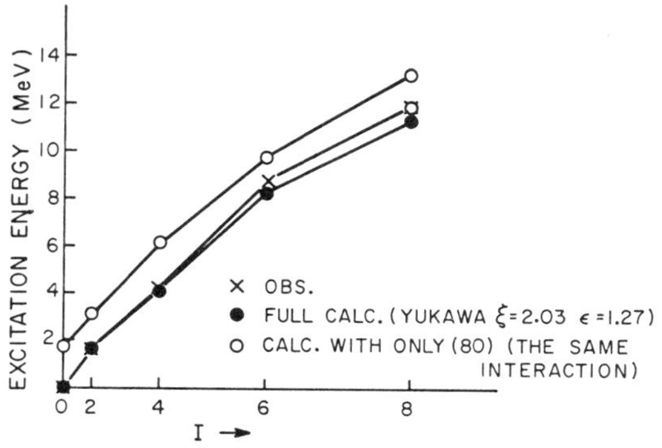

Figure 2
Rotational levels of ^{20}Ne. The excited-state energies are plotted as a function of I(I+1). In this figure I is indicated instead of I(I+1) in the abscissa.

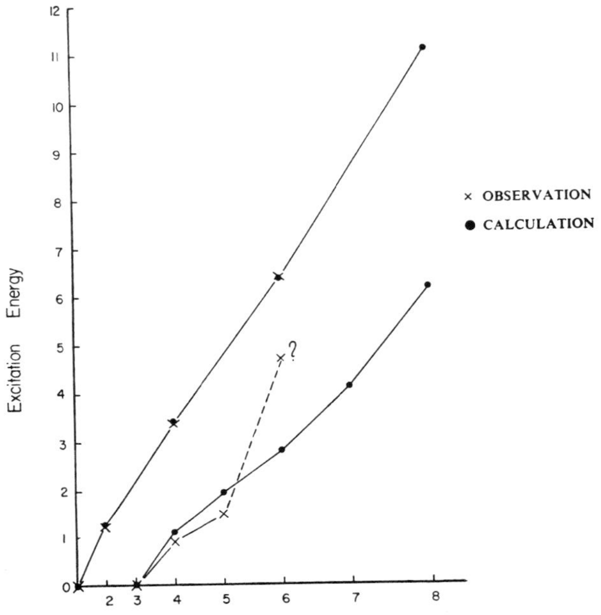

Figure 3
Rotational levels of ^{22}Ne and ^{22}Na.

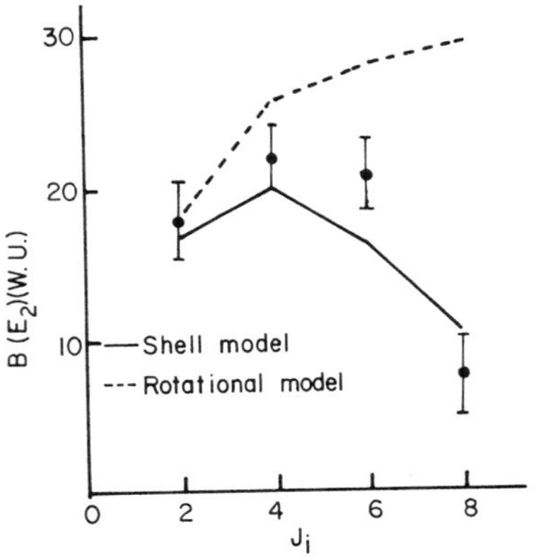

Figure 4
E2 transition probabilities in the ground state band of ^{20}Ne.

observed this level at 11.15 MeV.[3] This is a quite good agreement. Thus our prediction of kinks in ^{22}Ne was proved. The mechanism producing these trends was beautifully explained by M. Harvey and his collaborators.[4]

Now let me discuss the electric quadrupole transition probabilities in ^{20}Ne, summarized in Table 1. Old observed values have shown peculiar discrepancies with the calculated values which are shown in the fourth and fifth columns. Recently Hausser and his collaborators in Chalk River have remeasured the 4^+ to 2^+ life time and have found that the B(E2, 4→2) is 40% greater than previously reported.[5] Litherland, Dixon and their collaborators in Toronto obtained a smaller result for the B(E2, 6→4).[6] Furthermore the B(E2, 8→6) has been recently measured by the Chalk River group. The new result is extremely interesting, because it has clearly indicated the termination of the rotational band.

Table 1
E2 Transition Probabilities in the ground state band of ^{20}Ne

J_i	J_f	Exp	ASS	O.R.	Rot. Mod.	O+α
2^+	0^+	$17.8^{+2.9}_{-2.1}$	16.3	14.9	17.9	20.3
4^+	2^+	21.9±2.1	20.0	18.5	25.6	26.9
6^+	4^+	20.6±2.4	16.6	15.2	28.2	22.8
8^+	6^+	7.5±2.5	10.1	9.7	29.7	15.7

ASS: Arima, Sakakura and Sebe[1]
O.R.: Oak Ridge Group[2]
Exp.: Chalk River Group[5,6]
O+α: Cluster model by Horiuchi[27]

We have calculated the quadrupole moments Q of the ground band members. Using the simple formula

$$Q = -\frac{I}{2I+3} Q_o \quad ,$$

we can obtain the intrinsic quadrupole moments Q_o. The values of Q_o, Table 2, show the interesting feature that Q_o decreases as I increases; in other words, the deformation decreases as I increases. This indicates an anti-stretching effect in ^{20}Ne. The calculation shows moreover that the stretching phenomena can be found in ^{22}Ne.

The calculated shell model wave functions are complicated mixtures from the jj-coupling point of view. It is important to find a simple picture of them. Instead of the jj-coupling basis, we can use the basis provided by the SU_3 group[7] and supermultiplet theory.[8] Table 3 shows some of the calculated squared amplitudes.[1] The results demonstrate that, at least for ^{20}Ne, the SU_3 group method is very useful. Intriguing properties of the SU_3 wave functions will be discussed in the next section.

Table 2
Intrinsic quadrupole moments Q_0 in the ground state bands in ^{20}Ne and ^{22}Ne

	^{20}Ne		^{22}Ne	
	cal	obs[a)]	cal	obs[a)]
2^+	52.1	73.5	54.0	59.5
4^+	51.7		56.4	
6^+	51.1		58.8	
8^+	49.2		59.0	
10^+			43.6	

a) K. Nakai, A. Winther, F.S. Stephens and R.M. Diamond, private communication.

Table 3
Percentage analysis of the wave functions of low-lying states in ^{20}Ne terms of the SU_3 irreducible representations

		0_1	0_2	2_1	2_2	4_1	6_1	8_1
[4]	(80)	0.895	0.042	0.906	0.025	0.848	0.875	0.806
	(42)	0.016	0.828	0.018	0.717	0.061	0.008	
	(04)	0.025	0.053	0.004	0.105	0.001		
	(20)	0.000	0.004	0.000	0.002			
[31]	(61)	0.059	0.014	0.068	0.041	0.082	0.110	0.174
	(42)	0.000	0.000	0.000	0.019	0.002	0.001	
	(23)	0.002	0.043	0.001	0.063	0.001	0.000	
	(31)	0.001	0.009	0.001	0.014	0.002	0.000	
	(12)	0.000	0.002	0.000	0.002	0.000	0.000	
	(20)	0.000	0.000	0.000	0.000	0.000	0.000	
[22]		0.001	0.004	0.001	0.011	0.002	0.005	0.020
[211]		0.001	0.001	0.001	0.001	0.001	0.001	
[1111]		0.000	0.000	0.000	0.000	0.000	0.000	

II. SPECTROSCOPIC FACTORS AND ALPHA-TRANSFER REACTIONS

Table 4 shows the binding energies of several sd-shell nuclei relative to that of ^{16}O. This table shows that ^{20}Ne has a very large binding energy. This indicates the stability of a system with two protons and two neutrons, due to spatial symmetry. The two-body interaction is short ranged and therefore favors the relative s-state of the two-body system. Thus the spatially symmetric pair is favored by the two-body interaction. Each nucleon has, of course, four degrees of internal freedom: neutron spin up, neutron spin down, proton spin up and proton spin down. Therefore, for a system of no more than four particles, the spatial part of the wave function can be fully symmetric. This is the main reason why ^{20}Ne is very stable. The importance of spatial symmetry was recognized by Wigner who introduced the supermultiplet theory.[8]

Table 4
Binding Energies

	ΔBE^*	$\Delta BE/n$
^{16}O	0	0
^{18}O	3.8	1.9
^{18}F	4.8	2.5
^{19}O	3.6	1.2
^{19}F	11.2	3.7
^{20}O	7.0	1.7
^{20}Ne	24.0	6.0
^{21}Ne	26.0	5.2
^{22}Ne	32.2	5.4

$\Delta BE^* = BE(8+Z,8+N) - BE(^{16}O) - 4.2N - 0.6Z$
$n = N + Z$

Spatial symmetry is broken by the spin-orbit interaction, by other non-central interactions, and by Bartlet and Heisenberg central interactions. Table 5 shows how much this symmetry is broken in the sd-shell nuclei.[9] For the Rosenfeld interaction, the spatial symmetry is conserved very well, but the Kuo interaction badly breaks this symmetry above ^{20}Ne. However, the sum of intensities of the highest and next highest symmetries is always larger than 85% except for n=10 and 12 even with the Kuo interaction.

Table 5
Breaking of spatial symmetry.[9] n is the particle number. Representations are ordered according to the eigenvalue of the two-body Casimir (Majorana) operator; for example for n=8 [42], [431], [422], [332], [4211]

N	Rosenfeld	Kuo
4	92-8	70-27-3
5	90-9-1	75-20-4-1
6	87-9-4-1	73-17-5-5
7	80-19-1	50-45-3-1-2
8	79-19-1	36-49-11-1-2

The spatial symmetry, however, is not enough to completely determine the states in the sd-shell, the pf-shell and so on. For example we have two spatially symmetric states in ^{18}O which have spin 0:

$T = 1 \quad S = 0 \quad L = 0 \quad (1s)^2$

$\text{"} \quad \text{"} \quad \text{"} \quad (0d)^2.$

A residual interaction mixes these two states. If the residual interaction is approximated by the quadrupole-quadrupole interaction, the wave functions can be given by the SU_3 group.[7] The SU_3 wave functions are nothing but those projected from the intrinsic wave functions given by the anisotropic harmonic oscillator shell model. The Hamiltonian for a quadrupole-quadrupole residual interaction is given by

$$H = \sum_i (T_i + \frac{1}{2}m\omega^2 r_i^2) + \kappa \sum_{i,j} (q_i \cdot q_j).$$

Taking the Hartree-Fock procedure, we can approximate this Hamiltonian by the sum of single-particle Hamiltonians. This will be called the intrinsic Hamiltonian

$$H_{int} = \sum_i \{T_i + \frac{1}{2}m\omega^2 r_i^2 + \kappa Q_o \cdot q_{oi}\} ,$$

where $Q_o = \langle \sum q_{oi} \rangle$. Since $q_o \propto 3z^2 - r^2$, H_{int} is the sum of anisotropic harmonic oscillator Hamiltonians. Single-particle wave functions are given by products of three functions of the x,y and z coordinates, respectively. The quantum numbers which label those single-particle wave functions are n_x, n_y and n_z which are the numbers of oscillator quanta in the x,y and z directions. Any solution of H_{int} is provided by a Slater determinant of those single particle wave functions. Since ^{18}O has two nucleons in the (002) orbit, the intrinsic wave function χ_{int} is very simple:

$$\chi_{int} = N_{int}\frac{1}{\sqrt{2}} \begin{vmatrix} (z^2 e^{-\nu r^2/2} N\uparrow)_1 & (z^2 e^{-\nu r^2/2} N\downarrow)_1 \\ (\quad '' \quad)_2 & (\quad '' \quad)_2 \end{vmatrix} .$$

Because any solution of H should be an eigenstate of angular momentum, it must be projected out from the intrinsic state χ_{int}:

$$\Psi_{LM} = N_L \, P_{LM} \, \chi_{int},$$

where L is an angular momentum, and P_{LM} the projection operator. This projection is extremely easy in the present case. We can expand $z^2 \exp(-\nu r^2/2)$ in terms of 1s and 0d wave functions:

$$z^2 e^{-\nu r^2/2} = r^2 \cos^2\theta \, e^{-\nu r^2/2}$$

$$= \phi(0d\ m{=}0) - \frac{1}{\sqrt{2}} \phi(1s) + \alpha\phi(0s)$$

where

$$\phi(0d\ m{=}0) = \sqrt{\frac{\sqrt{\nu}\ \nu^3 \cdot 2^4}{\sqrt{\pi}\ 3 \cdot 5}} \, r^2 e^{-\nu r^2/2} Y_0^{(2)}(\theta\phi)$$

$$\phi(1s) = \sqrt{\frac{\sqrt{\nu}\ \nu \cdot 2}{\sqrt{\pi}\ 3}} (3 - 2\nu r^2) e^{-\nu r^2/2} Y_0^{(0)}(\theta\phi),$$

and α is a certain constant. Thus the intrinsic state may be written in the form

$$z_1^2\, e^{-\nu r_1^2/2}\; z_2^2\, e^{-\nu r_2^2/2}$$

$$\propto (\phi_1(0d\ m{=}0) - \frac{1}{\sqrt{2}}\phi_1(1s))(\phi_2(0d\ m{=}0) - \frac{1}{\sqrt{2}}\phi_2(1s))$$

$$= \frac{1}{2}|(1s)^2[2]S> - \frac{1}{\sqrt{2}}\{\phi_1(0d\ 0)\phi_2(1s) + \phi_1(1s)\phi_2(0d\ 0)\}$$

$$+\ \phi_1(0d\ 0)\phi_2(0d\ 0).$$

Here we have neglected the terms including $\phi(0s)$ because the 0s state is already occupied. The last term can easily be rewritten in terms of eigenstates of angular momentum by using Clebsch Gordan coefficients;

$$\phi_1(0d\ 0)\phi_2(0d\ 0) = \sum_{L=\mathrm{even}} (2200|L0)\,|(0d)^2L0>.$$

Thus we reach the following expression:

$$z_1^2 e^{-\nu r_1^2/2}\, z_2^2 e^{-\nu r_2^2/2}$$

$$\propto \frac{3}{2\sqrt{5}}\left[\frac{\sqrt{5}}{3}|(1s)^2S> + \frac{2}{3}|(0d)^2S>\right]$$

$$- \frac{3}{\sqrt{7}}\left[\frac{\sqrt{7}}{3}|1s\ 0d[2];D0> + \frac{2}{3}|(0d)^2;D0>\right] + \frac{3\sqrt{2}}{\sqrt{5.7}}|(0d)^2;G0>. \qquad (1)$$

The projection of an angular momentum eigenstate is now trivial. For example,

$$\Psi_{L=0} = N_0\, P_0\, \chi_{int}$$

$$= \frac{\sqrt{5}}{3}|(1s)^2S> + \frac{2}{3}|(0d)^2S>. \qquad (2)$$

These projected wave functions belong to the irreducible representation (40) of the SU_3 group. They have an interesting property concerning two nucleon transfer. According to Elliott and Skyrme,[10] the (spherical) harmonic oscillator shell model wave functions are products of the intrinsic wave functions and the 0s center of mass wave function, for example

$$\Psi_{HO}(^{16}O) = \Phi_{0s}^{cm}(\vec{R}_{16} - \vec{R}_{sh})\ \chi(^{16}O)$$

and

$$\Psi_{HO}(\text{two nucleon}) = \Phi_{0s}^{cm}(\vec{R}_2 - \vec{R}_{sh})\ \chi(0s)$$

where Φ_{0s}^{cm} is the wave function for the center of mass motion, R_{sh} means the center of the shell model field, and the χ's are the wave functions for the intrinsic motion.

Let $\Phi_{NLM}^{rel}(\vec{R}_2-\vec{R}_{sh})$ be the wave function of the two nucleons relative to the centre of the shell model potential. Let the quantity I be defined as follows: (Take $R_{sh}=0$ in the following)

$$I = \sqrt{\frac{18.17}{2}} \langle \Psi_{HO}(^{16}O)\chi(0s)\Phi_{NLM}^{rel}(\vec{R}_2) | \sqrt{\frac{2}{18.17}} A | \Psi_{HO}(^{16}O)(sd)^2 \times [2](40)LM\rangle$$

where all coordinates in ^{16}O are measured from the center of the shell model Hamiltonian, and A is the antisymmetrizer between 16 particles in ^{16}O and two nucleons in the sd shell. Using the Talmi-Moshinsky transformation,[11] we can rewrite the product $\Phi_{0s}^{cm}(\vec{R}_{16})\Phi_{NLM}^{rel}(\vec{R}_2)$ in terms of a linear combination of wave functions which are products of a wave function of the center of mass motion $\vec{R}_{18}$ and that of the relative motion between the center of ^{16}O and that of the two nucleons $(\vec{R}_2-\vec{R}_{16})$:

$$\Phi_{0s}^{cm}(\vec{R}_{16})\Phi_{NLM}^{rel}(\vec{R}_2) = \langle 0s(\vec{R}_{16})NL(\vec{R}_2);L | 0s(\vec{R}_{18})NL(\vec{R}_2-\vec{R}_{16});L\rangle \times \Phi_{0s}^{cm}(\vec{R}_{18})\Phi_{NLM}^{rel}(\vec{R}_2-\vec{R}_{16})$$

+ other terms including excited center of mass motion.

Inserting this into I, we obtain

$$I = \langle 0s(\vec{R}_{16})NL(\vec{R}_2);L | 0s(\vec{R}_{18})NL(\vec{R}_2-\vec{R}_{16});L\rangle \sqrt{S_L}$$

where

$$\sqrt{S_L} = \sqrt{\frac{18.17}{2}} \langle \chi(^{16}O)\chi(0s)\Phi_{NLM}^{rel}(\vec{R}_2-\vec{R}_{16}) | \int \Phi_{0s}^{*}(\vec{R}_{18})\Psi(^{18}O(40)LM)d^3R_{18}\rangle.$$

This S_L is the probability of finding the ^{16}O cluster in its ground state with the two nucleon cluster in the 0s and the relative motion between the two clusters in the NLM state.[12] In other words, S_L is the spectroscopic factor of the 0s two nucleon cluster in ^{18}O.

On the other hand, I can be easily rewritten as

$$I = \langle \Psi_{HO}(^{16}O)\chi(0s)\Phi_{NLM}^{rel}(\vec{R}_2) | (1-\sum P_{i17}-\sum P_{i18}+\sum P_{(ij)(17,18)}) | \Psi_{HO}(^{16}O)(sd)^2[2](40)LM\rangle$$

$$= \langle \chi(0s)\Phi_{NLM}^{rel}(\vec{R}_2) | (sd)^2[2](40)LM\rangle$$

where the exchange terms have no contributions because $\Psi(^{16}O)$ contains only 0s and 0p orbits. The last bracket is nothing but a Talmi-Moshinsky bracket. Using the wave functions obtained in (1) and (2), we evaluate I to be

$$I = \langle 0d0d;L=4|0s0G;4\rangle \qquad \text{for } L=4,$$

$$I = \frac{2}{3}\langle 0d0d;L=2|0s1D;2\rangle + \frac{\sqrt{7}}{3}\langle 0d1s;L=2|0s1D;2\rangle \qquad \text{for } L=2,$$

and

$$I = \frac{2}{3}\langle 0d0d;L=0|0s2S;0\rangle + \frac{\sqrt{5}}{3}\langle 1s1s;L=0|0s2S;0\rangle \qquad \text{for } L=0.$$

Using the explicit numerical values of these Talmi-Moshinsky brackets, we find

$$I = \frac{1}{2}\sqrt{\frac{3}{2}}$$

which is independent of L. The spectroscopic factors S_L thus do not depend on L and amount to

$$S_L = \frac{3}{8}\left(\frac{18}{16}\right)^4 \cong 0.6.$$

This value of 0.6 gives the probability of finding two nucleons in a relative 0s state in ^{18}O. This value is less than unity because of the Pauli principle.

We have wave functions Ψ^{ex}_{LM} which are orthogonal with the states $|(40)LM\rangle$; for example when L=0,

$$\Psi^{ex}_{o} = \frac{2}{3}|(1s)^2\ S\rangle - \frac{\sqrt{5}}{3}|(0d)^2\ S\rangle.$$

It is very easy to show that the spectroscopic factors for these excited states vanish. This means that only the $|(40)LM\rangle$ states can be excited by the two nucleon transfer reaction and the wave functions of these states can be given by the antisymmetrized cluster wave functions;

$$|\Psi_{HO}(^{16}O)(sd)^2[2](40);LM\rangle = N_L A\chi(^{16}O)\chi(0s)\Phi^{rel}_{NLM}(\vec{R}_2-\vec{R}_{16}) \times \Phi^{cm}_{os}(\vec{R}_{18}-\vec{R}_{sh}) .$$

This cluster wave function was introduced by Wildermuth.[13] The equivalence of the Wildermuth wave function to that of the SU_3 model was proved by Bohr and Bayman.[14]

The same argument as discussed above can be applied to ^{20}Ne. The wave functions of the ground state band can be projected out from the intrinsic state χ:

$$|(sd)^4[4](80)LM\rangle = N_L\ P_{LM}\chi$$

where

$$\chi = \frac{1}{\sqrt{4!}}\begin{vmatrix} (z^2e^{-\frac{1}{2}\nu r^2}N\uparrow)_1, & (z^2e^{-\frac{1}{2}\nu r^2}N\downarrow)_1, & (z^2e^{-\frac{1}{2}\nu r^2}P\uparrow)_1, & (z^2e^{-\frac{1}{2}\nu r^2}P\downarrow)_1 \\ (\qquad)_2, & (\qquad)_2, & (\qquad)_2, & (\qquad)_2 \\ - & - & - & - \\ (\qquad)_4, & (\qquad)_4, & (\qquad)_4, & (\qquad)_4 \end{vmatrix}$$

These states absorb all the strengths of alpha transfer reactions. The spectroscopic factor S_L of these reactions are expressed as

$$S_L = \frac{I}{\langle 0s(\vec{R}_{16})NL(\vec{R}_4);L|0s(\vec{R}_{20})NL(\vec{R}_4-\vec{R}_{16});L\rangle^2} I^2$$

where

$$I = \binom{20}{4}^{\frac{1}{2}} \langle \Psi_{HO}(^{16}O)\chi(\alpha)\Phi^{rel}_{NLM}(\vec{R}_4-\vec{R}_{sh}) \Big| \binom{20}{4}^{-\frac{1}{2}} A\,|\Psi_{HO}(^{16}O)$$
$$\times\ (sd)^4[4](80);LM\rangle,$$

and

$$\chi(\alpha) \propto e^{-\frac{1}{2}\nu_\alpha \sum_{i=1}^{4}(\vec{r}_i-\vec{R}_4)^2}$$

$$= e^{-\frac{1}{2}\nu_\alpha(\vec{r}_1-\vec{r}_2)^2}\ e^{-\frac{1}{2}\nu_\alpha(\vec{r}_3-\vec{r}_4)^2}\ e^{-\nu_\alpha\{(\vec{r}_1+\vec{r}_2)-(\vec{r}_3+\vec{r}_4)\}^2}$$

The integral I can be simplified to

$$I = \langle\chi(\alpha)\Phi^{rel}_{NLM}(\vec{R}_4-\vec{R}_{sh})|(sd)^4[4]\ (80)LM\rangle.$$

As an example, for L=8, consider the state

$$|(sd)^4[4](80)8M\rangle = |(0d)^4[4]8M\rangle.$$

Carrying out the Talmi-Moshinsky transformation three times, we can express I as the product of three Talmi-Moshinsky brackets:

$$I = \langle 0d0d;G|0G0s;G\rangle\langle 0d0d;G|0G0s;G\rangle\langle 0G0G;L|0L0s;L\rangle.$$
$$= \frac{3}{64}\sqrt{\frac{35}{2}}\ .$$

We thus obtain

$$S = \left(\frac{3}{64}\sqrt{\frac{35}{2}}\right)^2 \ \frac{20}{16}^{8} \ \cong 0.24$$

where we have used

$$\langle 0s(\vec{R}_{16})NL(\vec{R}_4);L|0s(\vec{R}_{20})NL(\vec{R}_4-\vec{R}_{16});L\rangle = \left(\frac{20}{16}\right)^4.$$

We can again prove that only these (80) states among the many states of the $(sd)^4$ configuration can be excited by the alpha transfer reaction on ^{16}O and that the S_L's are independent of L. Thus these (80) wave functions are equal to the Wildermuth wave functions consisting of the ^{16}O and alpha clusters.

Table 6 shows that the (^{7}Li,t) reaction on ^{16}O can excite only the ground state band members among the positive parity levels.[15]

Table 6
Results for reaction $^{16}O(Li^7,t)Ne^{20}$ [15]

Excitation energy (MeV)	J^{Π}	Required L	Relative $(d\sigma/d\Omega)_{max}$
0.00	0^+	0	100
1.63	2^+	2	272
4.25	4^+	4	171
4.97	2^-	...	...
5.63	3^-	3	39
5.80	1^-	1	82
6.72	0^+	...	...
7.03	4^-	...	...
7.17	3^-		
7.20	0^+	$0+3^b$	486
7.43	2^+	...	...

We can see from this table that a number of negative parity levels are also excited. They can be the (90) states, or in other words

$$\Phi_L^- = N\, \mathcal{A}\chi(^{16}O)\chi(\alpha)\Phi_{NL}^{rel}(^{16}O+\alpha),$$

where

NL = 4P, 3F, 2H, 1K, OM.

Expressed in terms of the shell model, the Φ_L^- are given by

$$\Phi_L^- = \alpha|(0s)^4(0p)^{12}\ (sd)^3(pf)[4^5](90)LM\rangle$$

$$+\ \beta|(0s)^4(0p)^{11}(sd)^5[4^5](90)LM\rangle,$$

where $\alpha^2>\beta^2$.

Recently the ^8Be transfer reaction has been carried out on ^{12}C by the Brookhaven group[16] and by the Pennsylvania group.[17] The former group used the $(^{14}N,^6Li)$ reaction on ^{12}C, which was also performed by the Heidelberg group,[18] whereas the latter group used the $(^{16}O,^{12}C)$ reaction. According to their experiments, we can observe that all members of the ground state band are excited and furthermore that the 9.08 MeV 4^+ and 12.19 MeV 6^+ states are strongly excited. The 4^+ and 6^+ states, however, are not strongly excited[15] by the alpha transfer reaction on ^{16}O. This indicates that these 4^+ and 6^+ states are 4h-8p states. This $(^{14}N,^6Li)$ reaction can not tell which 0^+ and 2^+ states are of a 4h-8p nature. Using the $(^{16}O,^{12}C)$ reaction, the Pennsylvania group concluded that the 7.20 MeV 0^+ and the 7.84 MeV 2^+ states are 4h-8p states.[17] This is very consistent with the systematics of alpha decay widths.[17]

There is a band with K=2$^-$, the members of which are strongly excited by the $^{12}C(^{14}N,^{6}Li)^{20}Ne$ reaction.[16,18] This fact is not contradictory to the usual assignment of 1h-5p(82) states. In order to obtain additional information about the negative parity bands with K=0$^-$ and K=2$^-$, it is desirable to study the one nucleon pick-up reaction on ^{21}Ne, such as $^{21}Ne(p,d)$ and $^{21}Ne(d,t)$, because this reaction can excite the 1h-5p states but not the states with the $(sd)^3(pf)^1$ configuration. If the picture of ^{20}Ne discussed above is correct, the pick-up reaction should excite the K=2$^-$(82) band strongly but the K=0$^-$(90) band only weakly.

III. QUARTET STATES

It is now apparent that ^{20}Ne has 4h-8p states which form the rotational band based on the 7.20 MeV, 0$^+$ state. In this region of the periodic table, $\hbar\omega$ is 15 MeV. Thus 4h-np states should have 60 MeV excitation. Why then do we have 4h-8p states as low as 7 MeV? This question was answered by G.E. Brown a number of years ago. According to him, a large energy gain is caused by a large deformation.[19] An alternative explanation will be discussed here.

If we excite four nucleons from the p-shell into the sd-shell, the four excited nucleons in the sd shell form a very stable structure as do the four nucleons in the ground state of ^{20}Ne. I will call this structure a "quartet" following Gillet and Danos.[20] Feshbach called it "fourness".[21] The single particle energy difference can be cancelled by the large internal energies of the quartets which amount to 25 MeV. Because we have an additional quartet in the sd shell and a quartet hole in the p shell, the sum of internal energies are roughly 25 x 2 MeV = 50 MeV. Therefore, we can expect the 4h-8p quartet states around 10 MeV (=4 x 15 - 25 x 2 MeV). Of course we must take into account the interaction between the quartet in the sd-shell and that in the p-shell. This can be estimated as follows. Suppose that the first excited state in ^{16}O is a one-quartet-one-quartet-hole state. Then the excitation energy of this state can be expressed by

$$\frac{Q_{sd}}{Q_{p\ hole}} - \underline{\qquad} = \frac{Q_{sd}}{} - \underline{\qquad} + \frac{\phantom{Q_{sd}}}{Q_{p\ hole}} - \underline{\qquad} + V_{p,sd}$$

$$^{16}O^{*} - {}^{16}O = {}^{20}Ne - {}^{16}O + {}^{12}C - {}^{16}O + V_{p,sd}\ .$$

The left hand side should be 6.05 MeV. Each term at the right hand side can be known from the binding energy of the corresponding nucleus. We can thus obtain

$$V_{p,sd} = 3.5\ \text{MeV (repulsive)}$$

Now we can easily estimate excitation energies of excited quartet states[23]; for example in ^{20}Ne,

$$\frac{Q_{sd}+Q_{sd}}{Q_{p\ hole}} - \frac{Q_{sd}}{} = \frac{Q_{sd}+Q_{sd}}{} - \frac{Q_{sd}}{} + \frac{1}{Q_{p\ hole}} - \frac{}{} + 2 \times V_{p,sd}$$

$$^{20}Ne^{*} - {}^{20}Ne = {}^{24}Mg - {}^{20}Ne + {}^{12}C - {}^{16}O + 2 \times V_{p,sd} \,.$$

The interaction energy between the two sd - quartets is automatically taken into account by using the observed binding energy of ^{24}Mg.[23] This formula predicts that the 8p-4h quartet state in ^{20}Ne should be at 6 MeV.

There are certain salient features of the quartet structures which are worth mentioning. A number of 0^+ states which are not explained by normal shell model calculations may be explained in terms of excited quartet states. Furthermore, we can speculate that the large alpha widths of some 0^+ levels may be attributed to quartet structure in the pf-shell. Some results are given in Table 7.

Table 7
Quartet 0^+ States

	Ex.	Ex.(th)	No.of Quartets			Γ_α(KeV)	θ^2_α(%)
			p	sd	pf		
^{12}C	0.0	0.0	2	0	0		
	7.7	10.0	1	1	0	<25	200
^{16}O	0.0	0.0	3	0	0		
	6.0	6.0	2	1	1		
	16.7	15.0	1	2	0		
^{20}Ne	0.0	0.0	3	1	0		
	7.2	5.0	2	2	0	4	1
	8.6	9.0	3	0	1	1300	70
^{40}Ca	0.0	0.0	3	6	0		
	3.4	3.4	3	5	1		
^{44}Ti	0.0	0.0	3	6	1		
	1.9[24]	2.3	3	5	2		

Using the $^{40}Ca(\alpha,\gamma)^{46}Ti^{*}$ reaction, Dixon and his colleagues recently found a 0^+ state at 1.9 MeV in ^{44}Ti which is not reproduced by the shell model with the $(pf)^4$ configuration.[24] We can expect the 4h-8p state in ^{44}Ti to be at 2.3 MeV. This state should not be excited by the transfer of an alpha particle to ^{40}Ca. Saclay and Argonne do not find the corresponding state by using the alpha transfer reaction.[25] This is consistent with the 4h-8p conjecture discussed above.

IV. SHELL MODEL AND CLUSTER MODEL OF ^{20}Ne

As discussed in section I, the shell model wave functions of lowlying states in ^{20}Ne are approximated with good accuracy by the (80) SU_3 wave functions which turn out to be equal to those of the Wildermuth cluster model. The root mean square of the distance between the two clusters in the (80) states can be estimated in the following way:

$$\langle R^2\rangle = \frac{1}{\nu_{Rel}}(2N + L + \frac{3}{2})$$

$$= \frac{1}{\nu_{Rel}}(8 + \frac{3}{2}),$$

where

$$\nu_{Rel} = \frac{16}{5}\nu ,$$

and

$$\nu = \frac{m\omega}{\hbar}$$

The factor $\frac{16}{5}$ comes from the reduced mass. Using the formula

$$\nu \cong A^{-1/3}\ \mathrm{fm}^{-2},$$

we obtain

$$\sqrt{\langle R^2\rangle} \cong 2.8\ \mathrm{fm}.$$

This value is much smaller than the sum of the radii of the two clusters

$$R_{^{16}O} \cong 3.0\ \mathrm{fm}$$

$$R_\alpha \cong 1.7\ \mathrm{fm}.$$

Thus, the overlapping between the two clusters is large and, therefore, the Pauli principle plays the important role of smearing the clusters in the shell model wave function.

Let us introduce a variational wave function which includes the SU_3 wave function as one of its limits. One of the variational parameters is the distance between these two clusters:

$$\Psi_{LM} = N_L\ P_{LM}\ \chi'$$

$$\chi' = N'\,A\,\{\chi(^{16}O)\Phi_{0s}(\vec{R}_{16}-\vec{R}_0)\chi(\alpha)\Phi_{0s}(\vec{R}_\alpha-\vec{R}_1)\}$$

We have three variational parameters $\nu_{^{16}O}$, ν_α and $D=|\vec{R}_1-\vec{R}_0|$. We assume $\omega_{^{16}O}=\omega_\alpha=\omega$ in order to reduce the number of parameters. This Ψ_{LM} tends to the (80) wave function when $D\rightarrow 0$. Using the wave function Ψ_{LM} and the Volkov potential, H. Horiuchi calculated the expectation value of the energy $E(\nu,D)$.[27,28] In figure 5,

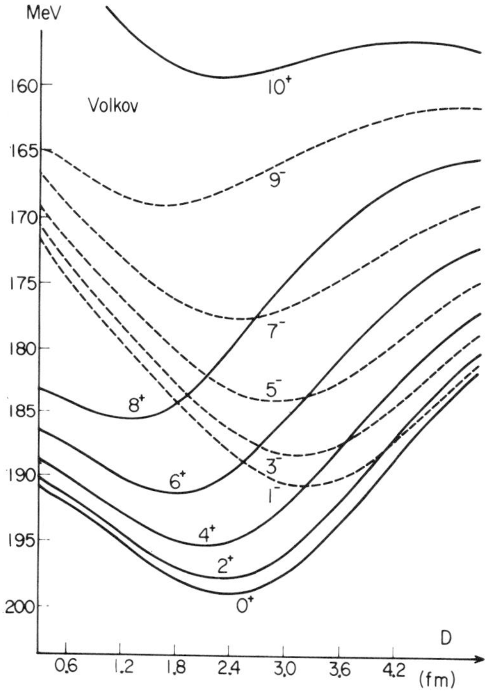

Figure 5
The expectation value $E(\nu,D)$ of the Hamiltonian for ^{20}Ne.

this $E(\nu,D)$ is plotted as a function of D. The value of ν is varied until $E(\nu,D)$ is minimized at each value of D.

We can find a few salient features in this figure:

(i) The value of D decreases as the spin increases. This means that the probability of finding an alpha cluster is larger in lower spin states.

(ii) The value of D is between 2.4 fm and 1.2 fm in the ground state band. Therefore the mean square distance between the ^{16}O and alpha clusters in the ground state band is roughly

$$\sqrt{\langle R^2 \rangle} \simeq \sqrt{(2.8)^2 + D^2} \leqq 3.5 \text{ fm}.$$

This value is still smaller than 4.7 fm which is the sum of the two radii.

(iii) The location of the 10^+ state which does not appear in the sd shell model, is discontinuously too high to belong to the ground state band.

(iv) The $K = 0^-$ excited band has larger D values than the ground state band. Therefore the alpha cluster has a larger probability of being found in the $K=0^-$ band than in the ground state band.

(v) Because of the nature of (i), the B(E2, $J_i \rightarrow J_f$) calculated by this model (Table 1) show a deviation from the predictions based on the rotational model. The results are very similar to those given by the shell model. However, in the cluster model, an effective charge is not necessary.
The cluster wave function is a very natural extension of the shell model one. The predictions of the cluster model are not very different from those of the shell model. The largest differences, however, can be found in alpha particle spectroscopic factors. The cluster model gives larger values than the shell model does. At the present moment, we do not have any clear experimental evidence which shows that alpha spectroscopic factors exceed shell model values and, therefore, except from the variational point of view, we have no explicit support of the cluster wave functions. This model, however, has the one clear advantage that the ground state band and the $K=0^-$ excited band are treated on the same footing.

It is highly desirable to carry out precise calculations in order to extract the spectroscopic factors for alpha transfer reactions.

The spatial symmetry breaking due to the strong spin-orbit interaction is neglected in the alpha cluster model. In order to take into account the strong spin-orbit interaction, a few techniques have been developed, they are discussed elsewhere.[28,29]

A final remark is that the relationship between quartet excited states and alpha cluster states should be made clear but this is still an open problem.

REFERENCES

1. Akiyama, Y., Arima, A. and Sebe, T. Nucl. Phys. A138, (1969) 273. Arima, A., Sakakura, M. and Sebe, T., Nucl. Phys. A170, (1971) 273.
2. Halbert, E.C., McGrory, J.B., Wildenthal, B.H. and Pandya, S.P., Advances in Nuclear Physics, Vol. 4, Ed. M. Baranger and E. Vogt (Plenum Press, New York, 1971).
3. Davies, W.G., Bronde, C. and Forster, J.S., Bull. Am. Phys. Soc. 15 (1970) 543.
4. Kalman, L.S., Bernier, J.P. and Harvey, M., Can. J. Phys. 45 (1967) 1297.
5. Hausser, O., Alexander, T.K., McDonald, A.B., Ewan, G.T., and Litherland, A.E., Nucl. Phys. A168 (1971) 17. Alexander, T.D., Hausser, O., McDonald, A.B., Ferguson, A.J., Diamond, W.T., and Litherland, A.E., to be published.
6. Rogers, D.W.O., Aitken, J.H., Litherland, A.E., Dixon, W.R. and Storey, R.S., Can. J. Phys. 49 (1971) 1397.
7. Elliott, J.P., Proc. Roy. Soc. A245 (1958) 128, 562; Elliott, J.P., and Harvey, M., Proc. Roy. Soc. A272 (1963) 557; Harvey, M. Advances in Nuclear Physics, Vol.1, Ed. Baranger, M. and Vogt, E. (Plenum Press, New York, 1968).
8. Wigner, E.P., Phys. Rev. 51 (1937) 106.

9. French, J.B., and Parikh, J.C., Phys. Lett., 35B (1971) 1.
10. Elliott, J.P., and Skyrme, T.H.K., Proc. Roy. Soc. (London) A232 (1955) 561.
11. Talmi, I., Helv. Phys. Acta 25 (1952) 185;
Moshinsky, M. Nucl. Phys. 8 (1958) 19; 13 (1959) 104;
Arima, A., and Terasawa, T. Prog. Theoret. Phys. 23 (1960) 115.
The following formula, derived by Arima and Terasawa, is very convenient:

$$\langle n_1\ell_1 n_2\ell_2;L|NL\ 0s;L\rangle = (-1)^{\frac{1}{2}(L+\ell_1-\ell_2)}\, 2^{\frac{1}{4}(\ell_1+\ell_2-3L)-N} \cdot\left[\frac{N!(2L+2N+1)!!(2\ell_1+1)(2\ell_2+1)}{n_1!n_2!(2\ell_1+2n_1+1)!!(2\ell_2+2n_2+1)!!(2L+1)}\right]^{\frac{1}{2}} (\ell_1\ell_2 00|L0),$$

for the case $m_1=m_2$, where m is a mass of each particle.
12. Rotter, I., Nucl. Phys. A122 (1968) 567; A135 (1969) 378.
13. Wildermuth, K. and Kanellopoulos, T., Nucl. Phys. 7 (1958) 150; 9 (1958)/59) 449.
14. Bayman, B.F., and Bohr, A. Nucl. Phys. 9 (1958/59) 596.
15. Middleton, R., Proceedings of the International Conference on Nuclear Reactions Induced by Heavy Ions, Heidelberg. Ed. by R. Bock and W.R. Hering (North-Holland Pub. Co., Amsterdam 1970). Middleton, R., Rosner, B., Pullen, D.J. and Polsky,L., Phys. Rev. Lett. 20 (1968) 118.
16. Nagatani, K., Levine, M.J., Belate, T.A., and Arima, A. to be published.
17. Middleton, R. private communication, to be published in Phys. Rev. Lett.
18. Marquardt, N., Von Oertzen, W. and Walter, R.L., Phys. Lett. 35B (1971) 37.
19. Brown, G.E., Proc. Paris Conf. on Nuclear Structure (1964); Brown, G.E., and Green, A.M., Nucl. Phys. 75 (1966) 401.
20. Danos, M. and Gillet, V., Phys. Rev. 161 (1967) 1034.
21. Feshbach, H., private communication.
22. Mattauch, J.H.E., Thiele, W. and Wapstra, A.H., Nucl. Phys. 67, (1965) 1.
23. Arima, A., Gillet, V. and Ginocchio, J., Phys. Rev. Lett. 25 (1970) 1043.
24. Simpson, J.J., Dixon, W.R. and Storey, R.S., private communication and to be published in Phys. Rev. C. (1971).
25. Morrison, G.C., Korner, H.J., Greenwood, L.R., and Siemssen,R.H. A.N.L. Phys.Research Monthly Report, June (1971).
26. Faraggi, H., Jaffrin, A., Lemaire, M.C., Mermaz, M.C., Faivre, J.C., Gastabois, J., Harvey, B.G., Loiseaux, J.M. and Papineau, A., Annals of Physics (de-Shalit Memorial issue) (1971).
27. Horiuchi, H., Ph.D. thesis. University of Tokyo (1970).
28. Arima, A., Horiuchi, H., Kubodera, K., and Takigawa, N., to be published in Adv. Nucl. Phys., Ed. M. Baranger and E. Vogt.
29. Abe, A., to be published in Prog.Theoret.Phys.

EXPERIMENTAL EVIDENCE FOR ALPHA PARTICLE CLUSTER STATES IN THE REGION ^{16}O to ^{20}Ne

A.E. LITHERLAND
Department of Physics
University of Toronto

The experimental evidence for alpha-particle cluster states in the region ^{16}O to ^{20}Ne has been reviewed recently by Ogloblin and by others[1]. First I would like to give a report on some recent new experimental evidence which has an important bearing on this problem and secondly I would like to review the evidence for alpha-particle cluster states in the region ^{16}O to ^{20}Ne in the light of the recent new evidence.

SOME ALPHA PARTICLE CLUSTER STATES IN ^{18}F

The situation in ^{18}F has only recently been clarified as a result of experiments by Rolfs et al.[2,3] Before these experiments it was not possible to distinguish between the weak coupling approach[4] and the strong coupling approach for the set of states which have been suspected for some time to be four-particle two-hole states (4p-2h). The situation in ^{18}F was unclear because of the lack of reliable spin and parity measurements. Unreliable evidence had led Middleton et al.[5] to propose, on the basis of their most valuable high-resolution measurements on the $^{14}N(^7Li,t)^{18}F$ alpha-transfer reaction, that the weak coupling picture might be valid for the (4p-2h) states in ^{18}F. On the other hand Bassichis et al.[6] had proposed that a (4p-2h) $K^{\pi} = 1+$ rotational band should exist in ^{18}F and their Hartree-Fock calculations suggested that the 1.70-MeV level might be the band head of such a band.

The experiments of Rolfs et al.[23] have now clarified the situation in ^{18}F and as the results are of considerable interest I would like to spend a little time discussing them in some detail.

The study of the 5298 and 6567 keV states in ^{18}F, using the $^{14}N(\alpha,\gamma)^{18}F$ reaction, by Rolfs et al.[2] has provided the most significant new information on the (4p-2h) states in ^{18}F.

These states were formed as resonances at 1135 and 2766 keV bombarding energy in the $^{14}N(\alpha,\gamma)^{18}F$ reaction. Both the resonances being studied were very weak ones and it was necessary to use the maximum alpha particle currents available and the best directly cooled TiN targets that could be made. TiN targets on Ta or W backings were bombarded by ^4He currents of 60 to 130 μA

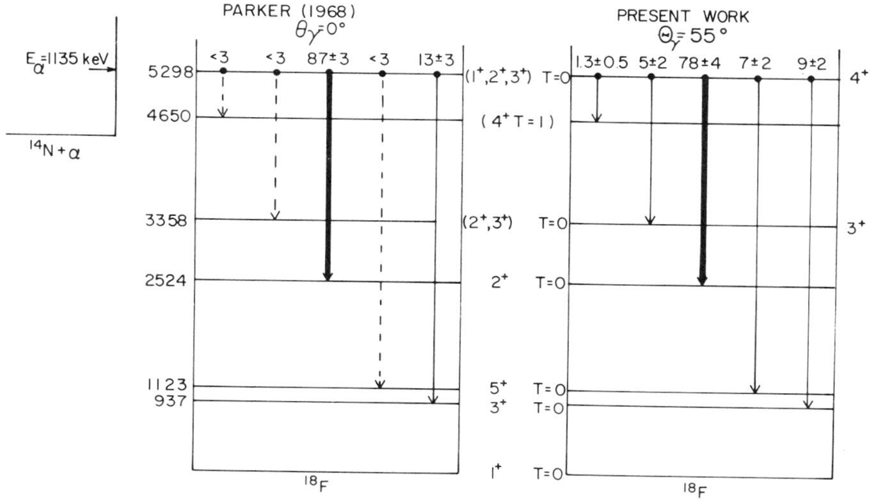

Figure 1
The gamma-ray schemes of the 5298-keV state in ^{18}F obtained by Parker[7] and Rolfs et al.[2] are compared.

from the KN-3 Van de Graaff at McMaster University. The capture gamma rays were observed with the help of 40 cc Ge(Li) detectors.

The right-hand side of figure 1 shows the results obtained following an analysis of the gamma-ray spectrum at the 1135-keV resonance which had previously been studied at $\theta_\gamma = 0^o$ by Parker[7]. In addition to the transition to the 3+, 937 -keV state in ^{18}F observed by Parker there is a prominent transition to the 5+, 1123-keV state in ^{18}F. This transition was not observed by Parker because of the unusually strong angular distribution. The data given by Parker[7] are shown on the left-hand side of Figure 1. Additional weak transitions to the states at 3358 and 4650 keV in ^{18}F are shown.

Some of the angular distribution data of Rolfs et al.[2] are shown in Figure 2. The analysis of the angular distribution of the primary gamma rays to the 5+, 1123 keV state in ^{18}F shows conclusively that the resonance has an angular momentum of 4 (solid curve). The angular distribution of the primary transition to the 3358 keV state shows conclusively that the 3358 keV state has an angular momentum of 3 and that there is a large admixture of quadrupole radiation in the transition.

The thick target yield of gamma rays from the reaction $^{14}N(\alpha,\gamma)^{18}F$ at the resonance gives $\omega\gamma$ = 19 ± 7 meV where

$$\omega\gamma = \frac{2J+1}{2I+1} \frac{\Gamma_\alpha \Gamma_\gamma}{\Gamma}$$

and the measurement of the lifetime of the resonant state by the Doppler Shift Attenuation Method gave a lifetime of 31 ± 6 fs.

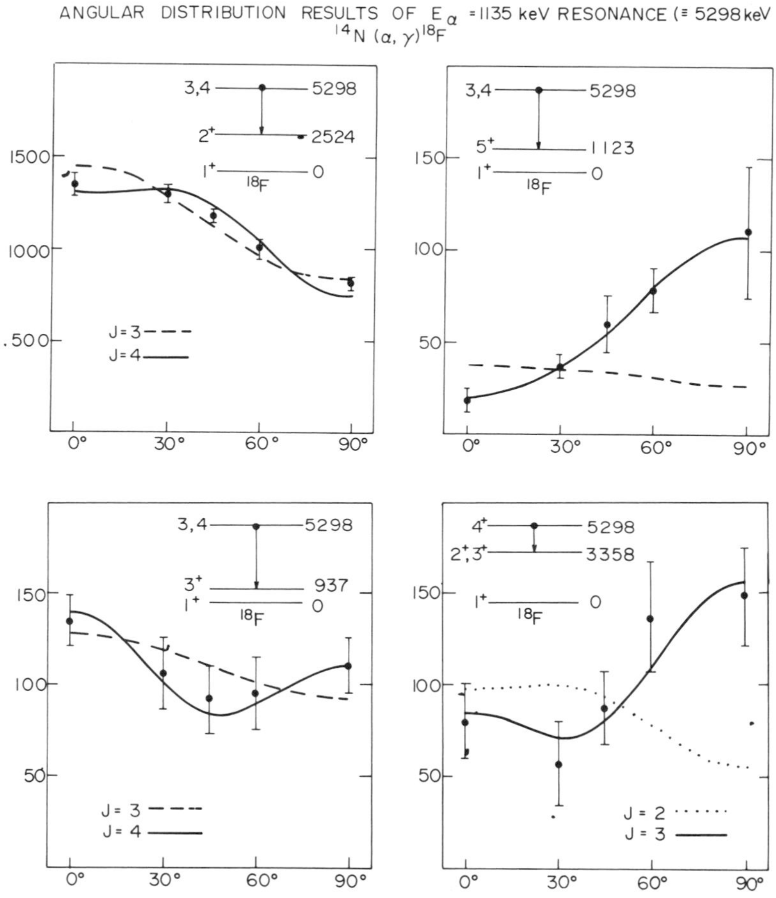

Figure 2
Angular distribution data[2] for the 5298 keV state in ^{18}F are shown.

It is worth noting at this point that there are very few charged particle resonances whose lifetimes can be measured by the Doppler Shift Attenuation Method and the 1135 keV resonance in $^{14}N(\alpha,\gamma)^{18}F$ is only the second to be measured in studies of the radiative capture of alpha particles. The reason for the rarity is due to the fact that Γ must be less than 50 meV for the method to succeed, otherwise the lifetime of the nuclear state would be too short to be measured by the Doppler Shift Attenuation Method. In the case being discussed the measured values of τ and $\omega\gamma$ give $\Gamma_\gamma = \Gamma_\alpha = 12 \pm 4$ meV. These values are in excellent agreement with the previous result of $\Gamma_\gamma/\Gamma_{total} = 0.5 \pm 0.1$, obtained by Gorodetsky et al.[8] The 78% branch to the $J^\pi = 2+$, 2524 keV level corresponds to an E2 or M2 strength of 25 or 580 Wu implying $J^{\pi}(5298) = 4+$.

The relevance of these results to the alpha cluster structure of states near ^{16}O is apparent when a value of the reduced alpha particle width for the resonance is deduced. Using an interaction radius of 4.8 fm a value of $\theta^2_{4+}(\ell = 4) = 0.5$ is obtained.

At this point in the study it was clear that we were dealing with a rotational band in ^{18}F. In addition the large yield of the $^{14}N(^7Li,t)^{18}F$ reaction to this group of states and the large value of $\theta^2_{4+}(\ell = 4)$ observed implied that all the members of the band probably had large alpha particle reduced widths.

To test these ideas further a study was made of the 2766 keV resonance in the reaction $^{14}N(\alpha,\gamma)^{18}F$. Previously Herring et al.[9] had shown this resonance to exhibit an $\ell = 4$ character, in the elastic scattering of alpha particles by ^{14}N, implying spin-parities of 3^+, 4^+ or 5^+. This resonance was suspected to be the 5^+ member of the $K^{\pi} = 1+$ rotational band because of its excitation energy and spin-parity possibilities. The E2 transition strengths observed for the lower members of the band suggested that the main decay of the 5^+, K = 1, state should be by an E2 transition to the 3^+, 3358 keV state in ^{18}F.

These expectations were fulfilled for the decay of the 2766 keV resonance. A spectrum of gamma rays from the resonance is shown in Figure 3 together with a decay scheme.

The strongest decay is as expected. The analysis of the angular distributions of the resonant gamma rays gives a unique J(6567) = 5 spin assignment. This 6567 keV state is unbound by 990 keV against proton decay to ^{17}O. However in a study of $^{17}O(p,\gamma)^{18}F$ it was found that $\omega\gamma < 0.25$ meV and consequently $\Gamma_p < 0.009\ \Gamma_\alpha$. From the observed $\omega\gamma$-width of 97 ± 20 meV in the $^{14}N(\alpha,\gamma)^{18}F$ reaction a width of $\Gamma_\gamma = 26 \pm 6$ meV can be deduced. In this case $\Gamma_\alpha >> \Gamma_\gamma$ or Γ_p, and so the E2 strength of the 6567 to 3358 keV gamma ray decay is 28 ± 6 Wu. The parity of the 6567 keV state is positive because the 3358 keV state has positive parity. The reduced alpha particle width for the 6567 keV state is $\theta^2_{5+}(\ell = 4) = 1.3$ assuming an interaction radius of 4.8 fm or $1.2(A^{1/3} + 4^{1/3})$ fm where A = 14 in this case. The reduced width in this case is clearly large and can be made equal to unity by an appropriate choice of interaction radius.

The situation in ^{18}F can be summarized with the help of Figure 4 as follows:

1. The 1702, 2524, 3358, 5298 and 6567 keV states are the first five members of a $K^{\pi} = 1+$ rotational band.

2. The reduced E2 transition strengths are large for transitions within this band and small for transitions between this band and other states in ^{18}F.

3. The E2 transition strengths are consistent with the strong coupling collective model[10].

4. The energies of the five members of the rotational band are consistent with the Hartree-Fock theory of Bassichis, Giraud and Ripka[6] although the predicted energy of the band head is about 2.5 MeV too high.

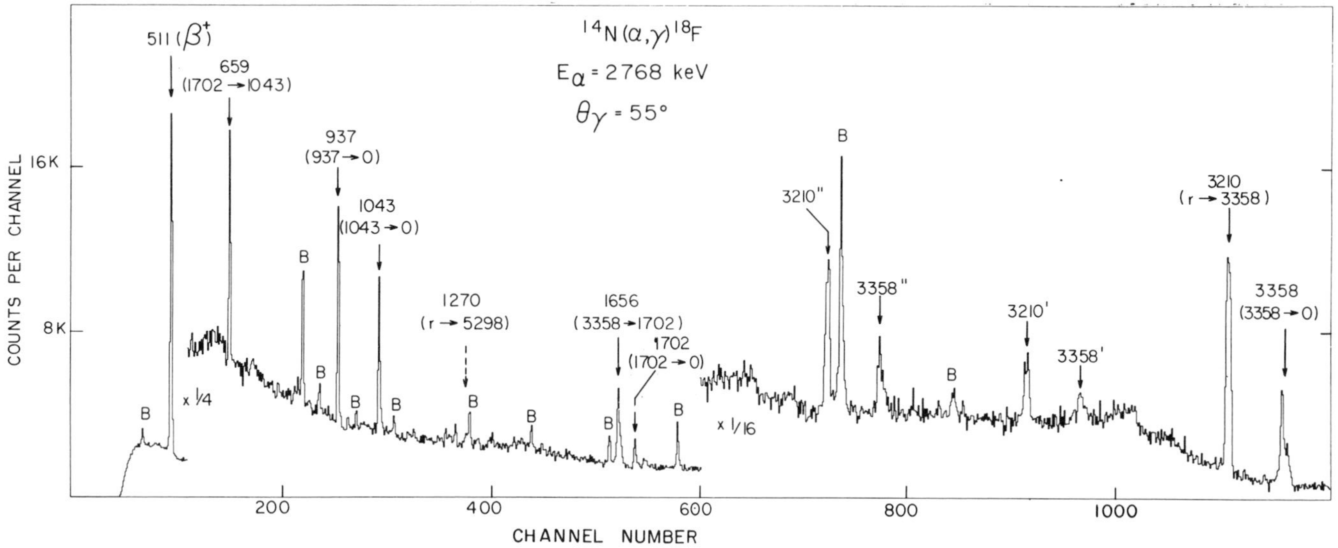

Figure 3a
Low energy portion of the gamma-ray pulse height spectrum obtained at the 2766 keV resonance in $^{14}N(\alpha,\gamma)^{18}F$.

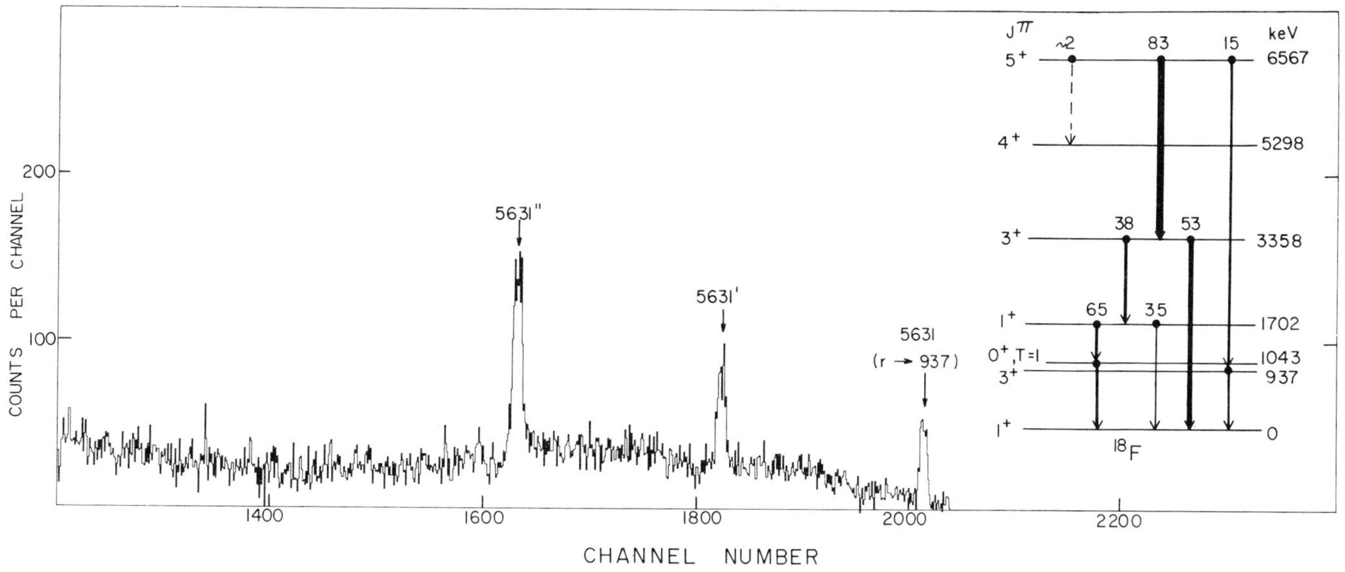

Figure 3b
High energy portion of the gamma-ray pulse height spectrum obtained at the 2766 keV resonance in $^{14}N(\alpha,\gamma)^{18}F$.

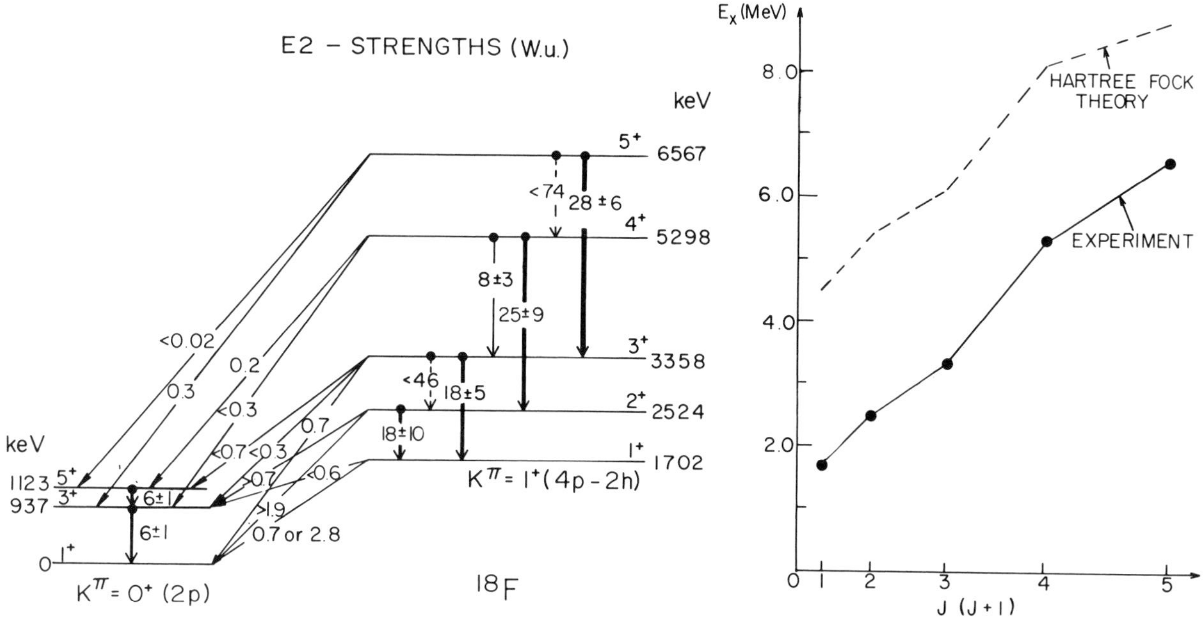

Figure 4
The reduced E2 transition probabilities in Weisskopf Units are shown for the transitions within and from the K^{π} = 1+ rotational band in ^{18}F.

5. The members of the rotational band seem to have large alpha particle reduced widths.

The weak coupling model is unsatisfactory for a number of reasons:

1. The energy spectrum looks too much like the spectrum of a rigid rotor.

2. The 1+ state associated with the 2+ and 3+ states either does not exist or else its properties are so different from those expected that it is unrecognizable. The 1+ state should decay to the 1+, 1702 keV state with an enhanced E2 transition. The 1+ states at 3724, 4363 and 4852 keV decay by M1 transitions with large reduced widths to states which are known to have dominant $(sd)^2$ configurations.

The problem of weak coupling versus strong coupling in ^{18}F will be discussed in a future publication but for the purpose of these lectures the main interesting property of these levels in ^{18}F is the large alpha particle reduced width of the rotational band members. At present this feature is not even approximately reproduced by any theories. However rotational bands of excited states with large alpha particle reduced widths are common between ^{16}O and ^{20}Ne so that a summary of this evidence will now be given.

A REVIEW OF ALPHA PARTICLE CLUSTER STATES IN THE REGION ^{16}O TO ^{20}Ne

This review is simply an extension of the excellent review given recently by Ogloblin[1] who discussed in some detail the results from lithium-ion induced nuclear reactions. I will concentrate on the region between ^{16}O and ^{20}Ne and I will compare and contrast the alpha particle reduced widths deduced from resonant reactions, induced by alpha particles, and those estimated in a very simple way from lithium-ion induced nuclear reactions.

Alpha Particle Reduced Widths from Resonant Reactions

The alpha particle reduced width θ^2 can be obtained from the measured alpha particle width Γ_α with the help of the formula[1]

$$\Gamma_\alpha = 2P \frac{\hbar^2}{mR^2} \theta^2$$

In this formula P is the penetrability of the coulomb and centrifugal barriers and $\hbar^2/mR^2$ is the so-called Wigner limit. R is known as the interaction radius and is often taken to be $1.2\ (A^{1/3} + 4^{1/3})$ for alpha particle induced resonance reactions.

In order to determine θ^2 from the measured value of Γ_α it is, of course, necessary to know the orbital angular momentum ℓ of the captured alpha particle in order to estimate correctly the value of P. The value of ℓ is usually deduced from angular distribution measurements.

The quantity Γ_α can be obtained in a number of ways:

1. If $\Gamma_\alpha > 500$ eV and if $\Gamma \sim \Gamma_\alpha$, then Γ_α can usually be determined by observing the width of a resonance in the elastic scattering of alpha particles. This limit is only an approximate one and depends upon the quality of the electrostatic accelerator and beam analysing system available to the experimental physicist.

2. If $\Gamma_\alpha < 500$ eV and if $\Gamma \sim \Gamma_\alpha$, then Γ_α can often be determined by measuring the area of the anomaly in the elastic scattering of alpha particles. In this case the value of Γ_α can be obtained with the help of the formula

$$2\pi^2(2J+1)\lambda\!\!\!^{-2}\Gamma_\alpha \quad = \quad \text{Area of Resonance}$$

When $\Gamma_\alpha < 0.1$ eV then other methods must be used to determine Γ_α. In addition the approximation $\Gamma \sim \Gamma_\alpha$ becomes a poor one and it is necessary to know all the other partial widths such as Γ_γ and in some cases Γ_p etc.

3. The total width Γ can be measured in the region 0.1 meV $<\Gamma<0.1$ eV by the Doppler Shift Attenuation Method. This method compares the slowing-down time of an excited atom with the lifetime $\hbar/\Gamma$ of the resonant state. An example of the use of this method was given previously for ^{18}F. The method is not of wide applicability, partly because the yield of the radiative capture reaction is also proportional to $\Gamma_\alpha \sim \Gamma$. In this case large ion beam currents are usually necessary in order to obtain enough statistics to measure the attenuation of the Doppler shift of the gamma rays from the de-excitation of the nucleus.

4. It is possible to obtain functions containing Γ_α by measuring the absolute yield of resonant reactions. For example it is possible to measure $\Gamma_p\Gamma_\alpha/\Gamma$, $\Gamma_p\Gamma_\gamma/\Gamma$ and $\Gamma_\alpha\Gamma_\gamma/\Gamma$ by studying (p,α), (p,γ) and (α,γ) reactions. If the only open channels are those for proton, alpha particle and gamma ray emission then it is possible to deduce Γ_p, Γ_α and Γ_γ from the above measured quantities provided some information is available from elastic scattering of alpha particles and protons. Examples were given in the previous lecture in which the 4^+, 1135 keV and 5^+, 2766 keV resonances were discussed. The value of Γ_p for the 4^+, 1135 keV resonance is zero because the channel is closed; $\Gamma_\alpha\Gamma_\gamma/\Gamma$, Γ_α/Γ, and Γ were measured in this case. The 5^+, 2766 keV resonance can be seen in elastic scattering and so provided $\Gamma_\gamma, \Gamma_p \ll \Gamma_\alpha$ the method given in section 2 can be used. In this case $\Gamma_\alpha\Gamma_\gamma/\Gamma$, $\Gamma_p\Gamma_\gamma/\Gamma$ are known and from these values it is clear that Γ_γ and Γ_p are $\ll \Gamma_\alpha$.

Alpha Particle Reduced Widths From Lithium Ion Reactions

At present there is no satisfactory theory of lithium ion alpha particle transfer reactions and so a quantitative discussion must be based upon gross oversimplifications in the interpretation of the data. However there are striking features in the experimental data which encourage one to believe that the development of a satisfactory theory is well worth the effort.

In the previous lecture I mentioned the large alpha particle reduced width obtained from the resonant reactions $^{14}N + \alpha$. In particular the 4^+, 5298 and 5^+, 6567 keV states in ^{18}F were shown to have large alpha particle reduced widths. The high resolution data of Middleton et al.[5] on the $^{14}N(^7Li,t)^{18}F$ with 15 MeV ^7Li ions show that those 4^+ and 5^+ states are populated very strongly. Other data, discussed by Ogloblin[1] for example, also demonstrate that nuclear states with large alpha particle reduced widths are populated most strongly in $(^6Li,d)$ and $(^7Li,t)$ reactions.

This situation is reminiscent of the results from (d,p) and (d,n) stripping reactions in which nuclear states with large neutron and proton reduced widths are most strongly populated. The deuteron is known to be a loosely bound structure of a neutron and a proton with a binding energy of 2.2 MeV. The ^6Li and ^7Li nuclei are also loosely bound structures and they are bound against break-up into α + d and α + t by only 1.47 and 2.47 MeV respectively. Both ^6Li and ^7Li are known to have a cluster structure in their ground states. Consequently by analogy with deuteron stripping reactions $(^6Li,d)$ and $(^7Li,t)$ reactions are expected to show strong preference for final states with large alpha particle reduced widths.

Reduced nucleon widths from deuteron stripping reactions are now obtained routinely as a result of many years of theoretical and experimental development. However a comparable theory of $(^6Li,d)$ and $(^7Li,t)$ alpha particle transfer reactions is still being developed[1]. Until a complete theory has been formulated the following summary[1] of the principal features of lithium-ion induced reactions is worth making:

1. The reactions are very selective and there is a definite correlation between alpha particle reduced widths and the intensity of the particle groups from lithium induced reactions. For example in ^{18}F the states in the $K^\pi = 1^+$ rotational band are strongly populated compared with other states. The relative intensities of the deuteron and triton groups to the same residual states in both the $(^6Li,d)$ and $(^7Li,t)$ reactions are approximately equal. This feature supports the concept of a directly transferred alpha particle cluster.

2. The selection rules

$$I_f = I_i + L, \quad \Delta\pi = (-1)^L$$

are obeyed quite well and they further support the concept of the transferred alpha particle cluster; I_i, I_f, L and π are the initial and final nuclear spins, L is the orbital angular momentum of the alpha particle and π is the parity of L. The isospin selection rule, $\Delta T = 0$, works not only for the $(^6Li,d)$ reaction but also for the $(^7Li,t)$ reaction. This again implies the transfer of a T = 0 alpha particle cluster.

3. The angular distributions have a form typical of a direct process.

I would like to conclude by giving examples of the first two of these features to illustrate how important is the development of a satisfactory theory of the lithium-ion alpha particle transfer reaction.

A most useful feature of the lithium-ion transfer reactions seems to be the correlation between the intensity of the deuteron or triton group and the angular momentum of the corresponding excited state in the residual nucleus[1]. Together with the correlation noted earlier between the alpha particle reduced width and the intensity of the deuteron or triton group, these two features suggest that one should try to correlate the data with the help of the formula

$$\frac{d\sigma}{d\Omega} = (2J_f+1)F\theta_t^2 \text{ mb/st}$$

The factor $(2J_f+1)\theta_t^2$ is suggested by experiment, where J_f is the angular momentum of the final state and θ_t^2 is the reduced alpha particle width obtained from the transfer reaction to distinguish it from θ_α^2 the reduced alpha particle width obtained from Γ_α.

The factor F is obviously a very complicated factor and must include such quantities as the Q-value of the reaction and the spin of the target nucleus and the projectile and reaction product. In addition it must contain the angle of observation, because $d\sigma/d\Omega$ is a differential cross section, and also the properties of the ground state of the lithium ions. Many other significant quantities are included and some guidance can be obtained from the plane wave theory of Neogy et al.[11] For the purposes of the remainder of this talk I am going to make the drastic assumption that for $(^7Li,t)$ reactions $F = 0.92(2J_i+1)^{-1}$. We then have

$$\frac{d\sigma}{d\Omega}(15^o) = \frac{2J_f+1}{2J_i+1}\ 0.92\ \theta_t^2\ \frac{\text{mb}}{\text{st}}$$

$F = 0.92$ is chosen so that $\theta_t^2 = 1$ for the 5^+, 6.548 MeV state in ^{18}F.

This expression will be assumed to apply irrespective of the incident ^7Li energy in the range 15-30 MeV. The choice of $\theta_t^2 = 1$ for the 5^+, 6.548-MeV state in ^{18}F is for the purposes of normalization only and is made because θ_α^2 for this state is known to be near the Wigner limit. As we will see the simple expression given above, with some reservations, shows an excellent correspondence between the θ_t^2 obtained from a variety of $(^7Li,t)$ work and the θ_α^2 obtained from resonance reactions.

The present situation with regards to alpha particle reduced widths in ^{18}F is summarized in Table 1. The $^{14}N(^7Li,t)^{18}F$ data were taken from the work of Middleton et al.[5] and the resonance

TABLE 1

FLUORINE - 18

	Jπ	E_x(MeV)	$\frac{d\sigma}{d\Omega}$ (15°) $\frac{\mu b}{st}$	θ_t^2	Γ	θ_α^2
T=0	1+	1.701	680	0.74	-	-
	2+	2.524	1,710	1.1	-	-
	3+	3.358	2,750	1.26	-	-
	4+	5.295	2.210	0.8	12-24 meV[2,12]	0.5-1.0
	5+	6.548	3,370	1*	700 eV[9]	1.32
T=0	1+	0	110	0.12		
	3+	0.937	140	0.09		
	5+	1,131	47	0.014		
T=1	0+	1.043	< 2		0	0
	2+	3.060	20			
	4+	4.651	18			
	0+	4.741	9		0	0
	2+	4.964	14			
	0+	6.139	200	0.22	0	0
T=1+0	1-	5.613	<84	<0.09	60 eV[13]	0.12
	1-	5.681	140	0.22	200 eV[13]	0.22
T=0	1+	4.852	150	0.16	93 μeV[12]	0.11
	4-	6.105	770	0.31	155 eV[13]	0.31

data were taken from the authors listed in the references. The agreement between the very different methods for obtaining θ^2 for the unbound states is quite striking and suggests that possibly the values of θ^2 for the bound states are being estimated reliably. However the results, from the $^{14}N(^7Li,t)^{18}F$ reaction leading to the 0+, T=1 states, indicate that caution is needed in interpreting the results. The yields of the reaction to the isospin one states are expected to be inhibited because of the ΔT=0 selection rule and in addition the yields of the 0+ states should be zero because of parity conservation. Consequently the 0+, T=1 states should not be excited by the $^{14}N(^7Li,t)^{18}F$ reaction provided the mechanism is an alpha particle transfer reaction. The 0+, T=1 states at 1.043 and 4.741 MeV are very weakly excited. However the recently assigned 0+, T=1 state at 6.139 MeV is excited rather strongly and this observation casts doubt on the predominance of the alpha particle transfer reaction mechanism at 15 MeV ^7Li ion energy

TABLE 2

NEON - 20

	Jπ	E_x(MeV)	$\frac{d\sigma}{d\Omega}(15^o)\ \frac{\mu b}{st}$	θ^2_t	Γ	θ^2_α
	0+	0	240	0.26	-	-
	2+	1.63	820	0.18	-	-
K=0+	4+	4.25	2,000	0.24	-	-
	6+	8.77	2,300	0.21	110 eV[14]	0.130
	8+	11.97	< 230	<0.02	35 eV[14]	0.017
	1-	5.80	1,450		-	-
	3-	7.19	5,000	0.77	8 keV[16]	1.48
$K=0^-$	5-	10.3	13,000		150 keV[15]	
	7-	15.6	5,000		470 keV[15]	
	2-	4.97	10		-	-
	3-	5.63	300	0.047	3.1 meV[14]	0.084
$K=2^-$	4-	7.02	15		-	-
	5-	8.46	15			
	6-	10.65			-	-
	7-	13.33			80 eV[14]	0.0012
	0+	6.72	300	0.3	19 keV[16]	0.68

on ^{14}N. This problem cannot of course be resolved at present but it is worth noting that the compound nucleus contribution to each triton group can fluctuate[1] greatly from group to group. The fluctuations are greatest for final states with low spin so that the result for the 0+, T=1, 6.139 MeV state at 15 MeV ^7Li energy may be an unusually large fluctuation in the compound nucleus contribution. The result does however suggest caution in interpreting values of $\theta^2 < 0.2$ obtained from the $^{14}N(^7Li,t)^{18}F$ reaction.

The situation in ^{20}Ne is shown in Table 2 and again there is a striking correspondence between the values of the alpha particle reduced widths obtained from resonance reactions and the $^{16}O(^7Li,t)^{20}Ne$ alpha particle transfer reaction. In this transfer reaction the excitation of states in ^{20}Ne with $\pi = (-1)^{J+1}$ is forbidden. The numbers in the table support this selection rule very well.

In Tables 3 and 4 the situation in ^{19}F and ^{16}O, ^{17}O and ^{18}O is shown and again there is good agreement between the few values of θ^2 obtained from resonance reactions and the alpha particle transfer reactions.

TABLE 3

FLUORINE - 19

Jπ	E_x(MeV)	$\frac{d\sigma}{d\Omega}(15^o)$ $\frac{\mu b}{st}$	θ_t^2
1/2-	0.11	500	0.55
5/2-	1.35	1700	0.61
3/2-	1.46	960	0.52
7/2-	4.00	840	0.23
9/2-	4.04	1100	0.24

Jπ	E_x(MeV)	$\frac{d\sigma}{d\Omega}(15^o)$ $\frac{\mu b}{st}$	θ_t^2
1/2+	0	400	0.43
5/2+	0.197	400	0.14
3/2+	1.56	800	0.43
9/2+	2.79	200	0.04
7/2+	4.39	200	0.05
13/2+	4.65	≤100	≤0.01

Jπ	E_x(MeV)	$\frac{d\sigma}{d\Omega}(15^o)$ $\frac{\mu b}{st}$	θ_t^2	Γ_α	θ_α^2
3/2+	3.92	200	0.11		
3/2-,5/2+	4.56,4.55	≤600	≤0.3	<1 meV[17]	<0.3
5/2-	4.68	≤100	≤0.036	2.1 meV[17]	0.026
1/2+	5.34	600	0.65		
3/2+	5.49	1400	0.77	4 keV[18]	0.87
5/2+	6.26	1000	0.36	3 keV[18]	0.43
7/2+	6.31	1200	0.33	3 keV[18]	0.43

In conclusion, it is clear the simple expression for $d\sigma/d\Omega$ given earlier shows that a detailed theory is well worth formulating. However it is also worth noting that more extensive high resolution data should be obtained at a variety of ^{7}Li ion energies at greater than say 15 MeV and preferably at energies greater than 30 MeV.

TABLE 4

$J\pi$	E_x(MeV)	$\frac{d\sigma}{d\Omega}(15^o)\ \frac{\mu b}{st}$	θ_t^2	Γ_α	θ_α^2
OXYGEN - 16					
0+	6.06	620	0.67	-	-
2+	6.92	6000	1.30	-	-
4+	10.35	6500	0.79	27 keV	
6+	16.2	-			
OXYGEN - 17					
5/2+	0	75	0.027		
1/2+	0.871	92	0.1		
1/2-	3.058	750	0.81		
5/2-	3.846	1400	0.51		
3/2-	4.551	1350	0.73		
OXYGEN - 18					
2+	1.98	190	0.04	-	-
0+	3.63	≤320	≤0.35	-	-
4+	7.10	3370	0.41	63 meV[19]	0.31

REFERENCES

1. A.A. Ogloblin, "Nuclear Reactions Induced by Heavy", edited by R. Bock and W.R. Herring (North Holland, 1970), p.231. See also the articles by R. Middleton, p. 263, and by K. Bethge, Annual Review of Nuclear Science 20 (1970), 255.
2. C. Rolfs, H.P. Trautvetter, R.E. Azuma and A.E. Litherland, to be published.
3. C. Rolfs, L. Keiser, R.E. Azuma and A.E. Litherland, to be published.
4. A.P. Zuker, Phys. Rev. Letters, 23 (1969), 983.
5. R. Middleton, L.M. Polsky, C.H. Holbrow and K. Bethge, Phys. Rev. Letters, 21 (1968), 1398.
6. W.H. Bassichis, B. Giraud and G. Ripka, Phys. Rev. Letters, 15 (1965), 980.
7. P.D. Parker, Phys. Rev. 173 (1968), 1021.
8. S. Gorodetsky et al., Phys. Rev. 155 (1967), 1119.
9. D.F. Herring et al., Phys. Rev. 112 (1958), 1210.
10. D.J. Rowe, "Nuclear Collective Motion" (Methuen, London, 1970).
11. H.E. Gove, "Resonance Reactions", in Nuclear Reactions, Vol. I, edited by P.M. Endt and Demers (North Holland, 1959), p.259.
12. R.G. Couch et al., Bull. Amer. Phys. Soc. 16 (1971), 511 and private communication.
13. E.A. Silverstein et al., Phys. Rev. 124 (1961), 868.
14. O. Hausser et al., to be published.
15. W.E. Hunt et al., Phys. Rev. 160 (1967), 782.
16. F. Ajzenberg-Selove and T. Lauritsen, Nuclear Physics
17. D.W.O. Rogers et al., to be published.
18. H. Smotrich et al., Phys. Rev. 122 (1961), 232.
19. H.E. Gove and A.E. Litherland, Phys. Rev. 113 (1959), 1078 and also L.F. Chase, Jr., et al., in "International Nuclear Physics Conference", edited by Becker, Goodman, Stelson and Zucker (Academic Press, 1967), p. 930.

PROMPT AND DELAYED PHOTOFISSION

J. GOLDEMBERG
University of São Paulo, São Paulo, Brazil
and
University of Toronto, Toronto, Ontario, Canada

In these two lectures on prompt and delayed fission I will review quickly the basis of the present ideas on the fission process, namely the Strutinsky double barrier model and results from two recent experiments done at the University of Toronto* on delayed photofission and at the University of São Paulo,** Brazil on prompt photofission. The Toronto experiment involves developments of a new technique to detect and locate in space fissioning nuclei.

1. THE LIQUID DROP MODEL (LDM)

The classical way[1] of describing the fission of heavy elements on the basis of the liquid drop model of Bohr and Wheeler has had an extraordinary success in describing the gross properties of fission. In this model one assumes that the incoming particle produces a symmetric deformation of the nucleus described by the coefficients in the expansion (Fig. 1)

$$r(\theta) = R[1 + \alpha_0 + \alpha_2 P_2(\cos\theta) + \alpha_4 P_4(\cos\theta) + \ldots] \qquad (1)$$

This deformation costs a certain amount of energy E_f and eventually carries the nucleus over a "saddle point" beyond which there is no return, leading to scission into two fragments (Fig. 2).

Pictorially one can describe what happens by considering a little ball with some initial potential energy in a potential well (Fig. 3): if the ball is released it will execute a complicated motion in space and eventually cross the saddle point.

Saddle point shapes for nuclei with different atomic numbers[2] can be seen in Fig. 4. The parameter x in this figure is the fissionability parameter

$$x = \frac{Z^2/A}{48.4} \qquad (2)$$

* In collaboration with L. Pai and A.E. Litherland.
** In collaboration with Olga Mafra and S. Kuniyoshi.

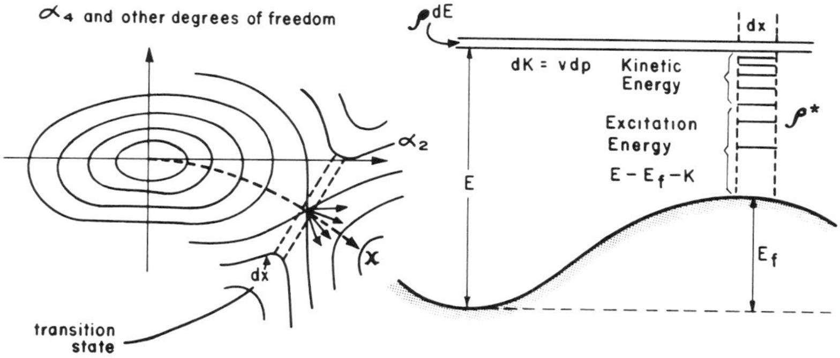

Figure 1
Potential energy contours for deformations specified by α_2 and α_4.

Physically x is related to the ratio of the electrostatic energy to the surface energy of the nucleus. For x = 1 the nucleus is unstable against fission.

The deformation of the nucleus is a slow process and one assumes in general that a compound nucleus is initially formed by the incoming particle or photon. The deformation (and subsequent fission) is therefore one of the possible ways[3] in which

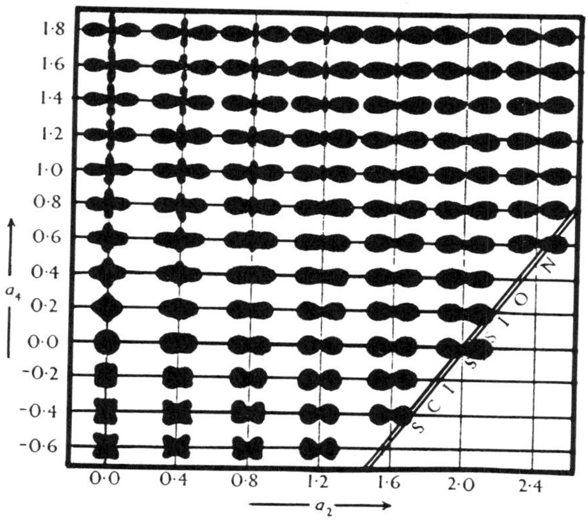

Figure 2
Shapes of nuclear surfaces on α_2 and α_4 deformations.

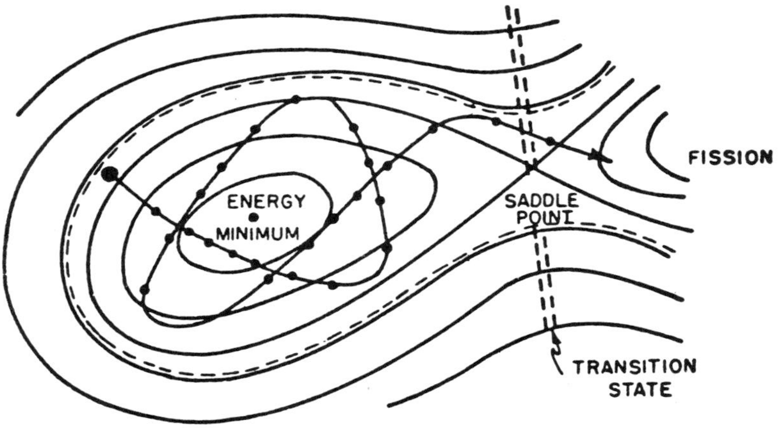

Figure 3
Motion of representative point of fissioning system in configuration space determined by α_2 and α_4.

the compound nucleus might evolve (Figure 5). Other processes, such as neutron emission, can compete with fission in the decay of the compound nucleus. Experimentally one finds that the branching ratio between neutron emission and fission for uranium and thorium is of the order of 1. Binding energies for the neutron and the height of the fission barrier are of course critical in determining this branching ratio. We will come back to this question later.

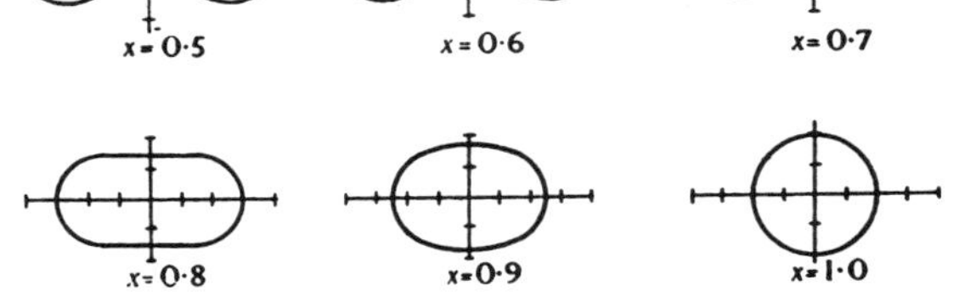

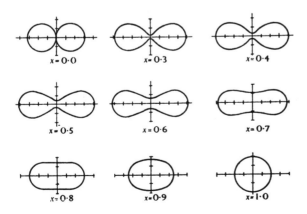

Figure 4
Saddle point shapes for various fissionability parameters.

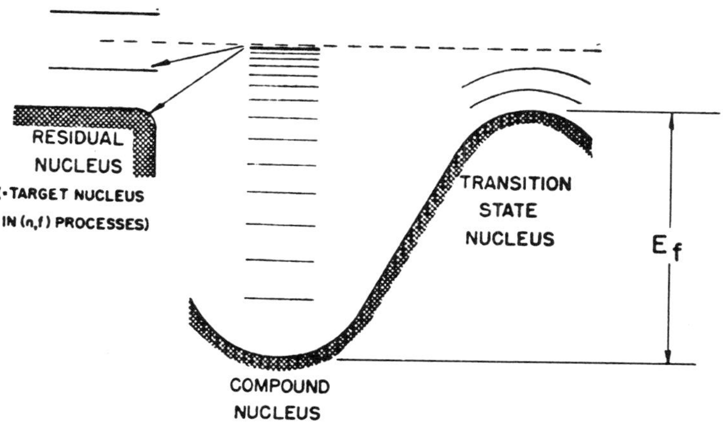

Figure 5
Levels in the compound nucleus, transition state nucleus and residual nucleus.

2. THE STRUTINSKY MODEL

Strutinsky[4] was the first to point out clearly that shells may be regarded more generally as a large-scale non-uniformity in the distribution of single-particle states; groupings of this kind may always be expected in finite-size nuclei. Usually one thinks of the shells as related to the degeneracy of single-particle states due to the sphericity of the nuclear shape and that they disappear or do not play an important part in highly deformed nuclei. The schematic Nilsson diagram of Fig. 6 shows how a shell can disappear for small deformations and reappear at larger ones.

The binding energy of the nucleus is modulated by the non-uniformities in the level density distribution (which changes with deformation). Figure 7 shows a Fermi gas distribution of levels in a LDM (Liquid Drop Model) nucleus and a more realistic level distribution: if the last occupied level (ε_F in the Fermi gas distribution) falls in a place corresponding to a low density, then there are many levels compressed below ε_F and

$\varepsilon_F < \Lambda$ (the nucleus is more bound than in the LDM).

If the last filled level falls between a group of levels (Fig.8)

$\varepsilon_F > \Lambda$ (the nucleus is less bound than in the LDM).

As a result of this the potential energy as a function of deformation is a curve that looks like Fig. 9, which gives the nucleus a double humped barrier against fission. The corresponding potential energy contours and saddle points are shown in Fig. 10.

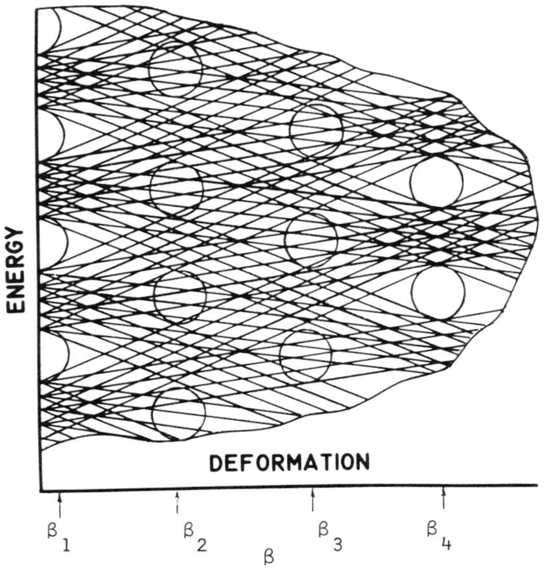

Figure 6
Simplified Nilsson diagram for large deformations.

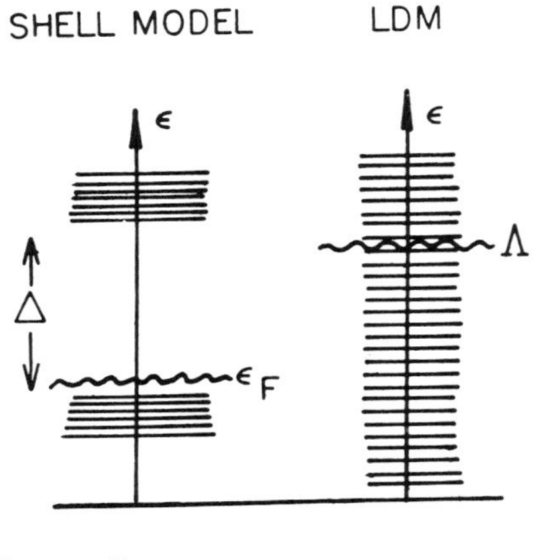

Figure 7
Last occupied shell level above closed shell clusters.

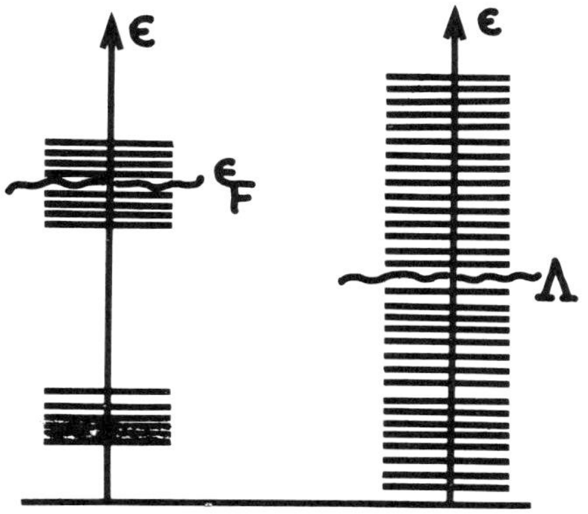

Figure 8
Last occupied shell level inside closed shell cluster.

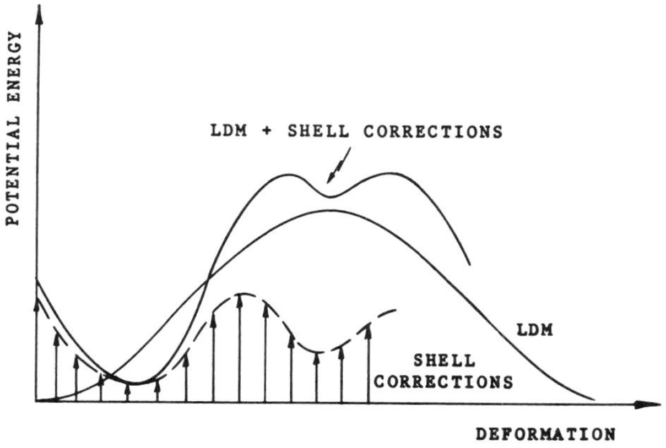

Figure 9
The single humped well in the LDM and the Strutinsky double humped well.

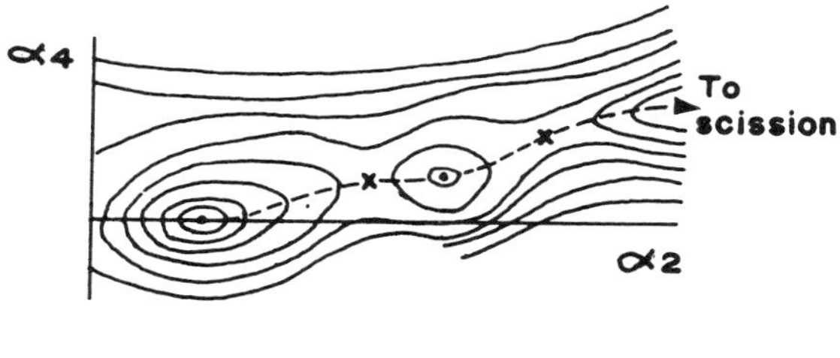

Figure 10
Potential energy contours for deformations in double humped well nucleus.

3. THE FRANCK-CONDON PRINCIPLE

The double hump barrier brings immediately to our mind ideas from molecular physics: states in the second well can be thought of as "quasimolecular" states in which the two fragments are already beginning to come apart. States of this kind have been investigated before in $C^{12} + C^{12}$ reactions, where one finds resonances that can be interpreted as "quasimolecular"[5] states in Mg^{24}.

If one thinks of the double humped barrier as a superposition of two potential energy curves, the Franck-Condon principle tells us a number of things regarding the photoexcitation of the original nucleus. The Franck-Condon principle in the case of molecules tells us in effect that (Fig. 12) the electron jump in a molecule takes place so rapidly in comparison to the vibrational period that immediately afterwards the nuclei still have very nearly the same relative position and velocity as before the "jump". In order to apply this principle let us consider Fig. 12(a), (b) and (c) in which are drawn the potential curves that were assumed by Franck and Condon for the upper and lower electronic states in the three typical cases of intensity distribution.

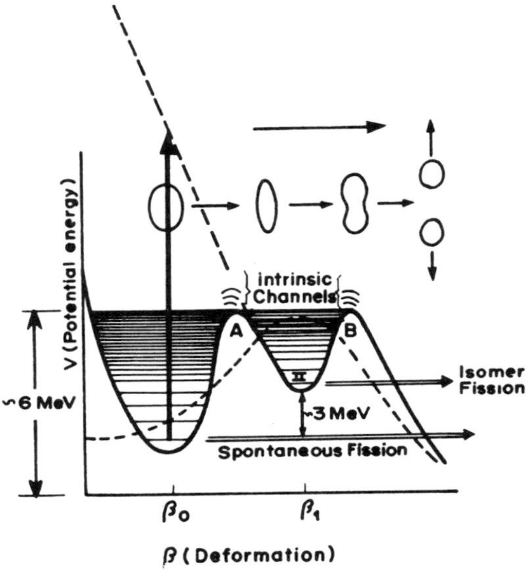

Figure 11
The second well nucleus as a "quasi-molecular" state.

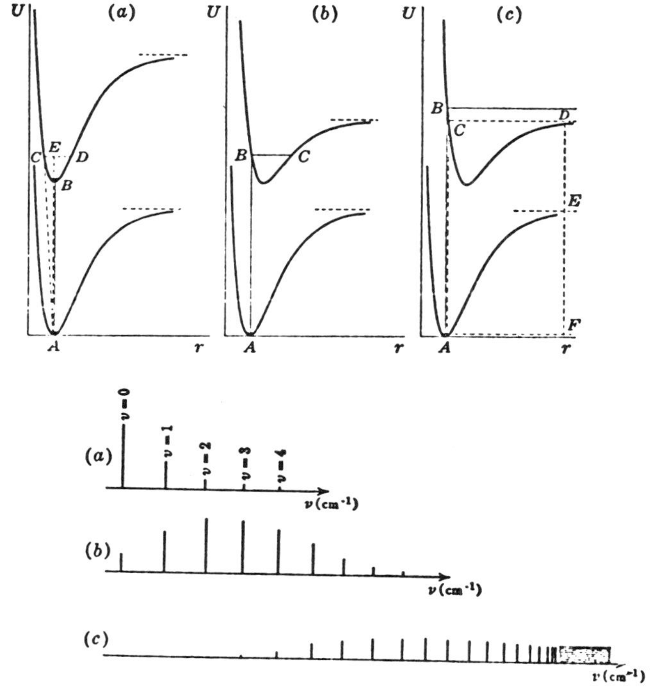

Figure 12
Illustration of the consequences of the Franck-Condon principle in molecular transition.

In Fig. 12(a) the potential curves of the two electronic states have been so chosen that their minima lie very nearly one above the other (equal internuclear distance). In absorption, the molecule is initially at the minimum of the lower potential curve, if we disregard the zero-point vibration. It can be seen that for a transition to the minimum of the upper potential curve the requirement of the Franck-Condon principle that the change of position and momentum be small is satisfied. On the other hand, a transition into a high vibrational state (CD in Figure 12(a)) would be possible only when, at the moment of the electronic jump, either the position (transition from A to C) or the velocity (transition from A to E) or both alter to an appreciable extent. At the point E, of course, the molecule has the amount of kinetic energy EB. Only at the turning point C or D are the velocity and the kinetic energy zero, as in the initial state at A. Thus, on the basis of the Franck-Condon principle, a transition to such a high vibrational level is forbidden or at least highly improbable. Thus we obtain an intensity distribution of the type illustrated in Fig. 12(a).

In Fig. 12(b) the minimum of the upper potential curve lies at a somewhat greater r value than of the lower. Therefore the transition from minimum to minimum is no longer the most probable, since the internuclear distance must alter somewhat in such a transition. The most probable of the transitions is that from A to B in Fig. 12(b) (vertically upwards). For this transition there is no change in the internuclear distance at the moment of the "jump" and no change of the velocity. Thus, immediately after the electron jump the two nuclei still have their old distance from each other and zero relative velocity. Since, however, the equilibrium internuclear distance has a different value in the new electronic state, the nuclei start to vibrate between B and C. The vibrational levels whose left turning points lie in the neighbourhood of B are the upper levels of the most intense bands. For still higher vibrational levels an appreciable change of the internuclear distance or velocity must take place, as a result of which the intensities of the bands decrease again. Thus the observed intensity distribution in the second case (Fig. 12(b)) is explained.

In Fig. 12(c) the minimum of the upper potential curve lies at a still greater internuclear distance. The Franck-Condon principle is strictly fulfilled for the transition AB. However, the point B on the upper potential curve lies above the asymptote of this curve and therefore corresponds to the continuous region of the vibrational spectrum of the upper state. After such an electron jump the atoms will fly apart. Transitions to points somewhat below B (that is, in the discrete region) and somewhat above B (that is, in the continuous region) are also possible. In this way the third case of intensity distribution (Fig. 12(c)) is explained. In the light of the Franck-Condon principle therefore one does not expect to excite states of the second well very readily; the absorption of a photon will lead preferentially to highly excited states above the two fission barriers so the excited state will fission promptly. If this is true one would not expect to form fission isomers by photon excitation. That this is indeed the case was confirmed in an experiment done recently at the University of Toronto.

4. EXPERIMENT ON THE PHOTOEXCITATION OF FISSION ISOMERS

In a photoexcitation experiment one has to start from a stable nucleus, the most convenient of which is U^{238}; one can try in this case to form the 195 ± 30 nanoseconds isomer.[6]

Bombardment by electrons is indicated for this type of experiment: this is equivalent to excitation by virtual photons of different multipolarities (a real photon beam is a plane wave). Fig. 13 shows the spectral composition of the virtual photon spectrum and also a typical bremmstrahlung spectrum produced in a converter of 0.01 radiation length (R.L.). This converter can be the target itself or an external converter, in which case some geometrical efficiency for intercepting the real photons

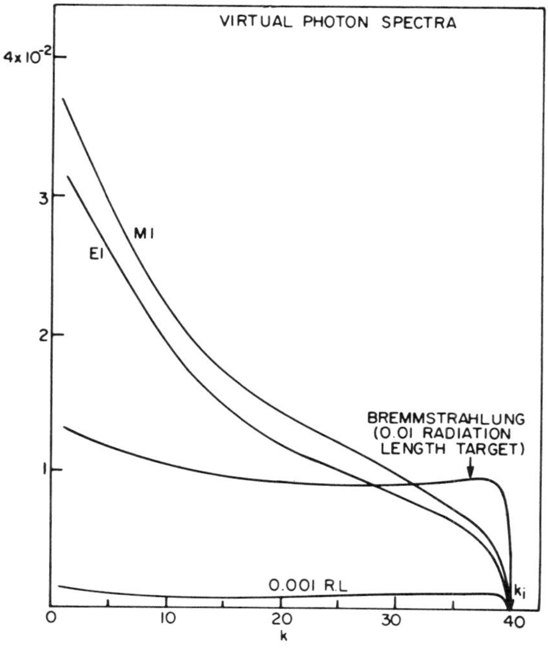

Figure 13
Virtual spectrum of photons from electrons hitting a nucleus and bremmstrahlung spectrum of photons produced in a 0.01 radiation length target.

will enter reducing the number of photons. One can see that for 0.01 R.L. the number of virtual and real photons is approximately equal; for thinner targets the number of virtual photons (per electron) is larger than the number of real photons.

When an electron of energy E_0 (E_0 = 40 MeV in these experiments) hits a uranium nucleus which absorbs an energy $h\nu$ and scatters the electron with energy $E = E_0 - h\nu$ to an angle θ, (Fig. 14), the nucleus recoils at an angle ϕ with an energy given by

$$\Delta E = \frac{h^2 s^2 c^2}{2Mc^2} \quad ; \tag{3}$$

$\vec{s}$ is the momentum transfer ($\vec{s} = \vec{p}_0 - \vec{p}$). It follows therefore that the target can be left with considerable speed (~10^7 cm/sec.).

The basic idea of the experiment is to observe fission fragments from uranium nuclei which recoiled (Fig. 15) in a region of space surrounded by detectors.

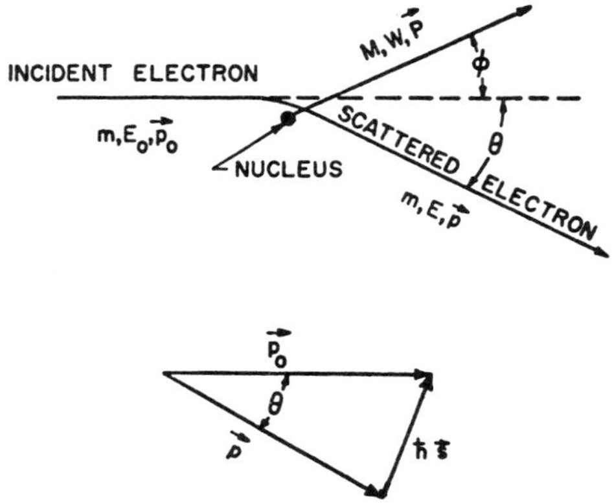

Figure 14
Kinematics of electron scattering

The recoil energies involved are of the order of a few kev, for which the ranges are quite small (Fig. 16). Therefore only the surface layer of the target, approximately $5\mu g/cm^2$, is used in effect. Even using such thin targets only ~ 1% of the recoiling uranium nuclei can get out of the target.

It is therefore essential to concentrate a large number of electrons in the target. This can only be done using electron beams because there are no photon beams of comparable intensities. About 10^{18} electrons (1/100 of a microgram of electrons) hit the target in our experiments.

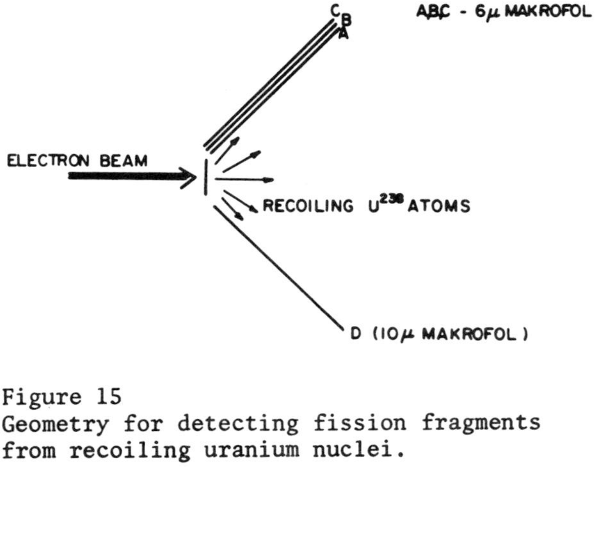

Figure 15
Geometry for detecting fission fragments
from recoiling uranium nuclei.

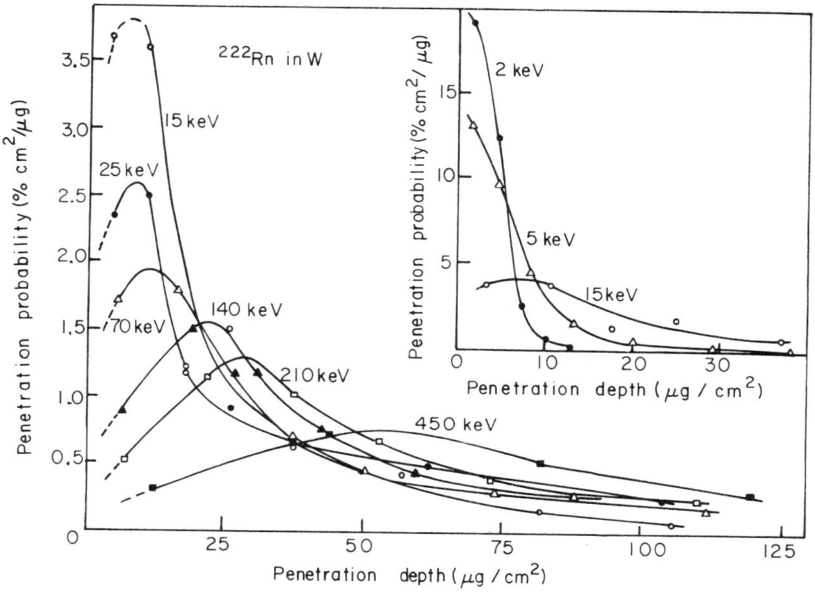

Figure 16
Penetration depths on Rn^{222} in tungsten, which corresponds approximately to uranium ions in uranium.

The detector used was a multi-foil makrofol arrangement which can be punctured by the fission fragments; the thickness of the makrofol is such (6 microns) that approximately 3% of the fission fragments puncture the 3 foils. When this occurs one can reconstruct the event in space and determine its origin and consequently the distance travelled by the fission isomer before disintegrating.

The detectors are totally insensitive to the sea of low energy photons, electrons, protons, neutrons, alphas, etc. present under the irradiation conditions.

The results indicate that the cross section for the electromagnetic excitation of the U^{238} isomer is smaller than $3 \times 10^{-30}cm^2$ which, compared to the known cross sections for prompt electrofission[7] (9×10^{-26} cm^2), gives a ratio R

$$R = \frac{\sigma(\text{isomer})}{\sigma(\text{ground state})} \lesssim 3 \times 10^{-5}.$$

One concludes from this experiment that there is little overlap between the ground state of U^{238} and the nuclear levels in the second well.

This leads us into the experiment done at São Paulo on prompt photofission.

5. PHOTOFISSION IN U^{238} AND Th^{232}

This problem has been intensively studied in the last few years using 3 types of γ-ray sources:

a) low resolution experiments (~1 MeV) - using bremmstrahlung radiation;

b) intermediate resolution experiments (~100 keV) - using Compton scattered γ-rays from incident monochromatic lines from neutron capture reactions;

c) high resolution experiments (~10 eV) - using Doppler broadened γ-rays from neutron capture reactions in a variety of elements.

The "channel theory"[3] in which fission experiments are usually analysed assumes that the fission mode associated with each channel is strongly mixed with the overall nuclear motion; this is the basis of statistical calculations using expressions of the type

$$\frac{\Gamma_f}{D} = \frac{N_f^*}{2\pi} \; ; \qquad D = \frac{1}{\rho(E)} \; . \tag{4}$$

where $\rho(E)$ is the compound nucleus density of levels before fission and N_f^* is the effective number of channels at the saddle point.

There is ample evidence in favour of a statistical treatment of fission.[8] For example, support comes from the study of the competition between fission and neutron emission for which an equation similar to the one above can be written:

$$\frac{\Gamma_n}{D} = \frac{N_n}{2\pi} \; . \tag{5}$$

In eq. (4) N_f^* is a monotonically increasing function of energy, although it may show a step-like behaviour if the levels at the saddle point are well separated. It is given by the expression

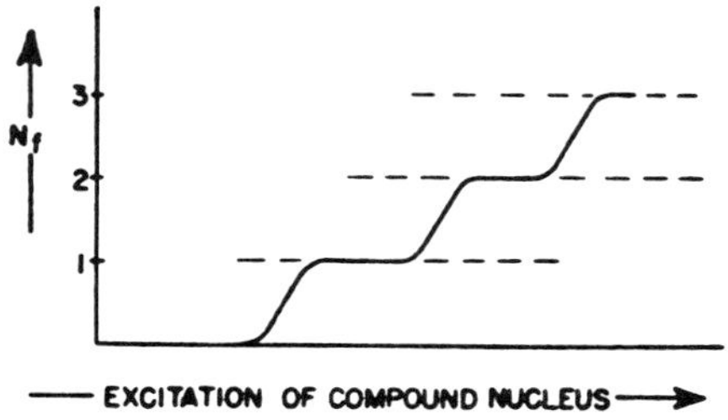

Figure 17
Number of effective fission channels as a function of excitation energy.

$$N_f^* = \int_0^{E-E_f} \rho^*(E-E_f-\varepsilon)\, d\varepsilon \qquad (6)$$

where $\rho^*(E-E_f-\varepsilon)$ is the density of levels at the saddle point, E_f is the effective fission threshold and E is the excitation energy. The effect of the double humped barrier for fission does not seem to be important in the calculation of Γ_f.[9] If we consider the two barriers we must specify the effective number of channels in the two wells (Fig. 18) and we get:

$$\Gamma_f = \frac{1}{2\pi\rho} \frac{1}{\frac{1}{N_A} + \frac{1}{N_B}} = \frac{1}{2\pi\rho} \frac{N_A N_B}{N_A + N_B} \qquad (7)$$

where N_A and N_B are the effective numbers of channels at the barriers, A and B. From experimental data we can see[9] that $N_A \cong 0.5$ and $N_B \cong 0.002$ so $N_A >> N_B$ and

$$T_f = \frac{1}{2\pi\rho} \frac{N_B}{1 + N_B/N_A} \cong \frac{1}{2\pi\rho} N_B \; ,$$

which is virtually the same expression as we have used without considering this correction.

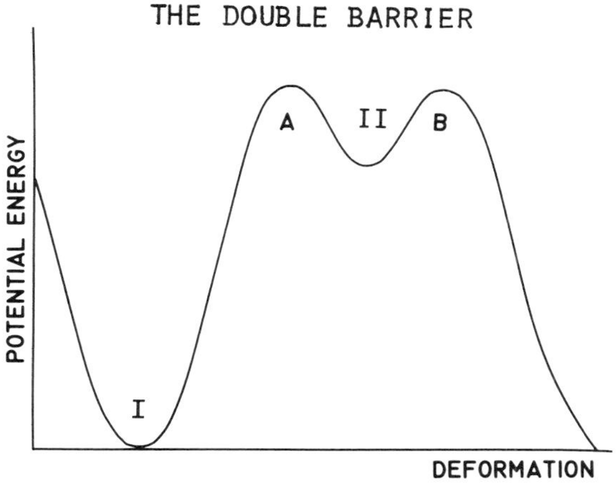

Figure 18
N_A and N_B are the numbers of effective fission channels at the two barriers A and B.

The neutron width was also calculated theoretically using Weisskopf's evaporation theory. According to this theory we can calculate the probability per unit time $\omega_n(\varepsilon)d\varepsilon$ for a nucleus A, excited to an energy E_A, to emit a neutron with kinetic energy between ε and $\varepsilon + d\varepsilon$, and become a nucleus B with excitation energy $E_B = E_A - B - \varepsilon$, where B is the neutron binding energy. We obtain

$$\Gamma_n = \hbar \int_0^{E-B} \omega_n(\varepsilon)\, d\varepsilon \qquad (8)$$

where

$$\omega_n(\varepsilon)\, d\varepsilon = \frac{\sigma(E_A,\varepsilon)}{\pi^2 h^3} \frac{gm\, \rho(E-B-\varepsilon)\varepsilon d\varepsilon}{\rho(E)} \qquad (9)$$

and where $\sigma(E_A,\varepsilon)$ is the inverse cross section; also:

m	is the neutron mass,
g=2	for neutrons,
$\rho(E)$	is the compound nucleus density of levels and
$\rho(E-B-\varepsilon)$	is the residual nucleus density of levels.

Making an hypothesis about the density of levels we can calculate the ratio Γ_n/Γ_f. We use $\rho = Ce^{E/T}$ (where T is the nuclear temperature) given by Huizenga[8] because we found that this is the formula which gives the best agreement with out experimental data. Performing the necessary integrations one obtains

$$\frac{\Gamma_n}{\Gamma_f} = \frac{2TA^{2/3}}{K_0} \frac{[-1-(E-B)/T + \exp[(E-B)/T]]}{[1- \exp[(E-E_f)/T]]} \qquad (10)$$

where $K_0 = 14.431$ MeV.

In Figs. 19 and 20 the theoretical results are plotted with our data. Above 9 MeV the ratio Γ_n/Γ_f seems to be constant but below this energy one sees clearly a variation of this ratio with energy.

One should notice that the data for uranium and thorium is best fitted for nuclear temperatures of 0.9 MeV and 1.5 MeV; this difference in temperatures is quite surprising and might be due to a deformation of the compound nucleus which would cool it prior to neutron emission.

In all of these experiments the cross section $\sigma(\gamma,f)$ was measured with a fission counter and the total neutron production cross section σ(neutrons) was determined by the use of a "long" neutron counter.

These cross sections are related by the expression

$$\sigma(\text{neutrons}) = \sigma(\gamma,n) + \nu\sigma(\gamma,f) \quad , \qquad (11)$$

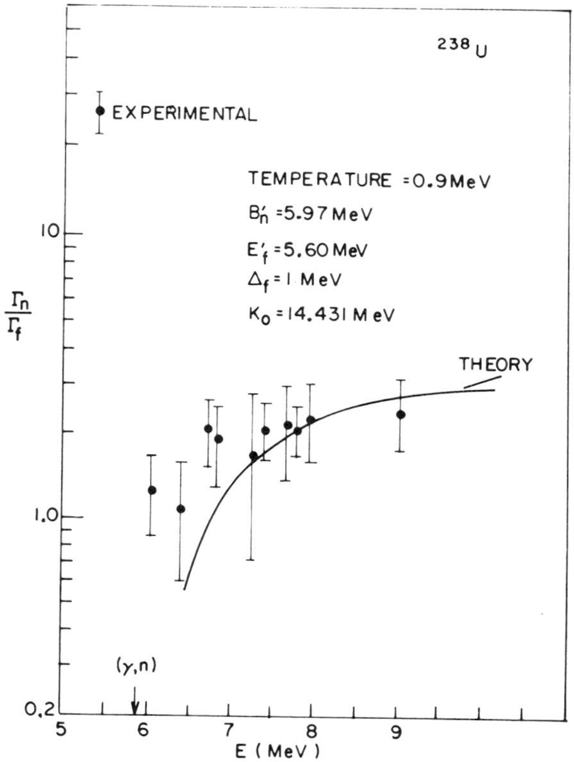

Figure 19
Experimental results for Γ_n/Γ_f as a function of excitation energy and results of statistical calculations for U^{238}.

(ν is the number of neutrons per fission) from which one can derive $\sigma(\gamma,n)$. The ratio $\sigma(\gamma,n)/\sigma(\gamma,f)$ is given by

$$\frac{\sigma(\gamma,n)}{\sigma(\gamma,f)} = \frac{\sigma_{abs}\ \Gamma_n/(\Gamma_n + \Gamma_f)}{\sigma_{abs}\ \Gamma_f/(\Gamma_n + \Gamma_f)} = \frac{\Gamma_n}{\Gamma_f} \quad . \qquad (12)$$

The existence of "intermediate" structure near the threshold for the (γ,f) reactions in ^{238}U and ^{232}Th has been clearly established by Knowles[10]; levels of ~200 keV width have been found while in (n,n) reactions only gross structure (~2 MeV in width) is observed. This indicates that the density of states available for energy dissipation in fission near the fission threshold is considerably smaller than the total density of states.

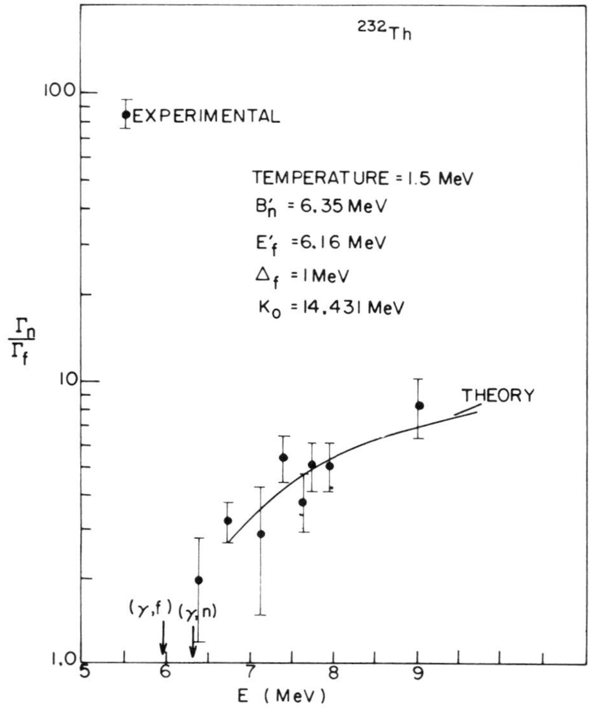

Figure 20
Experimental results for Γ_n/Γ_f as a function of excitation energy and results of statistical calculations for Th^{232}.

As pointed out by Knowles this is explained most conveniently by associating the vibrational states with the second well. Our data in general tend to support the findings of Knowles. However the same structure present in the (γ,f) cross sections (Figs. 21,22) is seen in the (γ,n) cross sections (Figs. 23,24); data derived from Lindner[11] is also plotted in Fig. 23. No (γ,n) results except the old bremmstrahlung results are available in the literature. This "intermediate" structure is rather surprising because in the (γ,f) cross section it has been attributed to the small number of states leading to fission; to find the same structure in the (γ,n) cross section indicates that the same states are involved in this process, which is somewhat unexpected. Pictorially one would be tempted to say that the neutrons are emitted mainly from the deformed nucleus on its way to scission and not from the excited undeformed nucleus; one should notice in connection with this point that in the photofission of ^{238}U and ^{232}Th the ratio Γ_n/Γ_f is of the order of 1 as discussed above, indicating that the mean lives for the (γ,n) and (γ,f) processes are of the same order of magnitude.

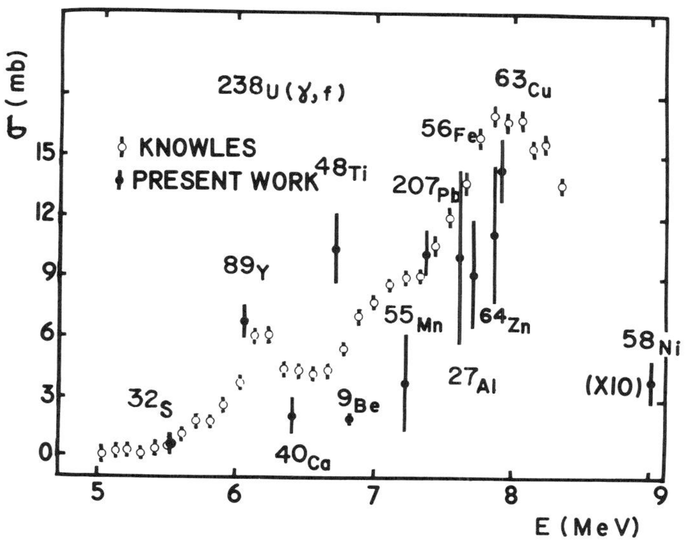

Figure 21
Cross section for the (γ,f) reaction in U^{238}.

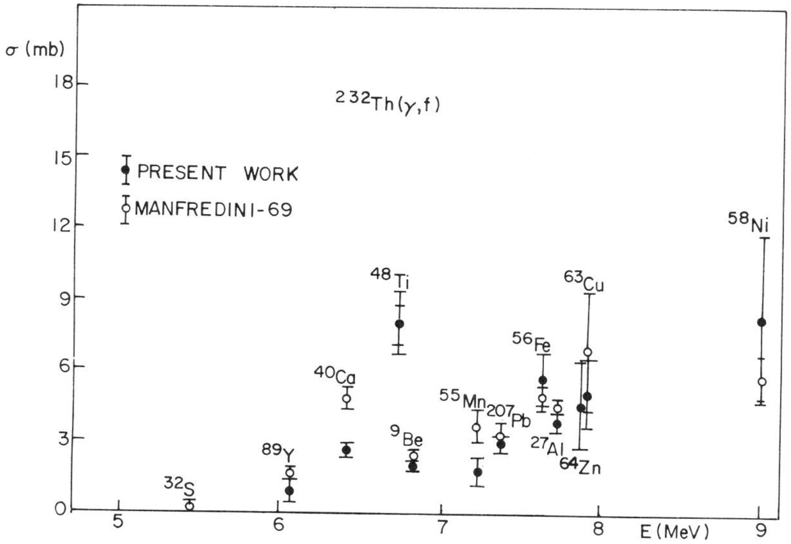

Figure 22
Cross section for the (γ,f) reaction in Th^{232}.

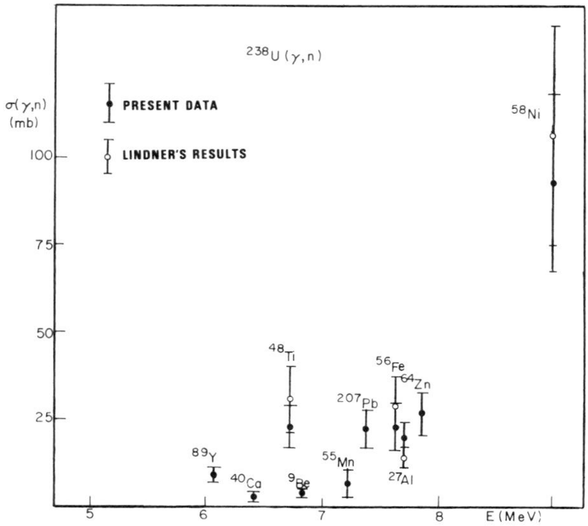

Figure 23
Cross section of the (γ,n) reaction in U^{238}.

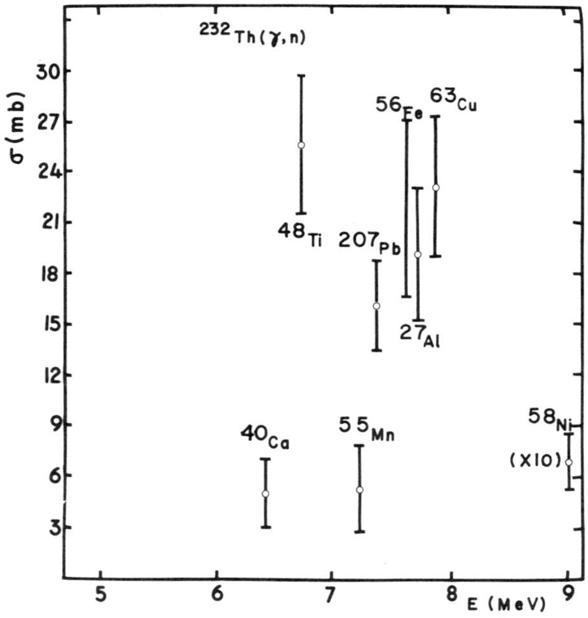

Figure 24
Cross section for the (γ,n) reaction in Th^{232}.

In addition to the peak at ~6.2 MeV which is well established and studied by Knowles our data indicates another peak at 6.73 MeV corresponding to the γ-ray emitted in the neutron capture of ^{48}Ti. Experimentally there seems to be no doubt that the cross section at this energy is much larger than the cross sections at the adjoining energies 6.83(^{9}Be) and 6.42(^{40}Ca). This peak is not seen by Knowles. It is conceivable that due to the "high resolution" nature of our measurements some levels with special properties are hit, a feature which is washed out in the "intermediate" or "gross structure" measurements. It is therefore possible that the low density of states in the second well is not the origin of the intermediate structure observed in photofission near the threshold.

REFERENCES

1. Bohr, N. and Wheeler, J.A., Physics Rev. 56, 249 (1939).
2. Wilets, L., Theories of Nuclear Fission, Clarendon Press, Oxford, 1964.
3. Wheeler, J.A., Fast Neutron Physics (Edited by J.B. Marion and J.L. Fowler, Interscience Publishers, New York, 1963).
4. Strutinsky, V.M., Nucl. Phys. A95, 420 (1967) and A122, 1 (1968).
5. Almquist, E., Bromley, D.A. and Kuehner, J.A., Phys. Rev. Lett. 4, 515 (1960) and Vogt, E. and McManus H., Phys. Rev. Lett. 4, 518 (1960).
6. Russo, P.A., Vandenbosch, R., Mehta M., Tesmer, J.R. and Wolf, K.L., Phys. Rev. 3C, 1595 (1971).
7. Moretto, L.G., Gatti, R.C., Thompson, S.G., Routti, J.T., Heisenberg, J.H., Middleman, L.M., Yearian, M.R. and Hofstadter, R., Phys. Rev. 179, 1176 (1969).
8. Huizenga, J.R. and Vandenbosch, R., Nuclear Reactions II (edited by P.M. Endt and P.B. Smith, North Holland, Amsterdam, 1962) p. 42.
9. Björnholm, S. and Strutinsky, V.M., Nucl. Phys. A136, 1 (1969).
10. Knowles, J.W. (to be published in Nucl. Phys. and private communication).
11. Lindner, M., Nucl. Phys. 61, 17 (1965).

EXPERIMENTS ON PHOTOFISSION IN HEAVY ELEMENTS

J.W. Knowles
Atomic Energy of Canada Limited
Chalk River Nuclear Laboratories
Chalk River, Ontario, Canada

Photofission cross section and angular distribution measurements of ^{238}U and ^{232}Th are described and the gross structure observed at the fission threshold is interpreted as a band of rotational levels similar to the low energy rotational band characteristic of heavy nuclei. The slope of the transmission factor at fission threshold is found to be twice as large in ^{232}Th as in ^{238}U. An asymmetric (octupole) deformation parameter which is larger in ^{232}Th than in ^{238}U can account for the difference in transmission factor and, in addition, can account for the difference in gross structure at the fission threshold.

INTRODUCTION

It is well known that nuclei between Th and Cf exhibit strong asymmetry in the fission mass distribution, the ratio of the most probable masses in the heavy and light group varying from 1.5 to 1.2[1]. Also stable asymmetric nuclear deformation is suggested by the systematic occurrence of low energy odd-parity rotational states in even-even nuclei[2]. It is the purpose of this talk to describe new measurements, in particular, the occurrence of odd-parity rotational levels at the fission threshold of ^{232}Th and ^{238}U, which can be interpreted in terms of asymmetric deformation.

Much theoretical calculation has been done on the nuclear structure and the stability of heavy and superheavy elements based on a modified oscillator model with quadrupole (P_2) and hexadecapole (P_4) deformations[3]. A deformed oscillator potential consisting of a series of terms with reflection and spatial symmetry has been sufficient for describing symmetric fission. But in order to describe asymmetric fission, asymmetric (P_3 and higher) terms are needed to describe the deformation. Strutinsky[4] pointed out that the mixing of single-particle states of opposite parity in a pear-shaped nucleus tends to stabilize such a deformation. Octupole (P_3) distortions of the nuclear potential, related to low-lying states, have been investigated by Lee & Inglis[5], Johansson[6] and recently by Vogel[7]. In particular Johansson concludes, for the very elongated shapes which are of interest in fission, that stable octupole deformation probably occurs and that the potential surface describing the fission barrier has two asymmetric valleys.

The gross structure in fission-yield spectra, at energies below the fission threshold, includes peaks which are less than 200 keV wide. These widths are at least ten times narrower than are expected for strongly coupled states and indicate that individual states which exist at the saddle point are not strongly mixed into the average nuclear motion. This situation can be described phenomenologically by assuming a double potential well[8].

To date, cross section measurements of fast-neutron induced fission, e.g. 0-2 MeV neutrons on ^{230}Th and ^{232}Th, and the energy variation of the angular distribution of fission products from (n,f), (d,pf) and (α,α'f) reactions have furnished the best evidence of gross structure[8]. Photofission yield measurements[9-13] and photofission fragment angular distributions[12,14-18] furnish additional but less conclusive evidence. It has been observed in photofission of even-even nuclei that the ratio of dipole to isotropic component in the angular distribution decreases uniformly with increasing energy between 5 and 8 MeV in support of the Bohr theory[19] and that angular momentum quantum numbers $K = 0$, $J^{\pi} = 1^{-}$, corresponding to low energy vibrational bands, are conserved during the fission process. More recently, the observation, in ^{238}U and other nuclei, of a relative increase of quadrupole to dipole component in the angular distribution near 5.0 MeV implies electric quadrupole absorption and thus the excitation of $J^{\pi} = 2^{+}$ rotational states[16,17]. However, these measurements do not give as clear a relationship between angular distribution and resonance structure as one would like because of either limited energy resolution or limited energy range. On the one hand the extraction of cross sections from photofission measurements[12-16,17] made with the bremsstrahlung continuum requires differential analysis which limits the energy resolution to about 10%. On the other hand measurements made with individual monochromatic γ-rays following (n,γ) reactions, while only a few eV wide, are spaced too far apart to be useful for structure width measurements ~200 keV wide[13], and those made with near monochromatic radiation following the $(p,\alpha\gamma)$ reaction[21] are restricted in range to a few hundred kilovolts.

PHOTOFISSION CROSS SECTION MEASUREMENTS

An improvement over earlier measurements is obtained by using a beam of effectively monochromatic γ-radiation of variable energy and moderate resolution. Such a beam is provided by the facility shown in figure 1. It includes a $^{58}Ni(n,\gamma)^{59}Ni$ source S, located in the thermal column of the NRU reactor, Chalk River. Source radiation of energy E_n, Compton scattered through angle θ_m, from all points of the curved aluminum plate C, is incident on target T with energy E_γ (MeV) according to the equation

$$E_\gamma^{-1} = E_n^{-1} + (0.511)^{-1} (1 - \cos\theta_m) \qquad (1)$$

providing that S, C and T are located on the circumference of a circle. The intensity incident on target T is about 1 quantum/eV-sec/cm height and is measured with a NaI detector behind a lead aperture. Beam energy calibration is made with respect to (n,γ) radiations of ^{59}Ni and ^{49}Ti using a Ge(Li) detector.

The energy, E_γ, of the incident radiation is changed by changing the position of the target with respect to the source and scatterer. As shown in figure 2 an increase in scattering angle θ_m causes the scattered (n,γ) spectrum to contract to lower energy. The advantage of this method is that about 65% of the incident intensity is in the two highest energy radiations, the 8.5 and 9.0 MeV γ-rays following neutron capture in ^{58}Ni. The energy resolution at the target which depends on the spatial extent of S, C and T is about 3% for the 1.7 cm and 5% for the 3.8 cm wide targets shown in figure 3.

Both absolute photofission cross section measurements and fragment angular distribution measurements have been made with this facility.

Fission ion chambers are used for the cross section measurements. The arrangement of targets in the ion chambers is shown in figure 3. The better geometry targets, type B, are 0.1 mg/cm^2 strips of fissionable material, coated on the inner surface of cylindrical aluminum tubes, Al, which form the outer wall of an ionization chamber filled with argon gas. Greater counting rates are obtained with cylindrical targets, type A, with 0.17 mg/cm^2 thick coatings and with three counters arranged in tandem. The total pulse height

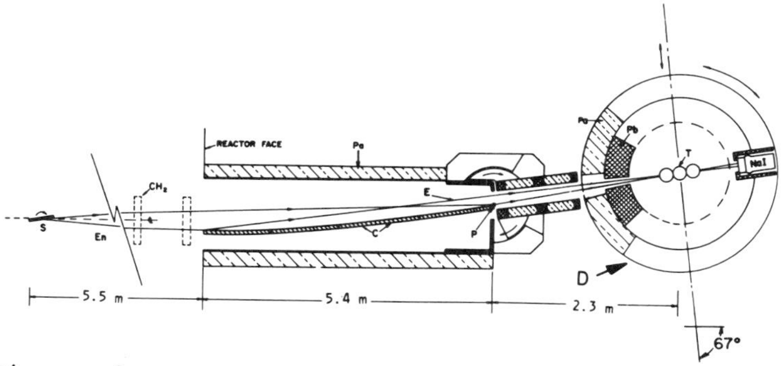

Figure 1
Variable energy γ-radiation facility; CH_2, fast neutron filter; Pb, shielding; Pa, paraffin shielding.

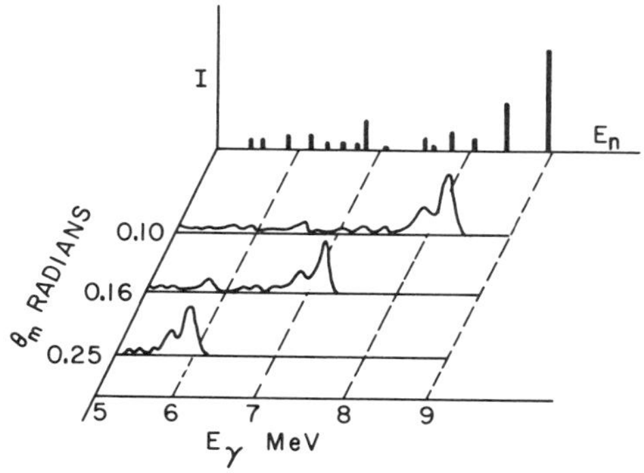

Figure 2
Incident intensity $I(E_\gamma, \theta_m)$.

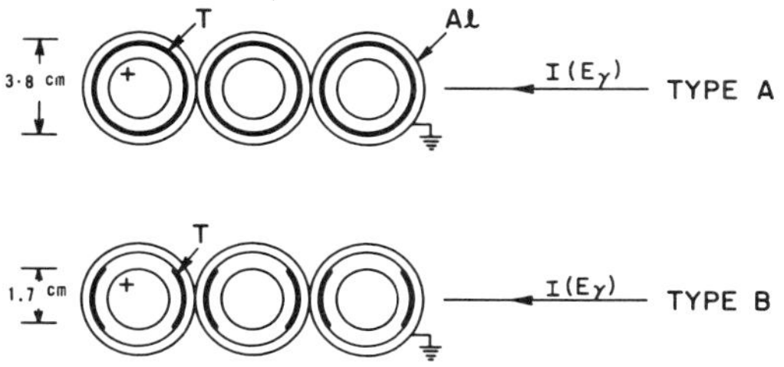

Figure 3
Photofission ion chambers; $I(E_\gamma)$ incident radiation; + 300 V anode.

distribution from the ion chamber above a fixed bias is chosen to exclude counts from spontaneous α-decay and noise, but not from spontaneous fission and photofission and possible neutron fission. Neutron-induced fission is negligible as demonstrated by the fact that the ion chamber count, with the incident γ-radiation beam, attenuated by a 15 cm thick lead absorber with the reactor at full power, is equal to the spontaneous fission count with the reactor at zero power.

The photofission yields of ^{232}Th and ^{238}U, measured with this apparatus, are not the result of a single incident γ-ray but of a spectrum of γ-radiations, dominated by the 9.0 and 8.5 MeV scattered radiation of the ^{59}Ni source as shown in figure 4 for ^{232}Th. Photofission cross sections $\sigma_F(E_\gamma)$ are obtained from the fission yield distributions Y_m at angular settings θ_m, using the equations

$$Y_m = \sum_n Y_{nm} = NG \sum_n I_{nm}\, \sigma'_{Fnm} \qquad (2)$$

$$\sigma'_{Fnm} = \int H(E-E_\gamma)\sigma_F(E_\gamma)\,dE, \quad \int H(E-E_\gamma)\,dE = 1 \qquad (3)$$

where N is the number of target atoms, G the detector efficiency and I_{nm} the incident intensity corresponding to a source γ-ray of energy E_n Compton scattered at angle θ_m. $H(E-E_\gamma)$, the energy distribution of a single incident γ-radiation, is measured with a thin Ge(Li) detector. NG is obtained from

$$NG = 1.44\ T\ d(NG)/dt,$$

$$T(^{238}U) = 0.884 \pm 0.026 \times 10^{20}\ \text{hr} \qquad (4)$$

where T is the spontaneous fission half-life[20] of ^{238}U and d(NG)/dt, the spontaneous fission rate measured in the narrow-strip uranium counters.

Absolute photofission cross sections $\sigma_F(E_\gamma)$ of ^{238}U and ^{232}Th as a function of incident γ-ray energy are shown in figures 5a and 5b respectively.

The ^{232}Th measurements, figure 5b, show a broad peak 400 keV wide at 6.4 MeV on both sides of the neutron separation energy $E(\gamma,n)$ followed by a deep valley at 6.9 MeV. A second small narrow peak is just visible at about 5.5 MeV. The ^{238}U measurements, figure 5a, show peaks at 5.05, 5.7 and 6.2 MeV, about 200 keV wide, and peaks at 7.2 and 8.0 MeV about 500 keV wide.

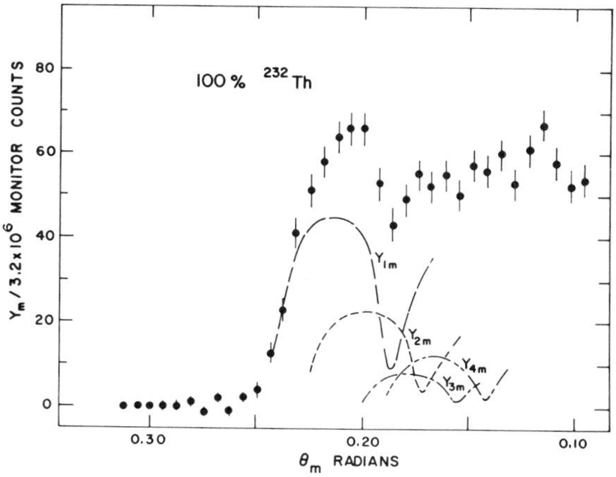

Figure 4
Photofission total yield $Y_m(\theta_m)$ and partial yields Y_{1m} to Y_{4m} of ^{232}Th.

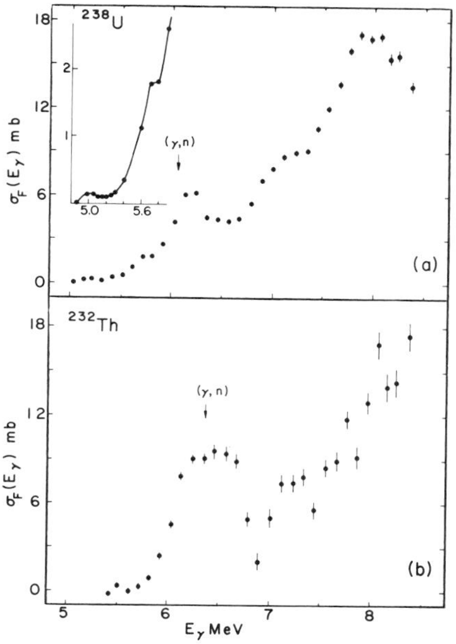

Figure 5
The photofission cross section $\sigma_F(E_\gamma)$ of (a) ^{238}U and (b) ^{232}Th.

It might be thought that the narrow peak at 6.2 MeV is not a fission spectrum resonance but exists because of the presence of the adjacent valley believed to result from neutron competition just above the neutron separation energy. However a quantitative estimate of neutron and γ-ray emission just above the (γ,n) threshold shows this explanation to be untenable.

Many attempts have been made to measure the photofission spectra of ^{232}Th and ^{238}U at threshold using bremsstrahlung and monochromatic γ-rays of neutron capture. Some of the most recent measurements on ^{238}U using bremsstrahlung, made by Rabotnov et al.,[14] are shown as a broken line in figure 6. It is difficult by this method, which uses differential analysis to obtain fission cross sections, to resolve to better than 10% structure on the flat and steep regions of the spectrum. However, overall agreement with our measurements is good. The ■ points and measurements of Manfredini et al.[13], using (n,γ) radiation are ≈10 eV wide. Large fluctuations are expected since the average level spacing of ^{238}U between 6 and 8 MeV is comparable to the average width of incident neutron capture γ-rays from ^{58}Ni. The Δ points are measurements of Huizenga[21] using Doppler broadened γ-rays, following the ^{19}F(p,αγ)^{16}O reaction. Two of the three measurements reported agree with present measurements within counting statistics.

Transmission Factors & Deformation Parameters

Photofission transmission factors T_F for ^{232}Th and ^{238}U are obtained from the photofission cross section measurements of figure 5 and photo-absorption cross sections σ_γ through the equation

$$T_F = (T_n + T_\gamma)/[(\pounds\ \sigma_\gamma/\sigma_F) - 1] \tag{5}$$

derived from compound nuclear statistical theory. Here £ is the statistical modification factor[22] (0.8<£<1.0, assuming a Porter-Thomas distribution of levels) and T_n and T_γ are the neutron and γ-ray transmission factors respectively, averaged over the many resonances which contribute to photo-absorption. Estimates of T_n, T_γ and σ_γ as a function of energy are based largely on experimental data and are discussed in detail elsewhere[23]. These parameters for ^{238}U are plotted as a function of photon energy in figure 7. It is evident that $T_n + T_\gamma$ increases rapidly at the neutron separation energy, $E_B(\gamma,n)$, but less rapidly at 200 keV above this energy. Curves 1 and 2 show that the variation of $T_n + T_\gamma$ with

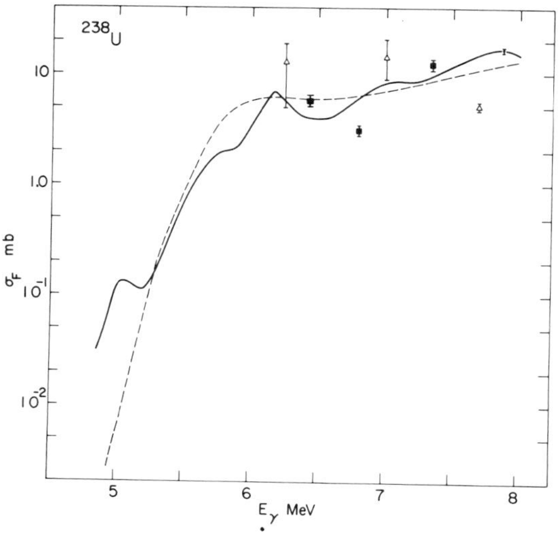

Figure 6
Photofission in ^{238}U, —— Khan and Knowles, --- Rabotnov et al., ■ Manfredini et al., Δ Huizenga et al.

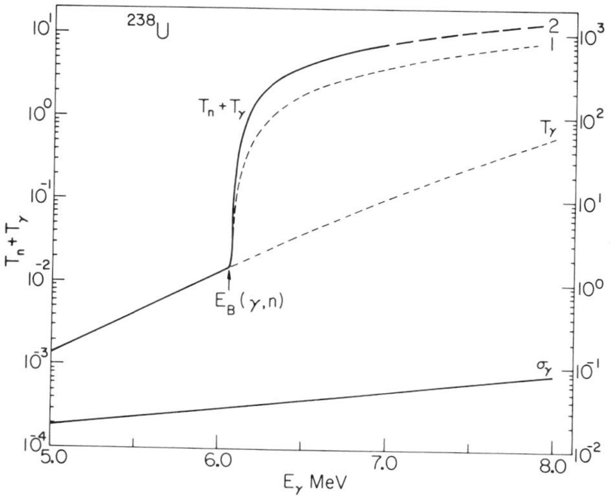

Figure 7
Estimates of $T_n + T_\gamma$, T_γ and σ_γ as a function of E_γ.

E_γ, caused by variations of T_n, is only moderately sensitive to a change in the nuclear model. Curve 1 is estimated for a black nucleus and curve 2 for a phenomenological model where the parameters of the potential are obtained from experiment[23]. Photofission transmission factors T_F for ^{238}U and ^{232}Th, obtained with the parameters just described, are shown as a function of energy in figures 8a and 8b respectively. The transmission factor of ^{238}U increases approximately exponentially with energy up to about 6 MeV and thereafter rises more steeply to a peak at 6.2 MeV. The transmission factor of ^{232}Th increases much more steeply at low energy than does ^{238}U and at higher energies exhibits several broad peaks with a pronounced dip at 6.9 MeV. The broken lines are transmission factors calculated from the Hill, Wheeler equation[24]

$$T_F^{-1} = 1 + \exp(-2\pi[E-E_F]/\hbar\omega) \tag{6}$$

in which single channel fission near threshold is assumed to occur from a real well of parabolic shape. The parameter $\hbar\omega$ and the fission threshold energy, E_F, are obtained by fitting to experimental values of T_F at low energies.

The transmission factors of ^{238}U and ^{232}Th at low energies, below the fission barrier, can be viewed in better perspective, as shown in figure 9, by including transmission factors for spontaneous fission which are near the bottom of the potential well. These transmission factors are obtained from the spontaneous fission half-lives $t = 1.01 \pm 0.03 \times 10^{16}$ yr and $t > 10^{21}$ yr for ^{238}U and ^{232}Th respectively[20]. Here T_F is calculated from equation $T_F = 2\pi\ \Gamma_F/D$ where Γ_F is the fission width corresponding to half-life t and D, the average level spacing near the ground state, is approximately 20 keV. The measurements of T_F just below the fission barrier are the solid lines shown in the upper part of figure 9. Interpolation between measurements at the fission barrier and at the bottom of the well is expected to follow approximately the broken line. The slope of this line at $E \approx 0$ is set within narrow limits by the transmission factor and by simple considerations about the form of the potential. According to WKB theory, the transmission factor T_F as a function of a single deformation parameter ε, at a depth > 0.5 MeV below the top of the fission barrier, is given by

$$T_F = \exp(-K) \tag{7}$$

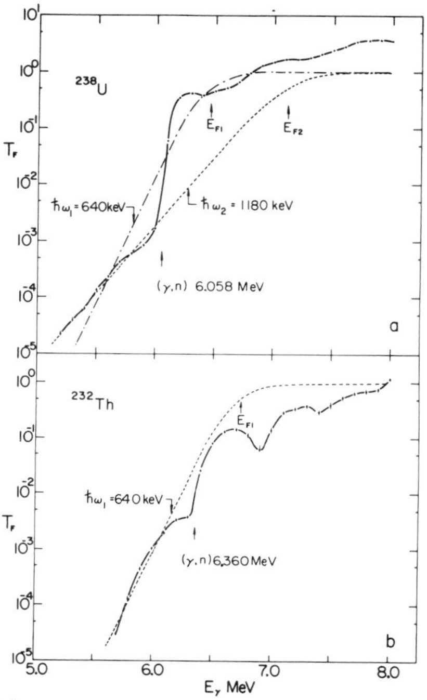

Figure 8
Photofission transmission factors $T_F(E_\gamma)$, (a) for ^{238}U, (b) for ^{232}Th.

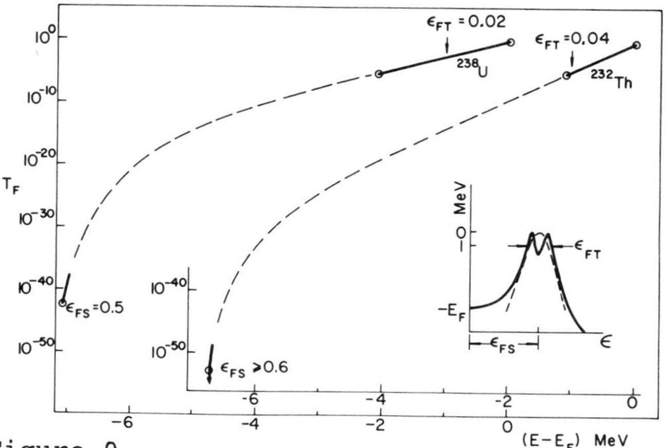

Figure 9
Transmission factors of ^{238}U and ^{232}Th at E≈0 and E_F-E = 1 MeV.

where

$$K = 2 \int_{\varepsilon'}^{\varepsilon''} \sqrt{2B(W(\varepsilon)-E)/\hbar^2}\, d\varepsilon\ . \qquad (8)$$

Here B is the inertial mass parameter associated with fission, $W(\varepsilon)$ is the potential energy surface of the fission barrier, E the excitation energy of the nucleus and $\varepsilon = \varepsilon' - \varepsilon''$ the change in deformation parameter on passage through the barrier. Parameters B and ε are defined in reference 3. Parameters ε_{FT}, 1 MeV below threshold, and ε_{FS} near the bottom of the well at $E \approx 0$, are obtained from equations (7) and (8) using the measured transmission factors T_F. For these calculations we use theoretical estimates[25] of the mass parameter B. We also assume that the potential $W(\varepsilon)$, which may have a composite shape at threshold, can be approximated over much of the deformation by a parabolic shape as shown in figure 9. It is found that ε_{FS} at the bottom of the potential well is only about 10-20 percent greater in ^{232}Th than in ^{238}U but that ε_{FT} near the fission threshold is twice as large in ^{232}Th. The values of ε_{FS} and ε_{FT} are listed in columns 6 and 7 of Table I.

It is interesting to compare the fission deformation parameters ε_{FS} and ε_{FT} with the equilibrium deformation parameters of ^{238}U and ^{232}Th obtained from inelastic proton scattering angular distribution and coulomb excitation measurements[26,28]. In order to compare parameters in Table I the β_2 and β_4 parameters listed by the authors[26] have been expressed in terms of the corresponding ε_2, quadrupole, and ε_4, hexadecapole, deformation parameters using the conversion graph of Nilsson et al.[3] Equilibrium ε_2 deformation parameters, obtained experimentally, are given in columns 2 and 3. These parameters are approximately equal for ^{232}Th and ^{238}U as predicted by theory. The corresponding ε_4 parameters, column 4, are different for the two nuclei, in disagreement with theory, column 5, but they contribute only a small amount to the total deformation. The parameters ε_{FS}, obtained from spontaneous fission, column 6, are appreciably larger than the equilibrium values. It is to be expected[3] that symmetric ε_2 and ε_4 parameters will be the main contributors to ε_{FS} but asymmetric parameters may contribute significantly[6]. The fact that ε_{FS} for ^{232}Th is slightly larger than for ^{238}U may be caused by an asymmetric deformation parameter, which is larger for ^{232}Th than for ^{238}U. Near the fission threshold, the values of ε_{FT}, column 7, show a relatively greater difference. A large difference is possible in theory since ε_{FT} is a measure, not of the

Table 1
Nuclear deformation parameters

	EQUILIBRIUM EXPERIMENT			THEORY	FISSION EXPERIMENT	
	ϵ_2*	ϵ_2**	ϵ_4	ϵ_4	ϵ_{FS}	ϵ_{FT}
^{232}Th	0.22	0.21 ± 0.01	0.022 ± .012	0.054	0.60	0.04 ± .01
^{238}U	0.25	0.25 ± 0.01	-0.003 ± 0.010	0.050	0.50	0.02 ± .005

* ELECTROMAGNETIC
** INELASTIC PROTONS

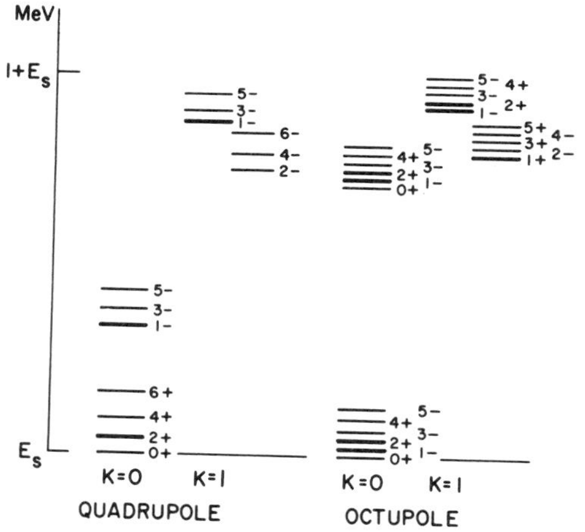

Figure 10
Energy levels of stable deformation.

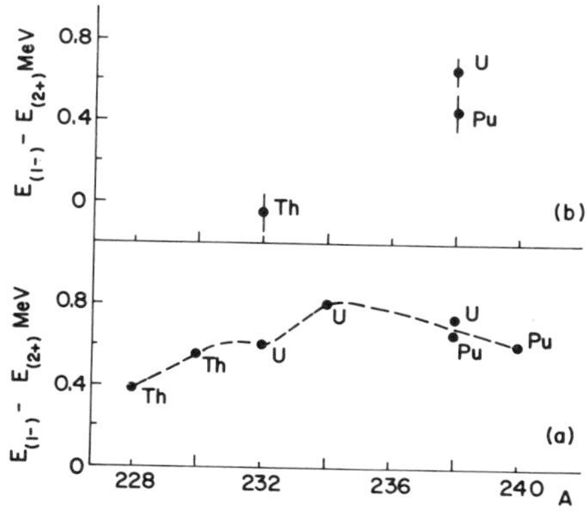

Figure 11
The $E_{(1-)} - E_{(2+)}$ difference of
(a) low energy collective states and
(b) fission threshold states as a
function of A of even-even nuclei.

total deformation for fission, but of the change of deformation near threshold.

It is reasonable to include asymmetric parameters in the description of the deformation at threshold because of the strong asymmetry in the fission mass distribution of ^{232}Th and ^{238}U. The belief that asymmetric parameters play an important part in fission near threshold is strengthened further by the presence of low-lying negative parity collective states in heavy even-even nuclei. It has been shown[4] that a pear shaped nuclear deformation can account for low energy negative parity states. Pear shaped oscillations mean that an inversion of the ends of the pear occurs by passage through a potential barrier caused by the higher energy of the spheroidal shape. As in molecular spectroscopy the effect of inversion is described by an even and odd set of eigenfunctions corresponding to rotational states J = 0,2,4,.. and J = 1,3,5,.. respectively. When the nucleus is very stable against pear shaped deformation, the odd states will be in their normal position between the even states as shown in figure 10. However, if the nucleus is soft, the odd states will be displaced upward by an amount which is inversely proportional to the strength of the deformation parameter. Referring to figure 11 which shows the energy difference of the low lying 1^- and the first 2^+ collective states of heavy nuclei between 228>A>242 it is apparent (since the energy of 2^+ states is nearly constant) that low A nuclei have low 1^- state. If the model is correct nuclei near ^{232}Th (the low lying 1^- state of ^{232}Th has yet to be observed) are expected to be significantly more stable than nuclei near ^{238}U.

ANGULAR DISTRIBUTION MEASUREMENTS

Near the fission threshold where the deformation is large we might expect to find strong evidence of asymmetric deformation. But in order to interpret the gross structure near the fission threshold, by analyses similar to that made near the ground state, we need to identify the spins and parities of the states. For this reason we have made photofission angular distribution measurements over the range of energies associated with gross structure close to the fission threshold.

Photofission angular distribution measurements between 5.4 and 7.4 MeV have been made, using the variable energy γ-ray facility, by placing a uranium target at position T, figure 1, inside the thin-walled aluminum vacuum chamber shown in figure 12. The target, a 20 mg/cm^2 thick film of U_2O_3 painted on a thin-walled aluminum cylinder 1.6 cm in diameter and 9.2 cm long, is located on the axis of a vacuum chamber 20 cm in diameter. The target is rotated about its axis in order to ensure uniform emission at all angles. Radiation I(E) from the γ-ray facility, incident on the target, is not attenuated appreciably by the thin aluminum walls of the vacuum chamber. The fission fragments are recorded on a plastic film (Makrofol) detector D placed on the inside wall of the vacuum chamber. For convenience in processing, eight flat films mounted in standard plastic frames and arranged as an octagon are a substitute for the cylindrical film.

The angular distribution measurements are corrected for the finite size of target and detector and are referred to the highest energy incident γ-ray using the method of analysis described previously. The angular distribution of the photofission differential cross section $d\sigma_F/d\Omega$, assuming only dipole and quadrupole photo-absorption, is given by

$$\partial\sigma_F/\partial\Omega = a(E) + b(E)\sin^2\theta + c(E)\sin^2 2\theta \qquad (9)$$

$$\sigma_F(E) = \sigma_a(E) + \sigma_b(E) + \sigma_c(E)$$

where

$$\sigma_a(E) = 4\pi a,\ \sigma_b(E) = (8\pi/3)b,\ \sigma_c(E) = (32\pi/15)c.$$

Coefficients a(E), b(E) and c(E) are obtained by least-square fitting of equation (9) to the angular distribution measurements of photofission fragments of ^{238}U. Since it will be of particular interest to compare the measurements with predictions, based on modified liquid drop models[6], the terms of equation (9) are regrouped as follows:

$$\partial\sigma_F/\partial\Omega = S(2 - \sin^2\theta) + D\sin^2\theta + C\sin^2 2\theta$$

$$\sigma_F(E) = \sigma_S(E) + \sigma_D(E) + \sigma_C(E) \qquad (10)$$

where σ_S, the cross section contributing to channel $K = 1$, $J^\pi = 1^-$, is characterized by angular distribution $(2 - \sin^2\theta)$, σ_D contributing to channel $K = 0$, $J^\pi = 1^-$ has angular distribution $\sin^2\theta$, and σ_C contributing to channel $K = 0$, $J^\pi = 2^+$ has angular distribution $\sin^2 2\theta$. The channel cross sections σ_S, σ_D and σ_C are obtained from equation (10) by integration over all

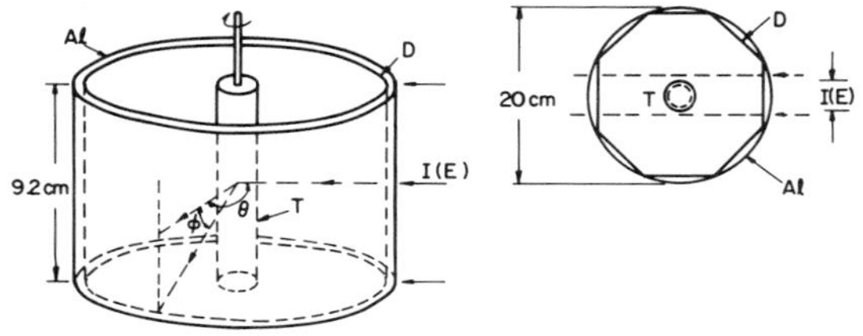

Figure 12
Vacuum chamber for photofission angular distributions.

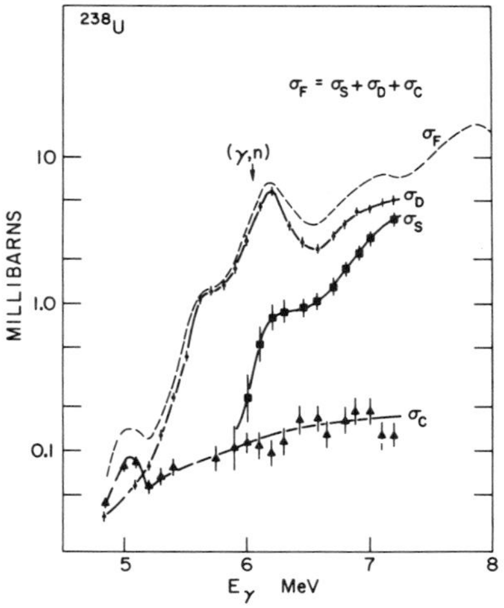

Figure 13
Total cross section σ_F and channel cross sections σ_S, σ_D and σ_C as a function of E_γ.

Table 2
Rotational states at the fission theshold

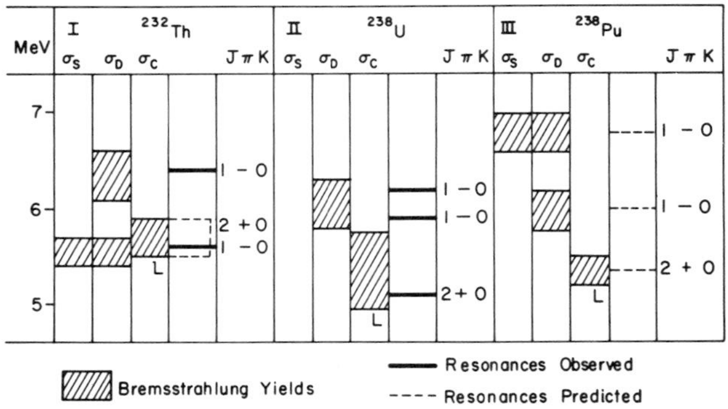

angles. The channel and total cross sections for photofission in ^{238}U between 5.0 and 7.5 MeV are shown in figure 13. Channel cross section σ_D, K = 0, $J^{\pi}=1^-$ has much of its strength in the peaks at 5.6 and 6.2 MeV. Channel cross section σ_S, K = 0, $J^{\pi} = 1^-$, shows a broad peak near 6.2 MeV which may not be a resonance since it can be accounted for by neutron competition just above the (n,γ) separation energy. Cross section σ_C has not been measured by us with sufficient precision near 5 MeV to resolve the peak at 5.05 MeV seen in the total cross section σ_T measurement. However measurements of σ_C and σ_T have been made by Rabotnov et al.[14] with poorer resolution, about 10%, but much better statistical accuracy. They find for ^{238}U that the ratio σ_C/σ_F increases substantially with decreasing energy near 5 MeV. Combining this ratio with our measurement of σ_F we conclude that the peak is associated with σ_C rather than with σ_D as shown in figure 13.

DISCUSSION

The results of the angular distribution measurements of ^{238}U and ^{232}Th are summarized in sections I, II and III of Table 2. The horizontal lines in column 4 of section II correspond to the narrow peaks in the photofission spectrum of ^{238}U, less than 200 keV wide, at 5.05, 5.7 and 6.2 MeV and have angular distributions which correspond to resonances in channels with K = 0, $J^{\pi} = 2^+$, 1^- and 1^- respectively. The corresponding measurements made by Rabotnov et al., made with 10% resolution, the σ_D and σ_C distributions shown as broad peaks or limits, L, in columns 2 and 3 of section II, overlap the narrow resonance observed by us. We have observed two peaks in ^{232}Th, at 5.6 and 6.4 MeV, both of which correspond to resonances with K = 0, $J^{\pi} = 1^-$ as shown in column 4, section I. Also a small σ_C contribution in ^{232}Th between 5.5 and 5.8 MeV has been observed by Rabotnov et al., which is evidence of a resonance in a channel with K = 0, $J^{\pi} = 2^+$. They also measured photofission angular distributions from ^{238}Pu, ^{240}Pu and ^{242}Pu. Their measurements on ^{238}Pu, interpreted as channel cross sections, show broad peaks or limits similar to those observed in ^{238}U and ^{232}Th. Thus there is a strong suggestion that narrow resonances are present in these nuclei also. For example, the broad peaks in the σ_S, σ_D and σ_C angular distributions of ^{238}Pu, shown in columns 1, 2 and 3, section III, have the spins and parities shown and probably indicate resonances at energies close to the broken lines shown in column 4.

Level distributions based on simple stable quadrupole and octupole deformations near the ground state and near the fission threshold have been investigated theoretically and shown[6,15] to have approximately the level sequences shown in figure 10. On the left side of figure 10 are shown, for quadrupole deformation, the rotational band, $K = 0$, $J^\pi = 0^+, 2^+, 4^+$ --- and at somewhat higher energies the vibrational band $K = 0$, $J^\pi = 1^-, 3^-, 5^-$ ---. On the right are shown, for octupole deformation, the rotational band, $K = 0$, $J^\pi = 0^+, 2^+, 4^+, 1^-, 3^-, 5^-$ --- and at higher energies the vibrational band $K = 0$, $J^\pi = 0^+, 2^+, 4^+, 1^-, 3^-, 5^-$. At still higher energies are the corresponding vibrational bands based on $K = 1$. However excitation with photons is very much simplified because only dipole and quadrupole photo-absorption have appreciable cross sections. Thus only states of low spin $J^\pi = 1^\pm$ and $2^\pm$ are excited appreciably from the $J^\pi = 0^+$ ground states of ^{238}U and ^{232}Th. These states are shown as heavy lines in figure 10. Thus for quadrupole deformation the most probable decay channels for fission are, for $K = 0$, two states, $J^\pi = 2^+$ and 1^-, and for $K = 1$, one state $J^\pi = 1^-$ and one $J^\pi = 2^-$. For octupole deformation, the fission channels are, for $K = 0$, one state $J^\pi = 2^+$ and two $J^\pi = 1^-$ states and for $K = 1$, one $J^\pi = 1^-$, one $J^\pi = 2^+$, one $J^\pi = 2^+$ and one $J^\pi = 1^+$ state. The $K = 1$, $J^\pi = 1^+$ and 2^+ states will be 6 and 20 times weaker respectively than the $K = 1$, $J^\pi = 1^-$ states. If we can observe all of the $J^\pi = 1^-$ resonances associated with the $K = 0$ band we can choose between the two models in figure 10. Since we observe two rather than one $K = 0$, $J^\pi = 1^-$ state close to the fission threshold in both ^{232}Th and ^{238}U we favour the octupole model.

Additional support for the octupole model is obtained from the relative distribution of levels in the $K = 0$ band. In this band, which includes plus and minus states, the minus states are not generally in their normal positions – those shown in figure 10 – but are displaced upwards with respect to the even states. To first order the displacement is inversely proportional to the strength of the octupole deformation parameter[6]. The energy separations, obtained from photofission measurements between $K = 0$, $J^\pi = 1^-$ and $K = 0$, $J^\pi = 2^+$ states at the fission threshold of ^{232}Th, ^{238}U and ^{238}Pu, as a function of A, are shown in figure 11a. For comparison the separations of 1^- and 2^+ states near the ground state of heavy even-even nuclei are shown in figure 11b. The energy difference increases with increasing A, the fission threshold measurements showing the largest increase. In terms of the octupole deformation this indicates a larger parameter for ^{232}Th than for ^{238}U, the parameter appearing stronger at the fission threshold than near the ground state. Theory suggests[6] that a large octupole deformation parameter can result in stable deformation which can be described by a double potential well. The narrow gross structure, <200 keV wide, observed at the fission threshold, can be explained in terms of interactions in a shallow second well. According to calculations by Strutinsky[29] a shallow second well should exist for many of the actinide nuclei, its position coinciding with the liquid drop model saddle point radius.

REFERENCES

1. E.K. Hyde, "The Nuclear Properties of the Heavy Elements", Vol. III, Prentice-Hall, Inc. (1964) 87-140.
2. C.M. Lederer, J.M. Hollander & I. Perlman, John Wiley & Sons, Inc. (1967) 418-437.
3. S.G. Nilsson, Chin Fu Tsang, A. Sobiczewski, Z. Szymanski, S. Wycech, C. Gustafson, I.L. Lamm, P. Möller & B. Nilsson, Nucl. Phys. A131(1969)1.
4. W.M. Strutinsky, J. Nucl. Energy 4(1957)523.
5. K. Lee & D.R. Inglis, Phys. Rev. 108(1957)774.
6. S.A.E. Johansson, Nucl. Phys. 22(1962)529.
7. P. Vogel, Phys. Lett. B25(1967)65.
8. J.E. Lynn, United Kingdom Atomic Energy Authority Research Group Report No. AERE-R5891 (1968).
9. E.J. Winhold & I. Halpern, Phys. Rev. 103(1956)990.
10. J.E. Gindler, J.R. Huizenga & R.A. Schmitt, Phys. Rev. 104(1956)425.
11. R.A. Schmitt & R.B. Duffield, Phys. Rev. 105 (1957)1277.
12. L. Katz, A.P. Baerg & F. Brown, Proc. Second United Nations Conference on the Peaceful Uses of Atomic Energy, Geneva, 15(1958)188.
13. A. Manfredini, M. Muchnik, L. Fiore, C. Ramorino, H.G. DeCarvalho, J. Lang & R. Müller, Nucl. Phys. 74(1965)377; Nuovo Cimento 44(1966)218.
14. N.S. Rabotnov, G.N. Smirenkin & A.S. Soldatov et al., Proc. Symp. on Phys. & Chem. of Fission, SM 60/81, Vienna (1965) 135; N.S. Rabotnov, G.N. Smirenkin, A.S. Soldatov, L.N. Usachev, S.P. Kapitza & Y.N. Tsipeniuk, Institute of Physics & Energetics report FEI-170 (1969), English trans. LA-4385-TR (1970).
15. E. Albertsson & B. Forkman, Nuclear Physics 70 (1965)209.
16. A.S. Soldatov, G.N. Smirenkin, S.P. Kapitza & Y.M. Tsipeniuk, Physics Letters 14(1965)217.

17. Kh.D. Androsanko et al., Second Symp. on Physics & Chem. of Fission, International Atomic Energy Agency, IAEA/SM/122/134(1969)419.
18. C.D. Bowman, G.F. Auchampaugh & S.C. Fultz, Phys. Rev. 133B(1964)676.
19. A. Bohr, Proc. Inter. Conf. on the Peaceful Uses of Atomic Energy, (1955) United Nations, New York, 2(1956)151.
20. E.K. Hyde, "The Nuclear Properties of the Heavy Elements", Vol. III, Prentice-Hall, Inc. (1964)74.
21. J.R. Huizenga, K.M. Clarke, J.E. Gindler & R. Vandenbosch, Nucl. Phys. 34(1962)439.
22. J.E. Lynn, The Theory of Neutron Resonance Reactions, Clarendon Press, Oxford (1968)226.
23. A.M. Khan & J.W. Knowles (to be published).
24. D.L. Hill & J.A. Wheeler, Phys. Rev. 89(1953)1102.
25. A. Sobiczewski et al., Nucl. Phys. A131(1969)67.
26. J.M. Moss et al., Phys. Rev. Lett. 26(1971)1488.
27. P. Möller et al., Phys. Lett. 26B(1968)418.
28. K.E.G. Lobner, M. Vetter & V. Honig, Nucl. Data, Sec. A7(1970)495.
29. V.M. Strutinsky, Nucl. Phys. A95(1967)420.

NUCLEOSYNTHESIS AND THE FORMATION OF SUPERHEAVY ELEMENTS

A.G.W. CAMERON
Belfer Graduate School of Science
Yeshiva University
New York, New York

Goddard Institute for Space Studies, NASA
New York, New York

PROCESSES OF NUCLEOSYNTHESIS

Cosmological and Stellar Nucleosynthesis

The discovery of the isotropic microwave background radiation, with an apparent blackbody spectrum corresponding to a temperature of 2.7°K, has had a profound influence on recent developments in cosmological theory. I believe that most cosmologists now accept the view that this radiation represents a remnant of a prior hot state of the universe, being emitted in its present form shortly after the radiation decoupled from matter, at a temperature near 3000 °K, when hydrogen atoms recombined. If one accepts this and asks what are the logical consequences, then one can project backwards in time to a point where nuclear reactions occurred very rapidly in the early history of the universe. At a sufficiently early time, the universe will be filled with many elementary constituents in thermodynamic equilibrium. These may include baryons and antibaryons, pions and other mesons, leptons and antileptons of all kinds, both charged and neutrinos, and photons. As the universe expands, the temperature falls. The baryons and antibaryons annihilate, leaving a small baryonic residue, at least in our part of space. The pions and muons disappear, leaving a portion of the energy, in the form of muon neutrinos and antineutrinos, out of interaction with the remainder of the particles. At still lower temperatures the electrons and positrons annihilate, leaving two other non-interacting fluids, the electron neutrinos and antineutrinos. There is a small residual charged lepton number which is equal to the residual baryon number.

Up until the time of disappearance of the electrons and positrons, the interaction of these particles with neutrons and protons was sufficiently fast to maintain neutrons and protons in statistical equilibrium with each other. When the electron pairs vanish, the neutron-proton ratio is frozen, leaving about one neutron to every 7 protons.

Upon further cooling, the neutrons can be captured by the protons and a series of further thermonuclear reactions occurs involving the light particles. Deuterium and tritium are formed and can interact to form ^{4}He. The net result of these other reactions is to convert the bulk of the neutrons into alpha-particles, so that alpha-particles constitute about one-quarter of the mass of the matter present. Small residual amounts of deuterium, tritium and ^{3}He remain.

The above description assumes a simple expanding Friedman universe. If the universe contains various possible pathologies, such as very high Fermi energies for the neutrinos or anti-neutrinos, or a rate of expansion several orders of magnitude faster than in a Friedman universe, then the above conclusions would be altered. However, I will not consider these cases here.

It is remarkable that nearly everywhere one looks in the universe where the hydrogen-to-helium ratio can be determined, the helium is about one-quarter of the mass. About the only exception to this approximate rule is the case of some very hot stars in advanced stages of evolution, where it is possible that the helium has diffused downwards from the surface layers and hence is deficient. Thus it appears that the bulk of the helium in the universe was cosmologically synthesized. If we should live in a closed universe, then that is about all that was synthesized. However, if we live in an open universe having an average mass density in space about equal to that which we see in the form of galaxies, then the ^{4}He will be slightly reduced compared to the closed case (24% compared to 29%), but the deuterium and ^{3}He will have abundances comparable to those found in solar system material. No other plausible mechanism has been suggested which gives the correct orders of magnitude for the abundances of deuterium and helium 3 in the solar system. Cosmological nucleosynthesis can also manufacture small amounts of ^{7}Li, via various minor nuclear reactions at the time of helium synthesis, but it does not appear that enough ^{7}Li can be made in this way to account for the observed solar system abundances.

Apart from the lithium, beryllium, and boron, which appear to be synthesized by cosmic ray bombardment of interstellar gas, the remaining elements all seem to have been synthesized in stellar interiors, either gradually or explosively. In this general picture, the material which is going to form the galaxy would be initially composed of hydrogen and helium. Stars form, heavier elements are then synthesized, and the synthesis products are ejected back into interstellar space. These are then incorporated in later generations of stars, and nuclear reactions make further transformations of the abundances before the material is once again ejected into space. In this way the interstellar medium is gradually enriched in the products of various classes of nuclear reactions. I shall briefly review the status of these various classes of reactions.

Hydrogen Burning

There are various complex sets of nuclear reactions which can convert hydrogen into helium under various stellar conditions ranging from about 10^7 to 5×10^7 ^{o}K. These reactions include the proton-proton chains, which have three distinct branches, and the carbon-nitrogen-oxygen bi-cycle, in which CNO isotopes act as catalysts in nuclear reactions in which hydrogen is converted into helium. These various reactions are reasonably well known and I shall not list them here. A great deal of effort has gone into the determination of the nuclear cross sections in the stellar energy region during recent years. It is believed that most of the important reactions have rates reasonably well determined by these efforts.

Current interest in these reactions stems from the efforts of Raymond Davis, Jr. to measure the neutrino flux from the sun. He utilizes a large underground tank of C_2Cl_4, which is commercial cleaning fluid. Neutrinos from the sun can convert ^{37}Cl into ^{37}Ar, which can be extracted from the underground tank with great efficiency and counted in a low-level counter. At present Davis has a marginally significant counting rate, and most of this is expected to be due to the neutrinos resulting from the decay of ^{8}B in the sun. At present his counting rate is a factor of 5 less than is expected on the basis of the best solar models, utilizing current estimates for the nuclear reaction rates as well as for the stellar opacities and other quantities which go into the calculation of a stellar model. This discrepancy has caused a great deal of scrutiny of the relative nuclear cross sections, and a revision of some of them, and it may well be that the remaining discrepancy is in some other area, such as the opacities in the solar interior. Nevertheless, until the discrepancy is quantitatively explained, doubt will remain that the nuclear cross sections have been completely adequately determined.

Helium Burning

We are concerned with only two major reactions in the helium burning stage. One of them forms carbon, the other destroys it:

$$3\,^{4}\mathrm{He} \rightarrow {}^{12}\mathrm{C}^{*} \rightarrow {}^{12}\mathrm{C} + h\nu$$

$$^{12}\mathrm{C}(\alpha,\gamma)^{16}\mathrm{O} \quad .$$

A year or two ago it was thought that the rates for these reactions were known sufficiently accurately for astrophysical purposes. However, recently a number of doubts about this have appeared. It is very important to determine the rates relative to one another, since if the formation of carbon is very inefficient, whereas the destruction of it should become very efficient, then we might never have any carbon left at all at the end of the helium burning process. However, nucleosynthesis seems to require the operation of an explosive carbon burning stage, and this in turn would appear to require significant amounts of ^{12}C to remain at the end of the helium burning process.

The triple-alpha reaction is in a sense the simplest, going through the second excited state of ^{12}C. However, recently the energy of this state was revised upwards by a small amount, which nevertheless had the effect of decreasing the carbon formation rate by a significant factor.

The destruction of carbon occurs mostly in the tail of the bound state of ^{16}O at 7.12 MeV. The radiation width is reasonably well known for this state, but the reduced alpha width is very much in dispute. Interference capture measurements have tended to indicate that the reduced alpha width approaches the single particle upper limit. On the other hand, ^{7}Li stripping experiments tend to indicate the reduced alpha width is much lower, in the range 0.02 to 0.05 of the single particle limit. It is important to settle this matter, since if the reduced alpha width approaches the single particle upper limit, then we can expect essentially no carbon to remain at the end of helium burning. On the other hand, if it is as low as 0.02 of the single particle limit, then we would have practically all carbon at the end of helium burning. One suspects that the truth ought to lie between these extremes.

Helium burning takes place at a temperature of the order of 2×10^8 °K and at densities of the order 10^3 or 10^4 grams per cubic centimeter.

Carbon and Oxygen Burning

If the temperature and density in a stellar interior can be substantially further raised, then, in turn, carbon and oxygen burn ing may take place. The basic character of these reactions is similar. Because of the difficulty of penetrating Coulomb barriers, the fastest thermonuclear reactions in a given astrophysical environment are likely to be those involving the lowest product of reacting charges. Thus, following helium burning, the lightest charge will be that of carbon, and the appropriate reactions will be those of ^{12}C with itself. These reactions lead to a compound ^{24}Mg nucleus. This compound nucleus may deexcite with an emission of an alpha-particle, forming ^{20}Ne, or with the emission of a proton, forming ^{23}Na. However, this is likely to take place at a temperature near 10^9 °K, and protons and alpha-particles cannot live an independent existence long, at least at these temperatures. They will be very quickly reabsorbed on some nucleus. This nucleus may be ^{12}C as long as some of it exists, giving rise to various interesting possibilities for the production of neutrons. Also, the protons and alpha-particles may be reabsorbed on the ^{23}Na and ^{20}Ne, forming ^{24}Mg. Once the latter nucleus is present in significant abundances, other reactions can take place leading to other minor constituents extending beyond ^{24}Mg. The basic reactions therefore consist of ^{12}C nuclei fusing and producing a series of nuclei in the general vicinity of ^{24}Mg.

Once the carbon has disappeared, the next reactions which can take place are those of ^{16}O with itself. This forms the ^{32}S compound nucleus, which can exoergically decay with the emission of neutrons, protons, and alpha-particles. These particles will be rapidly recaptured on other nuclei present, since the temperature will now be 2 x 10^9 ^{0}K, and the net results are likely to be a series of intermediate nuclei extending from ^{24}Mg through ^{32}S and perhaps slightly beyond (by this time ^{20}Ne has been destroyed by photonuclear reactions).

Of greater interest for nucleosynthesis is explosive carbon and oxygen burning. In this case the carbon and oxygen thermonuclear reactions are induced by the passage of a supernova shock wave through the material. As a result, the peak temperatures to which these nuclei are subjected may approach twice the normal temperature at which they would occur in a stellar interior. Thus carbon and oxygen thermonuclear reactions will partially go to completion on time scales of the order of a second. The incompletely burnt carbon produces good agreement with the relative abundances of most of the nuclides between ^{20}Ne and ^{27}Al. Explosive oxygen burning, again incomplete, gives rise to the correct proportions of several nuclides in the vicinity of ^{28}Si. We shall discuss the conditions giving rise to explosive nucleosynthesis later.

Silicon Burning

If the temperature should rise to around 4 x 10^9 ^{0}K during the normal course of stellar evolution, then a new type of nuclear transformation takes place. It might be thought that the next step would involve the fusion of two nuclei in the region of silicon. However, at these high temperatures, there is a significant population in the tail of the Planck photon spectrum at energies comparable to nucleon and alpha-particle separation energies. Thus all of the nuclei present in the stellar interior under these conditions will start to undergo photonuclear reactions. The self-conjugate nuclei such as ^{24}Mg, ^{28}Si and ^{32}S will mainly undergo (γ,α) reactions. Slightly more neutron-rich nuclei will easily lose a neutron in (γ,n) reactions. But once again these light particles do not lead an independent existence for very long in the stellar interior before being absorbed on other available nuclei. Thus a condition is established in the stellar interior in which there is a gradual approach to nuclear statistical equilibrium.

There is a tendency for the nuclei to flow toward those species with the largest binding energy per nucleon. If complete nuclear statistical equilibrium were to be established, there would be an abundance peak centered at ^{56}Fe. In fact, an abundance peak is observed in nature centered about this nuclide, and we call it the iron nuclear statistical equilibrium peak.

The nuclei in the vicinity of silicon will therefore have a tendency to approach iron. In order for them to do this, it is necessary that about half the nuclei be progressively broken down into neutrons, protons, and alpha-particles, so that these particles can be captured on the remaining nuclei in the silicon region, thus driving them up into the iron equilibrium peak region. This is essentially what happens, except that it happens via an approach to local statistical equilibrium at the same time that there are net flows through an extensive nuclear reaction network to produce both break-up of silicon-region nuclei and mass addition to such nuclei bringing them up toward the iron peak.

Once again, silicon burning is mainly of interest to nucleosynthesis in terms of explosive silicon burning. Explosive silicon burning can be initiated by explosive carbon or oxygen burning, if the density is high enough so that the energy released by the explosive carbon or oxygen burning raises the temperature beyond 4×10^9 °K, so that silicon burning can take place during the time scale associated with the passage of a supernova shock wave.

If explosive silicon burning is incomplete, which will occur in the vicinity of 4×10^9 °K, then an abundance distribution is set up extending from silicon to calcium, in which there is a local statistical equilibrium, and where the relative abundances of a great many of the nuclides agree excellently with the abundances in the solar system. Beyond calcium, a considerable amount of ^{56}Ni will be formed, which actually is the nucleus probably contributing most to the abundance of ^{56}Fe in nature, to which ^{56}Ni decays via ^{56}Co. There have been various speculations that large amounts of ^{56}Ni and ^{56}Co produced in supernova explosions by this process may be responsible for introducing very significant amounts of energy into the expanding supernova ejecta. It is conceivable that gamma-rays produced by the decay of these substances could be detected from supernovae at extragalactic distances.

In the vicinity of 5×10^9 °K, the actual abundance distributions in the vicinity of the iron equilibrium peak tend to reproduce rather well the abundance on the low mass side of the peak, but not those on the upper side. On the other hand, if the peak temperature should be raised to 7 or 8×10^9 °K then the ratio of helium to the iron equilibrium peak progressively increases, and when the temperature is lowered, much of this helium will be captured on iron peak elements, leading to a reasonable agreement between the nuclei produced on the upper side of the iron equilibrium peak and those observed in solar system material.

Formation of Heavy Elements

So far all of the processes discussed have involved the buildup of light elements toward the nuclear statistical equilibrium peak in the vicinity of iron. The heavy elements are not made by these processes, except through minor reactions which may take place in association with the main processes so far described. Indeed, if the temperature in a region containing heavy elements is raised to 4 or 5 x 10^9 ^oK, then the heavy elements would be broken down into material in the iron peak by a series of photodisintegration reactions.

Later on we shall discuss the problems of manufacturing superheavy elements. For this purpose we must concentrate on many of the details of one of the processes for manufacture of heavy elements, the r-process. In any case, it is best to consider the details of the processes manufacturing heavy elements in specific astrophysical contexts. Therefore, we consider at this stage only the broad aspects of these processes.

There are three principal processes for the manufacture of heavy elements. Two of these involve neutron capture; these can be divided into neutron capture taking place on a slow time scale, the s-process, and neutron capture taking place on a rapid time scale, the r-process. The third process makes nuclei which are on the neutron deficient side of the valley of beta stability, and which therefore cannot be made by neutron capture at all. This is called the p-process, where the p implies a rapid proton capture. It may also be possible to make p-process nuclei by photonuclear reactions, providing the temperature does not rise so high that the photonuclear reactions will quickly run all the way down to the vicinity of the iron equilibrium peak.

Two of these processes, the r-process and the p-process, are almost certainly associated with supernovae. The s-process can occur during advanced stages of stellar evolution, but before catastrophic end-points are reached since we see the products of neutron capture on a slow time scale in the spectra of certain red giant stars, of spectral class S. These stars usually have overabundances of the heavier elements of between 1 and 2 orders of magnitude. Since spectral lines of technetium are usually seen in these stars, and since the longest-lived technetium isotope which can be formed by neutron-capture is ^{99}Tc, with a half-life of 2 x 10^5 years, it is evident that the lifetime of stars in the red giant class S is relatively limited. We shall return to the mechanisms for the manufacture of these heavy elements after reviewing stellar evolution.

Lithium, Beryllium, and Boron

The isotopes of lithium, beryllium and boron are readily destroyed in stellar interiors, lithium at around 3 x 10^6 ^oK, beryllium at about 4 x 10^6 ^oK, and boron at about 6 x 10^6 ^oK. They undergo thermonuclear reactions with hydrogen, which results in the formation of helium. Thus these elements cannot have been formed in the normal course of stellar evolution in stellar interiors, since they are destroyed there, and the main energy generation processes jump across these elements via the triple-alpha reaction which occurs during helium burning.

It has been evident for some time, based upon the excellent measurements of high energy spallation cross sections carried out at Orsay, France, that the lithium, beryllium, and boron isotopes are for the most part consistent with spallation production cross sections in such targets as carbon, nitrogen, oxygen, and neon. The agreement is not precise, in particular the high ratio of ^{7}Li to ^{6}Li is much higher than corresponds to spallation production. In what follows we shall eventually find a separate way of manufacturing ^{7}Li. But the other isotopes of these elements will be ascribed to production by cosmic ray bombardment of the interstellar medium.

STELLAR EVOLUTION

The Hydrogen and Helium Burning Stages

A star is formed by the collapse and fragmentation of a cloud in the interstellar gas. Once a state of hydrostatic equilibrium has been established, we know from the virial theorem that half of the released gravitational potential energy will have been stored as internal thermal energy, and the remainder will have been either radiated away or utilized in dissociating the atoms in the stellar interior. As energy is radiated from the stellar surface, the internal thermal energy increases at an equal rate, and the star correspondingly shrinks. A temperature gradient is set up between the center and the surface sufficient to transport half of the released gravitational potential energy to the surface where it can be radiated away. The energy transport will take place by radiative transfer at lower temperature gradients, and by convective transport once the radiative temperature gradient necessary to transport the flux reaches the adiabatic temperature gradient.

As the star shrinks the center becomes hotter and denser, and eventually hydrogen thermonuclear reactions will commence, when the temperature reaches 10 to 20 x 10^6 ^{o}K. At this point the star settles down to a long period of stable operation as a "main sequence" star in a plot of stellar luminosity vs. surface temperature, which is called the Hertzsprung-Russell or HR diagram. Due to the high temperature sensitivity, the thermonuclear reactions providing energy generation in the star are more concentrated toward the center than are the gravitational potential energy release rates, so that as gravitational contraction ceases and nuclear energy burning takes over, the luminosity of the star declines slightly since the photons must travel through a longer path to reach the surface.

After the hydrogen burning has finished in the center, once again contraction begins in that region, and the entire central portion of the star is then increased in temperature. This is sufficient to ignite hydrogen burning in the fuel surrounding the helium core, and the star now obtains energy for a long period of time via the hydrogen burning shell source. In a massive star this gradually increases the mass of the helium core, which continues to contract and to increase its temperature until helium thermonuclear reactions are ignited at the center. Then the helium reactions will take over the burden of energy generation in the star until the helium has been exhausted, leaving a core composed of carbon and oxygen. The star will then have two energy burning shell sources, one of hydrogen and one of helium. Generally, these may be expected to progress outwards in the stellar mass, increasing the mass of the core that is left behind.

In a star of lesser mass, for example of about one solar mass, the sequence of events is somewhat different. In this case, the rate of addition of helium to the core from the hydrogen burning shell is very much reduced, since the luminosity of the star is much less, and therefore the rate of energy generation by contraction in the core becomes very small. The core stays essentially isothermal, but as contraction takes place it gradually becomes highly electron degenerate. The density may well reach 10^5 grams per cubic centimeter at the time that the core contains a good fraction of the stellar mass, of order 0.5 solar masses. At this time, the core is massive enough to start contracting fairly rapidly on its own, and the temperature is finally enabled to rise in the core, although it stays essentially isothermal in the highly conducting electron-degenerate central part. Contraction raises the temperature to about 10^8 ^oK, where the helium burning begins.

In a highly electron-degenerate gas, the pressure is a function essentially only of the density. Therefore the temperature can be raised a great deal with very little effect on the pressure. Thus when a thermonuclear reaction is ignited in an electron-degenerate core, the increasing rate of energy generation from nuclear reactions can boost the temperature rapidly in a thermal runaway. During the runaway the pressure in the core is very little affected until the thermal energy per particle ~kT, becomes of the order of the electron Fermi energy. At this point, the pressure in the core suddenly starts to rise rapidly, leading to an expansion of the core. The results are not more catastrophic than this because the hydrodynamic expansion time for the core becomes much less than the time required to complete thermonuclear burning of helium into carbon and oxygen.

During the time that the star of one solar mass has possessed an electron-degenerate core, the outer envelope has become very swollen, to a large radius, and the star is called a red giant. This is because, even though the luminosity may be very much greater than in the main sequence phase, the radius is so much greater that the surface temperature of the star is greatly lowered. With the onset of helium burning in the degenerate core (the helium flash) the expansion of the core is accompanied by a shrinkage of the outer hydrogen envelope back toward more normal radii. In massive stars the contraction of the helium core is also accompanied by an expansion of the outer hydrogen envelope, even though the core never becomes degenerate.

Also in the case of a star of one solar mass, after helium burning has gone to exhaustion in the core, once again a double-shell source is created, one of hydrogen and one of helium. It has been recently found that stars with double-shell sources tend to reach an unstable condition except in cases of very massive stars where the instability may be prevented by radiation pressure.

Helium shell flashes may occur, in which the thermonuclear reactions suddenly release a great deal of energy, thereby causing a local expansion in the intermediate layers of the star, and temporarily allowing the helium shell burning region to take over the full burden of energy generation in the star. Gradually the situation relaxes to normal with both shell sources again in operation. During the interval after the helium shell flash, the energy released can induce convection in the helium layer, and this convection may reach up close to the outer hydrogen burning layer. Indeed, calculations for the one solar mass case indicate that a mild amount of mixing of hydrogen into the helium layer can occur after a number of helium shell flashes have occurred. Investigations have not yet clarified the situation for more massive stars, but it is suspected that stars up to probably about 5 solar masses may also be able to mix hydrogen into the helium layer following helium shell flashing. As I shall discuss in more detail, I believe that this behaviour is intimately related to the production of neutrons and the formation of heavy elements by the s-process.

It takes a star like the sun about 10^{10} years to undergo hydrogen burning on the main sequence. Afterwards another 1 or 2 x 10^9 years may be occupied in helium burning stages. Thus during the age of our galaxy stars significantly less in mass than the sun are all still burning hydrogen on the main sequence. Stars of solar mass and greater can have gone through both hydrogen and helium burning phases. The later stages of evolution vary greatly in their general behaviour, depending upon the stellar mass. This forms the basis of three types of final evolutionary pathways that are open to the stars, leading to white dwarfs, neutron stars, or black holes. We shall now discuss these various pathways.

Formation of White Dwarfs

Among the lower mass stars, it appears that only hydrogen and helium burning reactions take place before the stars come to the end of their evolutionary lifetimes. Following helium burning, the helium burning products, carbon and oxygen, form a degenerate core which gradually increases in size as the two shell burning sources progress outwards in the mass of the star. As the two shell burning sources progress toward the surface of the star, the luminosity gradually increases, since there is less total opacity in the overlying layers for the photons to traverse. However, increasing luminosity of the star results in an increasing radiation pressure on the layers of the star near the surface. Eventually, the radiation pressure exceeds the gravitational forces, and matter is expelled from the star. Recent calculations by Paczynski[1] have indicated that, following the establishment of the two shell sources, the outer envelope will be expelled by radiation pressure in a remarkably short period of time, giving rise to a planetary nebula in the form of an expanding shell of gases, and a white dwarf remnant consisting of the degenerate core plus a thin overlying layer of non-degenerate material which forms the atmosphere.

Recent calculations by Paczynski show that the mass of the white dwarf remnant core gradually increases as the main sequence mass of the star increases. There is an upper limit to the mass which can form a stable white dwarf star, called the Chandrasekhar limit, which is about 1.4 solar masses. This arises because the addition of mass to a white dwarf star must compress the star in order to provide the additional pressure to support the increased mass, and there is an upper limit to the amount of mass that can be added this way. When the equation of state in the interior represents extremely degenerate electrons, then the compression of the material can only support the mass associated with the existing material, and not any additional material. Consequently, there is an upper limit to the mass of a white dwarf of the amount stated.

Paczynski found that the mass remaining in the degenerate core would reach the Chandrasekhar limit for a main sequence mass in the vicinity of 3.5 to 4 solar masses. The work by J.W. Truran and myself[2] on the chemical evolution of the galaxy and also the work of Gunn and Ostriker[3] on the types of stars likely to form pulsars within our galaxy, are consistent with this upper limit on the formation of white dwarfs at about 3.5 to 4 solar masses on the main sequence.

Even though stars forming white dwarfs probably never go beyond helium burning in their main sources of energy generation, they are not without interest for nucleosynthesis, as we shall discuss below.

Formation of Neutron States

Beyond about 4 solar masses on the main sequence, the amount of material remaining in the central core regions following any possible expulsion of matter from the surface by radiation pressure, will exceed the Chandrasekhar upper limit on the mass of a white dwarf. It follows that the central electron degenerate core of the star must contract and raise its temperature until a new set of nuclear reactions will be ignited. The next set of reactions is the carbon burning reactions. These will be ignited in a highly electron-degenerate condition, so that, as previously discussed for the helium flash in a stellar core, the nuclear reactions will produce a thermal runaway, leading to an increase of pressure only when the thermal energy per particle becomes comparable to the electron Fermi energy. However, it appears that there is a dramatic difference in kind between the thermal runaway which takes place in a helium flash, and the thermal runaway which might be termed a carbon flash. In the case of the helium flash, the time scale required to drive the helium thermonuclear reactions to completion exceeds the hydrodynamic expansion time of the core, so that the star is not disrupted when the core expands. However, the reverse situation obtains in a carbon flash. The carbon thermonuclear reactions will burn out to completion in a time scale that is short compared to the hydrodynamic time scale once the temperature has risen far enough to produce equality between thermal and degenerate energies. Consequently the temperature will rise to the point where the core would undergo a gentle hydrodynamic expansion, and then a great deal of additional energy will be dumped into the core before that expansion can take place. The result is a supernova explosion. The core violently expands, producing a very strong shock wave which races through the outer layers and ejects them. These processes are of great interest for a variety of kinds of nucleosynthesis.

The processes associated with explosive nucleosynthesis in the core have been discussed above. If the density is not too high, incomplete explosive carbon burning may produce elements in the vicinity of neon to aluminum. The rise in temperature produced by carbon ignition may cause explosive oxygen burning in a higher density range. If the oxygen burning is incomplete, then abundances of nuclides in the general vicinity of silicon are produced. The oxygen burning, in turn, may trigger explosive silicon burning. If this is incomplete, then the elements from the vicinity of silicon to calcium and parts of the iron equilibrium peak can be produced. If the silicon burning goes to completion, but the temperature never rises above about 5×10^9 $^{\circ}$K, then the iron equilibrium peak is produced with emphasis on the lower mass side of it. If the peak temperature should reach 7 or 8×10^9 $^{\circ}$K, then the upper side of the iron equilibrium peak is produced.

However, the range of densities which can be expected to be responsible for this range of reactions tends to be characteristic of the outer layers of the degenerate core. If the entire core were to explode and the products to be distributed in space, then we would expect to have a much higher abundance of material in the iron equilibrium peak than we would of such lesser elements as magnesium and silicon, which are the main components of the earth.

Thus it is comforting to know that the inner parts of the core of the star probably will not be completely exploded into space. If the thermonuclear ignition takes place at a density in excess of about 6×10^9 $^{\circ}$K, then the Fermi energy of the electrons approaches 10 MeV, and the rate of electron capture on the products in the equilibrium peak, particularly including ^{56}Ni, is especially rapid. Most of the pressure in the inner layers is due to the electrons. Hence the removal of the electrons by electron capture in any significant number causes a lowering of the pressure following the formation of the iron equilibrium peak, and hence a reimplosion of the central regions. The central region should then form a neutron star remnant.

There are major questions associated with these processes in connection with the r-process production of the heavier elements, which we will discuss in much greater length below.

As the supernova shock wave passes through the outer layers of the star, it can produce a number of other processes of interest for nucleosynthesis. In particular, the p-process products are expected to result when the shock wave passes through the lower layers of the hydrogen outer envelope. I will discuss this in due course.

The Formation of Black Holes

As the mass of the star on the main sequence increases, a point may be reached where the ignition of the carbon burning reactions takes place under non-degenerate conditions, or at least under conditions of sufficiently low degeneracy that an explosion does not result. In this case, carbon burning can be expected to go to completion in the core, and we may then have a configuration with three energy generation shells in it. Such a star would then form a degenerate core once again, and it would have another chance at a supernova explosion with the ignition of the oxygen burning reactions. However, in a sufficiently massive star, both carbon and oxygen burning may take place non-explosively in the star, and that will leave a degenerate core awaiting the ignition of silicon burning reactions. However, remarkably little energy is liberated during silicon burning reactions, even explosive ones, and it is very unlikely that a supernova explosion can be triggered by such reactions. Therefore, for very massive stars, it is likely that supernova explosions cannot occur because of the lack of sufficiently degenerate conditions during the ignition of major energy generation processes.

Under these conditions, the stellar interior has no choice but to reach increasingly extreme temperatures and densities. Energy must continue to flow toward the surface, and after nuclear energy generation resources of the star are exhausted, this energy must be supplied by gravitational contraction. The degeneracy of the core will become increasingly great, and if the mass is very large the temperature will become increasingly great. Both of these lead to a catastrophic end point of the stellar evolution.

If the density becomes sufficiently great, electron capture on nuclei present in the core will lead to the progressive removal of the electrons, which constitute the main source of pressure upholding the central regions. In this case, the contraction of the central regions will accelerate until free fall conditions are obtained. On the other hand, if the temperature becomes progressively larger, then any nuclei, which by now would be in the vicinity of the iron equilibrium peak, would no longer be stable in view of the very large number of high energy photons in the tail of the Planck spectrum. Nuclear statistical equilibrium predicts that the most abundant nucleus will shift from the iron region down to ^4He, and with still higher temperature, the helium will in turn be thermally dissociated into neutrons and protons. At the high densities which would necessarily be involved at this time, rapid electron capture on the protons would have to occur. A great deal of energy is absorbed in breaking the iron equilibrium peak into its constituent nucleons, all of the energy which has previously been released during the nuclear energy generation process of the star, and in order to supply this energy, a rapid release of gravitational potential energy is required. This also would lead to a collapse of the interior portions of the core.

At one time it was believed that this condition would lead to a supernova explosion. The idea was that the central regions of the core would collapse until nuclear densities were reached, raising the temperatures to values in excess of 10^{11} ^{o}K, whereupon under conditions of statistical equilibrium, very large numbers of neutrinos and antineutrinos would be created. These neutrinos and antineutrinos are expected to diffuse outwards in a time scale comparable to the hydrodynamic time scale, since at the high densities and temperatures the "optical" depth for the neutrinos at the center would be of order 10^2. According to a pioneering paper on supernova hydrodynamics published several years ago by Colgate and White[4], it was then expected that the outward diffusion of neutrino energy would cause a rapid heating of the descending outer envelope in the star, leading to the formation of a giant shock wave and an ejection of all the outer layers which were descending on the core. However, recent more detailed calculations by Wilson[5] have failed to substantiate this expectation. He has found that stars, with cores which collapse to densities in the vicinity of the nuclear density, will not succeed in exploding off the outer layers. Since we are discussing stars which have masses much in excess of the Chandrasekhar limit for white dwarfs, it follows that the mass of any core at nuclear densities, which is temporarily stabilized in approximate hydrostatic equilibrium, will progressively increase until a general relativistic upper limit is reached, following which general relativistic collapse will ensue.

This collapse will produce a black hole. It appears likely that the entire mass associated with the star at the time of formation of the black hole will disappear down toward the relativistic singularity. If the general theme of this analysis is correct, then it follows that the more massive stars do not contribute to nucleosynthesis in the galaxy.

ASTROPHYSICAL CONDITIONS FOR HEAVY ELEMENT NUCLEOSYNTHESIS

The s-Process

From our earlier discussion it is evident that heavy elements cannot be made under conditions approaching nuclear statistical equilibrium. Therefore it is necessary to search for neutron sources which can exist at temperatures much lower than those which produce nuclear statistical equilibrium. I shall discuss such a source within the context of a proposal recently made by W.A. Fowler and myself[6].

This proposal is based upon the helium shell flash mechanism of Schwarzschild and Harm[7] and others. In this mechanism, a star has advanced through hydrogen and helium burning in the core and possesses a double shell source surrounding an electron degenerate core. Following shell flashes, a great deal of energy is released at the base of the helium layer in the star, which produces convection throughout the helium layer, and leads to the mixing of some hydrogen from the surrounding layer into the helium layer. The hydrogen admixes in very small amounts in stars of one solar mass, but may admix in larger amounts in stars of around 2 to 4 solar masses.

Under these circumstances, hydrogen will first react with ^{12}C which has been formed by the triple-alpha reactions. The reaction is $^{12}C(p,\gamma)^{13}N(\beta^{+}\nu)^{13}C$. The resulting ^{13}C may then be convected to lower levels in the helium layer, eventually approaching temperatures high enough for the operation of the reaction $^{13}C(\alpha,n)^{16}O$.

The neutrons which are released are thermalized at a temperature of about 2×10^{8} °K, and can then be captured by the other elements which are present. This neutron energy corresponds to about 20 KeV. For such energies, all except the very lightest nuclei have neutron capture cross sections which effectively average over many resonances. As such, the neutron capture cross sections vary in a rather smooth way, but depend sensitively upon nuclear shell effects. Cross sections are very low at closed neutron shells. The cross sections are also high for nuclei with large level densities. This means that neutron capture cross sections are large for such heavy nuclei as rare earths and actinides, but are small for closed shell nuclei and the lighter nuclei.

The neutrons will be captured in proportion to the abundance of a given nuclide multiplied by an average neutron capture cross section. For an initial solar composition, this product has a peak centered on the iron equilibrium peak. Hence when the neutron capture starts, the capture takes place mainly within the iron equilibrium peak, building nuclei up toward heavier mass numbers by a chain of neutron captures interspersed, when necessary, by beta decays.

As the material initially in the iron peak moves up through heavier mass numbers, the abundance level in the heavy nuclei region may rise compared to solar values by factors in the range 10^3 to 10^4. The original iron peak becomes widely spread out among the heavier nuclei, since the neutron capture cross sections in the heavy nuclei are considerably larger than those in the iron peak itself. There are abundance maxima in the capture process at the neutron closed shell positions.

The capture becomes terminated in the vicinity of lead and bismuth. Capture beyond ^{209}Bi leads to nuclei which decay with short half-lives by alpha-particle emission, leading back toward lead and bismuth nuclei. This assures a natural mass number limit to the s-process. However, in nature this upper limit is not very important, since most of the nuclei formed in the s-process remain at mass numbers less than that of lead.

The s-process means a slow process. This means that when a nuclide which is unstable against beta decay is formed by neutron capture, ordinarily there will be time for that decay to take place before another neutron capture occurs in the same nucleus. However, this time interval between neutron captures may be as short as a few months without producing deviations from the observed abundance distribution corresponding to the s-process.

The extensive operation of the s-process releases a great deal of energy. The proposal by Fowler and myself was that this released energy would cause an extensive convection zone to be created extending upwards from the base of the helium layer, and joining onto the outer rather deep convection zone which should exist in such a star at this time, in order to be able to mix products of neutron capture throughout the upper level, thus producing a general mixing and a marked overabundance of the heavy elements at the surface. An enhancement of the surface abundance of carbon relative to oxygen may also take place. This would produce a red giant star in an advanced stage of evolution which is dominated by molecular lines either of oxides or carbides. In the first case, the spectrum would have characteristics of spectral class S, and in the second case it would be a carbon spectrum. Both types of red giant stars are known with considerable overabundances of heavy elements. One of the most striking features of the S stars is the presence of lines of technetium, probably due to the isotope ^{99}Tc, with a half-life of 2×10^5 years. It was the discovery of technetium lines in the spectra which originally led to the formulation of the idea of the s-process.

The r-Process

The rapid neutron capture process requires much more extreme conditions. It also requires a very short time scale. An early analysis of the process by Burbidge, Burbidge, Fowler, and Hoyle[8] assigned a time scale of the order of minutes for the operation of the r-process, and more recent analyses by some of my colleagues and myself[9] have suggested time scales as short as one second for the entire process to occur. Thus it is evident that the appropriate astrophysical environment is far from settled.

Two types of environment seem possible. We discuss the more extreme of these first.

It is evident that the r-process requires a very neutron-rich environment. Such an environment exists in the core of a star which undergoes a supernova explosion, following the collapse of the central regions to nuclear densities after the explosion has taken place. However, we have already seen that the reimploded material in such a neutronized core cannot be expected to be exploded away from the core after the collapse.

A possible way around this has emerged from two-dimensional supernova hydrodynamic calculations by LeBlanc and Wilson[10]. In these calculations an initial stellar configuration was taken to be rotating and to possess a seed magnetic field. Upon collapse, the rotation caused the material to settle into a fairly flat disk, with a great deal of differential rotation in the interior, so that the central regions rotated around the axis much faster than the peripheral regions. This differential rotation stretched out the magnetic lines of force and wrapped them into a toroid around the spin axis. This packed the lines of force closer together, leading to a very great enhancement of the magnetic field, with a corresponding large contribution to the total pressure from magnetic pressure. This had the effect of creating magnetic bubbles on the axis of rotation, which are buoyant because of the lower gas pressure in the regions of high magnetic pressure. These bubbles were then accelerated upward by the fluid and expelled from the collapsed configuration. If the material in the bubbles should have been neutronized, then this is one way in which neutronized material can be violently expelled into space.

If the neutronized material has been compressed to temperatures as high as 10^{10} ^{o}K, then during its subsequent expansion there will be no dependence upon the history of compression. If the peak temperature reached is less than this, then the degree of neutronization will in fact depend upon the history of the compression.

The reason for this is that at 10^{10} ^{o}K, weak interactions maintain a steady-state equilibrium between the abundances of the neutrons and the protons. If the material is electron degenerate, the degenerate electrons will supress the protons, leaving the neutrons in considerable excess.

As such material expands, the nuclear reactions are sufficiently rapid to bring the material to nuclear statistical equilibrium as long as the temperature exceeds 4×10^9 ^{o}K. As the temperature falls, the most abundant nuclei in statistical equilibrium will shift from nucleons or helium up into the heavy element region. For very large neutron excesses, the equilibrium peak no longer exists at ^{56}Fe, but rather it is shifted into the neutron-rich region and toward heavier mass numbers. A typical example of such a shift would be to make the nucleus ^{78}Ni most abundant. This has a double-closed shell of neutrons and protons, and consequently represents a local region of stability which becomes favored for nuclear statistical equilibrium under conditions of high density and neutron excess. Essentially all of the protons in the medium will be concentrated in the statistical equilibrium peak.

This will produce seed nuclei upon which rapid neutron capture can commence once the temperature falls to lower values. In the presence of a very large flux of neutrons, there remains an equilibrium between neutron capture processes and (γ, n) photodisintegration processes. Initially, at temperatures near 4×10^9 $^{\circ}K$, this equilibrium remains consistent with the nuclear statistical equilibrium peak. As the temperature falls lower, the photodisintegration rates weaken compared to the neutron capture rates, so that the equilibrium shifts toward larger neutron numbers. The more distant the point of equilibrium from the valley of beta stability, the more rapid the beta decay. Following a beta decay, a new equilibrium is quickly established with a capture of several neutrons beyond the mass number at which the beta decay occurred.

In this way a capture path can be defined for this version of the r-process, which exists far on the neutron-rich side of the valley of beta stability, close to the neutron drip line. We shall be interested in how far upward in mass number the capture path can go before it is terminated by fission. If fission occurs, the fragments become seeds for the continuation of the r-process, and if the original ratio of neutrons to seed nuclei was very high, then we might expect that a great deal of fission re-cycling will take place.

In any case, once the neutrons have become depleted, the r-process products must start to beta decay toward the valley of beta stability. This is a process in which considerable abundance redistribution can take place as a result of the delayed neutrons which accompany the beta decay for very neutron-rich nuclei. In the later states of the decay the mass number distribution is frozen, except insofar as any residual capture of delayed neutrons may alter it.

The expansion time scale likely to be associated with the jet ejection in the LeBlanc-Wilson picture would be about one second or less. In that case, the r-process would have to occur on a comparable time scale, and some of the detailed calculations which we have done, which will be discussed below, are in fact consistent with such a time scale.

A second possibility for the r-process may involve much milder conditions. This might involve a kind of thermal runaway in the helium burning layer of a star which has started to produce the elements by the s-process. According to the hypothesis outlined above, mixing of hydrogen into the helium layer can lead to neutron production, and, if the mean time between neutron capture is a matter of months, then heavy nuclei can be built up in the s-process. Now suppose that the temperature rises, so that the mean time between neutron captures is reduced to seconds or less. In this case, we have a very large number of seed nuclei extending throughout the heavy element region, and if a goodly number of neutrons can be produced, then several such neutrons could be captured per seed nucleus. This would certainly produce nuclei which would ordinarily be made by the r-process; it is an open question whether it would produce the right abundance distribution of the r-process products. Another feature with which one would have to be very concerned in such a proposal is the rapid rate of energy release, and the question as to whether this can lead to explosive conditions. The rate at which the energy can be released by rising temperature seems rather less than the rate at which the material is likely to expand on a hydrodynamic time scale, and consequently it is not clear that the necessary peak temperatures could be attained under realistic conditions to produce this process.

Because of our interest in the possible production of superheavy elements, we shall return to the discussion of the r-process later, since it is only by such a process that we can expect superheavy elements to be produced in nature.

The p-Process

There are two types of nuclear reactions which can lead to the production of heavy nuclei in the neutron deficient region on the neutron deficient side of the valley of beta stability. These are (p,γ) and (γ,n) reactions. In each case, the reactions would have to occur rapidly, and hence we must naturally associate the p-process with a supernova explosion which occurs rapidly.

It is evident that increasing Coulomb barriers will make the reaction rates for proton capture on heavy nuclei considerably slower than those on medium nuclei. However, since there are no barriers against the emission of neutrons, and since neutron binding energies along the valley of beta stability are somewhat less for heavy nuclei than for medium ones, we can expect the (γ,n) reaction rates will become slightly more rapid for heavy nuclei than for medium ones at a given temperature. This suggests that the proton capture reactions may dominate the situation for medium nuclei, but that photodisintegration reactions may become more important for the heaviest nuclei. Recent calculations by Truran and myself[11] confirm this general trend. However, the dynamics of the p-process have not yet been followed in detail.

In order that the p-process can operate on typical supernova hydrodynamic time scales, the peak shock temperature to which target nuclei are exposed must be of the order of 2.5×10^9 °K or higher. It is also necessary that the density should be at least one gram per cubic centimeter, preferably higher. Such densities in a hydrogen-rich layer are likely to be found only in the lower layers of the outer hydrogen envelope in the star whichis undergoing the supernova explosion.

The seed nuclei needed in the p-process must already be in

the heavy element region. These are likely to be the s-process and r-process products which have been manufactured in previous generations of stars, expelled into space, and reincorporated into the star which is undergoing the supernova explosion. Thus the p-process involves an extra degree of re-cycling of material between the interstellar medium and stars, and hence it is not surprising that the general abundance level of the p-process products is considerably less than the abundance level associated with the s-process and r-process products.

CHEMICAL EVOLUTION OF THE GALAXY

A Game and Its Assumptions

The discussion which we have carried out so far concerns a number of processes which we have discussed in an astrophysical and nuclear physical context. It is obvious that major quantitative uncertainties exist in the majority of these processes and in the details of the astrophysical environments associated with them. For these reasons, it may seem completely unjustified to attempt a quantitative grand synthesis of these processes into one overall picture. From the fundamental point of view, this is probably the case, but nevertheless it is instructive to see what effects certain combinations of assumptions concerning these processes will have on the production of the chemical elements in the galaxy and their relative abundances in the solar system.

A study of this kind has been carried out by J.W. Truran and myself.[2] We call it a game. The object of the game is to account for a great many observed aspects of our universe, and in order to accomplish this it is permissible to adjust a considerable number of astrophysical parameters, providing we do not assume anything we do not consider reasonable. The results obtained in the game are, of course, no more reliable than the assumptions that go into it, but at least we can determine what the consequences of certain assumptions will be.

The observable quantities that we would like to fit quantitatively in this game include the relative abundances of the different products of nucleosynthesis within the solar system, the relative abundances of the natural radioactivities in the solar system at the time of its formation, the distribution of the general level of abundances of the heavier elements among the stars in the galaxy, and the present rate of supernova activity.

The assumptions that we must make obviously must include the relative contributions of various ranges of stellar mass to the production of the different processes of nucleosynthesis. We must also make assumptions about the rate of star formation throughout the past galactic history, the relative numbers of stars of different mass which are formed (the stellar birth function), and whether or not the stellar birth function changes with time. We shall discuss these various assumptions in their turn.

The basic stellar composition assumptions made by Truran and myself are shown in Fig. 1. This figure is somewhat schematic, in the sense that we only needed to define average mass fractions over certain ranges of stellar mass of the products of nucleosynthesis that are formed. The basic astrophysical and nuclear physical reasoning which led us to these assumptions has been outlined in the earlier portions of these notes. The quantitative abundances shown in the figure result from the adjustments which we have made to fit the observable characteristics of the galaxy.

In the stellar mass range up to 4 solar masses, we have assumed that white dwarf stars are formed. A great deal of mass loss takes place during the red giant phase of stellar evolution, presumably by a stellar wind process, and hence for a star of about one solar mass we have assumed that a great deal of mass is lost by stellar winds. The remaining portion of the hydrogen layer and the helium layer will presumably be lost during the stage of formation of a planetary nebula. We do not need to specify how much of the mass is lost by stellar winds in the red giant phase and how much is lost during the formation of a planetary nebula. However, it is observed that stars which have about one solar mass on the main sequence later have only about

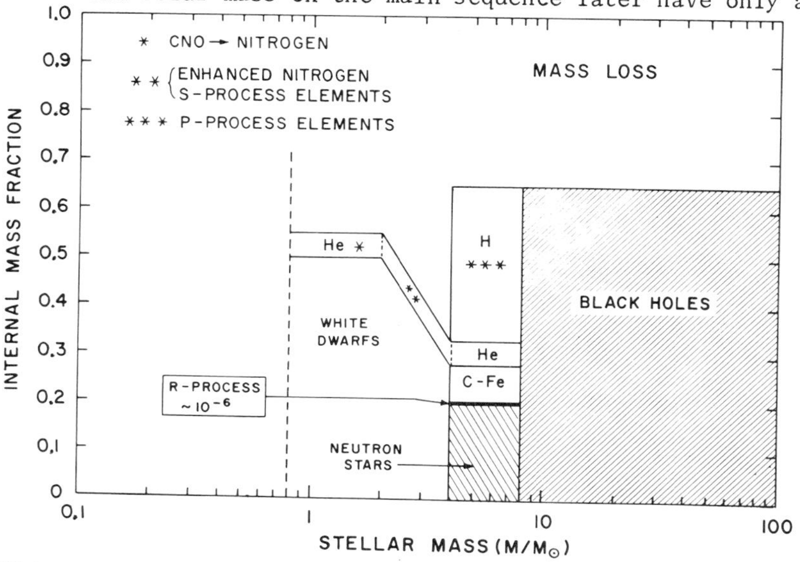

Figure 1
The adopted compositional structures for stars in their final stages of evolution are shown. The fractional stellar masses, both in the appropriate remnant and in various nuclear burning zones, are indicated as a function of main sequence mass.

0.5 solar masses after they have evolved past the red giant phase into the so-called horizontal branch position of the Hertzsprung-Russell diagram. We have been guided by this mass, which also seems to be a typical average mass for a white dwarf star, in drawing the mass of a white dwarf formed at the one solar mass position as half of the original mass of the star, as shown in Fig. 1. In the helium layer, which results from the previous operation of a hydrogen burning shell source, the original carbon, nitrogen, and oxygen in the layer will have been concentrated mostly into the form of ^{14}N by the operation of the carbon-nitrogen-oxygen bi-cycle of reactions. It is likely that in this mass range the operation of hydrogen mixing into the helium layer is very inefficient and that not much excess nitrogen would be produced as a result.

In the region from 2 to 4 solar masses, the mass fraction which we assumed to be left as a white dwarf turns down, in part because we cannot allow the Chandrasekhar mass limit to be exceeded, and in part because Paczynski has told us that the white dwarf masses should gradually approach the Chandrasekhar mass limit at about 4 solar masses. This is the region of stellar mass in which we assume that the main production of the s-process elements will occur as a result of hydrogen mixing into the helium layer, and we also assume that an enhanced production of nitrogen will occur by the $^{13}C(p,\gamma)^{14}N$.

We assume that the supernova process operates in the mass range 4 to 8 solar masses. I have discussed the various arguments which have led to the assignment of the lower mass limit for this, and at this point I can add another one: this limit is most consistent with an observed rate of about one supernova explosion per 25 years in galaxies like our own. The upper limit of 8 solar masses can be justified to a much lesser degree. In stellar evolution calculations the mass of the core seems to behave very similarly in the range 4 to 8 solar masses, and this is one possible justification for choosing the upper limit for the supernova range. However, our motivation was just as much to prevent the mass fractions of certain types of nucleosynthesis from becoming too small, on the average, in the stars undergoing supernova explosions. For example, if the upper limit to the neutron star region were raised a great deal, then the mass fraction of any one explosion contributing to the products of nucleosynthesis in the galaxy would have to be correspondingly lowered.

The mass fractions which appear to give agreement with solar system abundances include an assignment of some 20% of the mass to the neutron star remnant, an assignment of about 10^{-6} of the mass to the products of the r-process, 8% to explosive nucleosynthesis yielding the carbon-to-iron region, 5% to helium and some 32% to an outer hydrogen layer in which the p-process occurs. The remaining 35% of the mass fraction is assumed to be lost from the star by stellar winds in the red giant phase. Clearly this mass fraction is nothing more than a guess and is extremely unreliable.

Two points need comment at this stage. The mass fraction assumed to go into the r-process, 10^{-6}, is very much less than the mass fraction which was ejected in the magnetic bubbles of LeBlanc and Wilson, which was about 10^{-3}. This would thus seem to require that only a small part of the magnetic bubble can consist of neutronized material, or alternatively such ejection will only occur something like once in a thousand supernova events. The latter alternative is unattractive, because that would mean that there would be a large interval of time between the successive arrivals of fresh radioactive supernova debris at any given point in the galaxy, and it would be hard to understand why the abundances of the shorter-lived natural radioactivities found for solar system material agree so well with a more continuous rate of production, as will be shown later. On the other hand, if the r-process occurs under less extreme astrophysical environments, which was suggested as a possibility, then it may be easier to understand why such a small mass fraction should be required for the products of this process.

The second point concerns the mass fraction of the hydrogen layer which must be p-processed. We find that half of the layer shown must undergo p-processing. It might be assumed that the supernova shock wave should cause the entire hydrogen layer to be p-processed, in which case we would have assumed a too little stellar mass loss. However, our current indications are that the p-process is only effective at the higher densities in the hydrogen layer, and we believe that the processing of half the mass is physically a more reasonable picture.

Beyond 8 solar masses on the main sequence, we assume that about 35% of the mass will be lost in the red giant region, and that the remainder of the mass will form a black hole. The only contributions that this general process will make to the chemical evolution of the galaxy will be the destruction of light elements such as deuterium and ^{3}He in the outer envelope before the mass loss takes place in the red giant region.

I will only discuss the other assumptions very briefly. We must assume something about the stellar birth function. Here we take the observed stellar birth function which is characteristic of the galaxy now, and we attempt to apply this throughout as much of the past history of the galaxy as possible. This observed mass function is shown in Figure 2. One of the strong conclusions which we reach in our study is that it is impossible to satisfy all of our observational requirements if it is assumed that this mass function is unchanged throughout past time in the history of the galaxy. We therefore make an abrupt transition and assume that a first generation of stars will have a drastically different birth function, which contains no stars near one solar mass and many stars in the region of the supernova events. This alternative birth function for the first generation of stars is shown as dashed in Fig. 2.

We assume that the gas content of the galaxy decreases exponentially with time to a present level of about 5%. The age

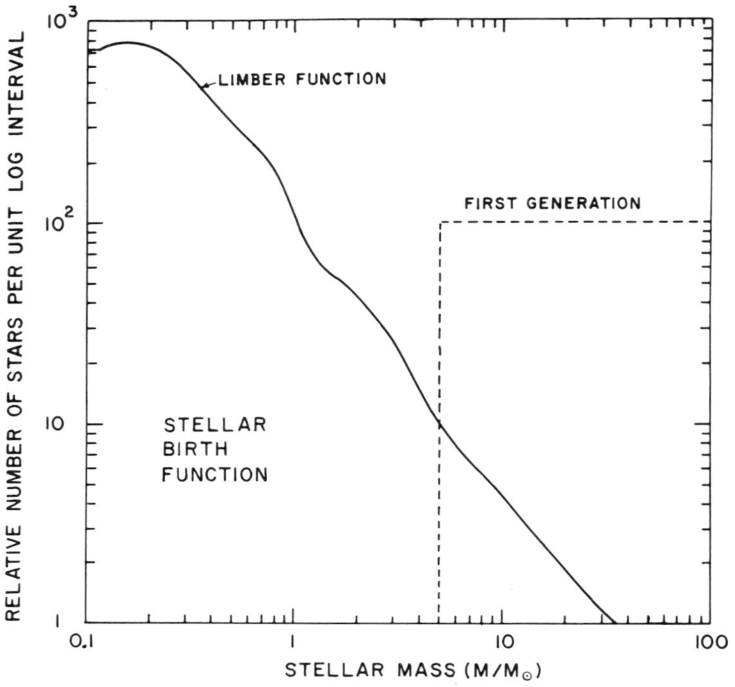

Figure 2
The adopted luminosity function. The relative number of stars per unit logarithmic mass interval is plotted as a function of main sequence mass. The dashed line indicates the stellar birth function adopted for the proposed pregalactic phase of star formation.

of the galaxy is chosen to produce agreement with the cosmochronology of the radioactive isotopes, and in our adopted model this amounts to some 12 x 10^9 years. The gas which is expelled from the stars is assumed to mix immediately and homogeneously with the interstellar medium, so that new generations of stars which are formed contain the appropriate compositions for the interstellar medium at the time they are formed. The remainder of the exercise is basically one of doing a lot of housekeeping with a computer program, forming stars out of the gas, changing the abundances in the interstellar medium as the stars evolve, keeping track of the stars which reach the end of their evolutionary lifetimes, and so forth. Such an exercise then generates a galactic history in terms of a changing star content, varied abundances of both of the stars and the interstellar medium, changing radioactive abundances, and a varying supernova rate.

Cosmochronology

With these basic assumptions established, it is possible to carry out numerical investigations of the chemical history of the galaxy. In the work described here the assumption has always been that the gas content of the galaxy decreases exponentially with time, to become 5% of the initial value at the time which is identified as the present in any given galactic cosmochronology.

It is first of interest to see what will occur if the Limber birth function is used throughout the entire history of the galaxy. The results of galactic evolution calculation in which such an assumption is made are shown in Fig. 3. In this figure, the mass fractions formed in different stellar mass ranges have been adjusted to produce an approximate equality of all the abundance ratios at a time to be associated with the formation of the solar system on the basis of cosmochronology. The ratios are very similar to the ones shown in Fig. 1. This figure shows the buildup of various products of nucleosynthesis relative to the abundances present in solar system material as

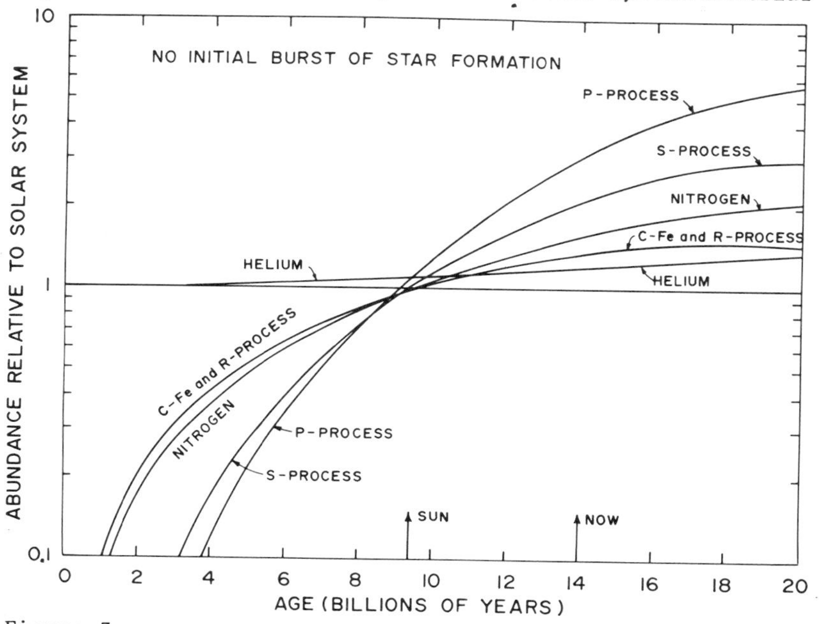

Figure 3
The abundances of nuclei formed by the various mechanisms of nucleosynthesis, relative to their solar system values, are shown as a function of galactic age when only the Limber function is used. The time of formation of the sun is specified by the concordance of the $^{235}U/^{238}U$ and $^{232}Th/^{238}U$ cosmochronological ratios.

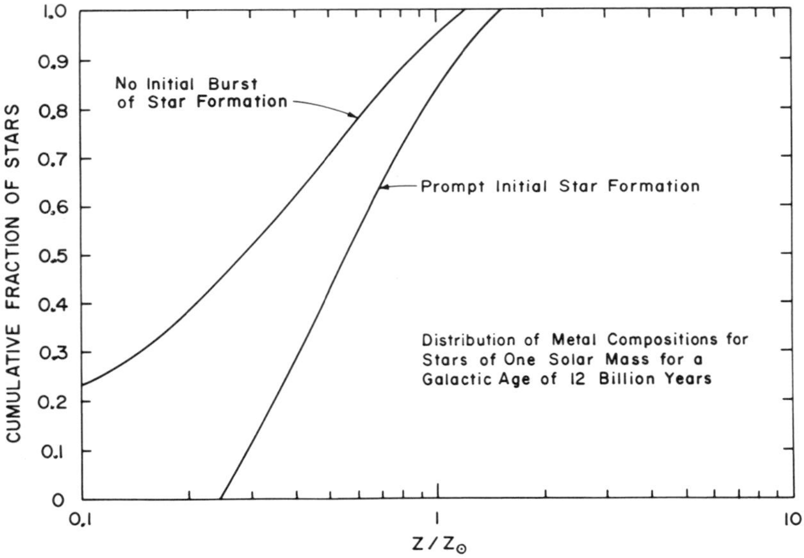

Figure 4
The distribution of metal compositions for stars of one solar mass in the galaxy is shown for models with and without prompt initial star formation.

a function of galactic age. Ideally, all the lines should pass through unity at the time associated with the formation of the sun and solar system. An approximate adjustment of assumptions has been made which very nearly brings this about. From the point of view of the buildup of the products of nucleosynthesis, this galactic history is quite satisfactory. However, it runs into great difficulties with regard to the distribution of abundances in stars of slightly less than one solar mass in the galaxy which have never evolved off the main sequence.

The distribution of heavy element abundances in such stars is shown in Fig. 4. It may be seen, for the curve marked "no initial burst of star formation" that some 23% of all the stars near one solar mass in the galaxy should have less than 10% of the solar system abundances of the heavy elements. This is clearly contrary to observation, as has been emphasized by Schmidt in 1963[12]. Stars with less than 10% of the solar system abundances of heavy elements are in fact exceedingly rare in space.

Because of this difficulty, we have gone over to a galactic history which is based upon a special stellar birth function for a first generation of stars, which is succeeded by the Limber birth function for all subsequent generations of stars. The special birth function is the first generation birth function shown in Fig. 2. This birth function starts at about 5 solar

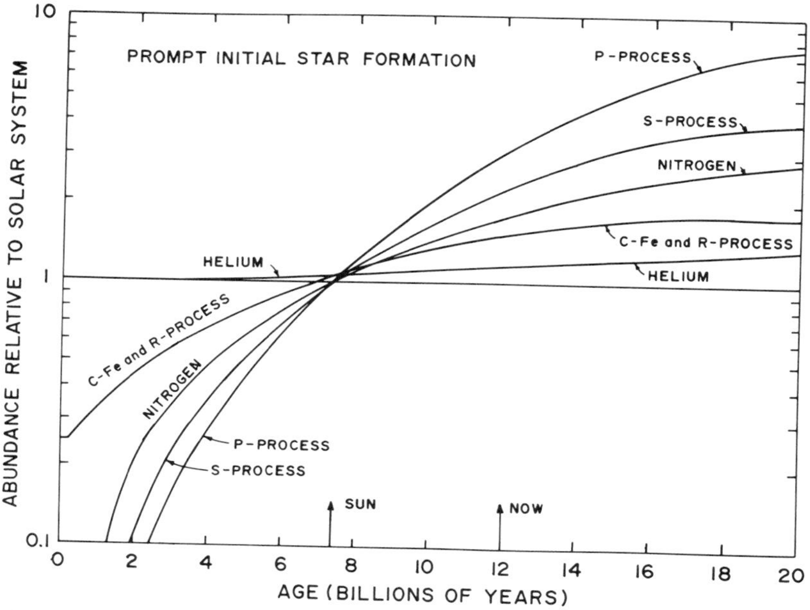

Figure 5
The abundance of nuclei formed by the various mechanisms of nucleosynthesis, relative to their solar system values, are shown as a function of galactic age for our adopted galactic model with prompt initial star formation.

masses and extends out to very large stellar masses. However, all that is really required is that there be a negligible number of stars near one solar mass formed in this birth function, and a large number of stars in the region giving rise to supernova explosions and explosive nucleosynthesis.

The results of the change in birth function after the first generation are also shown in Fig. 4. The curve marked "prompt initial star formation" results from this assumption, combined with making the initial stellar generation produce one-quarter of the solar abundances of heavy elements. The distribution of abundances shown in this curve is more in accord with the observed distribution of such abundances in stars in the galaxy. It does not include the small number of stars with very small contents of heavy elements, but such stars can be accounted for with the present approximations if it is assumed that some of them result from an inhomogeneous mixing of the first generation of abundances into the gases which give rise to stars, contrary to the assumptions of a homogeneous mixing which is inherent in the general framework of our postulates.

The buildup of the products of nucleosynthesis in the resulting galactic history is shown in Fig. 5. This bears a general resemblance to Fig. 3, but it may be noted that the cosmochronology now favors the formation of the sun at a galactic age of about 7.4 billion years, which makes the present occur at a galactic age of 12 billion years. It may be noted that this galactic model predicts a moderate buildup of secondary and tertiary products of nucleosynthesis in the interstellar medium, but not by very large factors. The primary products, such as the carbon-to-iron region, are produced in only very small overabundances in the interstellar medium in the interval between the time of formation of the sun and the present, so that not much general change in abundance levels can be expected. Of course, inhomogeneities in mixing produce larger abundance fluctuations in actual stars.

The relative abundances of long-lived radioactive elements are shown in Fig. 6. In this figure, the ratios of $^{232}Th/^{238}U$, $^{244}Pu/^{238}U$, $^{235}U/^{238}U$, and $^{129}I/^{127}I$ are shown. The actual cosmochronology which has been adopted in the preceding discussion is based upon the fact that the relative isotopic ratios pass through the original solar system values at a time of 7.4 x 10^9 years, for the abundances of the uranium and thorium isotopes.

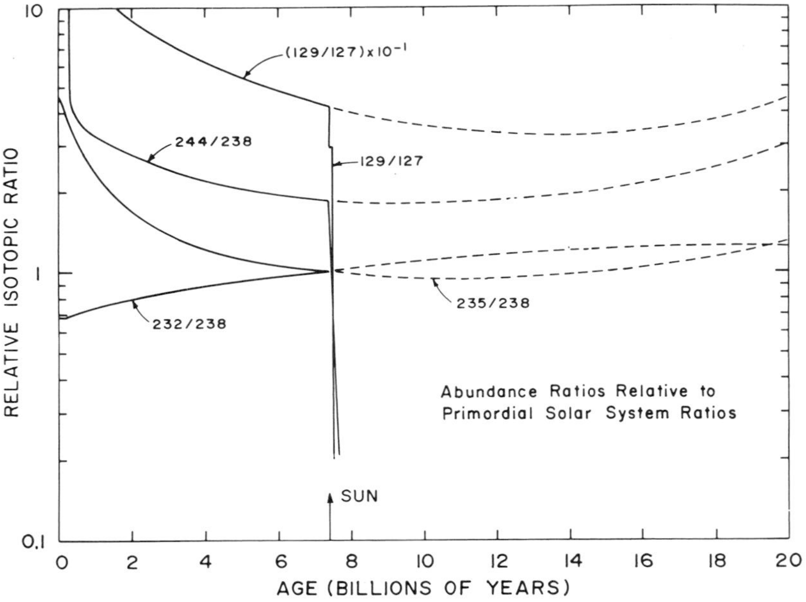

Figure 6
Radioactive ratios of cosmochronological importance, relative to the primordial solar system ratios, are shown as a function of galactic age. The solid lines show the production termination at time of formation of the solar system; the dashed lines show the subsequent ratios in the interstellar medium.

The abundance ratios of the shorter-lived ^{244}Pu (82 million years half-life), ^{129}I (17 million years half-life) are overabundant at this time. The decay of these radioactivities establishes some sort of formation interval during which the material that is going to form the solar system becomes isolated from the processes of galactic nucleosynthesis, and during this interval the shorter-lived radioactivities decay until the eventual xenon decay products are trapped and preserved in meteorites.

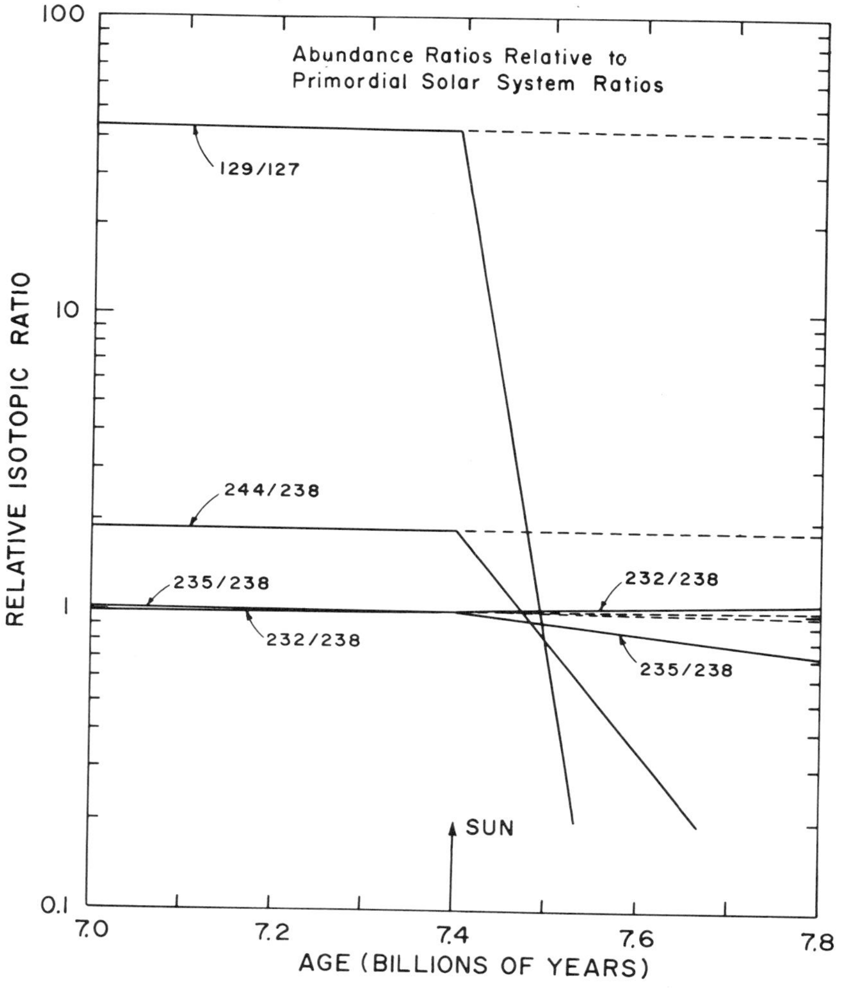

Figure 7
The region of concordance of the cosmochronological ratios is shown. An isolation time of approximately 100 million years is consistent with both the $^{129}I/^{127}I$ and the $^{244}Pu/^{238}U$ abundance ratios.

The cosmochronological situation is shown in greater detail around the time of formation of the solar system in Fig. 7. Here the artificial assumption has been made that galactic nucleosynthesis has been terminated at the time of formation of the sun. One would then expect that all of the relative isotopic ratios should be unity at some time shortly thereafter. It may be seen that this is approximately the case, but the worst discrepancy may amount to a slight overabundance of ^{129}I. Thus the formation interval to be associated with the presence of extinct radioactivities in meteorites is of the order of 80 million years. This is an interval in which the continuous galactic production rate of such radioactivities has been cut off and the radioactivities are able to decay freely before the product can be preserved in the meteorites. This principle will be applied later when we discuss the possible evidence for the preservation of the decay products of extinct radioactivities from the superheavy region in meteorites.

The Light Elements

The light elements, deuterium, ^3He, lithium, beryllium and boron, are all destroyed in stellar interiors. Therefore it is not surprising that they do not have a large abundance in nature. However, they do have a finite abundance, and it is a challenge to the theory of nucleosynthesis to account for this. I discuss very briefly the predictions made by the current model on the abundances of these nuclei.

I do not know of any way in which the deuterium and ^3He in solar system material can have been produced by cosmic ray activity, as I shall presently invoke for the lithium, beryllium, and boron. The most probable source of these two light nuclides is cosmological nucleosynthesis itself, which has been responsible for the bulk of the helium synthesis in our model. However, the synthesis of deuterium and ^3He is only possible within the context of an open universe, subject to no cosmological pathologies such as high Fermi level of the neutrinos or antineutrinos, or an expansion rate for the universe which is orders of magnitude different from that predicted by the Friedman models. If we indeed live in an open universe having a mean density equivalent to that which one would deduce from the matter which is seen in the form of galaxies in our universe, then the results shown in Fig. 8 are obtained. In this figure, there is an initial destruction of deuterium and ^3He which is caused by the assumptions about our first generation of stars, in which some 25% of the solar abundances of the products of explosive nucleosynthesis are made. Following this, the deuterium is destroyed by matter in the interstellar medium passing through stellar envelopes, where it will have been subjected to temperatures in excess of 10^6 $^{\circ}K$, and is then ejected by stellar winds. The only nuclear reaction of any importance in the interval 1 to 10 million degrees is the destruction of deuterium. Above 10 million degrees, the

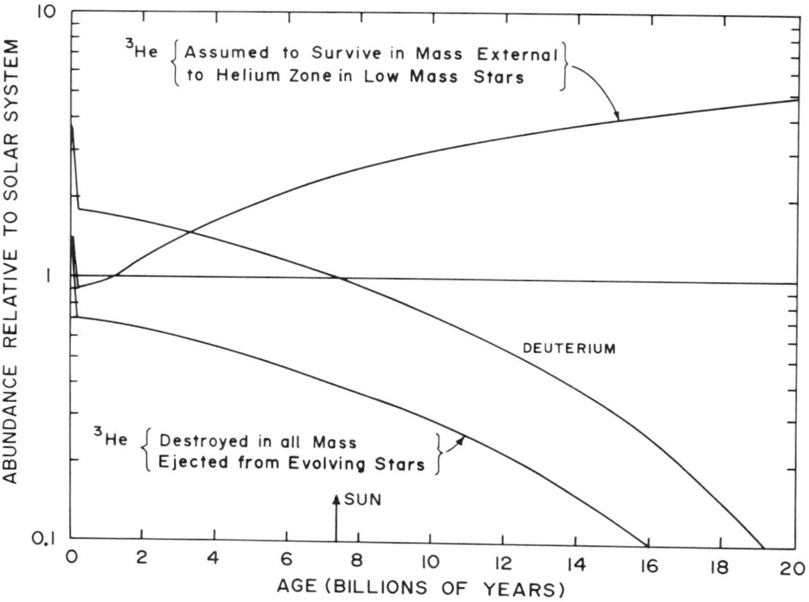

Figure 8
The abundances of ^{2}D and ^{3}He in the interstellar medium relative to solar system matter, are plotted as a function of galactic age. The two curves for ^{3}He correspond to limiting assumptions concerning its destruction in stellar envelopes.

^{3}He is also destroyed. It may be seen that at the time of formation of the sun the deuterium in the interstellar medium has been destroyed to the extent of about 50% of the initial abundance in the galaxy following the first generation of stars. This is indicative of the fact that about half of the interstellar medium at the time of formation of the solar system had been circulated through stellar interiors. It may be seen that an additional 50% of the remaining material has been circulated through stellar interiors since the sun has been formed, to judge by the continued lowering of the deuterium level as shown in Fig. 8.

The ^{3}He may be depleted or enhanced in the interstellar medium depending upon the behavior of the storage mechanism for the ^{3}He in stars. The destruction of deuterium produces ^{3}He, and if it is assumed that this nuclide is preserved in stellar envelopes at temperatures below 10 million degrees up until the time such material in stellar envelopes is lost in stellar winds, then there is a net enhancement of the ^{3}He in the galaxy with increasing age. On the other hand, if there should happen to be circulation currents in stars which will bring surface material through regions with temperatures in excess of 10 million degrees during the lifetime of stars on the main sequence or thereafter, then there is a net destruction of ^3He in stellar interiors and a depletion of it in the interstellar medium which resembles the depletion of deuterium. The truth presumably lies between the two types of behavior, and indeed there can also be an enhancement of the ^3He abundance in the interstellar medium as a result of ejection from stars which have formed some of it by the basic proton-proton reaction in the outer parts of the envelope. For our present purposes it is sufficient to note that the abundance of ^3He does not necessarily constitute a problem in nucleosynthesis, but it will require detailed consideration of various astrophysical mechanisms to determine what the present abundance level should be.

It has recently been suggested by Reeves, Fowler, and Hoyle[13] that the lithium, beryllium, and boron in the solar system may be produced by cosmic ray spallation on carbon, oxygen, nitrogen and neon in the interstellar medium throughout past galactic history. It is possible to test this prediction within the context of the present model. The abundances of the target elements are predicted as a function of time and we can assume that the fluxes of cosmic rays throughout past galactic history have been proportional to the supernova rate throughout this history. Its rate is shown in Fig. 9. Also involved are calculations of the spallation cross sections for production of the isotopes of these three light elements, based partly upon measurements of the cross sections made mainly at Orsay in France, and partly upon the extrapolation (demodulation) of the cosmic ray flux in the solar system to interstellar space. This demodulation calculation mainly produces uncertainties in the flux of low energy cosmic rays, and it may affect the predictions concerning the relative abundances of beryllium and boron relative to lithium.

In addition, it is not to be expected that the cosmic ray spallation processes will produce sufficient ^7Li. The measured spallation rates do not enhance ^7Li relative to ^6Li by nearly large enough a factor. On the other hand, the process of mixing of hydrogen into a helium layer by helium shell flash processes which may have been responsible for the s-process, can also bring ^3He down to high enough temperature levels where it can burn with ^4He to produce ^7Be. If a general outer convective zone is produced shortly after this event, as Fowler and I have suggested, then the ^7Be will be transported, in part, near the surface before electron capture takes place. Material may then be preserved in the surface layers as ^7Li for a sufficient length of time to be present in these layers when they are ejected in stellar winds.

The predictions of the model are shown in Fig. 10. The abundance of ^7Li shown in this figure is that which is a normalized value, and hence is not a prediction of the model. It is, however, quite reasonable in terms of the suggested process. The other curves are predicted values. The abundance of ^6Li

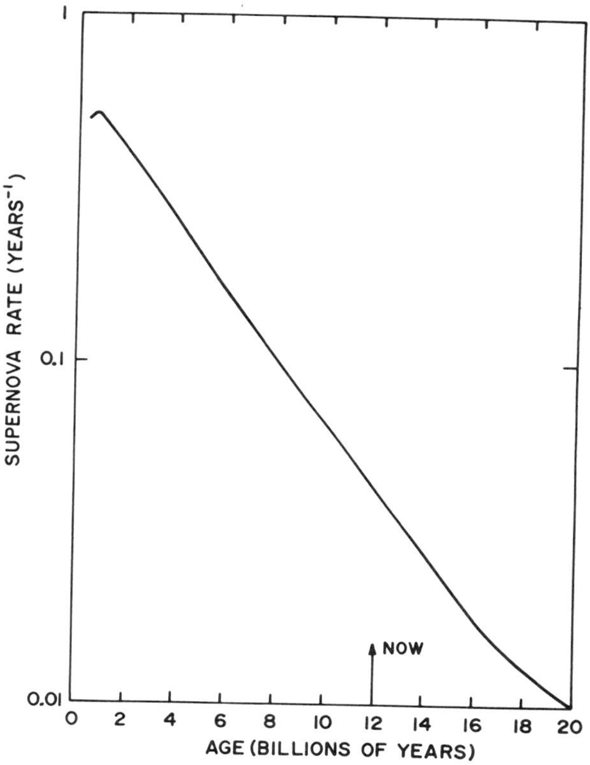

Figure 9
The rate of occurrence of supernovae (events per year) in the galaxy is plotted as a function of galactic age.

exceeds the observed one by only a factor of 1.2, which is much better agreement than can be expected from such considerations. There are two curves representing the sum of the abundances of the beryllium and boron isotopes. These curves are analogous to those shown for the preservation or destruction of ^{3}He in Fig. 8. One curve corresponds to preserving these isotopes in outer stellar envelopes, and the other to destruction of them. It does not make a great deal of difference which is chosen, because in each case the predicted production for solar system material exceeds that observed by about a factor of 3. Considering the various uncertainties in the parameters which enter into this calculation, this cannot be considered as disagreement between observation and theory. Thus the general indication is that these abundances are essentially correctly predicted by the process of cosmic ray spallation in the interstellar medium.

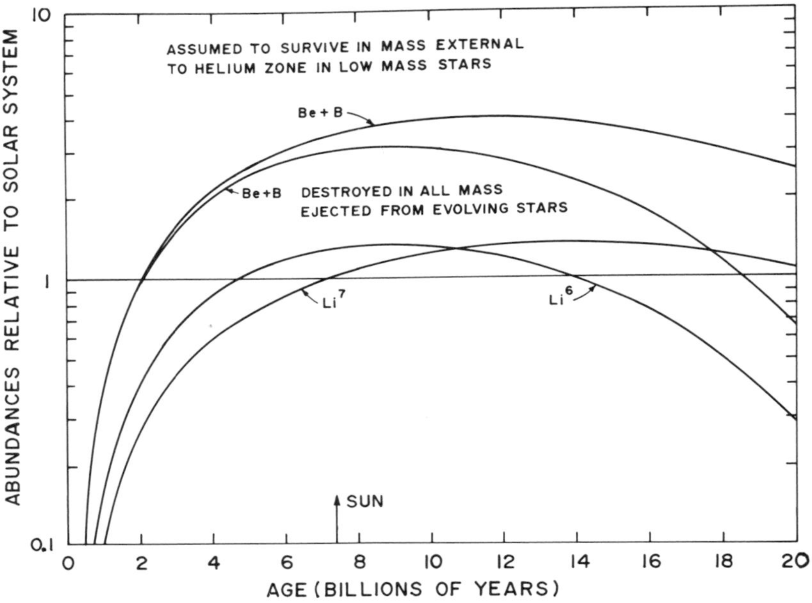

Figure 10
The abundance of ^{6}Li, ^{7}Li and the isotopes of beryllium and boron in the interstellar medium relative to solar system matter, are plotted as a function of galactic age. The two curves for Be + B correspond to limiting assumptions concerning its destruction in stellar envelopes.

THE NUCLEAR MASS FORMULA

In the remainder of these notes we shall concentrate on the problems associated with the r-process and with the manufacture of superheavy elements in nature by the r-process. The dynamics of the r-process depends very much upon the properties of nuclear matter in the neutron-rich region far away from the valley of beta stability. The properties of the nuclear mass formula which is usually derived to obtain masses along the valley of beta stability, are poorly determined for extrapolation far away from the valley of beta stability. Therefore J.W. Truran and I have spent some time during the last year or two attempting to determine more refined criteria which will improve the extrapolations away from the valley of beta stability of the mass formula. More recently, such extrapolations have also become of considerable interest for the prediction of the properties of nuclei near the base of the outer crusts of neutron stars, in neutron-rich environments approaching normal nuclear densities. Langer, Rosen Cohen, and Cameron[14], and also Bethe, Borner and Sato[15], determined the expected composition of such neutron star crusts based upon the use of an extremely simple basic form of mass formula as originally introduced by von Weizsacker. More recently, Baym, Bethe, and Pethick[16] have revised the mass formula and re-discussed the problem, also taking into account bulk properties of the matter in neutron star crusts such as the lattice Coulomb energy. Their results are very different from those which were obtained using the simple method by the first mentioned groups of authors. Thus an improved mass formula designed for validity in the neutron-rich region is of considerable interest both for r-process nucleosynthesis and for the structure and properties of neutron stars.

The basic form of the mass formula is as follows:

$$(M-A) = 8.07144\ A - 0.78245\ Z + E_{vol} + E_{surf} + E_{coul} + E_{wig} + S + P \qquad (1)$$

where (M-A) is the mass excess in MeV, the first two terms on the right hand side are the mass excess of the neutrons and protons, E_{vol} is the volume energy, E_{surf} is the surface energy, E_{coul} is the Coulomb energy, E_{wig} is the Wigner energy, S is the shell correction energy, and P is the pairing correction energy.

The volume energy can be written in the following form

$$E_{vol} = \beta_0 A + \beta_2 I^2 A + \beta_4 I^4 A + \beta_6 I^6 A \qquad (2)$$

where the coefficients multiply ascending even powers of the neutron excess I, defined by

$$I = \frac{A - 2Z}{A} \quad . \qquad (3)$$

Of the terms in this energy, the first, $\beta_0 A$ is the normal volume energy in the mass formula, and the second, $\beta_2 I^2 A$ is the normal volume symmetry energy. The higher terms are included in order to take into account the departure from a parabolic form of the volume energy with increasing asymmetry between the neutrons and protons.

The surface energy is given as follows:

$$E_{surf} = \gamma_0 A^{2/3} + \gamma_2 I^2 A^{2/3} \quad \gamma_4 I^4 A^{2/3} + E_{curv} \quad . \qquad (4)$$

Of the terms in this expression, the first $\gamma_0 A^{2/3}$ is the normal surface energy of the nucleus, and the second, $\gamma_2 I^2 A^{2/3}$, is the surface symmetry energy which has often been introduced into mass formulae in the last few years. The third term on the right hand side is an additional symmetry correction, and the last term E_{curv}, is a surface curvature energy which is given by the following expression:

$$E_{curv} = \alpha_c A^{1/3} \qquad (5)$$

where the coefficient α_c represents the sum of any energy terms which have an $A^{1/3}$ dependence.

The Coulomb energy consists of the following terms:

$$E_{coul} = \frac{0.864}{r_a}\frac{Z^2}{A^{1/3}} - \frac{0.864}{r_a}\left[\frac{5}{6}\frac{(1.49)^2}{r_a^2}\right]\frac{Z^2}{A} - \frac{0.660}{r_a}\frac{Z^{4/3}}{A^{1/3}} - \frac{1}{4}\frac{A^2I^2}{Z^2}\frac{E_{coul}^2}{E_{sym}}\frac{f(1-f^{2/3})}{(1-f)} \quad . \qquad (6)$$

The first term is the basic Coulomb energy which is applicable to a charged distribution having a sharp outer edge. The second term on the right hand side is a correction to this energy for the diffuseness of the surface. The third term is the Coulomb exchange energy, which takes account of the fact that two protons are unlikely to approach each other at very short distances, and the fourth term is a redistribution term which allows for the change in the Coulomb energy resulting from the Coulomb expansion of the proton fluid inside the nucleus, which is opposed by the tendency of the nuclear symmetry energy to keep the neutron and proton numbers as closely equal as possible in the interior.

In these expressions the term r_a is the mean particle spacing in the interior of the nucleus; we write r_a and f in the form

$$r_a = r_o(1+I)^{1/3} \qquad (7)$$

$$f = \frac{r_aA^{1/3}-1.5}{r_aA^{1/3}} \quad . \qquad (8)$$

In expression (7) r_o is taken to be a basic particle spacing for the deep interior of the nucleus, for normal nuclear matter in which A = 2Z. Such matter has a density of about 4×10^{14} grams per cubic centimeter. However, in the limit of the pure neutron gas, the neutrons have about their partial nuclear density in normal nuclear matter at which density the nuclear forces between the neutrons switch over from being basically attractive to being basically repulsive. The corresponding density at which this happens is about 2×10^{14} grams per cubic centimeter. The expression shown in Eq. (7) makes allowance for the decreasing density of the nuclear matter as the nucleus becomes more neutron-rich. This has very little effect on nuclei near the valley of beta stability, but is a feature which must be included to obtain reasonable results in the limit of a pure neutron gas.

The fourth term on the right hand side of Eq. (6) is the Coulomb redistribution energy. It should be noted that the entire expression for E_{coul} appears in this expression, so that in fact one should iterate to obtain a proper value for the total expression. Note also that the symmetry energy terms appear in this, as defined in Eq. (9):

$$E_{sym} = \beta^2I^2A + \beta^4I^4A + \beta^6I^6A + \gamma_2I^2A^{2/3} + \gamma_4I^4A^{2/3}. \qquad (9)$$

The term E_{wig} which appears in Eq. (1) is an empirical correction term introduced to improve the fit of the mass formula to the lighter nuclei. In much of our work we have adopted an expression given by Myers and Swiatecki for this term as follows:

$$E_{wig} = \alpha_w \exp(-6|I|) \tag{10}$$

where α_w is a coefficient to be specified.

We write the shell correction function S as the sum of two independent corrections which are separate functions of the proton and neutron numbers:

$$S = S(Z) + S(N) \ . \tag{11}$$

It may be noted that it is always necessary to specify one among the set of shell corrections, since any arbitrary number may in principle be subtracted from S(Z) and added to S(N). The shell corrections are in practice determined by subtracting from measured nuclear masses the fitted reference mass formula, and then fitting the residuals by the expression of Eq. (11), in which the values are necessarily over-determined by the experimental data. Thus a least-squares fit is required to obtain the values of the separate shell correction functions for the protons and the neutrons. The pairing correction P is intended to take into account the odd-even oscillation of the nuclear mass surface. The basic mass formula is defined as being applicable to nuclei having odd numbers of neutrons and odd numbers of protons. If either or both of the neutrons and protons become even in number, then the mass surface is depressed with respect to that for odd-odd nuclei. Thus the pairing correction is written as the sum of corrections for the proton and neutron numbers, as shown in Eq. (12),

$$P = P(Z) + P(N) \tag{12}$$

where P(Z) and P(N) take on non-zero negative values only for even values of Z and N.

It must be remembered that the nuclear masses, as expressed by the mass formula, are small differences between very large energy terms. It is traditional to fit nuclear masses by least-squares adjustments of the coefficients of the major terms. It is also important to realize that the reference mass formula will be fitted to nuclear masses without the inclusion of the shell and pairing correction terms. Under these circumstances, most nuclei will have large shell-induced departures from the reference mass formula, so that the least-squares adjustment of the coefficients in the mass formula merely gives the line that runs through some average trend of a function with considerable fluctuations in it. It is possible to add all sorts of terms to the reference mass formula and still to obtain about as good a least-squares fit, as judged by mean square deviations from the reference formula, as in just a standard simple case. Under these circumstances it is clear that many subsidiary considerations must be introduced to justify the goodness of fit not only of the basic form of the mass formula but also for the values of any secondary terms which are introduced into it, of which there are a great many in Eqs. (1), (2), (4), and (6). The coefficients which we determine by least-square fit to masses include β_0, β_2, γ_0, and r_0. All the other terms, with the exception of S and P which are fitted to mass excess residuals must be justified on the basis of physical reasoning and subsidiary tests of the goodness of fit of a postulated reference mass formula to the experimental masses.

One subsidiary condition which can be imposed is that the mass formula should yield the proper energy for a neutron gas in the limit where I becomes unity. For such an energy, we may take the volume energy per nucleon of a pure neutron gas as calculated by many-body techniques by Siemens[17]. This gives rise to Eq. (13),

$$\beta_0 + \beta_2 + \beta_4 + \beta_6 = 13 \text{ MeV} \qquad (13)$$

which is the applicable form of Eq. (2) in the neutron gas limit. The surface energy is normally positive, expressing the decrease in binding energy of a nucleon in the surface layer of the nucleus compared to the deep interior. If we imagine a lump of pure neutrons, there would still be attractive forces in the interior even though, as shown by Eq. (13), the volume binding energy is positive. Thus there would remain a residual, though artificial, positive surface energy in the neutron gas limit for a hypothetical clump of neutrons, and this also can be estimated on the basis of the Siemens calculations for a pure neutron gas, taking into account the depth of the attractive potential in that limit compared to ordinary nuclear matter. This leads to Eq. (14)

$$\gamma_0 + \gamma_2 + \gamma_4 = 11 \text{ MeV} \qquad (14)$$

in which the sum of the coefficients should equal 11 MeV in the limit of a pure neutron gas.

The second volume symmetry energy coefficient, β_4, has been estimated on the basis of a Thomas-Fermi calculation by Myers and Swiatecki. We have generally used values for this coefficient about equal to their estimate of - 25 MeV. A similar magnitude for this term and the indicated sign has been advocated for many years by myself, in order to account for the relatively slow rate of decrease of neutron binding energies for heavy nuclei.

One of the major requirements that Truran and I have imposed upon the mass formula is that the individual shell correction functions S(Z) and S(N) should be well-behaved functions. The rationale for this is that their sum S, fluctuates about zero throughout the table of masses as one goes from light nuclei to heavy ones. We require that each of its components should do

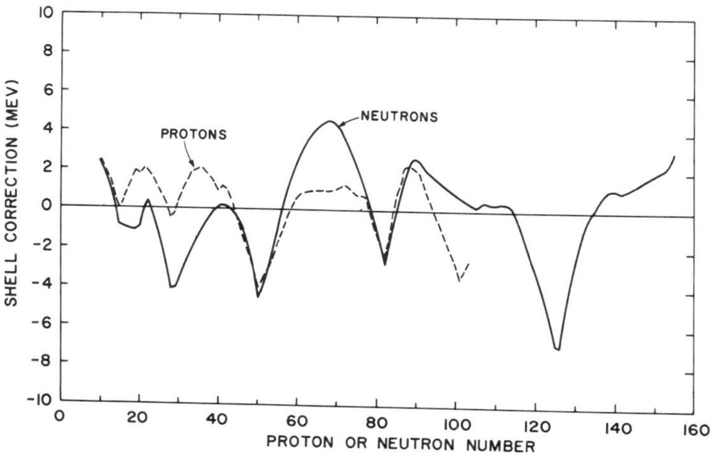

Figure 11
Proton and neutron shell corrections

likewise. Any failure of this requirement would lead to untrustworthy shell correction functions for nuclei far away from the valley of beta stability. These conditions impose strong constraints on the surface symmetry energy γ_2, and upon the coefficient of the Wigner energy, α_w. A detailed demonstration of this conclusion may be found in our discussion of the mass formula in the Proceedings of the Leysin Conference[18]. What appears to be required is that $\gamma_2 \simeq -100$ MeV and $\alpha_w \simeq -15$ MeV.

A reasonably well-behaved set of shell correction functions is shown in Fig. 11. Note that each curve has an oscillation about its mean value. Note also the energy scale on the ordinate.

If the Wigner energy coefficient is set equal to zero, the shell correction terms deteriorate badly as shown in Fig. 12. The ordinate scale here is three times that of Fig. 11.

If the surface symmetry energy coefficient γ_2 is set equal to zero, a much worse deterioration in the shell correction terms occurs, as shown in Fig. 13.

Another major check which can be made is that the nuclear radius parameter r_o should be approximately equal to the value measured by Stanford electron scattering, 1.03 fermis. However, a straightforward fit to the masses gives a value of r_o of about 1.07 or 1.08 fermis. One of our major concerns has been to resolve this descrepancy.

We have found one way to change the fitted value of r_o very considerably with very little effect upon the good behavior of the shell correction functions. This is to adjust the coefficient of the surface curvature energy α_c. This coefficient was estimated by Myers and Swiatecki from a Thomas-Fermi calculation of surface properties; they found it to be about 7 MeV. Since

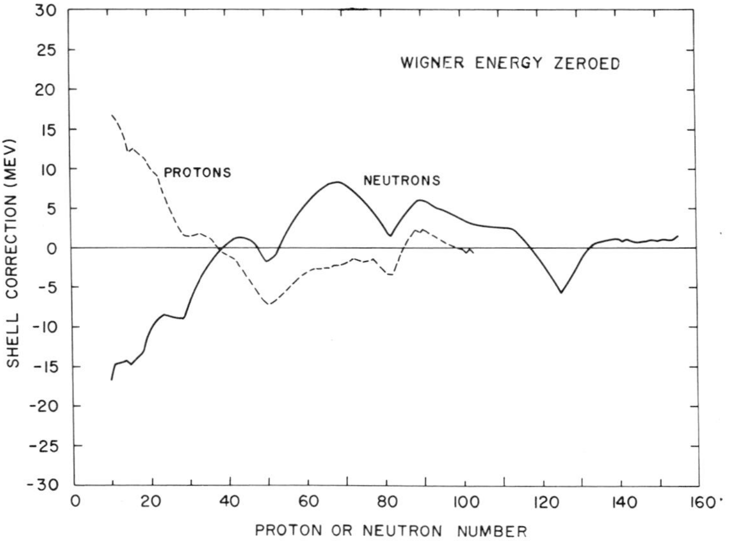

Figure 12
Proton and neutron shell corrections calculated for the choice: $\alpha_w = 0$.

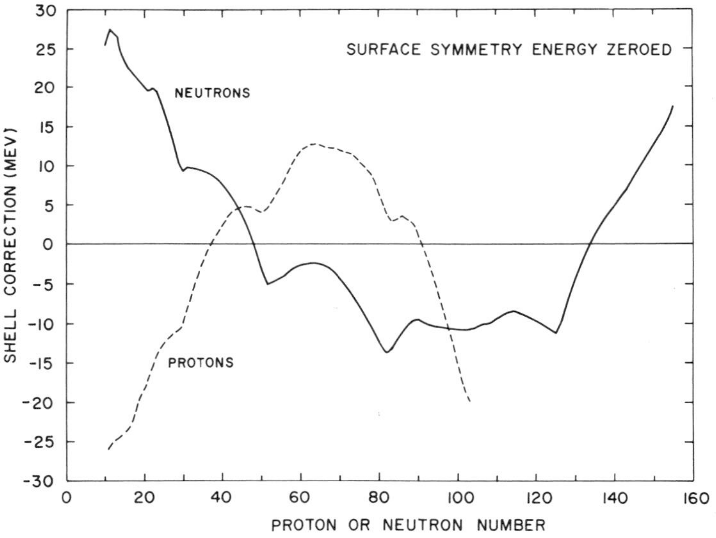

Figure 13
Proton and neutron shell corrections calculated for the choice: $\gamma_2 = 0$.

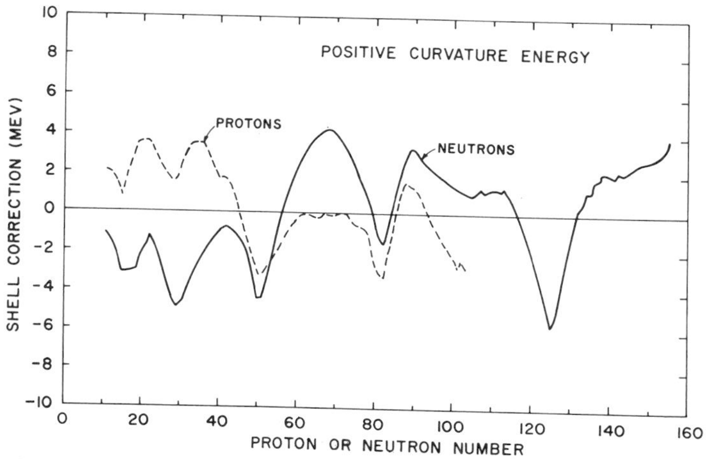

Figure 14
Proton and neutron shell corrections calculated for the choice: $\alpha_c = +10$ MeV.

this value was a difference between two larger terms, its value cannot be regarded as very well established. In addition, the difference between the Fermi energy in a real nuclear potential with shell effects and between the Fermi surface energy of neutron and proton gases contained within a spherical box, is an energy which may vary like $A^{1/3}$. As such, it could appear as an energy term having the properties of a surface curvature energy, but which would change the values of the surface curvature energy coefficient considerably. Thus we have maintained that we lack a very good prediction for the value of α_c.

The unit radius r_0 can be reduced to about the Stanford value if one takes $\alpha_c \simeq -20$ MeV. Compared to a fit in which $\alpha_c = 0$, all of the mass adjusted coefficients are larger. However, there is little effect on the shell correction functions, as may be noted from Fig. 14, with $\alpha_c = 10$ MeV, or 20 MeV higher than used in Fig. 11. However, there are profound effects upon the fissionability parameter of the nucleus. The fissionability parameter of a nucleus, which helps to determine the height of the fission barrier, is given by

$$x = \frac{E_{coul}}{2\ E_{surf}} \cdot \qquad (15)$$

In order to account for the measured fission barrier heights of nuclei in the vicinity of uranium, such nuclei should have values of the fissionability parameter of about $x \simeq 0.8$. However, with $\alpha_c \simeq -20$ MeV, it turns out that $x \simeq 0.6$. This would predict that the fission barrier heights for nuclei near uranium are far too high. Hence the choice of a negative surface energy coefficient as a technique for depressing the value of r_0 must be rejected on the basis that it leads to impossible fission parameters for nuclei. We prefer a value of α_c near the Myers-Swiatecki value of 7 MeV, which gives the value $x \approx 0.8$ for nuclei in the uranium region.

This leads to an apparent paradox. How do we reconcile the large value of r_0 with the considerably smaller value indicated by Stanford electron scattering measurements? We suggest the difference is more apparent than real, and may be indicative of a departure from a uniform proton density distribution in the nuclear interior. Such departures are suggested by typical Hartree-Fock calculations for the distribution of the densities of nucleons in the nuclear interior. Thus, if the protons are somewhat more concentrated toward the surface of a nucleus than would be indicated by a constant density, the Coulomb energy would be lower, and this would increase the characteristic radius r_0 in a fit to the Coulomb energy. For example, suppose that ρ_c is the density of protons in the nuclear interior, and ρ_s is the density of protons nearer the nuclear surface, for a sharp-edged nucleus of radius R and a dividing line between ρ_c and ρ_s at R_c. Then setting $h = \rho_s/\rho_c$ and $g = R_c/R$, we obtain

$$\frac{1}{r_0'} = \frac{1}{r_0} \left[\frac{h^2 - (h^2 - 1)\ g^5}{(1-h)^2 g^6 + h^2 + 2h\ (1-j)\ g^3} \right] \tag{16}$$

It may be noted that the quantities g and h may be adjusted in various ways to make the real radius parameter, r_0', differ from the apparent one obtained by mass fitting, r_0, by several per cent. We prefer this interpretation of the radius difficulty.

One other test which can be applied to the goodness of fit is the curvature of the valley of beta stability. This involves the second derivative of the reference mass surface with respect to the nuclear charge number Z. This test can be made by subtracting from the nuclear masses the fitted shell correction and pairing functions and then comparing the differences between the corrected masses for fixed mass number with the similar differences in the reference mass formula. We have found that discrepancies can be reduced if we replace the ordinary symmetry parameter I by a more elaborate expression given by Blatt and Weisskopf for the symmetry parameter:

$$I^2 \rightarrow I^2 + \frac{8|I|}{A} + \frac{32}{A^2} \tag{17}$$

This is a substitution for I^2 which should be made in all symmetry energy terms. In principle, this substitution should accomplish much the same effect as the original Wigner term, and we originally introduced it in place of the inclusion of the Wigner term. However, we found that we are unable to obtain properly behaving shell correction functions without the form of the Wigner term given by Eq. (10).

This is a progress report on the state of our investigations of the nuclear mass formula. We have not settled on a final form of it yet, and the final one that we do use will also be tested against its ability to give reasonable values for parameters in a nuclear level density formula and in a nuclear radiation width formula. However, it is very likely that we shall utilize most of the experience described in this section in making this final adjustment.

THE R-PROCESS

Probably the major mystery associated with the processes of nucleosynthesis is the site of the r-process. It is possible that several environments are involved. It is clear that if there is a rapid thermonuclear production of neutrons, then any existing heavy nuclei in that environment will be able to capture several neutrons, and the result will be typical products normally ascribed to the r-process. Many r-process nuclei may indeed be formed in that fashion. However, it is clear that the most prominent r-process features cannot be formed in this almost casual way.

There are prominent abundance peaks at mass numbers 130 and 195 which are due to the r-process. In these abundance peaks, there is a great diminution of the usual odd-even fluctuations in abundances. These fluctuations arise in a neutron capture process owing to the larger neutron capture cross sections that are ordinarily associated with nuclei with odd mass numbers compared to those which have even numbers of both neutrons and protons. It will also be true that under conditions of partial nuclear statistical equilibrium the even nuclei will be more abundant than the odd nuclei, often by much larger factors than in the case of the fluctuations in neutron capture cross sections. In a neutron capture chain, a steady flow is approached in which each of the successive members of the chain become subject to the condition that the product of cross section times abundance is roughly constant. This leads to abundance inversely proportional to the cross section.

Since the odd-even abundance fluctuations are greatly reduced in the vicinity of the r-process peaks at mass numbers 130 and 195, it is evident that a new condition has been responsible for the establishment of the actual abundances in these peaks. There is also a region of r-process production in the vicinity of mass number 164 where the odd-even fluctuations are somewhat reduced, although the abundance peak there is not nearly as pronounced as at 130 and 195. The only way in which we have found it possible to reduce the odd-even fluctuations is to subject nuclei to such a large neutron flux that the neutron capture path for the products lies far to the neutron-rich side of the valley of beta stability, so that the beta decay energies of the product nuclei are very high, and so that neutrons will be emitted following such beta decays. The emission of such delayed neutrons tends to smear over odd-even fluctuations in abundances. This provides a clue that extremely large neutron fluxes are needed, so that any given heavy seed nucleus subjected to these fluxes must capture a large number of neutrons during the r-process that produces the abundance peaks at mass numbers 130 and 195.

Such extreme neutron fluxes may arise if material is compressed to extremely high densities. This compression takes place if one supposes that material is compressed to about nuclear densities, whereupon the high Fermi energies of the electrons will convert most of the protons in the system into neutrons. This process is also assisted by the very high temperatures that will be produced in the system by such compression, which will break up nuclei and lower the threshold for electron capture on the resulting protons to the point where the capture can take place extremely rapidly. If such material is subsequently ejected by a hydrodynamic process, then an environment suitable for the r-process would certainly be produced. The difficulty lies in finding the precise conditions in which the compression and re-expansion will take place, and to find the conditions taking place in a fraction of the stellar material that is consistent with the approximately 10^{-6} of the stellar mass in the supernova event which produces the r-process.

Let us put aside our concern with the astrophysical location of such a process, and consider the nuclear physics of it. The environment is taken to be at a very high temperature, but also at a sufficiently high density so that the electrons are degenerate. Under these conditions, at temperatures in excess of 10^{10} °K, an equilibrium is established between the neutron and proton abundances by weak interactions. Bombardment of these nucleons by electrons and positrons interconverts the neutrons and protons, with appropriate emission of neutrinos and antineutrinos. If the electrons were not degenerate, so that electrons and positrons were present in equilibrium with the radiation in nearly equal numbers, then the protons would be produced in excess over the neutrons owing to the smaller mass of the proton. However, because of the electron degeneracy, positrons are suppressed relative to electrons in the equilibrium with the radiation field, and neutrons will be in considerable excess over protons. As the material expands on a hydrodynamic time scale of a fraction of a second, the weak interactions will freeze out at a temperature of about 10^{10} °K, leaving the neutrons in excess over the protons.

The resulting nuclear reactions which take place upon further expansion have been investigated by Truran, Arnett, Tsurata and Cameron[19], and in a different context by Delano and Cameron[20]. These investigations showed that as the temperature falls between 10^{10} °K and 5×10^{9} °K, there will be a very rapid and efficient capture of neutrons by protons, followed by secondary nuclear reactions to form helium, and then reactions of the helium with itself and with the neutrons to form heavier nuclei. Indeed, the system approaches nuclear statistical equilibrium on a time scale which is several orders of magnitude shorter than the hydrodynamic time scale.

At low densities and temperatures, the nuclear statistical equilibrium produces the maximum abundance in the peak centered about ^{56}Fe. At higher densities, this equilibrium peak becomes shifted, since one is no longer concerned just with producing the maximum nuclear binding energy per particle in the system, but the minimum energy of the entire system. This can be accomplished at very high densities by reducing the number of electrons at the expense of making the nuclei more neutron-rich than is true of those in the iron equilibrium peak. Thus, at considerably higher densities than those which establish the iron equilibrium peak in explosive nucleosynthesis, nuclear statistical equilibrium will be centered about the nucleus ^{78}Ni. This nucleus has 28 protons and 50 neutrons, and it is therefore particularly stable owing to the double-closed shell.

As the temperature falls lower, the charged particle reactions are no longer able to take place rapidly enough to maintain a full statistical equilibrium in the system. However, capture of neutrons and loss by (γ,n) reactions can still take place very rapidly. In a system with a very large neutron excess, such as is being discussed here, the neutron capture processes will continue as the temperature falls lower and lower. Indeed, the photonuclear reactions will become progressively less important as the temperature falls, so that the point of equilibrium at which capture and loss of neutrons is in balance will shift progressively toward lower neutron binding energies. Eventually, at low enough temperatures, the nuclei will shift out to a very neutron-rich position close to the neutron drip line, at which the neutron binding energy falls to zero.

The equilibrium which is established between the capture and loss of neutrons between adjacent nuclei of common charge Z is given by the following statistical equilibrium expression:

$$\ln \frac{n(Z,A+1)}{n(Z,A)} = \ln \frac{w(Z,A+1)}{w(Z,A)} + \ln n_n - 78.45 - \frac{3}{2} \ln T_9 + \frac{11.61}{T_9} Q_n(Z,A+1) \qquad (18)$$

where $Q_n(Z,A+1)$ is the binding energy of the last neutron in the nucleus $(Z,A+1)$, T_9 is the temperature measured in units of 10^9 °K, and $w(Z,A)$ and $w(Z,A+1)$ are the appropriate nuclear partition functions. For a given temperature and specified free neutron number density, n_n, the most abundant isotope for each value of Z can be determined from this relationship.

The neutron binding energies for very neutron-rich nuclei must be determined from the use of a mass formula. I shall report on calculations which were made about a year ago using the interim mass formula developed by Truran, Cameron, and Hilf[18].

The equilibrium indicated by Eq. (18) will be very rapidly established in the kind of astrophysical system under discussion. Nuclei will remain in the relative abundances predicted by this formula until beta decay takes place. Following a beta decay, additional neutron capture will bring the nuclei, which now have one more proton, into the corresponding abundance pattern predicted for that new value of Z. In this way, characteristic flow paths can be defined in the schematic nuclide chart.

The flow paths defined by the positions of the most abundant nuclei for each Z predicted by Eq.(18) are illustrated in Fig. 15 for a temperature of 2 x 10^9 °K and for two choices of the adiabat upon which the material is taken to be expanding:

$$\rho = 5 \times 10^4 T_9^{\ 3}$$
$$\rho = 5 \times 10^2 T_9^{\ 3}$$

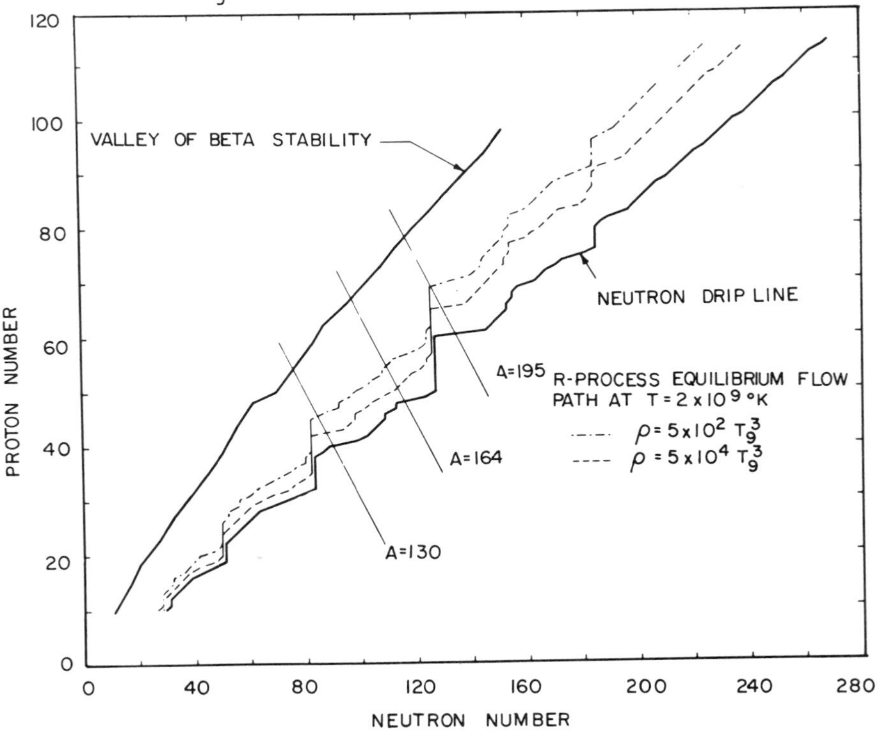

Figure 15
R-process paths in a nuclide chart under steady flow conditions at T_9 = 2, for two different densities. Note that the ordinate and abscissa scales differ by a factor two.

In this figure the free neutron density is taken to comprise 80% of the mass. The experimentally determined valley of beta stability and the neutron drip line predicted by the mass formula are shown on the figure. In each case, the flow paths follow the neutron drip line rather closely in the light element region, but not so closely for heavier nuclei. However, the flow paths come very close to the drip line at the positions of closed neutron shells. This figure basically shows the effect of density upon the position of the flow path.

The effect of temperature can be seen in Fig. 16, where the flow paths defined by the positions of the peak nuclei in the equilibrium condition at each Z-chain are shown for the intermediate adiabat $\rho=5 \times 10^3 T_9{}^3$ for temperatures of $T_9 = 1,3$, and 5.

These variations in flow path positions indicate that the actual dynamics of the r-process must be followed in some detail in order to understand where the flow paths should lie at critical times during the process in a supposed astrophysical environment. The dynamics will depend crucially upon the rates of beta decay in the system. In calculating the beta decay rates, account must be taken of the fact that the beta decay energies are very

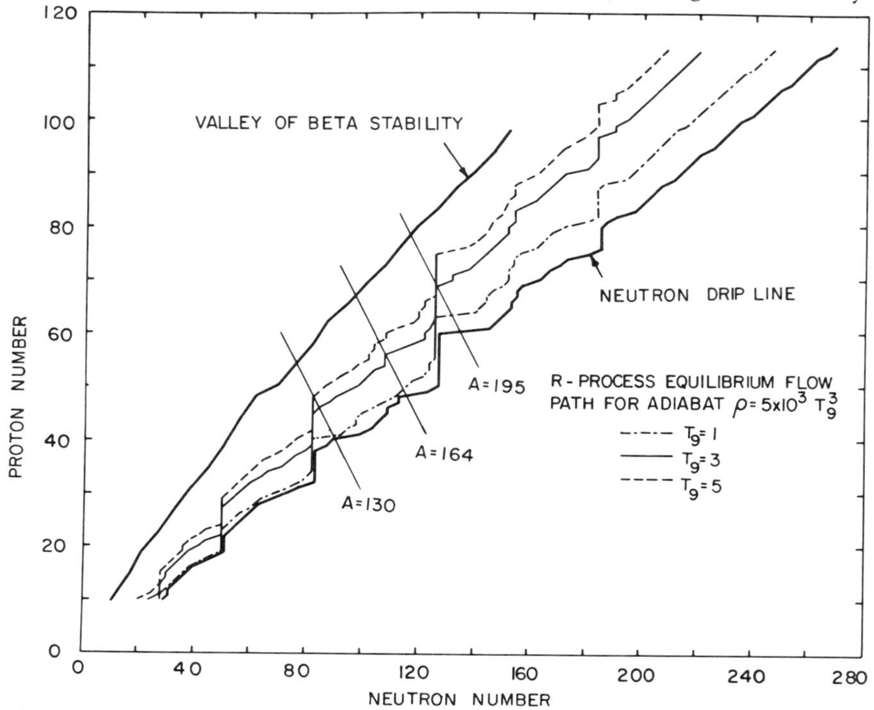

Figure 16
R-process paths in a nuclide chart under steady flow conditions at several different stages of an adiabatic expansion. Note that the ordinate and abscissa scales differ by a factor two.

large, so that a very large number of final states in the daughter nucleus is accessible to the decay process. The characteristic ft values which are well known for transitions between low-lying states of the parent and daughter nuclei will certainly be appropriate to use for the decay to the lower-lying states in the daughter nuclei. However, there may be thousands of millions of accessible excited states at higher energies. The single particle transition probabilities for the beta decay will become mixed into large numbers of these states, so that the ft values for any particular transition are likely to become inversely proportional to the density of nuclear states in the higher-lying region. Since the transition probability must be summed over all these states, which involves multiplying a single transition probability by the density of states, the density of states cancels out, and one is left with the transition probability per unit energy interval in the daughter nucleus. This is known as the beta strength function, which is now beginning to come under experimental attack.

In the work shown here, we have taken a very simple-minded approach to the beta strength function. We have assumed that the usual ft values apply to the beta decay to the first 100 excited states in the daughter nucleus, or at least to that fraction of the states which is likely to have an appropriate combination of spin and parity for an allowed transition, and above the 100th level the beta strength function has been assumed to be constant at the value which was attained in a vicinity of the 100th level.

The mean chain half-life is plotted as a function of proton number in Fig. 17 for the adiabat $\rho = 55 \times 10^3 T_9{}^3$ and a temperature of 2×10^9 °K. The longest mean half-lives, approaching 10^{-2} seconds, are those corresponding to Z-chains for which the most abundant isotope contains a magic number of neutrons (N=82 or 126). However, it may be seen that most of the beta decays take place with half-lives between 10^{-3} and 10^{-4} seconds. This indicates that the rate of progress of a nucleus along the r-process flow paths can become quite fast compared to the hydrodynamic time scale once the temperature has become low enough.

Cameron, Delano and Truran[9] have calculated the flow of nuclei in a nuclear reaction network for a variety of adiabatic conditions, starting with the seed nucleus ^{78}Ni, and a hydrodynamic expansion time scale. This expansion time scale was such that the temperature fell from $T_9 = 4$ to $T_9 = 2$ in 3.8×10^{-2} seconds, and from $T_9 = 2$ to $T_9 = 1$ in about a tenth of a second. The dramatic increase in the beta decay rate in the system with falling temperature is shown in Fig. 18, which corresponds to an adiabat $\rho = 4 \times 10^4 T_9{}^3$ and an initial ratio of neutrons to seed nuclei of 122. The depletion of the nuclei as the temperature falls may be seen by noting how this ratio changes along the curve in Fig. 18 The rapidity of buildup of abundances during the expansion can be judged from Fig. 19 which shows three snapshots of the abundance confined to a time of the first tenth of

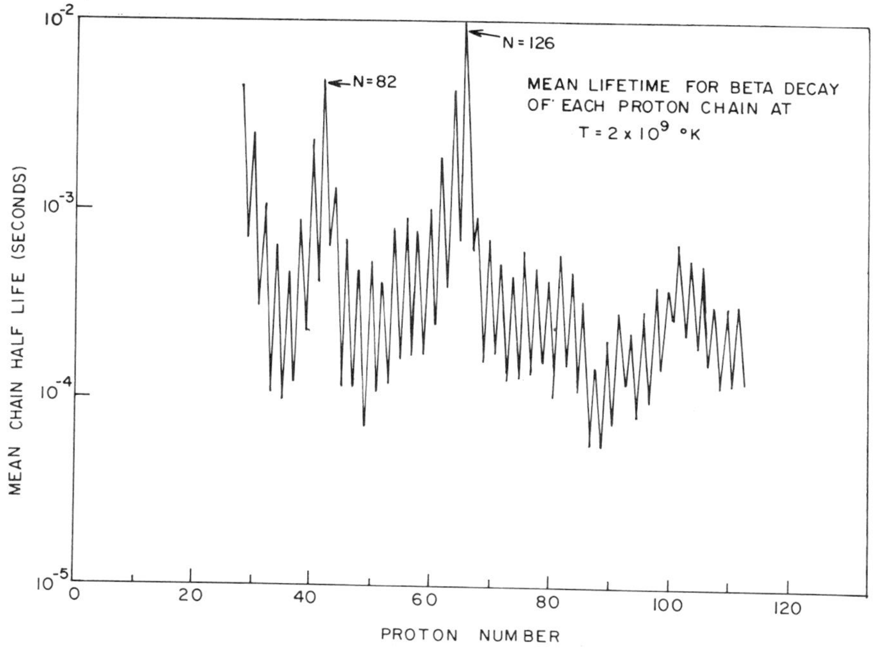

Figure 17
The average beta decay lifetime of the nuclei with given proton number at $T_9 = 2$ and 4×10^4 gm./cm.^{3} with 80% free neutrons by mass.

a second. It may thus be seen that starting from a single seed nucleus at mass 78, a widespread buildup of r-process abundances extending beyond mass number 300 can be obtained very quickly under the hypothetical astrophysical conditions we are considering.

When the neutrons are exhausted, a complex set of abundance readjustments takes place. The nuclei will emit neutrons following beta decay, at least until the beta decay paths take them somewhat closer to the valley of beta stability, and this tends to lower their mass number. On the other hand, the neutrons are recaptured on the nuclei, and this tends to increase some mass numbers. It is evident that the net gain and loss of mass numbers will be zero, but this does not prevent significant readjustments of abundance peaks in the system. In particular, those nuclei which decay fastest toward the valley of beta stability will acquire the highest neutron capture cross sections, and hence they will be most susceptible to capturing the neutrons released by the slower decaying nuclei, particularly those in the closed shell peaks. We have not succeeded in matching any of the abundance peaks observed in the r-process with our sample calculations so far carried out. On the other hand, it is evident that we have not yet put in all of the necessary physics for a fully realistic calculation.

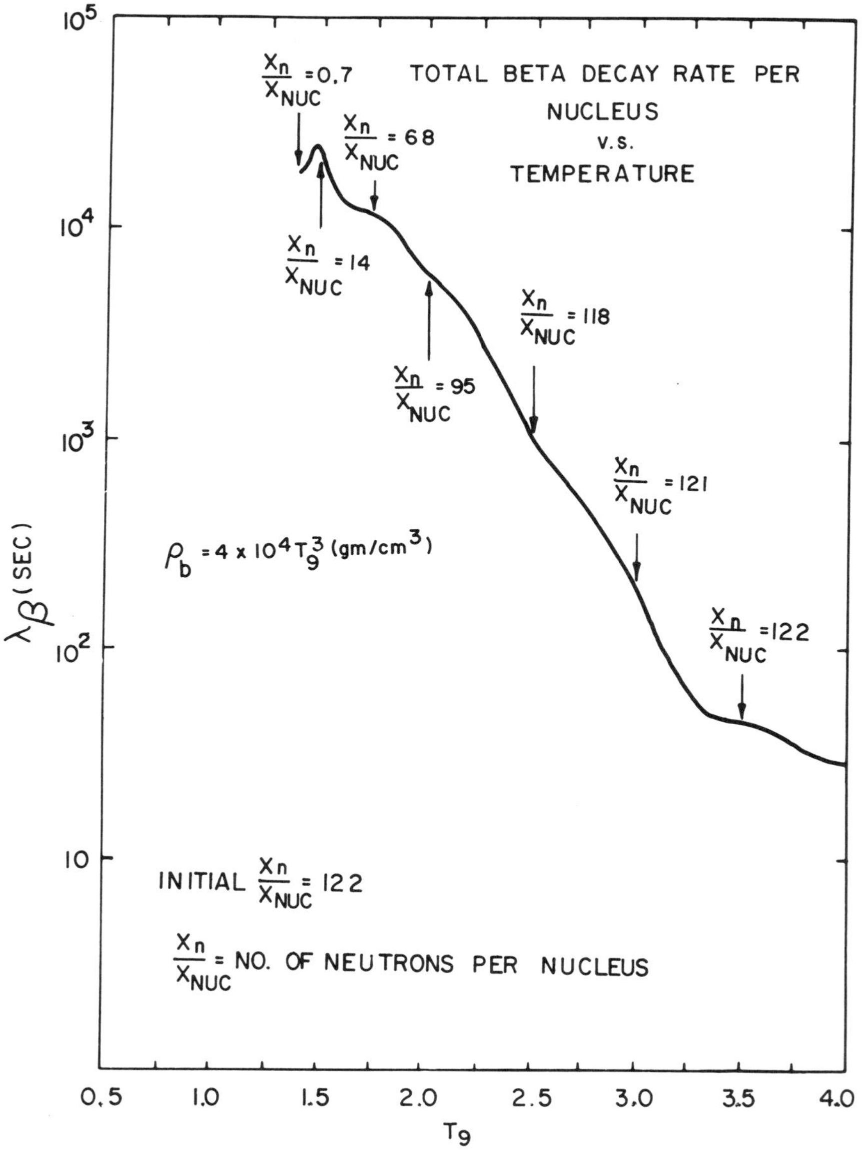

Figure 18
Change in the total beta decay rate per nucleus during expansion of a low adiabat for an initial ratio of neutrons to ^{78}Ni nuclei of 122.

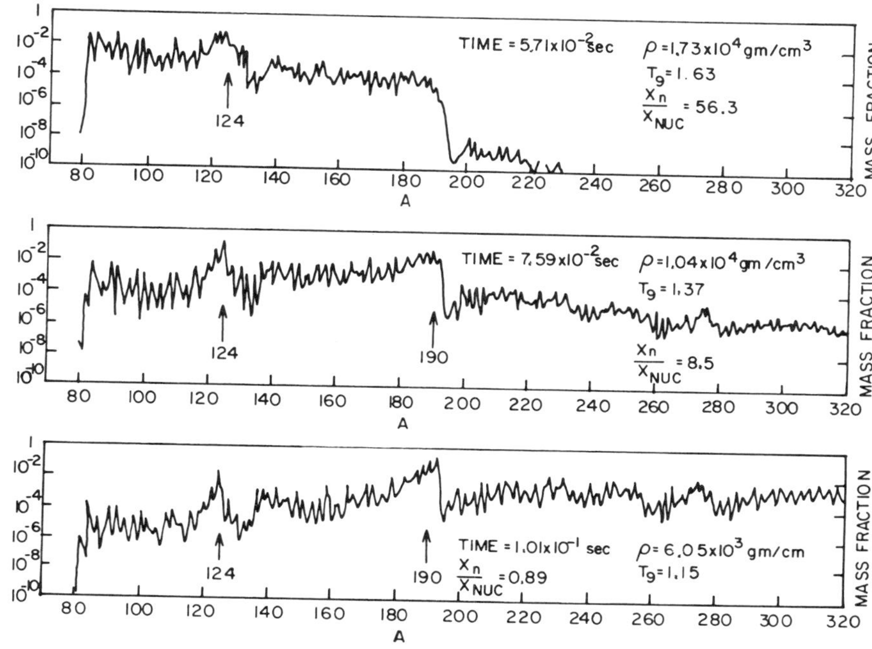

Figure 19
Evolution of the r-process abundances during the expansion of a low adiabat.

One feature which needs considerable improvement is our treatment of the beta strength functions. The formula that we have been using for computing the beta decay rates was established by us for use several years ago, and it should now be possible to improve upon the formulation of beta strength functions on the basis of experimental evidence, at least for electron capture processes.

A major process which was not included in the r-process calculations is that of fission. The buildup toward heavy nuclei beyond mass number 300 can take place exceedingly rapidly, and at some point the nuclei should become fissionable, either following neutron capture or following beta decay. We have so far not attempted to determine where the fission will take place. Following fission, we would have two seed nuclei, each with about half of the mass number of the fissioning nucleus, and the increase of seed nuclei in the system would then very rapidly lead toward depletion of the neutrons.

The fissionability of very neutron-rich nuclei depends critically upon the properties of the adopted mass formula. At very large neutron excesses, the surface energy tends to decrease, since the leading coefficient for the surface symmetry energy, γ_2, is very much larger than the leading surface coefficient,

γ_0. Of particular interest is whether the capture path can reach mass numbers in the vicinity of 300 prior to the fission process, in order that there may be some chance that the r-process will form nuclei in the hypothetical island of stability near Z = 114 and N = 184. The flow path tends to pass near mass number 300 in the vicinity of Z = 90, and some of our trial mass formulas indicate that the fissionability parameter x of Eq. (15) is still less than 0.9. Under these conditions, it seems quite possible that the r-process would be able to build nuclei as high in mass number as the hypothetical island of stability. Whether such nuclei could withstand fission during the beta decay process is yet another question which we have not investigated.

A further piece of physics which must be introduced is the use of true adiabats. In the exercises that we have carried out, we have adopted a simple relationship between temperature and density which assumes a gas with a ratio of specific heats of 4/3, as is appropriate to a medium dominated by radiation pressure, but for a real adiabat one must be prepared to calculate the ratio of specific heats due to a mixture of particles and radiation, and also to take into account the change in the internal energy and its influence upon the adiabat. A great deal of nuclear energy is released during the neutron capture process, and this will tend to keep the temperature raised during the capture process while the density is decreasing significantly.

SUPERHEAVY NUCLEI IN NATURE?

If superheavy nuclei with membership in the suspected island of stability near Z = 114 and N = 184, exist in nature, then they must almost certainly have been produced by the r-process. I shall conclude these notes with a brief discussion of the situation in this regard.

There has been dispute over whether superheavy nuclei might have been discovered in the cosmic rays. There have been a couple of occasions in which very heavy cosmic ray nuclei have passed through plastics and nuclear emulsions being exposed at high altitude in balloons. In the best documented case of such a heavy nucleus, which passed through interleaved sheets of plastic and nuclear emulsion, the charge as determined by the plastics was about 92, whereas that estimated from the nuclear emulsions was about 103. However, the experimenters have agreed that the charge Z = 96 is consistent with their data within the limits of error[21]. Since such transuranic elements can easily be produced by the r-process, and since several of them have half-lives in excess of 10^5 years, where they can easily reach the earth, then there is no present indication that products from the island of stability are involved.

However, the charges found by the plastic surveys of very-very-heavy cosmic rays tend to indicate an abundance distribution more characteristic of the r-process than of any other distribution. This means that if superheavy nuclei are produced by the r-process, they should certainly exist in the cosmic rays, and we should certainly find them when the surveys of very-very-heavy nuclei in the cosmic rays are pushed to a low enough flux level.

In our discussion of the chemical evolution of the galaxy, we showed the results of cosmochronology calculations, which showed that quite short-lived radioactivities can in favorable instances be detected through the traces of their daughter products which they leave in meteorites. It appears that the meteorites were formed very promptly in the early stages of formation of the solar system, and that they cooled quickly thus tending to preserve a record of very early conditions. In particular, since the ^{129}Xe which results from the beta decay of ^{129}I (half-life 1.7×10^7 years), is present in considerable abundance, then it is evident that if any nuclei in the superheavy island of stability have half-lives of the order of 10^7 years or longer, evidence for them should be found in some meteorites. All such nuclei should eventually decay by fission before alpha decays can take them all the way down to the region of uranium.

The xenon extracted from meteorites has indicated a great deal of information about the early history of the solar system[22]. This xenon not only contains varying amounts of ^{129}Xe from the decay of ^{129}I, but also usually a certain amount of xenon in the form of the isotopes 131, 132, 134, and 136 all of which result from the fission of ^{244}Pu. We have seen from the cosmochronology discussion that this isotope has a half-life of 8.2×10^7 years, and that significant amounts of it are therefore around at the time of formation of the solar system. It might be supposed that the presence of xenon due to the decay of a known fission radioactivity might spoil the search for the fission decay products of a hypothetical superheavy element. It is true that this complicates the situation, but it does not prevent the search for a superheavy element.

During the last few years it has become apparent that there are two different abundance patterns in the fission xenon present in meteorites. One abundance pattern has now been definitely established to correspond to ^{244}Pu, as a result of laboratory measurements of the resulting fission pattern from the nuclide. The abundance ratios from ^{244}Pu are as follows:

$$131 : 132 : 134 : 136 = (0.31 \pm 0.08) : (0.97 \pm 0.08) : (0.93 \pm 0.08) : 1.0.$$

This fission abundance pattern is present in most meteorites at a low level, corresponding to the fact that plutonium, like all actinides, is a very nonvolatile element. On the other hand, the second fission abundance pattern is quite different, as follows:

$$131 : 132 : 134 : 136 = (0.20\pm0.1) : (0.37\pm0.1) : (0.70\pm0.05) : 1.0.$$

This abundance pattern can be quite high in abundance in certain selected meteorites, such as carbonaceous chondrites, which have a high abundance of the more volatile elements. It may be seen that the abundance pattern here is quite different from that of the ^{244}Pu and also that it is quite different from the abundance pattern produced by either the spontaneous or neutron-induced fission of any known elements in the actinide series. Thus it is a very good candidate to be fission products from a superheavy element in the hypothetical island of stability.

The elements mercury, thalium, lead, and bismuth are very volatile and are usually greatly underabundant in ordinary classes of meteorites. If the elements immediately beyond bismuth happened to be stable, they would also be underabundant in meteorites owing to the fact that they also are volatile. The elements in the suspected island of stability are mostly chemical analogues of these volatile elements, and hence it is suspected that they should be volatile. $Z = 112$ is the chemical analogue of mercury, and hence this charge number or any charge number above it would be a possible candidate for the suspected superheavy extinct radioactivity which has given rise to the second fission abundance pattern in the meteorites. Since the indicated half-life for the superheavy elements is of order 10^7 years, it must be stable against both alpha decay and fission for a half-life of that order of magnitude. It is possible that this could favor the nuclide $^{297}113$.

It is evident that many more investigations of meteorites and of their various phases will be necessary in order to tie down the definite existence of this superheavy element and to determine its properties well enough to allow an assignment of the charge number. Personally, I find the evidence for the one-time existence of this superheavy extinct radioactivity in meteorites to be rather compelling. This raises a number of challenges to the nuclear physicist concerning the stability of the flow path in the r-process against fission until higher mass numbers than about 300 have been reached, and stability of the beta decay products against fission along the path leading into the island of stability. The abundance of the second fission pattern in those meteorites which contain it, is sufficiently great to indicate that there cannot have been a major depletion of the material that has been formed on the r-process flow path, during the beta decay phase into the island of stability. This challenge must dominate our further research on the r-process.

REFERENCES

1. Paczynski, B. Acta Astr. 20, 47, 1970.
2. Truran, J.W. and Cameron, A.G.W. Astrophys. Space Sci., in press, 1971.
3. Gunn, J.E. and Ostriker, J.P. Astrophys. J. 160, 979, 1970.
4. Colgate, S.A. and White, R.H. Astrophys. J., 143, 626, 1966.
5. Wilson, J.R. Astorphys. J., 163, 209, 1971.
6. Cameron, A.G.W. and Fowler, W.A. Astrophys. J., 164, 111, 1971.
7. Schwarzschild, M. and Harm, R. Astrophys. J., 150, 961, 1967.
8. Burbidge, E.M., Burbidge, G.R., Fowler, W.A. and Hoyle, F. Rev. Mod. Phys., 29, 547, 1957.
9. Cameron, A.G.W., Delano, M.D. and Truran, J.W. in Proceedings of the International Conference on the Properties of Nuclei Far From the Valley of Beta-Stability, Vol. 2, CERN, 1970.
10. LeBlanc, J.M. and Wilson, J.R. Astrophys. J., 161, 541, 1970.
11. Truran, J.W. and Cameron, A.G.W., to be published.
12. Schmidt, M. Astrophys. J., 137, 758, 1963.
13. Reeves, H., Fowler, W.A. and Hoyle, F. Nature, 226, 727, 1969.
14. Langer, W.D., Rosen, L.C., Cohen, J.M. and Cameron, A.G.W. Astrophys. Space Sci., 5, 259, 1969.
15. Bethe, H.A., Borner, G. and Sato, K., Astron. Astrophys. 7, 279, 1970.
16. Baym, G., Bethe, H.A. and Pethick, C.J., Nucl. Phys., in press, 1971.
17. Siemens, P.J. Nucl. Phys., A141, 225, 1970.
18. Truran, J.W., Cameron, A.G.W. and Hilf, E. in Proceedings of the International Conference on the Properties of Nuclei Far from the Valley of Beta-Stability, Vol. I, CERN, 1970.
19. Truran, J.W., Arnett, W.D., Tsuruta, S. and Cameron, A.G.W. Astrophys. Space Sci., 1, 129, 1968.
20. Delano, M.D. and Cameron, A.G.W., Astrophys. Space Sci., 10, 203, 1971.
21. Price, P.B., reported at Copenhagen cosmic ray conference, March, 1971.
22. Cameron, A.G.W., Comments Astrophys. Space Phys., 2, 18, 1970.

ELASTIC AND INELASTIC ELECTRON SCATTERING MEASUREMENTS, EXTENDED NILSSON MODEL AND PROJECTED HARTREE-FOCK

T.E. DRAKE
Department of Physics
University of Toronto

INTRODUCTION

Electron scattering experiments were first attempted in the early 1950's with the extracted electron beams from betatrons and synchrotrons[1]. With the development of the electron linear accelerator, a laboratory was established at Stanford where physicists could use this technique to measure the properties of atomic nuclei[2,3]. The resolution used in the early experiments was only 3% in incident electron energy, or 3 MeV FWHM for 100 MeV incident electrons. Present day electron scattering facilities are capable of 0.08% resolution but targets of thickness ~50 mgm cm^{-2} are used to reduce the data collection time and the overall resolution is seldom better than 0.15%. This limitation of resolution is illustrated by the $(e,e')^{24}Mg$ data of Fig. 1 [48]. Recent developments in accelerator technology should give a resolution of 0.01%, and with greater average current and lower background the target thickness used may be lowered to ten μgm cm^{-2}.[47]

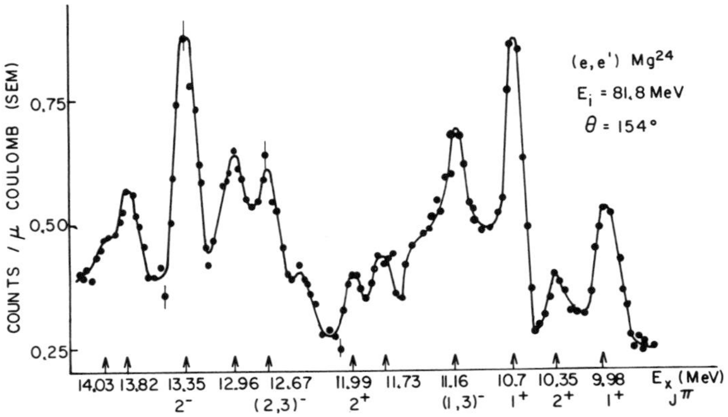

Figure 1
Raw data $(e,e')^{24}Mg$.

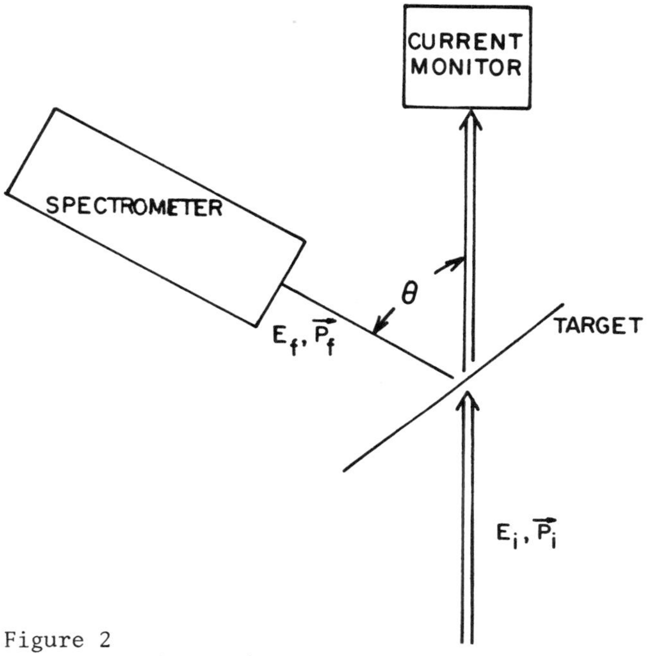

Figure 2
Schematic of (e,e')

In Fig. 2 the incident electron beam is seen to approach the target which is in transmission to give minimum ionization straggling, and to pass on through it into a current monitor. A small number of electrons in the beam interact with the target nuclei and they are scattered into a magnetic spectrometer placed at an angle θ to the direction of the incident beam. These electrons are momentum analysed by the spectrometer which focusses them into detectors positioned in the focal plane beneath several tons of shielding. The cross section $d^2\sigma/d\Omega dE_f$ is measured as a function of energy E_f at fixed angle θ. Integration over E_f for a peak in the measured spectrum gives the cross section $d\sigma/d\Omega$ differential only in solid angle Ω. A detailed derivation of the cross section can be found in references 3, 4, 5, and the discussion here will be limited to the plane wave Born approximation for which

$$\frac{d\sigma}{d\Omega} = \frac{d\sigma}{d\Omega}_{Mott} \eta_R \left|F(q,\theta)\right|^2$$

where $\vec{q} = \vec{p}_i - \vec{p}_f$ is the momentum transferred to the nucleus, $d\sigma/d\Omega_{Mott}$ is the cross section for electrons scattered from an infinitely heavy point nucleus, η_R is a factor which is due to the recoil of real nuclei and $|F(q,\theta)|^2$ contains the information about the spatial distribution of the nuclear source densities; charge, current and magnetization[4].

$$|F(q,\theta)|^2 = |F_{coul}(q)|^2 + (0.5 + \tan^2\frac{\theta}{2})|F_{TRAN}(q)|^2$$

and a plot of $|F(q,\theta)|^2$ against $(0.5 + \tan^2 \frac{\theta}{2})$ at fixed q gives a straight line of slope $|F_{TRAN}(q)|^2$ and intercept $|F_{coul}(q)|^2$ which are the fourier transforms of the nuclear current density and the charge density respectively:

$$|F_{coul}(q)|^2 = \sum_\lambda |F_{c\lambda}|^2$$

$$|F_{TRANS}(q)|^2 = \sum_\lambda (|F_{m\lambda}|^2 + |F_{E\lambda}|^2) .$$

Therefore for each nuclear excited state two diffraction patterns are measured, one of the charge density and one of the current density. Since the virtual photon transferred to the nucleus is spacelike, then $|\vec{q}| = 2\pi/\lambda \geq Ex/\hbar c$ for a level at excitation energy Ex, and with high energy electron accelerators the inner nature of the nuclear surface can be probed.

Models of the nucleus have made use of the Hartree-Fock method. The Hamiltonian is

$$H = \sum_{i=1}^{A} KE_{(i)}^{(1)} + \sum_{i>j=1}^{A} PE_{ij}^{(2)} , \qquad (1)$$

$$E = \frac{\langle\psi|H|\psi\rangle}{\langle\psi|\psi\rangle} \qquad (2)$$

is the energy and

$$\psi = \frac{1}{\sqrt{A!}} \begin{vmatrix} \phi_1(\vec{r}_1) & \ldots\ldots & \phi_1(\vec{r}_A) \\ \phi_A(\vec{r}_1) & \ldots\ldots & \phi_A(\vec{r}_A) \end{vmatrix} \qquad (3)$$

is a Slater determinant of the single particle orbitals $\phi(\vec{r})$. The substitution of eq. (3) into (2) and the application of the variational technique to minimize E will give the Fock-Dirac equation[6]. An iterative solution of this equation provides a self-consistent set of orbitals with their energy eigenvalues ε_i. In terms of these self-consistent orbitals the Hamiltonian can be written as

$$H = \sum_{i=1}^{A} H_{av}^{(1)}(i) + \sum_{i<j=1}^{A} V_{ij}^{(2)R} .$$

If the one-body Hamiltonian $H_{av}^{(1)}$ is successful, the remaining influence of the residual interaction which cannot be included in the average field is negligible and the observed properties may be explained by the average field alone, which might have symmetries such as good orbital parity and good orbital m_j (axial symmetry) for example. Of course the Slater determinant ψ does not, in general, have good angular momentum and a state of good angular momentum must be projected from it at some state in the process, either before variation (PHF) or after variation (HFP).

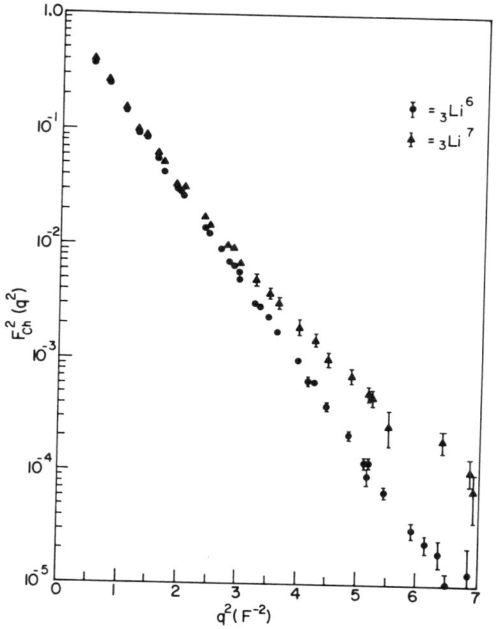

Figure 3

Ground state Coulomb form factors of ^{6}Li and ^{7}Li

^{6}LI AND ^{7}LI

^{6}Li has a ground state spin and parity of 1^+ and therefore only coulomb monopole C0, magnetic dipole M1 and coulomb quadrupole C2 terms can contribute to the coulomb and transverse ground state form factors[52].

The measurements of the coulomb form factors for the ^{6}Li and ^{7}Li ground states are given in Fig. 3 [7]. In the analysis of elastic electron scattering data it has been the custom to assume the orbitals of the average field to be the eigenfunctions of the spherical harmonic oscillator potential[7]. Suelzle et al., have shown that for ^{6}Li the measured coulomb form factor could be fitted with a phenomenological charge distribution to give a charge rms radius of 2.54 ± 0.05 fm. However, when harmonic oscillator orbitals were used no satisfactory fit to the data could be achieved with equal oscillator parameters for the 0s and 0p orbitals. For momentum transfers q less than 1.7 fm^{-1} ($q^2 \leq 3fm^{-2}$) a reasonable fit giving a charge rms radius of 2.43 fm, was obtained for different oscillator parameters b_s = 1.632 fm, b_p = 1.98 fm for the 0s and 0p orbitals respectively. It was impossible to fit the data over the entire range of 0 to 7 fm^{-2}.

The situation was not improved when Elton and Lodhi[8] attempted to include the effects of the residual interaction in LS coupling. They assumed

$$H = \sum_i \left(-\frac{\hbar^2}{2m} \nabla_i^2 + \frac{1}{2} m\, \omega^2\, r_i^2\right) + \sum_{i<j} V_{ij}^R$$

$$- V_{12}^R = \left(\frac{3}{4} V_t + \frac{1}{4} V_s\right) + \left(\frac{1}{4} V_t - \frac{1}{4} V_s\right) \vec{\sigma}_1 \cdot \vec{\sigma}_2$$

where

$$V_s^t = \begin{Bmatrix} 50 \text{ MeV} \\ 32 \text{ MeV} \end{Bmatrix} \exp - \left(r/\left(\begin{matrix} 1.36 \\ 1.76 \end{matrix}\right)\right)^2 .$$

Ground state correlations were introduced retaining the h.o. orbitals

$$\psi_{gs} = |(0s)^4(0p)^2 v=0) + g_{1p}|(0s)^4(0p)(1p)v=2) + g_{2p}|(0s)^4(0p)(2p)v=4)$$
$$+ g_{1s}|(0s)^3(0p)^2(1s)v=2) + g_{2s}|(0s)^3(0p)^2(2s)v=4)$$

with $g_{nl} = \langle\psi_0|V_{12}^R|\psi_{nl}\rangle/(\nu\hbar\omega)$ only to first order in the perturbation expansion. To the first order it can be shown that the correlations which they have calculated correspond simply to a change of single particle basis. However, they could not simultaneously reproduce the elastic charge form factor, the electric quadrupole moment and the magnetic dipole moment of the ^{6}Li ground state.

Wong and Lin[9] have calculated the coulomb form factor of ^{6}Li using intermediate coupling ground state wave-functions obtained from three different residual p-shell interactions: i) Brueckner, Gammel, Thaler potential with the spin orbit parameter ξ varied between 1 and 7 MeV, ii) an effective interaction from the best fit to p-shell nuclei obtained by Cohen and Kurath, and iii) an effective p-shell interaction calculated with the Hamada-Johnston potential with ξ from 3 to 5 MeV. The effects of short range correlations were included in a manner described by Da Providencia and Shakin[10]. In all cases the calculated coulomb form factors did not fit the data at large momentum transfer q without giving a ground state quadrupole moment at least 10 times the measured value of -0.08 e fm^2. (See Fig. 4).

Clearly there is no physical reason why the spherical harmonic oscillator orbitals should in any way resemble the orbitals of the average field within the nucleus. The success of the fermi charge distribution in reproducing the measured ground state coulomb form factors indicates that, if the nuclear surface and the field equipotential surfaces are the same, the orbitals have a Saxon-Woods radial dependence[11]. However the harmonic oscillator orbitals do form a complete set and Bouten et al[12] have expanded the s and p orbitals in LS coupling as

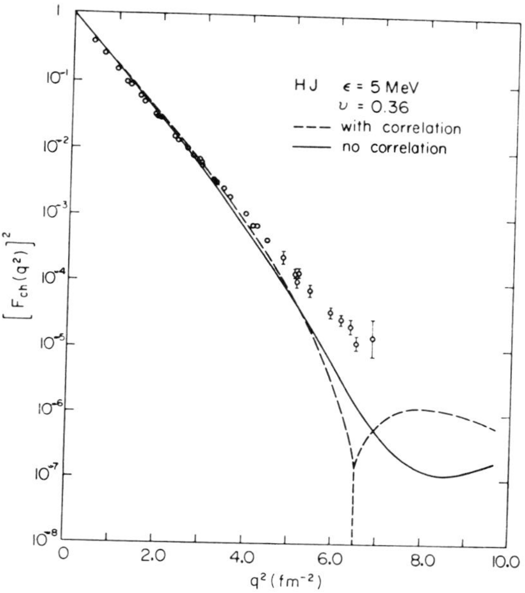

Figure 4
Ground state Coulomb form factors of ^6Li calculated with correlation effects for HJ force with $Q_{cal} = -0.086$; experiment and theory of ref. 9.

$$|e_0\rangle = \sqrt{1-\alpha^2-\beta^2}\,|0s0\rangle + \alpha\,|1s0\rangle + \beta\,|0d0\rangle$$

$$|\theta_0\rangle = \sqrt{1-\gamma^2-\delta^2}\,|0p0\rangle + \gamma\,|1p0\rangle + \delta\,|0f0\rangle$$

retaining the parity and axial symmetry. The coefficients and the oscillator parameter b are varied until a minimum is obtained for the energy

$$E_L = \langle\psi_L|H|\psi_L\rangle / \langle\psi_L|\psi_L\rangle$$

with

$$\psi_L = \underline{P}_L\phi$$

$$\underline{P}_L = \{(2L+1)/8\pi^2\}\int d\Omega\; D^L_{00}(\Omega)\;\underline{R}_\Omega \;:$$

$\underline{P}_L$ is the Peierls-Yoccoz projection operator[13]. ϕ is the Slater determinant of the orbitals and the centre of mass kinetic energy is subtracted from H. No assumption is made about the average field potential which is essentially included in the orbital coefficients α, β, γ, δ, and therefore higher order terms than r^2Y_{20} of the Nilsson potential are implicitly contained within this approach.

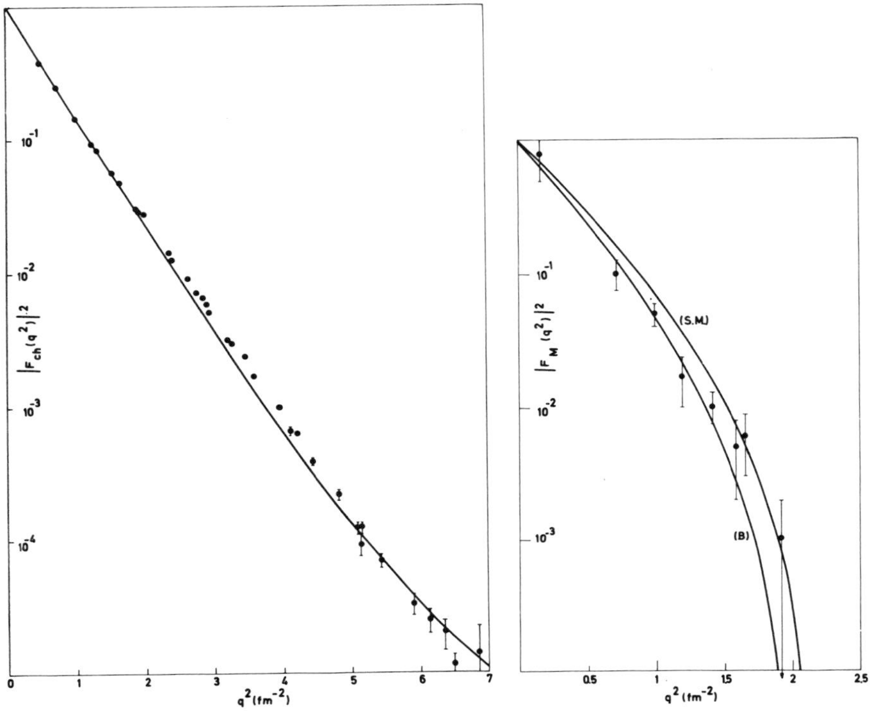

Figure 5
Ground state Coulomb and transverse form factors of ^{6}Li; experiment and theory of ref. 12.

For the ^{6}Li ground state the agreement with experiment is shown in Fig. 5. The oscillator parameter is b = 1.60 fm with $\alpha = 0.015$, $\beta = 0.291$, $\gamma = -0.141$, $\delta = 0.221$ and the "equivalent" Nilsson deformation parameter $\delta_{Nils} = 0.760$ for the $|e_0\rangle$ orbital and $\delta_{Nils} = 0.488$ for the $|\theta_0\rangle$ orbital. The agreement between the calculated quadrupole moment and experiment is better, but the distortion of the ^{6}Li ground state is extremely severe especially in the even "^{4}He" core orbital $|e_0\rangle$.[54] When the second member of the ground state rotational band at 2.18 MeV is projected out of the Slater determinant ϕ the agreement with experiment is good, see reference 12. On a rotational model the spectroscopic quadrupole moment from the BE2 value of 32 e^2fm^4 is -4.6 efm^2 which is fifty times larger than the measured ground state moment of -0.08 efm^2. This suggests a large dynamic change in deformation between the ground state and 2.18 MeV level. The level width is 25±1 keV and it decays entirely to ^{4}He+d. This is undoubtedly on the Wigner limit but the Wigner limit is difficult to estimate accurately using bound state orbitals. The parameters of the 2.18 MeV level are b = 1.60 fm, $\alpha = 0.034$, $\beta = 0.193$, $\gamma = -0.140$ and $\delta = 0.139$ [12] with the Nilsson $\delta_{Nils} = 0.49$ for $|e_0\rangle$ and

δ_{Nils} = 0.2 for $|\theta_0>$, so a large change in deformation does in fact occur between the ground state and 2.18 MeV level.[12] The differences between theory and experiment for ^{6}Li probably are due to the use of truncated expansions in bound harmonic oscillator functions for the orbitals, rather than say Saxon-Woods functions. The electron scattering form factors have recently been measured up to 14 fm^{-2} by Li et al[14] but the region from 7 to 14 fm^{-2} has not yet been compared with theory.

The situation in ^{7}Li is very much the same with the 4.63 MeV level exhibiting a level width of 93 ± 8 keV for decay to ^{4}He + t. The PHF coulomb form factors agree very well with experiment, see reference 15, and the success here is not immediately obvious since the equivalent Nilsson parameters are δ_{Nils} = 0.828 for $|e_0>$ and δ_{Nils} = 0.487 for $|\theta_0>$ orbitals.

^{9}BE

For ^{9}Be Bouten et al[16] have used intermediate coupling with the PHF method to calculate the ground state and 2.43 MeV level form factors which are compared to experiment in Figs. 6, 7. With a spin orbit splitting of the p orbitals of 2.5 MeV they predict the measured ground state electric quadrupole moment of 5.8 efm^2 and magnetic dipole moment of -1.177 n.m. The equivalent Nilsson deformation parameter of the ground state of ^{9}Be is about one-half of that of ^{6}Li for all orbitals and it is not surprising to find good agreement with experiment.

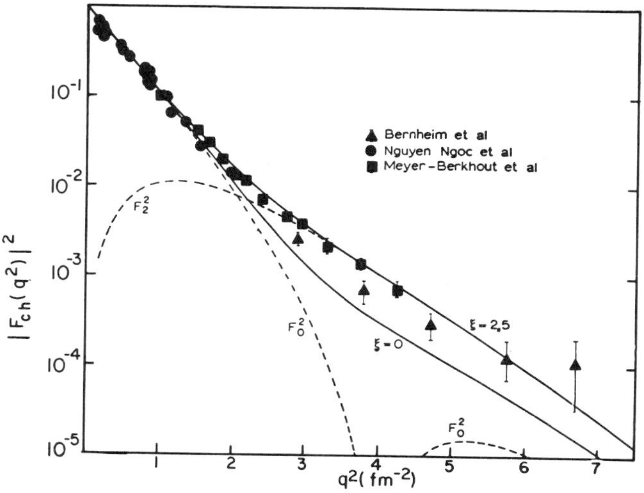

Figure 6
Coulomb form factor for the ground state of ^{9}Be; experiment and PHF of ref. 16.

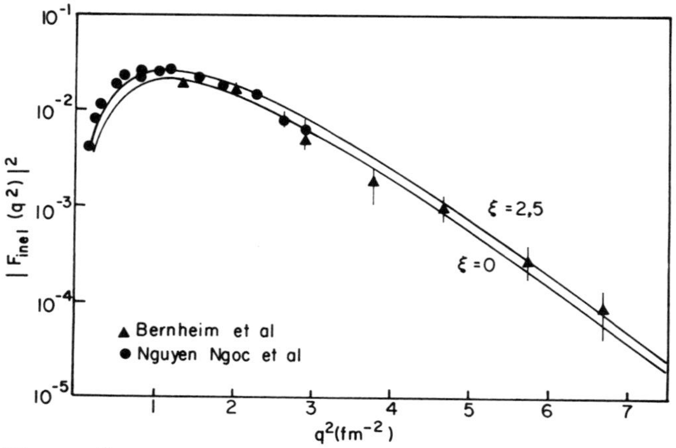

Figure 7
Coulomb form factor for the 2.43 MeV level of ^9Be; experiment and PHF of ref. 16.

In his original paper Nilsson[17A] proposed that both the nuclear surface and the equipotential surface of the average field on a nucleon were ellipsoidal in shape. This potential was written as the deformed harmonic oscillator

$$V = \frac{1}{2} m\, \omega^2 r^2 \{1-\frac{8}{3}\sqrt{\frac{\pi}{5}}\, \delta Y_{20}\} - 2K_1 \vec{\ell}\cdot\vec{s}\, h\omega_0 + D\, \vec{\ell}\cdot\vec{\ell}$$

δ is the deformation parameter, K_1 is a measure of the spin orbit strength and D is zero for the p shell. The Nilsson orbitals were expanded as linear combinations of spherical harmonic oscillator functions within the same shell. In this form the model was incapable of producing reasonable ground state charge form factors and intrinsic quadrupole moments in the p shell. Vinciguerra and Stovall[17B] have shown that the inclusion of higher shells in the Nilsson orbitals will produce an enhancement of the quadrupole moment and Drake[49] has extended the mixing to 7 $\hbar\omega$ to produce the orbitals of Fig. 8. The extended Nilsson calculation gives a minimum ground state energy for a deformation η = 5.3 where

$$\eta = \{\delta_{Nils}/K_1\} \times \{\frac{\omega(\delta_{Nils})}{\omega_0}\}$$

and

$$\omega(\delta_{Nils}) = \omega_0(1-\frac{4}{3}\,\delta^2_{Nils} - \frac{16}{27}\,\delta^3_{Nils})^{-\frac{1}{6}}$$

The oscillator parameter b = 1.606 fm reproduces the experimental rms radius of the charge distribution of 2.46 fm, and the fit to the measured form factor is good with a spectroscopic quadrupole moment of 4.1 efm^2, see Fig. 9. In Fig. 10 the agreement with the measured transverse form factor is comparable to that of Bouten[16] for a magnetic dipole moment of -1.15 n.m. and a

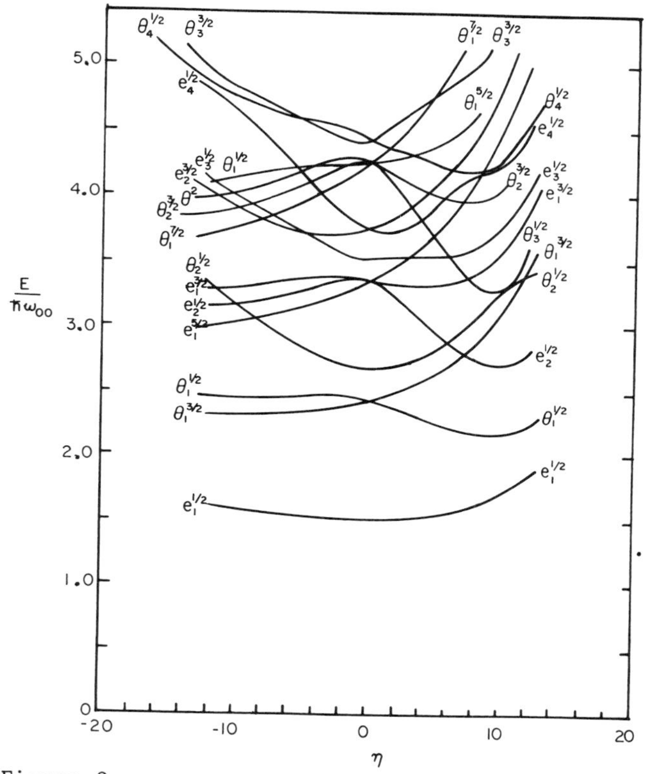

Figure 8
Nilsson orbitals

ratio of magnetic octupole to magnetic dipole moments = Ω/u = 3.3 fm^2. Both of these form factors were obtained with a Nilsson model spin orbit parameter K_1 = 0.08 which is in agreement with Bouten's spin orbit strength. In fact the ^9Be ground state appears to be well described by the simple ellipsoidal r^2Y_{20} deformation of the Nilsson potential.

For the 2.43 MeV level of ^9Be the longitudinal form factor calculated with the ground state deformation parameter η = 5.3 is shown in curve A of Fig. 11. The ground state configuration for ^9Be gave a minimum total energy for a prolate deformation, η = 5.3, which is consistent with a ground state J^π of $3/2^-$, but the disagreement between theory and experiment for the 2.43 MeV level is ~ 2 to 1. If we attempt to enhance the quadrupole moment by increasing the deformation, η ~ 7.8 then the $1/2^+$ orbital originating in the $|0d_{5/2}\rangle$ spherical state becomes the ^9Be ground state, see Fig. 8. Therefore the necessary enhancement cannot be obtained in any consistent manner with this ellipsoidal type of deformation. Since Bouten et al. get very good agreement with experiment we conclude that the simple

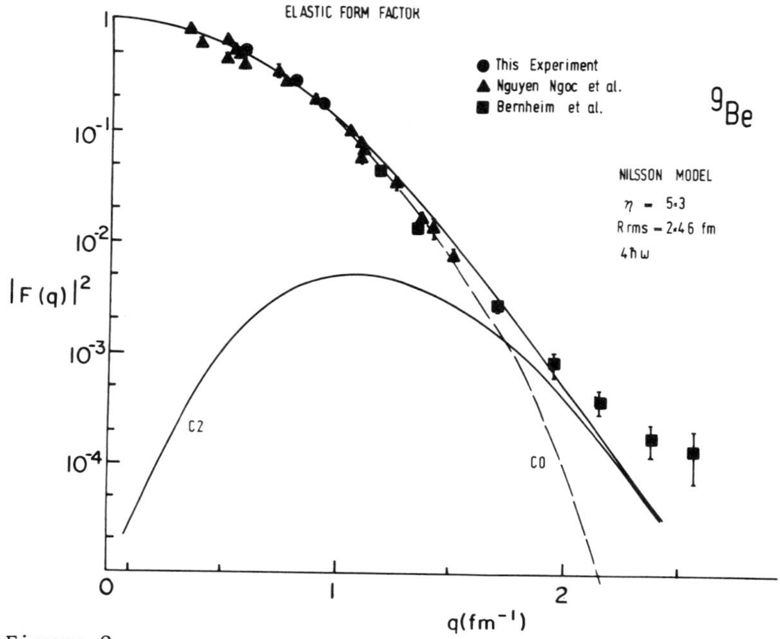

Figure 9
Ground state coulomb form factor of ^{9}Be with extended Nilsson model.

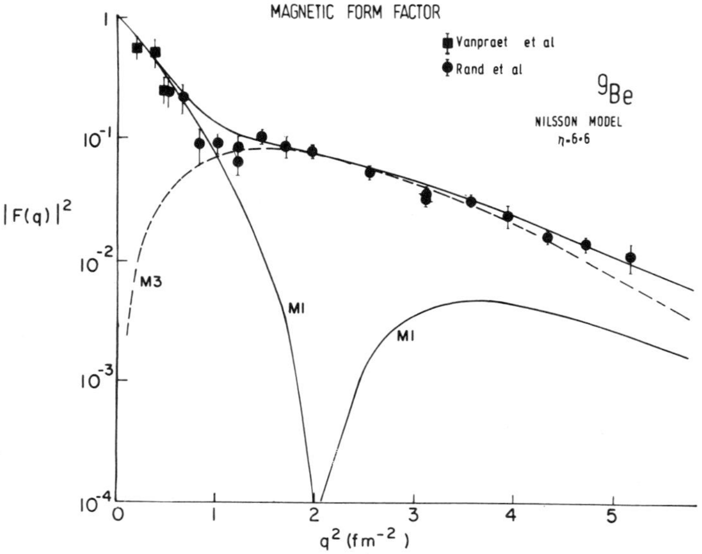

Figure 10
Ground state transverse form factor of ^{9}Be with extended Nilsson model.

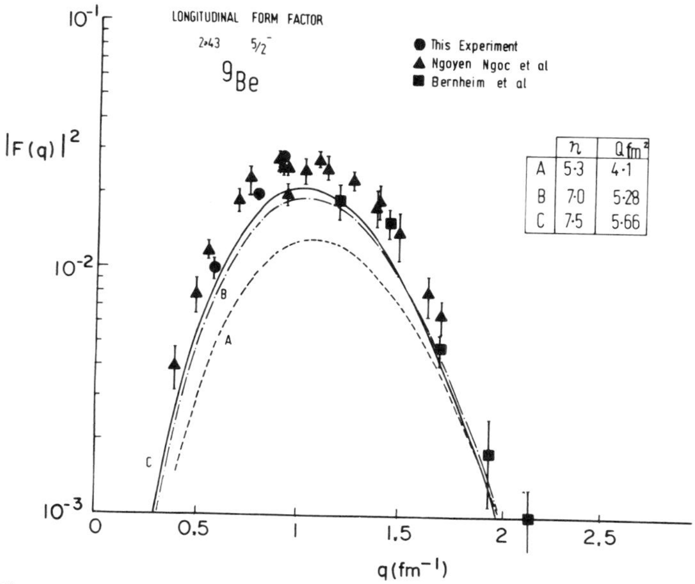

Figure 11
Coulomb form factor for the 2.43 MeV level of ^{9}Be with the extended Nilsson model.

Y_{20} deformation is not sufficient to describe the properties of this state and higher order terms $r^\lambda Y_{\lambda\mu}$ must be added to the Nilsson potential to fit this data. Some enhancement is also needed in the transverse Nilsson form factor[23].

The only deformation which one can envisage for this rotational state is a dumb-bell of two α particles (two centred α-α deformation of the ^{8}Be core). This interpretation is reinforced by the fact that the total width of this level is near the Wigner limit, and it is 90% due to p-wave neutron emission to the 2^+ state of ^{8}Be [19]. The remaining 10% of the level width is due to both decay by f-wave neutron emission to ^{8}Be ground state and also to radiative decay. Cocke and Christensen[20] have established the branching ratio to the ^{8}Be ground state to be 7.5±1.5%. Purdom et al[21] measured the ratio $\Gamma_\gamma/\Gamma = (1.16\pm.14) \times 10^{-4}$. Clerc et al[22] have established $\Gamma_\gamma = (91\pm10) \times 10^{-3}$ ev by electron scattering which implies a total level width of 0.78±.15 keV and a width of 59±20 eV for f-wave neutron emission. Since higher shells are mixed into the Nilsson orbitals one should be able to calculate the reduced width for f-wave neutron emission from the 2.43 MeV level to the g.s. of ^{8}Be, but as stated above, this is made difficult by the bound nature of the orbitals.

1.7 MeV Resonance

Inelastic proton scattering from ^{9}Be reveals a sharp final state interaction near the neutron emission threshold to the ground state of ^{8}Be [19,23,24]. The peak has also been observed in many other experiments such as (e,e'), (d,d'), (α,α') and $^{11}B(d,\alpha)^{9}Be$ [25] The large radiative width of this transition which implies an E1 character, and the isotropic distribution of the emitted neutrons suggest a $1/2^+$ assignment. Salyers[24] has analysed the (γ,n) cross section near threshold using a Dirac model of two α particles and a neutron for ^{9}Be, and he has shown that a $1/2^+$ resonance is necessary to obtain agreement with the measured (γ,n) cross section. In Fig. 8 the second even 1/2 orbital moves down in energy for prolate deformation, and above η=6.2 it drops below the second odd 1/2 orbital. In the ground state the neutron is in the odd 3/2 orbital and we would expect an excited $1/2^+$ state when the neutron jumps up into this second even 1/2 orbital. A deformation of η=6.6 would put the $1/2^+$ state at 1.7 MeV and since the valence neutron in the ^{9}Be ground state is in an odd 3/2 orbital, which will not overlap with either of the two core orbitals, it is not possible to connect the ground state to the $1/2^+$ state with the coulomb operator and still maintain axial symmetry in the Nilsson potential.

Slight et al[23] have observed this resonance by inelastic electron scattering, see Fig. 12 , and in Fig. 13 the transverse and coulomb form factors at 1.7 MeV are given. The transverse form factor agrees with the calculated Nilsson model prediction for a final state resonance between the neutron and the ^{8}Be core. The very large coulomb contribution has a diffraction maximum at ~0.6 fm^{-1}, which shows a transition radius much larger than expected if the valence neutron is given an "effective charge".

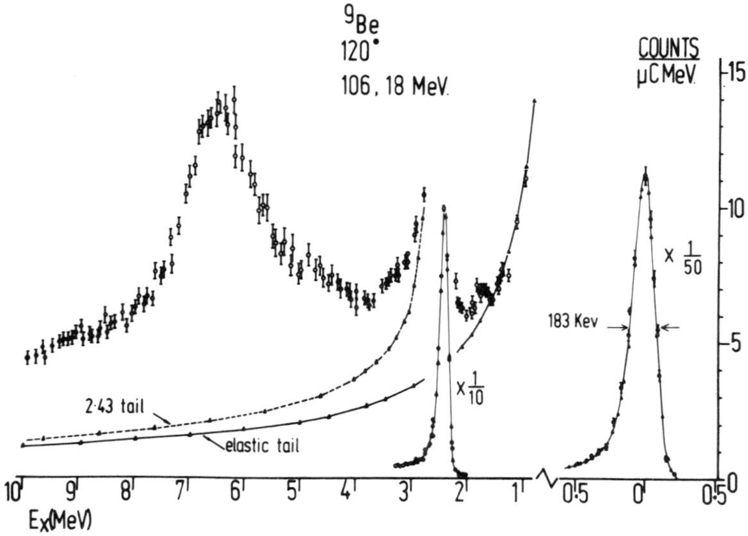

Figure 12
Raw data $(e,e^1)^9Be$.

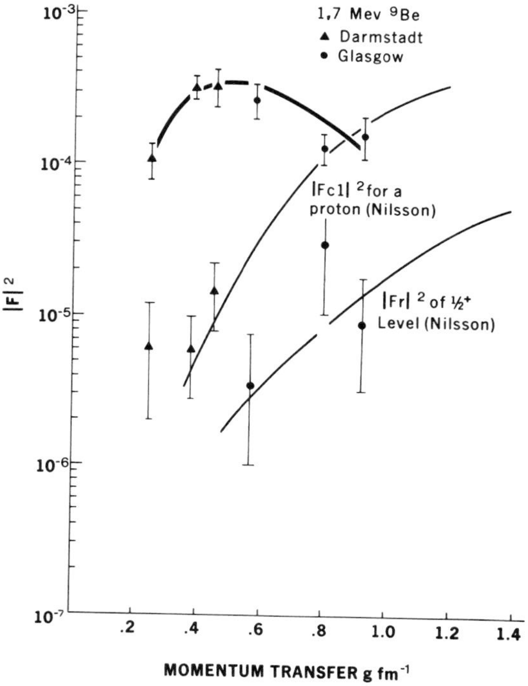

Figure 13
Coulomb and transverse form factors with extended Nilsson model for the 1.7 MeV resonance of ^{9}Be.

Therefore the coulomb contribution is not simply due to band mixing or to mixing via the residual interaction in the ^{9}Be average field and the concept of neutron effective charge has no meaning. The coulomb contribution must be due to the stretching of the two α particles in the core.[19] The form factor is consistent with a coulomb C1 interaction. Consider the argument of Tang et al.[26] for $e + {}^9\text{Be} \rightarrow e' + {}^8\text{Be} + n$. Let k be the neutron wave number, R the interaction radius and suppose that there is no final state resonance between the ^{8}Be core and the neutron. For small $kR \ll 1$, the s-wave neutron phase shift is $\delta_S(k) = kR(f^{-1} - 1)$ where f is the logarithmic derivative at the boundary. This phase shift gives a cross section increasing linearly with k near the threshold where $kR \ll 1$. However, for $kR \sim 1$ an enhanced cross section is observed and although the threshold for n + ^{8}Be is 1.67 MeV, the threshold for n + 2α is 1.57 MeV. If the ^{8}Be core is stretching there is extra kinetic energy available to the neutron which will allow a C1 contribution near ^{8}Be + n threshold. A dipole interaction can contribute to the ^{9}Be(γ,n) reaction, however a tiny neutron yield is observed below 1.67 Mev excitation in ^{9}Be.[24] Therefore the coulomb contribution is due to the ½$^+$

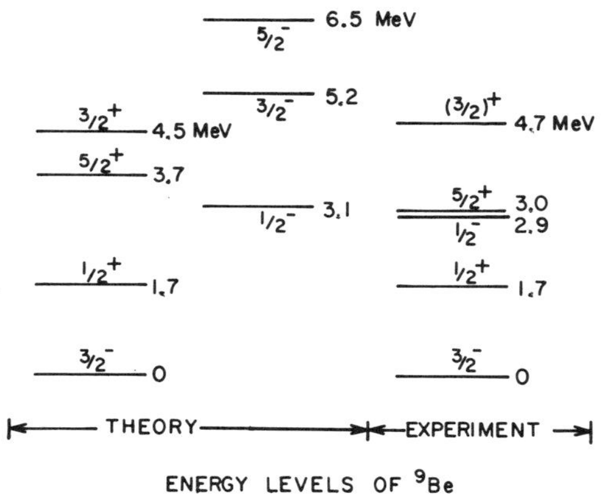

Figure 14
Energy levels of ^{9}Be

resonance of ^{9}Be, and the transverse part also displays a large transition radius.

With the ground state moment of inertia, the energies of the rotational band built upon this $1/2^+$ level and the rotational band built upon the $1/2^-$ level are shown in Fig. 14. The $1/2^-$ state corresponds to a valence neutron jump from the odd 3/2 orbital to the second odd 1/2 orbital of Fig. 8. The β decay of ^{9}Li has confirmed the existence of a $1/2^-$ level at 2.8 MeV with a width of 0.9 MeV[27] and further evidence for this level is given by Slight et al[23].

SUMMARY

The problem of E2 enhancement of p-shell nuclei has been treated by Kurath[28] using first order perturbation theory, (i.e., assuming that the deformation of the orbitals is small). In recent years the development of self consistent calculations, such as those of Bouten et al. described above, has upset the confidence in this shell model picture. It was found that by allowing the shell model field to deform, the minimization of the energy automatically lead to large deformations which gave the correct enhancement of the E2 data.

^{12}C has been studied with PHF[29] and with alpha cluster models[30] and the inclusion of higher order shells ($6\hbar\omega$) extends the "shell model cut-off" of the ground state rotational band.

S-D SHELL

Throughout the 1s - 0d shell from ^{16}O through ^{40}Ca and in the 1p - 0f shell there are collective low-lying negative parity states from 4 to 8 MeV excitation. Near the closed shells for ^{16}O and ^{40}Ca the thresholds for α emission are considerably lower (~ 4.5 MeV for ^{19}F and ^{20}Ne), than half way through the shell, (~ 10 MeV for Mg isotopes). Therefore one would expect that the average H.F. field would deviate from the ellipsoidal shape (r^2Y_{2m}) and tend to distort as the α particle pulls away from the residual nucleus. Onishi and Sheline[31A] have considered the 4 α cluster with tetrahedral deformation in ^{16}O and they point out that the lowest rank of the spherical harmonics generating the tetrahedrally deformed field is Y_{3m}.[31B] Therefore an average field of r^2Y_{3m} may account for the 3^- vibrational mode which gives 3^- states at ~4 to 8 MeV in the 1s - 0d shell and at 6.13 MeV in ^{16}O, while $r^2/2! \{\partial^2/\partial r^2 \rho(r)\} Y_{3m} Y_{3m}$ terms in the Taylor Series expansion of the transition charge density may enhance the observed BE2 values near the closed shells, in ^{18}O and ^{20}Ne.

With this in mind the Nilsson model with configuration mixing up to 7 $\hbar\omega$ has been used to calculate the ground state properties of nuclei from ^{16}O to ^{50}Cr [32]. The results are given in Table I. The quadrupole moments of ^{22}Ne, ^{24}Mg, ^{25}Mg, ^{26}Mg are predicted very well but as one moves through ^{20}Ne and ^{18}O toward ^{16}O, the agreement becomes progressively worse. A similar situation is observed in the 1p - 0f shell where the agreement is good for the stable Ti and Cr isotopes but it is poor as one moves toward ^{40}Ca. Near the closed shells a large enhancement to the intrinsic quadrupole moment is needed, and presumably a higher order deformation than Y_{2m} such as Y_{3m} etc. will suffice just as it did for the 2.43 MeV level of ^{9}Be mentioned above.

The basic argument here is that one can either define a poor average field $H^{(1)}_{av}$, $(H = H^{(1)}_{av} + V^{R}_{12})$, and rely on the diagonalization of the residual interaction within a large basis to provide the needed E2 enhancement, or one can properly average over the two body force as Bouten et al. have done and neglect the residual interaction. This latter approach allows us to think in terms of deformations. Each state of the nucleus has its own average field, or deformed shape for the nuclear surface, which can be quite different from that of the ground state as we saw in ^{6}Li. It would appear that the nucleus assumes the Nilsson ellipsoidal shape modified by the distortions due to the possible channels into which it can disintegrate, with the shell structure maintained by expressing the state functions as Slater determinants of the average field orbitals. This approach might also reproduce the deformations due to α clustering in light nuclei.[56]

The 0^+ state at 6.06 MeV in ^{16}O would not be accessible by this method if the orbital parity were preserved, as the average field would have 4 nucleons in even parity orbitals and 12 nucleons in odd parity orbitals, and an even number 2, 4, etc. of 1 $\hbar\omega$ jumps where the nucleons change parity, would not be included.[33] However for transitions such as the M1 transitions at 15.11 MeV in ^{12}C, 11.4 MeV in ^{28}Si,^{11C} 9.98 MeV and 10.74 MeV in ^{24}Mg.[48] etc. which are believed to be predominantly one particle excitations of $P_{3/2} \rightarrow P_{1/2}$ and $d_{5/2} \rightarrow d_{3/2}$ respectively, and for giant resonance states 1^-, 2^- in ^{12}C, ^{16}O ^{11}C, ^{24}Mg [48] this method might work rather well. To the author's knowledge it has not yet been attempted. However, the 3^- state in ^{16}O at 6.13 MeV has been calculated in this way by Rowe et al.[51] Reference 51 also compares thoroughly this PHF method with the Random Phase Approximation (RPA), the Tamm-Dancoff Approximation (TDA) and the Equations-of-Motion technique.

MAGNESIUM ISOTOPES

Coulomb form factors for the ground states and first excited states of the g.s. rotational band have been measured at Glasgow[35] and they are shown in Figs. 15, 16, 17, 18. To compare these form factors with those calculated in Born approximation using our extended Nilsson model program, the measured data were fitted using the Rawitscher-Fischer phase shift code to a

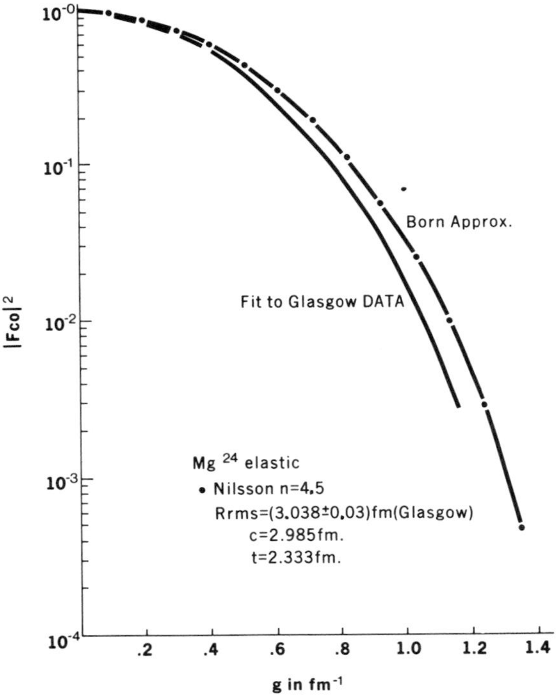

Figure 15
Coulomb form factors for the ground states of ^{24}Mg and ^{26}Mg.

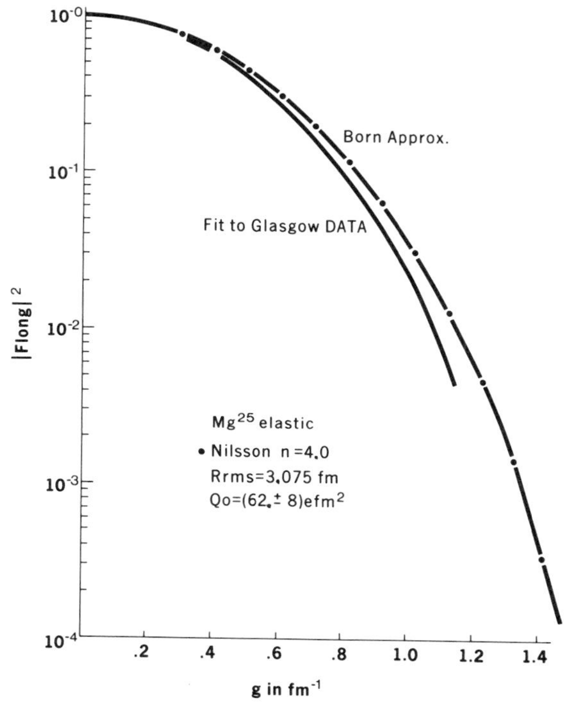

Figure 16
Coulomb form factor for the ground state of ^{25}Mg.

phenomenological charge distribution with parameters determined by the method of least squares. Then this charge distribution was used to evaluate the coulomb form factor in Born approximation for comparison with theory, see Figs. 15 to 18. This analysis gives the intrinsic quadrupole moments Q_0 of the magnesium isotopes from the coulomb form factors of the excited states at 1.37 MeV, 1.61 MeV, 1.81 MeV in ^{24}Mg, ^{25}Mg and ^{26}Mg respectively as well as the intrinsic quadrupole moment of ^{25}Mg ground state from the g.s. coulomb form factor. The ^{25}Mg g.s. has $Q_0 = 62 \pm 8$ efm^2 and the 1.61 MeV coulomb form factor gives 62 ± 5 efm^2. The transition to 1.37 MeV in ^{24}Mg gives $Q_0 = 65 \pm 4$ efm^2 and to 1.81 MeV in ^{26}Mg gives $Q_0 = 58 \pm 5$ efm^2. The rms radii of the Mg isotopes have been measured to be 3.038 ±.03 fm for ^{24}Mg, 3.075 ± .03 fm for ^{25}Mg and 3.037 ± .03 fm for ^{26}Mg in agreement with the extended Nilsson model prediction, see Table I [35]. These measurements suggest that the magnesium quadrupole moment does not change very much with rotation, at least not for the first excited state of the g.s. bands. This contradicts the large intrinsic quadrupole moment of 85 ± 10 efm^2

TABLE 1: Ground state radii and intrinsic quadrupole moments for nuclei in 2s-1d and 2-1f shells

Nucleus	Z	PRESENT CALCULATION RMS Radius Charge	Matter	Neutron	Q_{20}	Q_{20} Experiment*	η	Q_{20} HF (ref.53)	Charge RMS Radius Experiment*
^{16}O	8	2.667	2.667	2.667	0.0	0.0	0.0	0.0	2.667±.033 2.674±.022 2.670
^{18}O	8	2.663	2.760	2.836	+ 9.5	20.5	+2.75	-	2.727±.020 2.723
^{20}Ne	10	2.861	2.861	2.861	+49.5	60 ± 7	+4.5	25.5	3.040±.025 2.99 ±.04
^{22}Ne	10	2.894	2.943	2.984	+50.7	52 ± 10	+4.5	-	2.969±.021
^{24}Mg	12	3.024	3.024	3.024	+65.3	62 ± 10 65 ± 4g)	+4.5	31.4	3.038±.03
^{26}Mg	12	3.057	3.093	3.122	+66.8	57.9±5g)	+4.5	-	3.037±.03
^{28}Si	14	3.161	3.161	3.161	-68.8	35 ± 12	-6.4	-37.5	3.06 ± .05
^{30}Si	14	3.193	3.213	3.230	-54.6		-4.0	-	
^{32}S	16	3.263	3.263	3.263	0.0	50 ± 15	0.0		3.184±.05
^{34}S	16	3.306	3.327	3.346	-24.8		-1.50		

TABLE 1 (Continued)

^{36}Ar	18	3.380	3.380	3.380	+ 39.6	46 ± 5	-2.0		
^{38}Ar	18	3.416	3.431	3.446	- 26.2	46 ± 5	-0.75		
^{40}Ar	18	3.424	3.488	3.541	+ 38.8	62.0	+2.0		3.473±.045 3.41 ±.03
^{40}Ca	20	3.487	3.487	3.487	0.0	0.0	0.0	0.0	3.487
^{42}Ca	20	3.490	3.536	3.577	+ 18.0	60 ± 17	+1.30	20.8	3.517
^{44}Ca	20	3.494	3.584	3.610	+ 28.6	57 ± 12	+2.0	35.8	3.515
^{46}Ca	20	3.512	3.632	3.721	+ 21.2		+1.50	31.6	
^{48}Ca	20	3.525	3.676	3.782	- 24.8	28.5 ± 4	-2.0	0.0	3.476
^{46}Ti	22	3.584	3.622	3.666	+ 93.6	113 ± 40	+2.90	97.8	3.585±.03
^{48}Ti	22	3.602	3.670	3.724	+ 92.0	102 ± 23	+2.75	92.2	3.585±.03 3.55 ±.04
^{50}Ti	22	3.614	3.713	3.789	+104.2	68 ± 14	+3.25	15.0	3.585±.03
^{50}Cr	24	3.682	3.713	3.742	+141.0	127 ± 25	+3.50	75.2	

* References to the measurements of charge radii and quadrupole moments are given in reference 32.

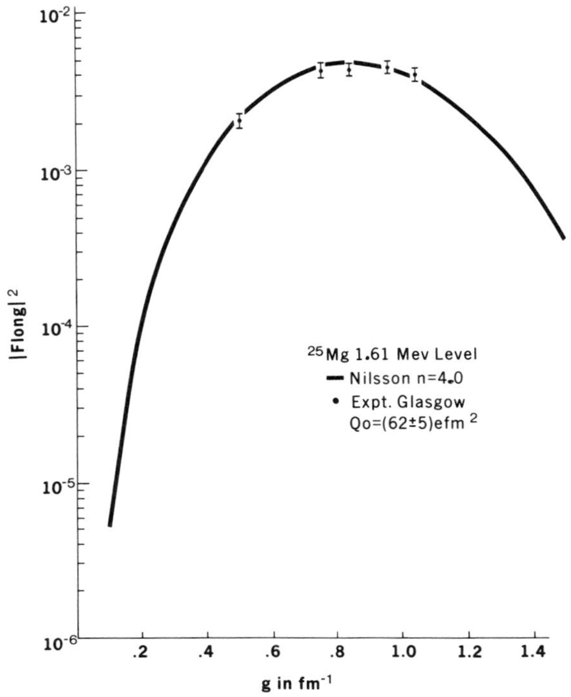

Figure 17
Coulomb form factor for the 1.61 MeV level of ^{25}Mg.

measured by Hauser et al.[36] using the reorientation effect for the 1.37 MeV level of ^{24}Mg, but this measurement has recently been revised and it is now in agreement with the above.[55]

A comparison of the ground state quadrupole moments for ^{19}F, ^{21}Ne, ^{23}Na with those calculated from BE2 values for transitions from the g.s. to first excited states of g.s. bands would be interesting, however the K = 5/2 g.s. band is mixed with the K = 3/2 band ($\Delta K = 1$) and it is difficult to separate the effects of band mixing and the large deformation due to the tendency toward α decay. The observation of large transition radii for the coulomb form factors for the valence neutron transitions (as for 1.7 MeV level in ^{9}Be) might help to clarify the situation here. These measurements have already been made by inelastic electron scattering on ^{25}Mg at Glasgow.[50]

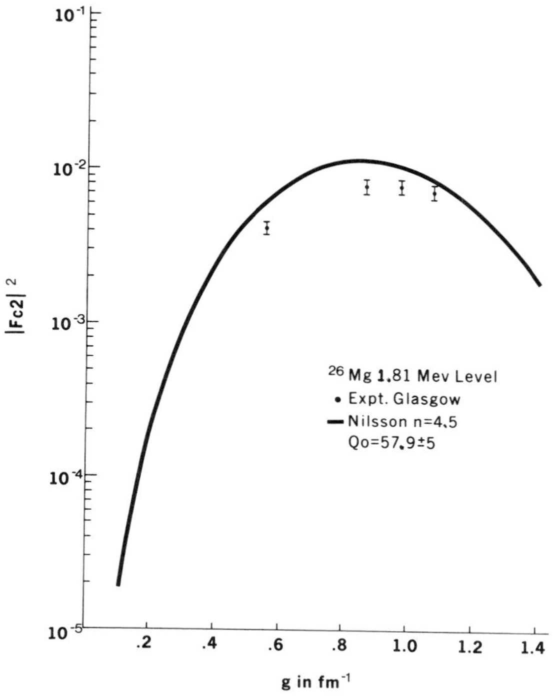

Figure 18
Coulomb form factor for the 1.81 MeV level of ^{26}Mg.

RADII OF THE NUCLEAR GROUND STATE

Since the coupling constant α is small for the electromagnetic interaction, the electron does not appreciably polarize the nuclear charge distribution[40] and elastic electron scattering has been used to measure the radii of the ground state charge distributions. Although the absolute measurement of radii is limited to ~ 2 per cent, the differences in the radii among isotopes or isotones has been accurately established by measuring the coulomb form factors beyond the first diffraction minimum. The Stanford group[37] measured the change in the charge rms radius between ^{40}Ca and ^{44}Ca. In Fig. 19 a plot is given of the ratio of the cross sections $\{\sigma_{44} - \sigma_{40}/\sigma_{44} + \sigma_{40}\}$ for momentum transfer q. Note how the surface thickness t and the half density radius c must both be changed to fit the data. In Glasgow the isotones ^{89}Y, ^{90}Zr and ^{92}Mo [38] have been studied and the same features are observed in the ratios of their cross sections, Figs. 20, 21. In the analysis of the data the charge distribution was assumed to be a static spherical fermi distribution. However the

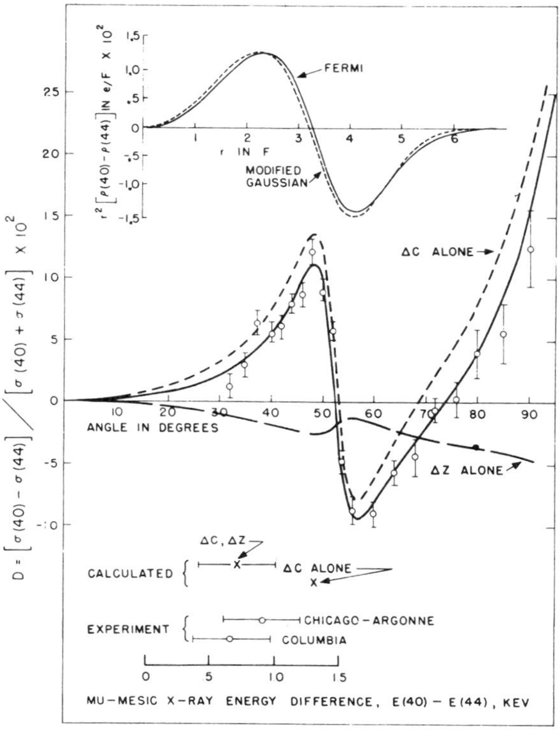

Figure 19
The ratios of the cross-sections for elastic (e,e') on the ^{44}Ca and ^{40}Ca isotopes.

polarizability of the nucleus is not negligible in the diffraction minimum and Sick et al.[39] have claimed to have measured a 10% enhancement in the cross sections at the first diffraction minima of ^{12}C and ^{16}O. This agrees with the dispersion correction calculations of Brown and Kujawski[40], who also present a good review of previous attempts to calculate this effect by others. The nuclear polarizability will certainly alter the nuclear surface thickness parameter t, Madsen et al.[41]

The measured charge radii of ^{40}Ca, ^{42}Ca, ^{44}Ca and ^{48}Ca do not follow an expected $A^{1/3}$ dependence, and in fact the radius of ^{48}Ca is smaller than that of ^{40}Ca by 0.31%, whereas the $A^{1/3}$ dependence predicts an increase of 6.26%. Similar deviations from the $A^{1/3}$ relation have been observed in the groups of isotopes ^{16}O, ^{18}O;[45] ^{20}Ne, ^{22}Ne;[43] ^{24}Mg, ^{25}Mg, ^{26}Mg [43] and ^{46}Ti, ^{48}Ti, ^{50}Ti, ^{52}Ti.[43] An attempt to include deformation in the shell model in a consistent manner was made by Drake

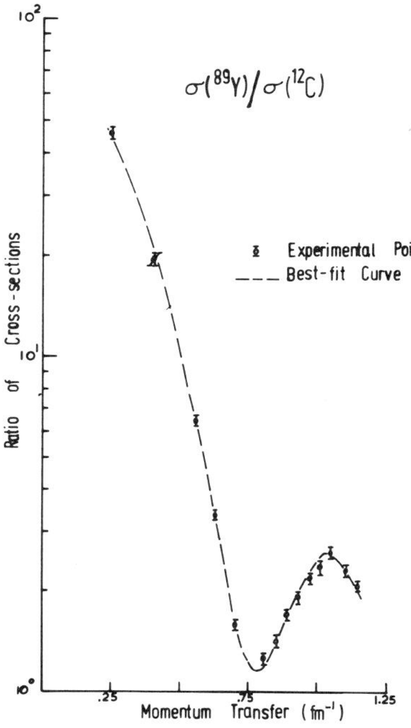

Figure 20
The ratio of the cross sections for elastic (e,e') on ^{89}Y and ^{12}C

and Singhal[32]. The extended Nilsson model was used to calculate the radii and the intrinsic quadrupole moments of even nuclei from ^{16}O to ^{50}Cr, see Table I. The nuclear volume of a uniform sphere of radius R is $4/3\pi R^3 = 4/3\pi r_0^3 A$. But $R^3 = 2\langle r^3 \rangle$ and so

$$2\frac{b_0{}^3}{A} \langle gs| \left(\frac{r^3}{b^3}\right) |gs\rangle = C\ A \qquad (4)$$

with $r_0 \simeq (C)^{1/3}$. For the spherical nuclei ^4He, ^{16}O, ^{40}Ca, ^{90}Zr the oscillator parameters b_0 were chosen to give the measured rms radii, (second moment $\langle gs|r^2|gs\rangle$), of the ground state charge distribution. The third moment $\langle gs|r^3/b^3|gs\rangle$ was evaluated with the extended Nilsson model and the parameter C was calculated. C is related to the compressibility and is given in Fig. 22. The oscillator parameters of the other nuclei of Table I were determined by an interpolation for C on Fig. 22, and equation 4 above. Therefore the model has <u>no</u> arbitrarily chosen parameters[32].

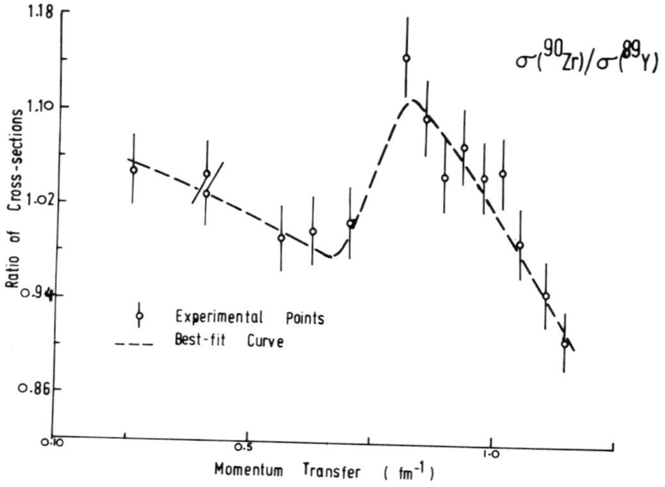

Figure 21
The ratio of cross sections for elastic (e,e') on the isotones ^{90}Zr and ^{89}Y.

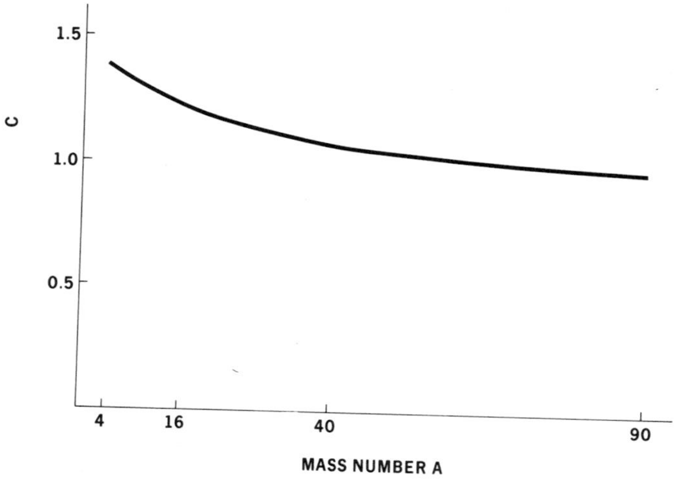

Figure 22
Determination of the oscillator parameter for the Nilsson model.

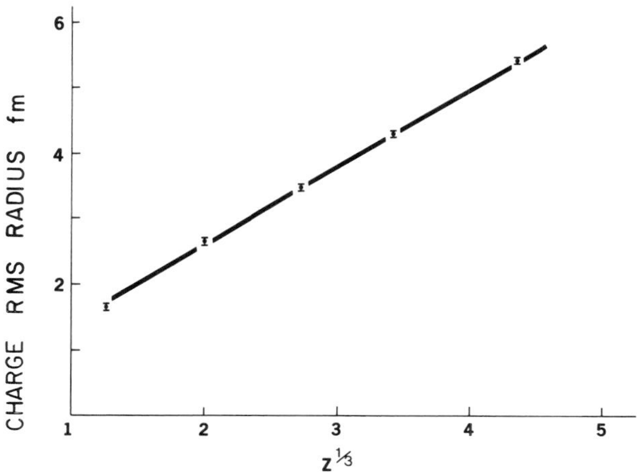

Figure 23
The R rms = 1.19 $Z^{1/3}$ + 0.27 relation for the spherical nuclei ^{4}He, ^{16}O, ^{40}Ca, ^{90}Zr, ^{208}Pb.

Accurate measurements are not available for the matter radii. The matter radii are seen to deviate from an $A^{1/3}$ prediction by approximately one per cent over the mass range 16 to 50. The rms radii for neutrons are larger than the corresponding proton radii because the valence nucleons are neutrons and they extend the neutron distribution beyond the "core", which contains an equal number of protons and neutrons. In the literature it is general practice to compare the measured charge radii with the $A^{1/3}$ relation. However, the above considerations suggest that while the $A^{1/3}$ relation is almost valid for matter radii, a Z-dependent relation might be more appropriate for charge radii. From Fig. 23 it can be seen that charge rms radii of the spherical nuclei ^{4}He, ^{16}O, ^{40}Ca, ^{90}Zr, ^{208}Pb all follow an approximate $Z^{1/3}$ dependence, R rms = 1.19 $Z^{1/3}$ + 0.27, and those of deformed nuclei will be altered by the deformation of the core[32].

(e,e' X) REACTIONS

The electron linear accelerators at Saclay and M.I.T. were designed for duty cycles of 2 and 5 percent respectively to measure coincidence cross sections. The M.I.T. machine has not yet been completed and we are still awaiting (e,e'p) measurements from Saclay. The cross section for the ^{12}C(e,e'p) ^{11}B reaction measured with the Frascati synchrotron[44] has been compared to theoretical cross sections calculated by Boffi et al.[45] using the PHF approach. They find that the coincidence cross section is not sensitive to the deformation of the ^{12}C ground state, which has a very large oblate deformation both from PHF and from the extended Nilsson model. Boffi et al.[45] conclude that there is no significant difference between the (e,e'p) cross section calculated in PHF with a highly deformed ^{12}C ground state and that calculated with a spherical ^{12}C ground state. Improved measurements are needed.

One might expect that the coincidence cross section which is most sensitive to this deformation would be that channel which is causing the deformation. It would be very interesting indeed to measure (e,e'α) cross sections and (e,e'x) cross sections where x may be ^{8}Be, ^{12}C, ^{16}O, etc. This would require an electron accelerator of ~ 0.1% resolution with a duty cycle of 1 and high currents. Targets would be ~ 10 micrograms cm^{-2} thick.

REFERENCES

1. Lyman, E.M., Hanson, A.O. and Scott, M.B., Phys. Rev. 84, 626 (1951).
2. Hofstadter, R., Fechter, H.R. and McIntyre, J.A. Phys. Rev. 92, 978 (1953).
3. A Collection of Reprints "Electron Scattering and Nuclear and Nucleon Structure", edited by R. Hofstadter, W.A. Benjamin (1963).
4. Willey, R.S., Nucl. Phys. 40, 529 (1963); Bishop, G.R. (Scottish Universities Summer School", edited by N. MacDonald (1964).
5. de Forest, T. and Walecka, J.D., Adv. in Phys. 15, 1 (1966).
6. Messiah, A., "Quantum Mechanics" 2 North Holland (1962).
7. Suelzle, L.R., Yearian, M.R. and Crannell, H. Phys. Rev. 162, 992 (1967).
8. Elton, L.R.B. and Lodhi, M.A.K., Nucl. Phys. 66, 209 (1965); Lodhi, M.A.K., Nucl. Phys. 80, 125 (1966).
9. Wong, S.S.M. and Lin, D.L., Nucl. Phys. A101, 663 (1967).
10. da Providencia, J. and Shakin, C.M., Nucl. Phys. 65, 548 (1965).
11. A: Hodgson, P.E., "The Optical Model of Elastic Scattering" Clarendon Press, Oxford. Appendix B. (1963).
    B: Elton, L.R.B., "Nuclear Sizes", Oxford University Press, (1961).
    C: Drake, T.E., Tomusiak, E.L. and Caplan, H.S., Nucl. Phys. A118, 38 (1968).
12. Bouten, M., Bouten, M.C. and Van Leuven, P., Phys. Lett. 26B, 191 (1968); Nucl. Phys. A100, 105 (1967); Bouten, M., Bouten, M.C., Van Leuven, P., Depuydt, H. and Schotsmans, L., Nucl. Phys. A100, 90 (1967).
13. Lamme, H.A. and Boeker, E., Nucl. Phys. A111, 492 (1968).
14, Li, G.C., Sick, I., Whitney, R.R. and Yearian, M.R. Nucl. Phys. A162, 583 (1971).
15. Bouten, M., Bouten, M.C. and Van Leuven, P., Nucl. Phys. A102, 322 (1967); Nucl. Phys. A111, 385 (1968).
16. Bouten, M., Bouten, M.C. and Depuydt, H., Nucl. Phys. A131. 385 (1969); Nucl. Phys. A127, 177 (1969).

17. A: Nilsson, S.G., Mat. Fys. Medd. Dan. Vid. Selsk 29, (1955).
B: Vinciguerra, D. and Stovall, T., Nucl. Phys. A132, 410 (1969).
18. Rand, R.E., Frosch, R. and Yearian, M.R., Phys. Rev. 144, 866 (1966).
19. Spencer, R.R., Phillips, G.C. and Young, T.E. Nucl. Phys. 21, 310 (1960).
20. Cocke, C.L. and Christensen, P.R., Nucl. Phys. A111, 623 (1968).
21. Purdom, P., Seeger, P.A. and Kavanagh, R.W., Nucl. Phys. 83, 513 (1966).
22. Clerc, H.G., Wetzel, K.J. and Spamer, E., Nucl. Phys. A120, 441 (1968).
23. Slight, A.G. and Drake, T.E., Electron Scattering from ^{9}Be, submitted to Nuclear Physics.
24. Salyers, A., Phys. Rev. 2C, 1653 (1970).
25. Phillips, Thomas W., Report No. MIT-208-330 of Mass. Inst. of Techn. and references contained therein. (1967).
26. Tang, Y.C., Khanna, F.C., Heindon, R.C. and Wildermuth, K. Nuc. Phys. 35, 421 (1962).
27. Chen, Y.S., Tombrello, T.A. and Kavanagh, R.W., Nucl. Phys. A146, 136 (1970); Macefield, B.E.F., Wakefield, B. and Wilkinson, D.H., Nucl. Phys. A131, 250 (1969).
28. Kurath, D., Phys. Rev. 140, B1190 (1965).
29. Bouten, M., Bouten, M.C. and Van Leuven, P., Nucl. Phys. A168, 438 (1971); Nucl. Phys. A94, 687 (1967); Ann. of Phys. 43, 421 (1967).
30. Takigawa, N. and Arima, A., Nucl. Phys. A168, 593 (1971); Das Gupta, S., Hocguengham, J.C. and Giraud, B., Nucl. Phys. A168, 625 (1971).
31. A: Onishi, N. and Sheline, R.K., Nucl. Phys. A165, 180 (1971).
B: Krappe, H.J. and Wahsweiler, H.G., Nucl. Phys. A104, 633 (1967).
32. Drake, T.E. and Singhal, R.P., Submitted to Nuclear Physics.
33. Brown, G.E. and Green, A.M., Nucl. Phys. 75, 401 (1965).
34. Rowe, D.J., Contribution to the Mont Tremblant International Summer School, (1971).
35. Curran, C.S., Drake, T.E., Johnston, A., Gillespie, W.A., Lees, E.W., Singhal, R.P. and Slight, A.C. , Submitted to Nuclear Physics.
36. Häusser, O., Hooten, B.W., Pelte, D. and Alexander, T.K., Can. J. Phys. 48, 35 (1970).
37. Hofstadter, R., Noldeke, G.K., van Oostrum, K.J., Suelzle, L.R. and Yearian, M.R., Phys. Rev. Lett. 15, 758 (1965).
38. Singhal, R.P., Curran, C.S., Drake, T.E., Gillespie, W.A., Johnston, A. and Lees, E.W., Submitted to Journal of Physics.
39. Sick, S.I. and McCarthy, J.S., Nucl. Phys. A150, 631 (1970).
40. Brown, W.D. and Kujawski, E., Ann. of Phys. 64, 573 (1971).
41. Madsen, D.W., Cardman, L.S., Legg, J.R. and Bockelman, C.K. Nucl. Phys. A168, 97 (1971).
42. Myers, W.D. and Swiatecki, W.J., Ann. of Phys. 55, 395 (1969).

43. References contained in reference 32 above.
44. Amaldi, Jr., et al., Phys. Rev. Lett. 13, 141 (1964); Phys. Lett. 35B, 24 (1967).
45. Boffi, S., Bouten, M., Ciofi Degli Atti, C. and Sawicki, J., Nucl. Phys. A120, 135 (1968).
46. Hogg, G.R., Slight, A.C., Drake, T.E., Johnston, A. and Bishop, G.R., "The Glasgow Electron Scattering Facility", Submitted to Nucl. Insts. and Methods.
47. Drake, T.E. et al. Report to the National Research Council of Canada, Study Committee for "An Eastern Regional Facility", November, 1971.
48. Johnston, A. and Drake, T.E., "Inelastic Electron Scattering from ^{24}Mg". To be submitted to Nuclear Physics.
49. Drake, T.E., Programs to evaluate transverse and Coulomb form factors and ground state moments with shell model, Extended Nilsson Model and P.H.F. wave functions. University of Glasgow, Kelvin Laboratories.
50. Curran, C.S., Drake, T.E., Johnston, A., Gillespie, W.A., Lees, E.W. and Singhal, R.P., "Valence Neutron Transitions in ^{25}Mg", to be submitted to Nuclear Physics.
51. Rowe, D.J., Ullah, N., Wong, S.S.M., Parikh, J.C. and Castel, B., Phys. Rev. C3, 73 (1971).
52. Pratt, R.H. and Walecka, J.D., Nucl. Phys. 64, 682 (1965) (Appendix).
53. Castel, B. and Parikh, J.C., Phys. Rev. C1, 990 (1970).
54. Cheon, Il.-T., Phys. Lett. 30B, 81 (1969).
55. Häusser, O., private communication. August 1971.
56. The H.F. approach has been extended to the sd shell recently by Lee et al. H.C. Lee, private communication, Mont Tremblant International Summer School. August 1971.

GAMMA SPECTROSCOPY WITH HEAVY IONS

O. Häusser
Atomic Energy of Canada Limited
Chalk River Nuclear Laboratories

The usefulness of heavy ion reactions for the measurement of nuclear electromagnetic matrix elements is demonstrated. The effects associated with gamma emission from a swiftly-moving heavy ion are mentioned. It is shown how these effects can be exploited experimentally to deduce static and transition moments for states with mean lifetimes shorter than 10^{-11} sec. The techniques covered in some detail include the recoil-distance method, the Doppler-shift attenuation method, radiative capture using heavy-ion beams and the reorientation effect in Coulomb excitation.

Although the emphasis is on a discussion of these methods, examples of experimental work are given and their relevance to nuclear structure theory is briefly discussed.

1. INTRODUCTION

The interaction between the electromagnetic field and the nucleus is weak, allowing the use of perturbation theory, i.e.

$$H(t) = H_o + H_{int}(t).$$

H_o is the nuclear model Hamiltonian and the usual non-relativistic ansatz for the interaction is

$$H_{int}(t) = -\beta\sum\{2g_\ell^{(i)}\ \vec{P}_i\cdot\vec{A}(r_i)+g_s^{(i)}\ \vec{S}_i\cdot\vec{H}(r_i)\} \qquad (1)$$

where $\beta = eh/2mc$, g_ℓ and g_s are the orbital and spin g-factors of the nucleon, $\vec{P}_i$ and $\vec{S}_i$ its momentum and spin vectors, $\vec{A}(r)$ is the classical vector potential, $\vec{H}(r)$ = curl $\vec{A}(r)$ and the sum extends over the individual nucleons. A multipole expansion of $\vec{A}(r)$ and use of the long wavelength approximation ($kr << 1$) yields the usual multipole operators[1]

$$G^{el}_{LM} = \frac{eg_\ell}{E_\gamma}\ [H_o, r^L C_{LM}] \qquad (2a)$$

$$G^{mag}_{LM} = \beta(2g_\ell/L+1\ \vec{\nabla}(r^L C_{LM})\cdot\vec{\ell} + g_s\vec{\nabla}(r^L C_{LM})\cdot\vec{S}) . \qquad (2b)$$

The electric multipole operator (2a) is more general than the special form of the interaction (1) might suggest. According to Siegert's theorem[2] G^{el}_{LM} remains unchanged even if nuclear forces are assumed to have an exchange character or a velocity dependence.

The magnetic matrix elements, on the other hand, depend on, in addition to the structure of the eigenfunctions of H_0, nuclear current contributions which are not included in (1), for example two-body contributions. Recent calculations by Chemtob[3] indicate that meson exchange between two nucleons contributes up to 0.5β to the magnetic moments of single-particle states near closed shells. In practice it is however difficult to identify such contributions from any measured matrix element, since we do not know whether g_ℓ and g_s in a heavy nucleus are identical to those of the free nucleons. This point has been emphasized by de Shalit in his summary talk of the Heidelberg Heavy Ion Conference.[4] We can either use phenomenological values of g_ℓ and g_s or, preferably, deduce renormalized g-values from a perturbation calculation which takes detailed configuration mixing into account.

It is apparent from this example that one can only have faith in such procedures if a large number of diagonal and off-diagonal matrix elements in the same nucleus are accurately known for comparison with theory. Such a program poses severe problems to experiment because the lifetime of the states may vary by many orders of magnitude, requiring a variety of different techniques. During the last few years substantial progress has been made in this area mainly through the use of heavy ion reactions combined with gamma-ray detection by Ge(Li) spectrometers. It is the purpose of this progress report to present selected examples of the uses of heavy ions to measure electromagnetic matrix elements for states with mean lives shorter than 10 ps. The majority of the experimental information presented was obtained with the MP Tandem accelerator at Chalk River, mainly in collaboration with Drs. T.K. Alexander and A.B. McDonald.

2. GAMMA-EMISSION AT HIGH VELOCITIES

2.1 Kinematic Effects

In this section we briefly mention a number of kinematic effects associated with the emission of gamma rays from swiftly moving ions. These effects are easily observable with present Ge(Li) detectors ($\Delta E/E \lesssim 0.002$) and have to be taken into account in the analysis of most gamma-decay studies with heavy ions.

Doppler shift. The relativistically correct expression for the energy of a gamma ray, E_γ, observed after emission from an excited nucleus travelling with velocity v, is

$$E_\gamma = E^o_\gamma \frac{\sqrt{1-(v/c)^2}}{1-\frac{v}{c}\cos\theta} = E^o_\gamma \left(1 + \frac{v}{c}\cos\theta + \left(\frac{v}{c}\right)^2 \left(\cos^2\theta - \frac{1}{2} + \ldots\right)\right) \quad (3)$$

where E_γ^o is the gamma-ray energy in the nucleus rest frame and θ is the angle between the gamma ray and the nucleus in the observer's frame. The last term in (3) is the second order Doppler effect predicted by the special theory of relativity. An experiment designed to measure the quadratic effect for $E_\gamma \approx 10$ MeV and $v/c \approx 0.05$ is presently being undertaken at CRNL. Previous tests of (3) using the Mössbauer effect were restricted to rather low recoil velocities ($v/c \approx 10^{-6}$) and gamma-ray energies (< 10 keV).

Aberration. If θ' is the angle of emission in the rest frame of the moving nucleus, the observation angle in the lab system is obtained from the relation

$$\cos\theta = \frac{\cos\theta' + v/c}{1 + \frac{v}{c}\cos\theta'} \quad . \tag{4}$$

An important consequence of this "aberration" effect is that the relativistically correct transformation of solid angles is given by

$$d\Omega^{TN} = d\Omega^{LAB}\,\frac{1-(v/c)^2}{(1-\frac{v}{c}\cos\theta)^2} = d\Omega^{LAB}(1+2\frac{v}{c}\cos\theta + \ldots) \tag{5}$$

This equation implies large forward-backward asymmetries even for nonpolarized nuclei. The solid angle correction is sizable even for light particle reactions such as $(p,\alpha\gamma)$ measurements on light targets, and changes the values of mixing ratios deduced from angular distribution measurements obtained only in the forward hemisphere.

Doppler broadening. The gamma-emitting recoils following a nuclear reaction have generally a distribution in velocity correlated with the recoil direction. Because the three directions of beam, recoil and gamma detector are involved, we require the knowledge of the cross section $d\sigma(\theta)/d\Omega_p^{LAB}$ and a normalized angular distribution tensor $W_{k\kappa}$ to calculate the energy distribution of gamma rays. The Doppler shift is in first order

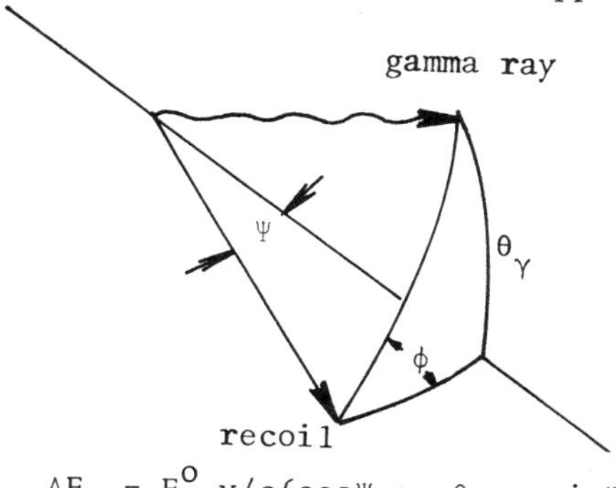

$$\Delta E_\gamma = E_\gamma^o\, v/c(\cos\Psi\,\cos\theta_\gamma + \sin\Psi\,\sin\theta_\gamma\,\cos\phi) \tag{6}$$

and the gamma-ray lineshape is

$$\frac{dN(E_\gamma)}{dE_\gamma} \sim \int \frac{dE}{dE/dx} \int d\Omega_p^{LAB} \frac{d\sigma(\theta)}{d\Omega_p^{LAB}} \sum_{k\kappa} \int \frac{dW_{k\kappa}(\theta,\Omega_\gamma^{TN},v)}{d\Omega_\gamma^{LAB}} \frac{dv}{dE_\gamma} d\Omega_\gamma^{LAB} . \qquad (7)$$

To provide an example equation (7) has been integrated to obtain the lineshape of the $2^+ \rightarrow 0^+$ transition in ^{28}Si following Coulomb excitation with 60 MeV ^{32}S ions. Calculated lineshapes for various angles and two different values of the static quadrupole moment Q are shown in Figure 1. The vanishing yield at E_γ^0 of the 90° lineshape is caused only by the properties of the angular distribution tensor $W_{k\kappa}$ for a $2^+ \rightarrow 0^+$ transition. At this observation angle the lineshape is completely insensitive to Q, whereas the influence of Q is largest at $\theta_\gamma = 0°$. The main cause for the sensitivity of the lineshape to the quadrupole moment at $\theta_\gamma = 0°$ can be traced to the difference in the particle angular distribution $d\sigma(\theta,Q)/d\Omega_p$ rather than to the particle angular distribution tensor.

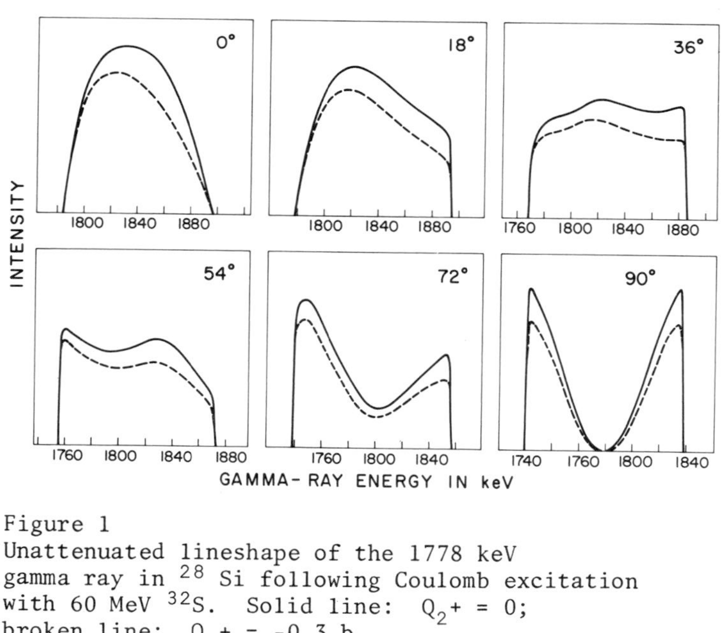

Figure 1
Unattenuated lineshape of the 1778 keV gamma ray in ^{28}Si following Coulomb excitation with 60 MeV ^{32}S. Solid line: $Q_{2^+} = 0$; broken line: $Q_{2^+} = -0.3$ b.

2.2 Deorientation

The distribution of magnetic substates following a nuclear reaction — which in turn determines the gamma-ray angular distribution — can change under the influence of external or internal fields. This can lead to annoying corrections in gamma-ray distribution measurements provided the nuclear lifetime is long enough for the perturbations to be effective. If, on the other hand, we know the strength and time-dependence of the perturbing fields, we can deduce nuclear moments from the data.

The large recoil velocities associated with heavy-ion reactions can be used to implant excited nuclei into host lattices with internal magnetic and electric fields. Large internal magnetic fields up to 2 MGauss have been observed in magnetized Fe hosts and have been used to measure g-factors of excited states with $\tau \gtrsim 10$ ps. An excellent review on details of this so-called IMPAC technique has been given by Grodzins.[5]

Much larger magnetic fields are obtained if excited nuclei are recoiling into vacuum. An atom recoiling out of a solid target is generally highly ionized. In medium heavy nuclei the charge state distribution is approximately Gaussian, typically with a full width of about four charge states and an average value $q \approx 10$. The outer electrons are initially expected to be distributed over many excited states. The unpaired electrons produce fields at the nucleus which fluctuate at a time-scale τ_c determined by the atomic transition rates. The fields are on the average isotropic, producing an attenuation of the gamma-ray angular distribution, i.e.

$$W(\theta) = \sum_{K=0}^{K_{max}} A_K G_K P_K(\cos\theta) \qquad (8)$$

where G_K are the time-integrated attenuation factors. If one assumes that the fields are fluctuating rapidly during the nuclear lifetime τ, one obtains[6,7]

$$G_k = 1/\tau \int_0^\infty G_K(t) e^{-t/\tau} dt = [1 + P_K \omega^2 \tau\ \tau_c K(I,J)]^{-1} \qquad (9)$$

where for a 2^+ state

$P_2 = 2$, $P_4 = 20/3$, $\omega = \mu_N gH/\hbar$ for magnetic dipole interactions;

$P_2 = 34/5$, $P_4 = 4$, $\omega = eQ/\hbar\ \partial^2 V/\partial Z^2$ for electric quadrupole interaction.

The factor K(I,J) depends on the relative magnitude of the nuclear spin I and the average electronic spin J. K is unity for I << J and < 1 for $I \gtrsim J$. It is evident from (7) that the multipolarity of the interaction can be determined from the ratio of G_2 and G_4. The pioneering work at the Weizmann Institute[7] showed that the interaction is almost entirely magnetic dipole. The dominance of the dipole interaction is confirmed in Figure 2, which was

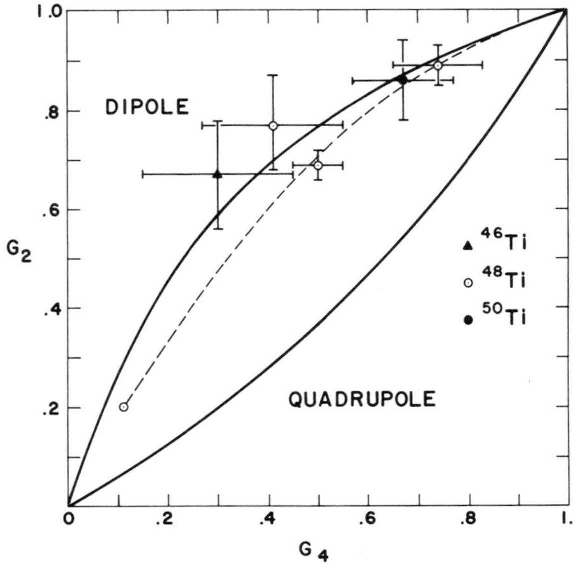

Figure 2
Attentuation factors G_2 and G_4 from Coulomb excitation of titanium isotopes. The solid line is calculated assuming a rapidly fluctuating perturbation, the dashed line is for a random static perturbation.

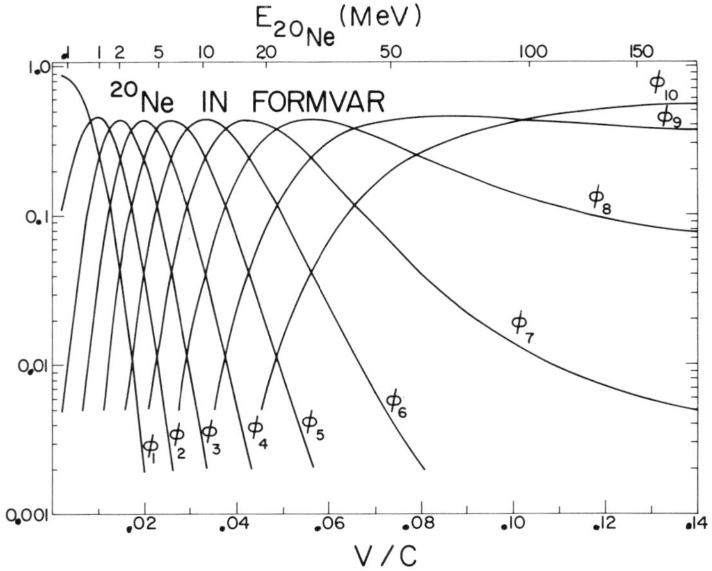

Figure 3
^{20}Ne charge state distributions calculated from ref.9.

obtained at Chalk River as a by-product of reorientation experiments in even titanium isotopes.[8] The use of attenuation coefficients for deducing absolute g-factors in medium heavy nuclei requires however a more detailed knowledge of the atomic processes than is presently available.

The situation appears somewhat more hopeful in light nuclei where only a few electronic configurations are involved whose fields can be accurately calculated. Calculated ionic charge state distributions for ^{20}Ne recoiling out of Formvar[9] are shown in Figure 3. Above $v/c \sim 0.06$ the probability for charge state 9^+ exceeds 30%. The lifetimes for allowed electronic transitions in the hydrogen-like configuration[10] are much shorter that 1 ps, which then implies a static perturbation for states with nuclear lifetimes of the order of 1 ps arising from the innermost (1s) electron. The magnetic field produced by a (1s) electron at the nuclear site is[10]

$$H(0) = 0.167\ Z^3 \text{ megagauss}$$

and the hyperfine splitting is

$$\Delta E_{HF} = \frac{g\beta(2I+1)}{\hbar} H(0) \quad . \qquad (10)$$

The attenuation factors for a static isotropic hyperfine interaction are[11]

$$G_k(t) = \frac{1}{2J+1} \sum_{FF'} (2F+1)(2F'+1)\left\{\begin{matrix} F & F' & k \\ I & I & J \end{matrix}\right\} e^{-i\omega_{FF'}t} \qquad (11)$$

where F is the total angular momentum and the $\omega_{FF'}$ are the hyperfine frequencies. For a 1s electron (11) simplifies and one obtains the time-integrated attenuation coefficients[12]

$$G_k(1s) = 1 - \Delta G_k = 1 - \frac{k(k+1)}{(2I+1)^2} \frac{(\omega\tau)^2}{1+(\omega\tau)^2} \quad . \qquad (12)$$

Since the magnetic field produced by (1s) electrons exceeds for example those of the $(1s)^2(2s)$ electrons by about a factor of 10, the attenuation factors for short-lived states (~ 1 ps) are almost entirely determined by the probability for the (1s) configuration. Accurate measurements of G_k can then be used to derive absolute magnetic moment measurements provided $\omega \approx 1$, i.e.

$$\tau \approx \frac{1.25}{(Z/10)^3(2I+1)g} \text{ ps} \quad .$$

This restriction on the lifetime is very severe; however a few magnetic moments may be measured this way. Such measurements would be very valuable since <u>relative</u> g-factors can be determined if ions are recoiling into gas.[7] In the recoil-into-gas method the characteristic collision time τ_c (eq. (9)) can be controlled by varying the gas pressure and a knowledge of the electron configuration is not essential.

3. RECOIL-DISTANCE METHOD

One of the most accurate methods for measuring lifetimes in the 10^{-12} - 10^{-9} sec range is the recoil-distance method. Its present form as developed from the early application of Alexander and Allen[13] is shown schematically in Figure 4. The decay of the excited state is measured directly, the variable delay being the time taken by recoils to cross a vacuum gap of adjustable length. The velocity distribution of the excited nuclei and the number of nuclei surviving for a time t can be measured, provided adequate gamma-ray resolution is available, because the Doppler effect separates nuclei decaying in flight from those decaying at rest in the metal stopper. The method is applicable for lifetimes as short as the slowing-down time of the ion in the stopper, i.e. $\approx 5 \times 10^{-13}$ sec. The practical problems of placing target and stopper at very small distances ($\geq$ 0.0002") and monitoring these distances have been solved[14,15] and a lifetime less than 1 ps (0.7 ps for the first excited state in ^{26}Mg) has already been measured.[16]

The large mean distances obtainable with heavy ion reactions are of course most welcome for this type of measurement. Figure 5 shows as an example three spectra obtained[17] at different target-

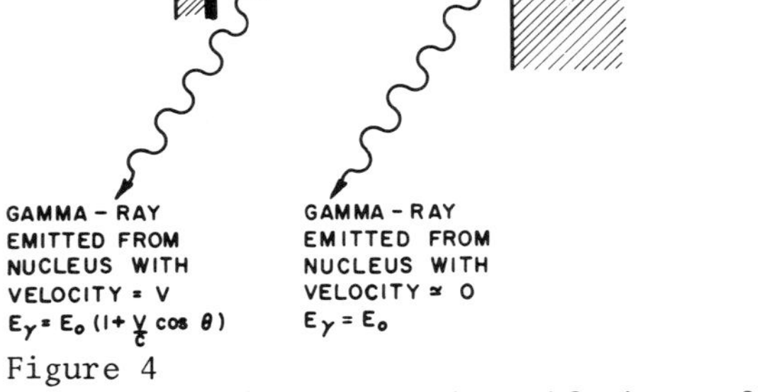

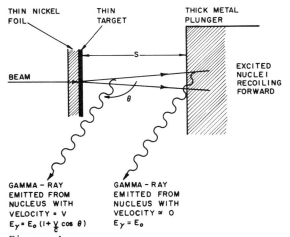

Figure 4
Recoil method of measuring lifetimes of excited states. For nuclei with lifetime τ; the number of nuclei decaying in flight is $N_1 = N_0(1-e^{-t/\tau})$, $t = S/V$, and the number of nuclei decaying in the plunger is $N_2 = N_0 e^{-t/\tau}$.

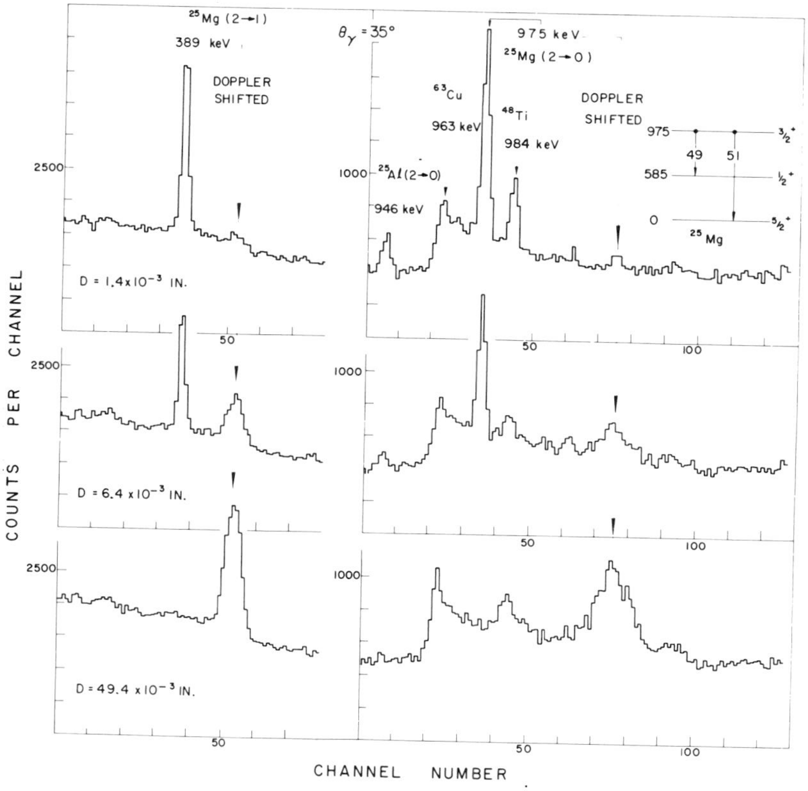

Figure 5
Spectra from the $^{2}H(^{24}Mg,p\gamma)^{25}Mg$ reaction obtained at three plunger distances.

plunger distances with the reaction $^{2}H(^{24}Mg,p\gamma)^{25}Mg$. In this case we use for a projectile what would be conventionally the target. The large recoil velocity ($v/c \sim 0.055$) and the small opening angle of the recoil cone (< 5°) makes the Doppler-splitting of the 389 keV line from the 975 → 585 transition in ^{25}Mg easily observable. The corresponding decay curves for two separate observation angles are shown in Figure 6. Needless to say, all corrections mentioned in section 2 have been applied to the intensities of stopped and moving peaks before fitting the data. The measured lifetime of the 975 keV level of ^{25}Mg corroborates the interpretation of the low-energy states in terms of the Nilsson model.

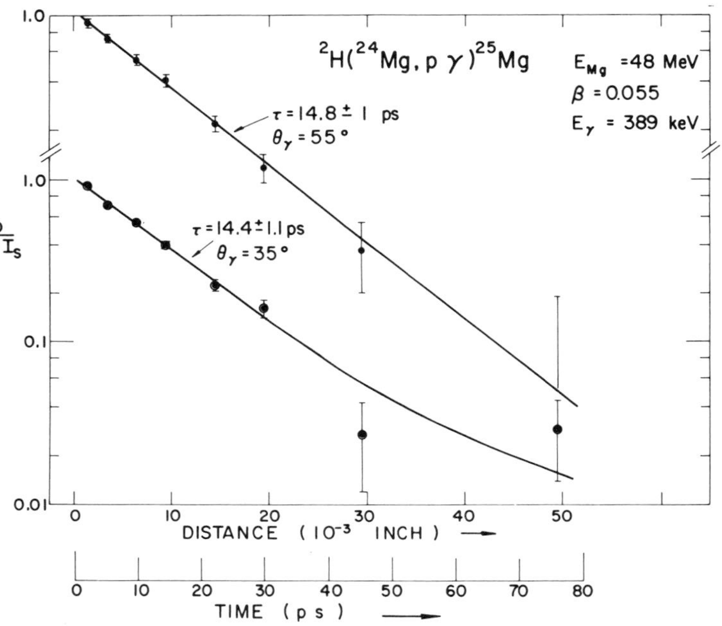

Figure 6
Decay curves for the 975 → 585 keV transition in ^{25}Mg.

4. DOPPLER-SHIFT ATTENUATION METHOD (DSAM)

In the DSAM as illustrated in Figure 7 the experiment determines either the projected velocity of the recoiling excited nuclei averaged over their lifetime, or the complete projected velocity distribution. Typical slowing-down times in solids are 10^{-14} - 10^{-11} sec and thus define the range of measurable lifetimes. The measurements can usually be made with good accuracy; however their interpretation requires the knowledge of the full history of the slowing-down process. In the majority of the analyses the stopping power is calculated from the theory of Lindhard[18] and the change of direction due to large-angle nuclear scattering is treated according to Blaugrund[19] by taking only the cosine of the average scattering angle, $\langle\cos\phi(t)\rangle$, into account. The uncertainties in this procedure can introduce systematic errors in the deduced lifetime of almost a factor of two as has been emphasized by Broude.[20] We shall briefly restate the source of these uncertainties:

a) at low velocities ($v/c < 0.2\%$) the stopping cross section is dominated by nuclear scattering. Measurements of the stopping power of various ions in argon have shown large deviations from Lindhard's predictions based on the Thomas-Fermi screened potential. Also the Blaugrund treatment of the average scattering angle yields poor agreement with observed lineshapes if light

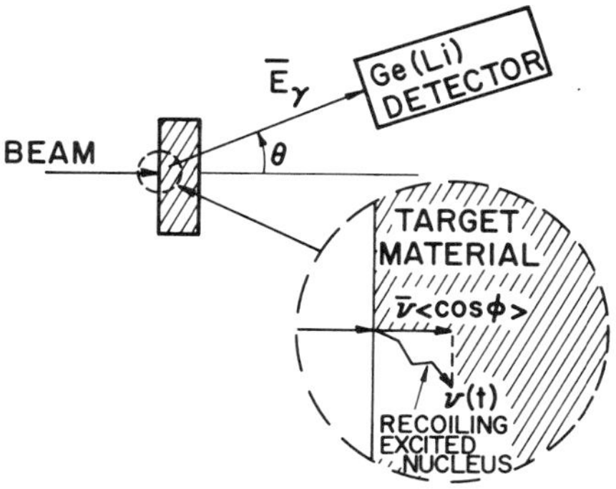

$$E_\gamma(t,\theta) = E_0\left[1 + \frac{\nu(t)}{c}\cos\theta\right]$$

$$\overline{E}_\gamma(\theta) = E_0\left[1 + F(\tau)\,\frac{\nu_0}{c}\cos\theta\right]$$

$$\text{where: } F(\tau) = \frac{\overline{\nu_z}}{\nu_0} = \frac{\int_0^\infty \frac{\nu(t)}{\nu_0} < \cos\phi(t) > \frac{dN}{dt}\,dt}{\int_0^\infty \frac{dN}{dt}\,dt}$$

or

$$F(\tau) = \frac{1}{\nu_0\tau}\int_0^\infty \nu(t) < \cos\phi(t) > e^{-t/\tau}\,dt$$

Figure 7
Definition of the F-factor in the DSAM.

nuclei recoil into heavy backings.[21] In these cases detailed Monte-Carlo calculations should be used. Recent measurements at low recoil velocities by Bister et al.[22] in which a variety of different backing materials were used show about a ±20% variation in the deduced lifetimes.
b) at intermediate velocities ($0.2\% < v/c < 2\%$) the electronic stopping cross section, which is proportional to the ion velocity, becomes dominant. Periodic fluctuations with projectile number, Z_1, about the theory of Lindhard with deviations of ±20% have been observed.[23] Lifetime measurements by Currie et al.,[24] obtained with ^{30}Si recoils of similar velocity but with different backing materials, showed deviations of much the same magnitude.

Measurements at initial recoil velocities $v/c > 2\%$ promise to yield somewhat more reliable lifetimes. The poorly understood and inaccessible region of nuclear stopping plays only a minor role in the slowing-down process. The stopping power of ions at high velocities depends mainly on the effective charge of the ion and a fairly reliable compilation exists.[25] The stopping power can also be measured fairly easily with new techniques[26] provided a low intensity beam of ions with nuclear charge Z_1 can be produced with an accelerator.

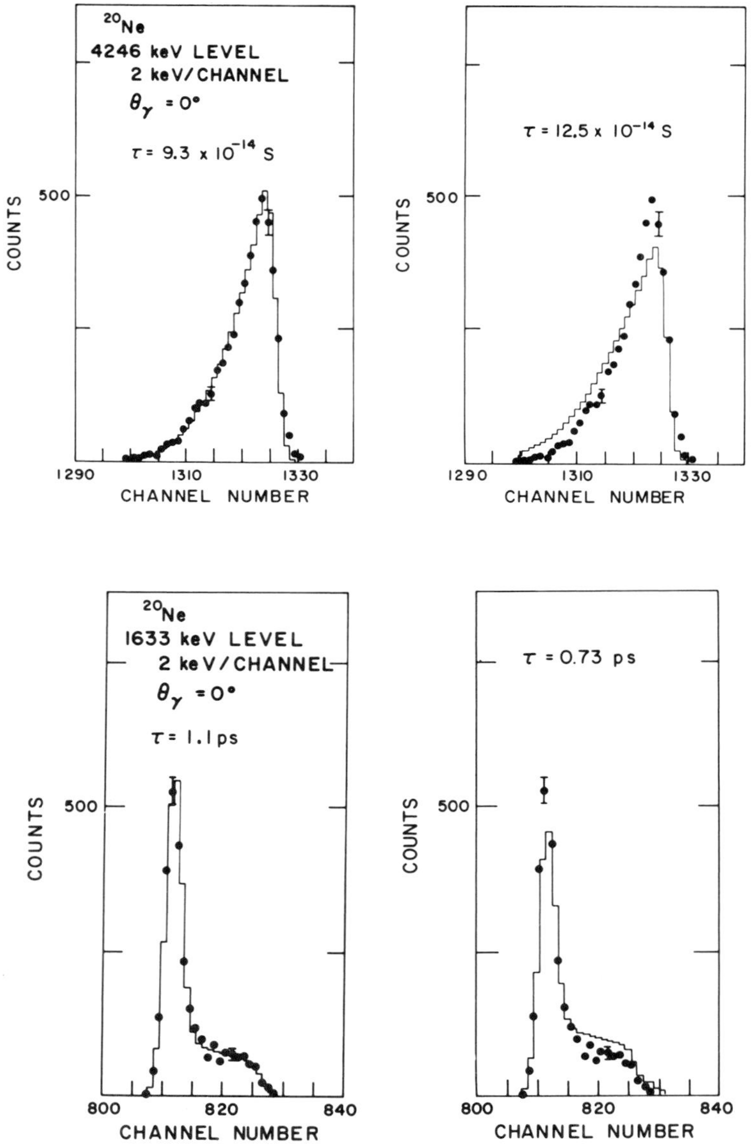

Figure 8
Lineshapes of the $4^+ \rightarrow 2^+$ and $2^+ \rightarrow 0^+$ gamma-ray transitions in ^{20}Ne, obtained with a well-defined initial velocity.

As an example Figure 8 shows lineshapes of transitions in ^{20}Ne fitted[27] with experimentally determined stopping cross sections for ^{20}Ne slowing down in Ni. The size and direction of the recoil velocity was defined by observing alpha particles from the $^{12}C(^{12}C,\alpha\gamma)^{20}Ne$ reaction near $\theta_\alpha = 0°$. The deduced lifetimes should in this case be fairly free of systematic errors.

The direct observation of gamma ray lineshapes following Coulomb excitation provides an example where the direction of the recoils is poorly defined. As pointed out before (eq. (7)) a calculation of the gamma-ray lineshape requires the knowledge of the cross section and the angular distribution tensor as a function of the particle scattering angle in addition to the slowing-down information. Figure 9 shows fitted lineshapes observed after Coulomb excitation of the first excited state in ^{28}Si by 60 MeV ^{32}S ions.[28] The lifetime was fitted at $\theta_\gamma = 90°$, where the lineshape is insensitive to the static quadrupole moment. In spite of the strong attenuation the lineshape at $\theta_\gamma = 0°$ has a slight sensitivity to the quadrupole moment favouring an oblate deformation of the 2^+ state in ^{28}Si.

The same technique has been applied recently[29] to Pb-isotopes. Figure 10 shows a line shape for a $3^- \to 2^+$ E1 transition in ^{206}Pb. Reasonable fits have been obtained at three observation angles although the somewhat high bombarding energy of 80 MeV and the poor statistics precluded a measurement of the 3^- quadrupole

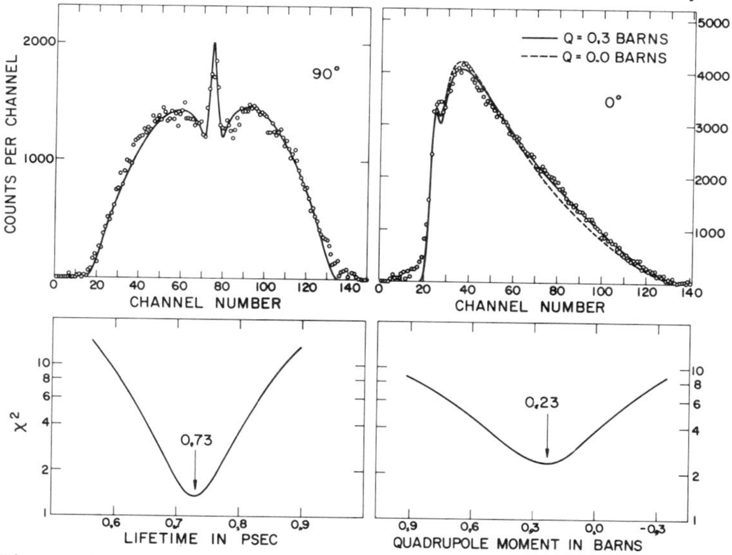

Figure 9
Gamma-ray lineshapes following Coulomb excitation of a thick ^{28}Si target with a 60 MeV ^{32}S beam. The $\theta_\gamma = 0°$ lineshape favours an oblate deformation of the 1778 keV first excited state in ^{28}Si.

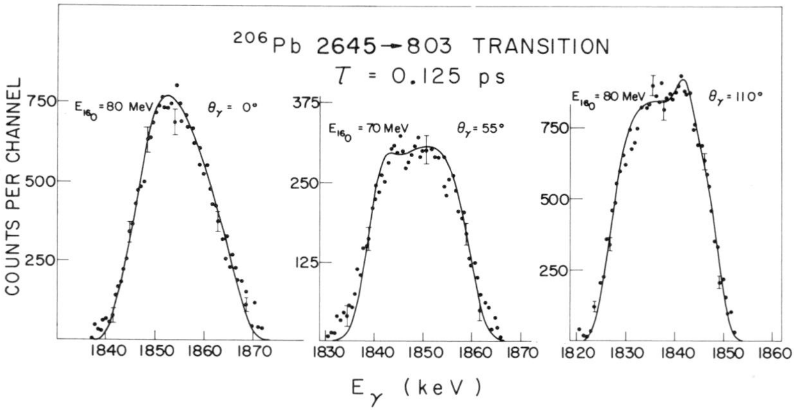

Figure 10
Gamma-ray lineshape of the $3^- \rightarrow 2^+$ E1 transition in ^{206}Pb obtained after Coulomb excitation with ^{16}O ions.

moment. Similar measurements of M1 matrix elements between single-particle states in ^{207}Pb and ^{209}Bi [29] indicate that magnetic moments and M1 transitions cannot be reproduced with a single set of effective g_ℓ and g_s for proton and neutron, respectively.

5. RESONANCE-CAPTURE REACTIONS WITH HEAVY ION BEAMS

The relatively large particle decay widths of unbound nuclear states precludes in most cases a direct measurement of the lifetime by the methods described previously. The gamma-decay widths of these states can however be determined with good accuracy by resonant capture, preferably using the particles of the predominant decay channel as target and projectile. The analyses of the traditional resonance reactions have been reviewed by Gove[30] and their study with the excellent resolution of Ge(Li) detectors has led recently to a renaissance of low-energy accelerators.

The small cross section of resonance-capture reactions makes their study difficult at high incident energies ($E_\alpha \gtrsim 5$ MeV, $E_p \gtrsim 3$ MeV) and only a very few experiments have been done at these energies. The prohibitive factors in alpha capture, for example, are the very large cross section for $(\alpha,n\gamma)$ reactions from the ubiquitous contaminants ^{13}C and ^{17}O, and the onset of $(\alpha,\alpha'\gamma)$ and $(\alpha,p\gamma)$ reactions.

The use of windowless differentially pumped gas targets, large NaI(Tℓ) detectors and the inversion of the conventional role of target and projectiles has now enabled the study of weak resonance-capture reactions at high CM energies.[31] The impurities in a gas can be kept extremely low and the build-up of hydrocarbons, a common problem with solid targets, is avoided. The advantage of using heavy ion beams is demonstrated in Figure 11 for the ^{4}He(^{16}O,γ)^{20}Ne reaction which is taken from the original work of

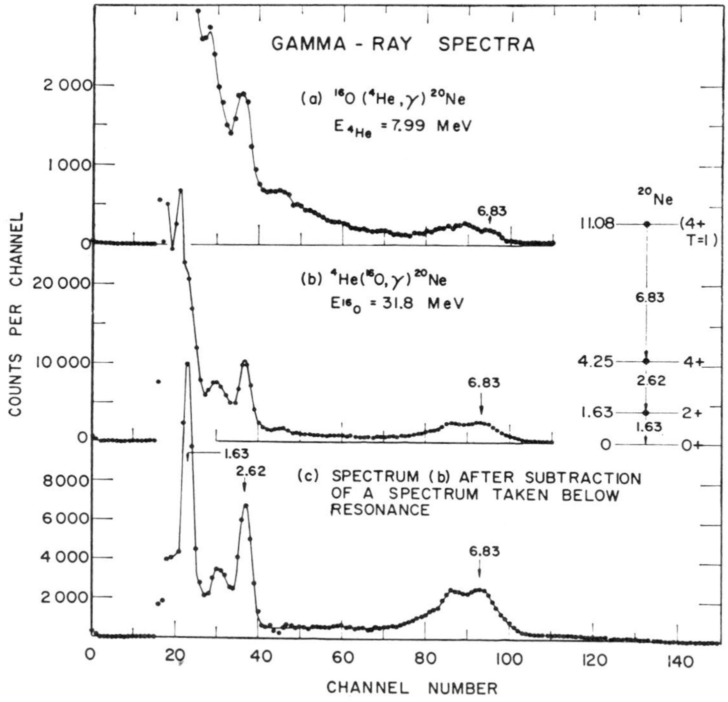

Figure 11
NaI(Tℓ) spectra from the decay of the 2^+, T=1 level at 11.08 Mev in ^{20}Ne. The spectrum obtained with an ^{16}O beam and a ^{4}He gas-target (b) is cleaner than the one where the roles of the target and projectile are interchanged (a).

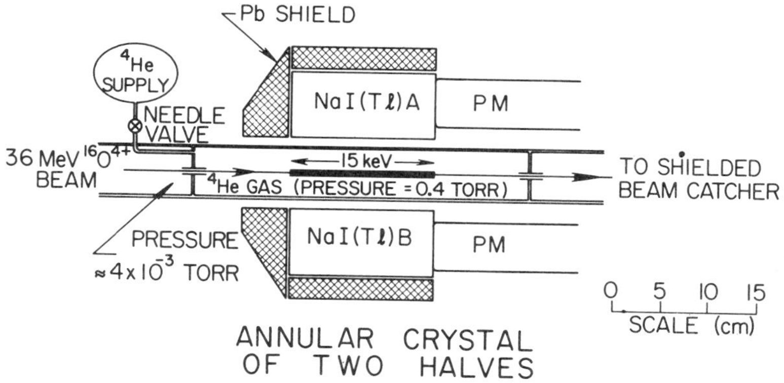

Figure 12
Schematical view of the differentially pumped Chalk River gas target.

Litherland and co-workers.[32] The background at low energies is improved considerably, mainly because activation of the beam lines by the Fe(α,n) reaction is avoided. The set-up of the present Chalk River gas target is shown in Figure 12. The differentially pumped gas cell is about 35 cm long. The pressure is adjusted to confine the resonance to within a 6" long annular NaI(Tℓ) detector, which surrounds the target and has a high efficiency and a high add-up probability for cascade gamma rays.

In the following, some recent measurements on states within the ground state band of ^{20}Ne [27,31,32] will be briefly discussed. A yield curve for gamma rays from the $J^\pi = 6^+$ resonance at 8.78 MeV is shown in Figure 13. The solid lines are the result of a calculation assuming a Gaussian energy profile of the beam and the E2 transition probability for the $6^+ \rightarrow 4^+$ transition of 21 Weisskopf units (W.u.). A similar yield curve for the $J^\pi = 8^+$ resonance state at 11.95 MeV in ^{20}Ne is shown in Figure 14. An excess of counts appears within 5 keV (1 keV CM) of the 8^+ resonance energy predicted from the elastic (α,α) scattering work.[33] Spectra obtained on resonance are displayed in Figure 15. The raw spectrum is shown on the top, the middle spectrum has an off-resonance spectrum subtracted, and a smoothed version of the

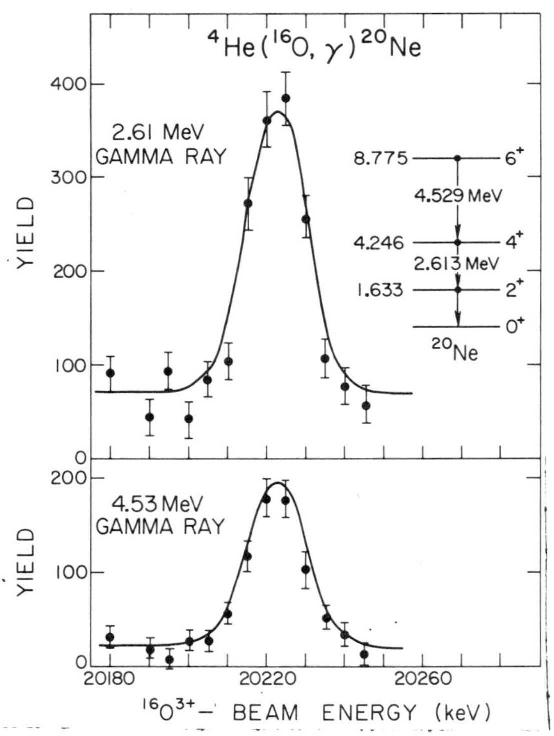

Figure 13
Yield curves obtained for the $J^\pi = 6^+$ resonance level at 8.775 MeV in ^{20}Ne.

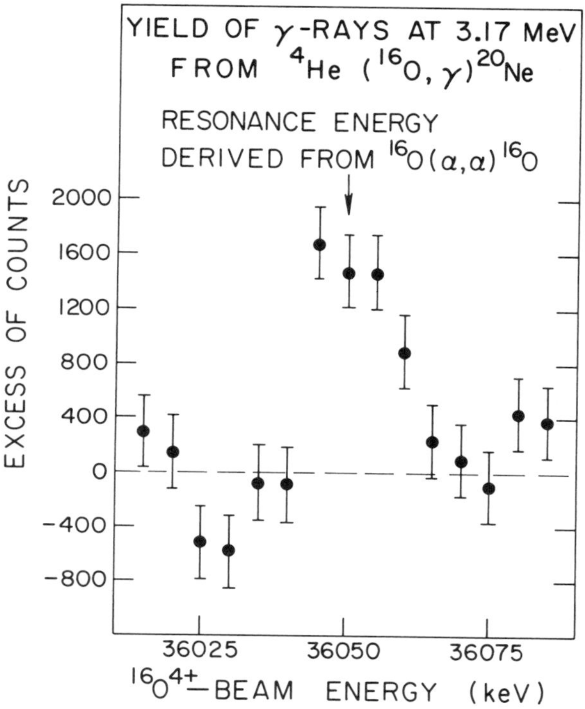

Figure 14
Yield curve of the 3.17 MeV gamma ray resulting from the $8^+ \rightarrow 6^+$ transition in ^{20}Ne.

latter spectrum (below) shows as the highest gamma ray in the spectrum the 3.17 MeV gamma ray from the $8^+ \rightarrow 6^+$ transition.

A summary of all measured E2 transition probabilities in ^{20}Ne [27,31] is shown in Figure 16 together with the theoretical predictions. The dashed lines are obtained from the adiabatic rotational model assuming a constant moment-of-inertia and neglecting Coriolis effects. The solid line was calculated by Harvey[32] in the SU_3 model which is representative of the shell model with four valence particles. The data are in good agreement with the shell model assuming a constant effective charge (0.43 for neutrons and 1.43 for protons).

A further indication that the spherical shell model provides an appropriate description of the lowest $K = 0^+$ and $K = 2^-$ bands in ^{20}Ne comes from recently measured alpha decay widths[30] (see Figure 17). The small values of the reduced widths in the high angular momentum states indicate a low probability for alpha-cluster formation. The decrease of the reduced widths from the $J^\pi = 6^+$ to the $J^\pi = 8^+$ state has been qualitatively explained as a decrease in the distance of the alpha-cluster from the ^{16}O core as mentioned in the lecture of Professor Arima.

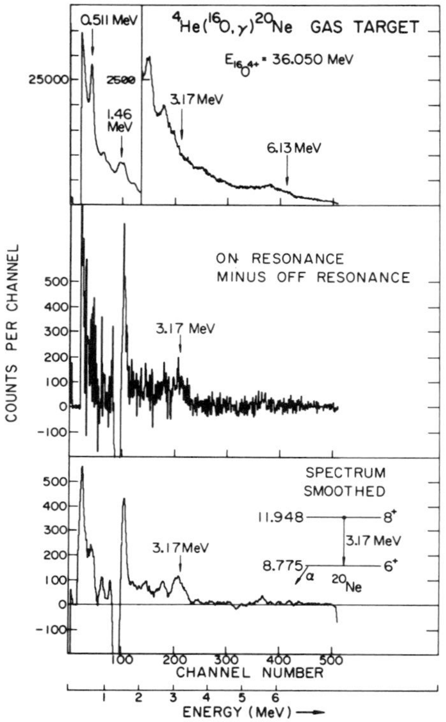

Figure 15
Gamma-ray spectra obtained from the 11.948 keV $J^{\pi} = 8^{+}$ resonance level in ^{20}Ne.

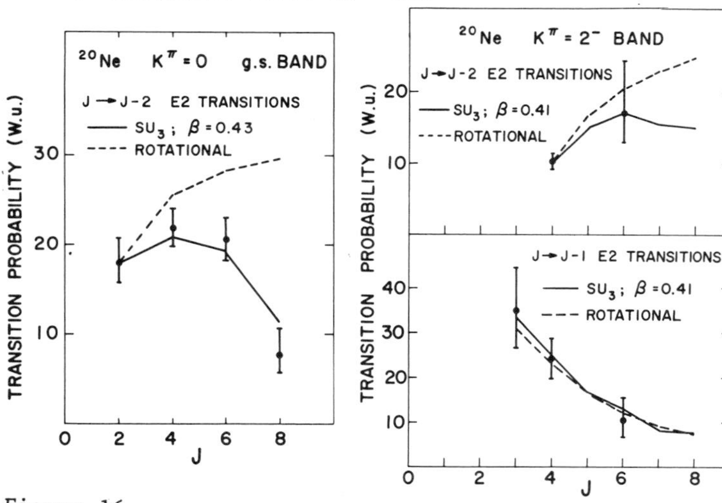

Figure 16
Summary of E2 transition probabilities in ^{20}Ne.

E_{exc}	J^{π}	Γ_{CM} (keV)	$\theta_{\alpha}^2 * 10^3$
8775 ± 3	6^+	0.110 ± 0.020	85 ± 15
11948 ± 5	8^+	0.035 ± 0.010	10 ± 3
5618 ± 3	3^-	$(3.1 \pm 0.7) * 10^{-6}$	57 ± 13
13333 ± 6	7^-	0.080 ± 0.030	0.8 ± 0.3

Figure 17
Alpha decay widths in ^{20}Ne.

6. COULOMB REORIENTATION IN LIGHT NUCLEI

One of the more obvious and fruitful applications of heavy ion beams in gamma spectroscopy is their use in Coulomb excitation experiments. One of the previous talks was in fact devoted entirely to the methods used to extract transition matrix elements from this reaction. I will therefore restrict my remarks to measurements of static quadrupole moments, in particular to the taming of the reorientation effect in light nuclei.

In all cases considered here the deBroglie wavelength is small compared with nuclear dimensions ($\lambda \lesssim 1$ fm). The localization of the projectile "orbit" then allows the use of semiclassical approximations. It has been suggested by Breit et al.[35] that the electric field gradients produced by the heavy ions at the target are sufficiently strong to modify measurably cross sections and gamma-ray angular distributions via their interaction with the quadrupole moment of the target. The effect on the excitation cross section can best be understood in second-order perturbation theory as indicated in the short notation of Figure 18. Each vertical line represents an Eλ matrix element —

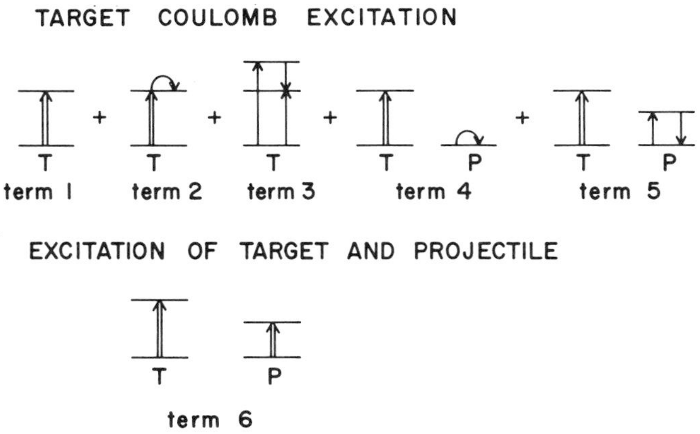

Figure 18
Processes contributing in a perturbation expansion of the cross section to Coulomb excitation of the target.

multiplied by appropriate orbital integrals — occurring in a perturbation expansion of the cross section. The first three terms result from the interaction of the projectile monopole with multipoles of the target and can be taken into account by the de Boer - Winther computer code.[36] The interference term 2 allows a determination of the sign and magnitude of the diagonal matrix element. Since the collision time is short ($\lesssim 10^{-21}$ sec) off-energy-shell contributions to term 3 can be fairly important. It has until recently been believed that the major contribution of this type results from virtual excitation of the 2nd lowest 2^+ level. We shall see later that excitation via the giant dipole resonance can also be very significant. Terms 4 - 6 come from projectile reorientation and simultaneous excitation of target and projectile and have been shown to be fairly small in most situations.[37]

All reorientation measurements so far have made use of the fact that the particle angular distribution is more sensitive to the reorientation term at backward angles. A typical set-up for particle-gamma coincidence measurements is shown schematically in Figure 19. ^{32}S particles scattered from a thin ^{112}Cd target are detected near 70° and 180° in two surface-barrier detectors. Inelastic events are selected by coincidence with several NaI(Tℓ) detectors which allow to determine empirically deorientation effects. Coincidence gamma-spectra are shown in Figure 20. The highest line results from projectile excitation of ^{32}S; the other lines originate from excitation of ^{112}Cd. A detailed analysis yields a prolate deformation of the 2.23 MeV state in ^{32}S, $Q = -0.175 \pm 0.050$ b, a large value in view of the vibrational features of the lowest five states. In Figure 21 the results of an anharmonic model are shown in which one- and two-phonon 2^+ states are allowed to mix with an amplitude X.

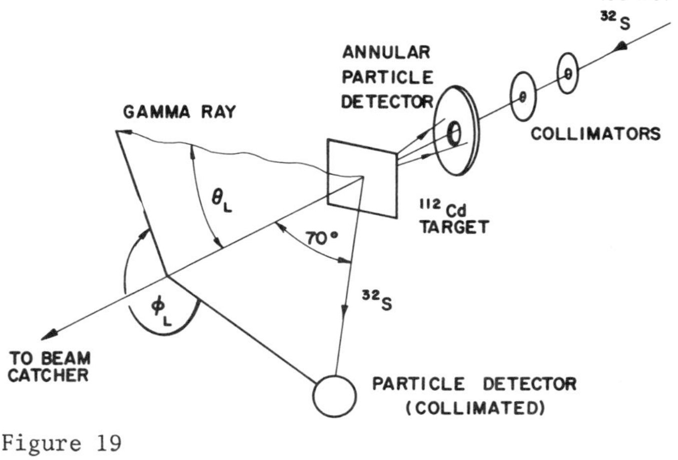

Figure 19
Schematic diagram of the experimental arrangement used to study projectile excitation of ^{32}S.

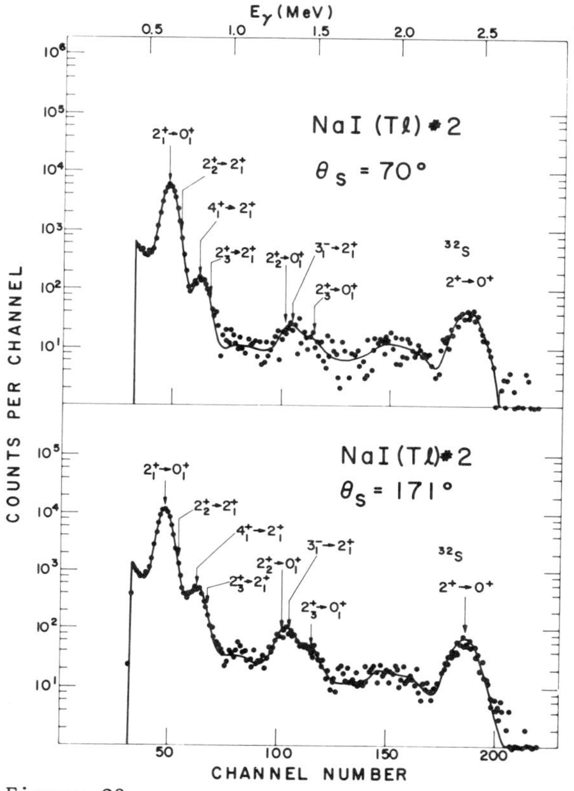

Figure 20
Two gamma-ray spectra obtained with a NaI(Tℓ) detector in coincidence with ^{32}S scattered at θ ~ 70° and θ ~ 171°. The solid lines were obtained from a fitting program using known lineshapes.

A large amount of mixing is needed to obtain only approximate agreement with experiment. A transitional model combining both vibrational and rotational features seems to be required to describe ^{32}S.

A summary of static quadrupole moments measured in the (sd) shell is presented in Figure 22. The intrinsic moments Q_0 are evaluated from Q_{2^+} and $B(E2;\ 0^+ \rightarrow 2^+)$ using the relations of the rotational model:

$$Q = Q_0 \frac{3K^2 - J(J+1)}{(J+1)(2J+3)} \qquad (13)$$

and

$$B(E2,\ J_i \rightarrow J_f) = 5/16\pi\ (J_i 2K0|J_f K_f)^2 Q_0^{\,2} \qquad (14)$$

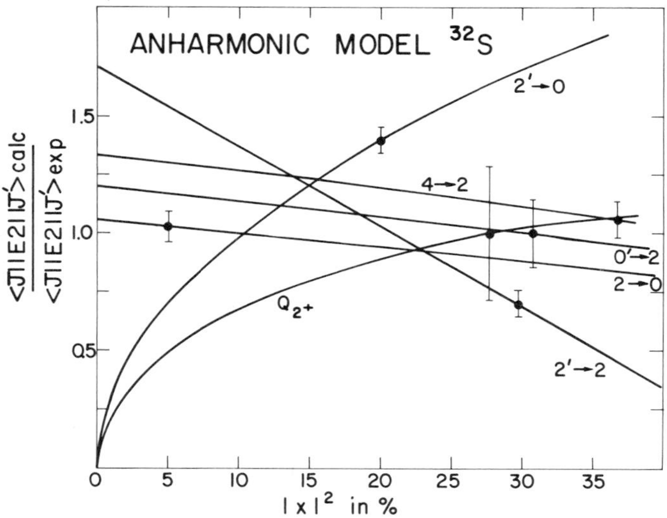

Figure 21
Ratios of theoretical and experimental reduced matrix elements in ^{32}S. Each line signifies a band with a width determined by the experimental errors. The errors are shown at arbitrary values of mixing parameter $|X|^2$.

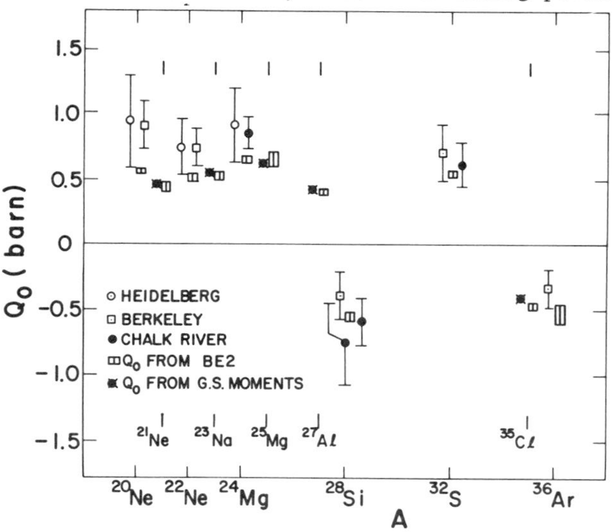

Figure 22
Summary of presently known intrinsic quadrupole moments in sd-shell nuclei derived from experimental static moments and BE2 values in the framework of the rotational model. Effects of the giant dipole resonance have been ignored in the interpretation of the reorientation measurement.

It should be noted that the rotational model is not strictly applicable to the low-energy levels in nuclei beyond mass 25. The marked change of sign of Q_0 between ^{24}Mg, ^{28}Si, ^{32}S and ^{36}Ar is qualitatively understood by Hartree-Fock calculations including explicitly up to five major shells.[39]

Another striking feature for mass 20, 22 and 24 nuclei consists in the 30-50% difference between Q_0 as deduced from the static moments and Q_0 deduced from BE2. In contrast to this behaviour neighbouring odd-A nuclei show excellent agreement between Q_0 from static moments and Q_0 from $B(E2, J_{g.s.} \rightarrow J_{g.s.}+1)$. It appeared therefore as a possibility that the reorientation measurements had not been correctly interpreted.

The same conclusion was also obtained from a recent study of projectile excitation of ^{18}O using ^{208}Pb as a target.[40] This experiment illustrates a simple and accurate technique applicable for light nuclei. Figure 23 shows a particle spectrum in

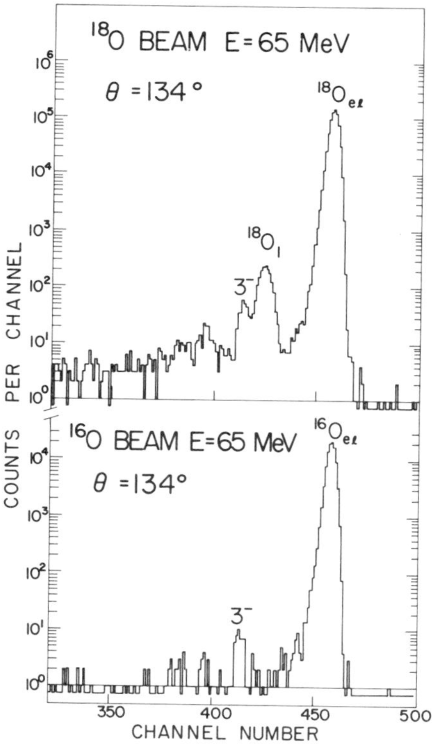

Figure 23
Spectra obtained with 65 Mev ^{18}O and ^{16}O beams and a ~20 $\mu g/cm^2$ thick ^{208}Pb target. The inelastic ^{18}O line is wider than the elastic ^{18}O line because of the Doppler shift of the unobserved $2^+ \rightarrow 0^+$ gamma-ray in ^{18}O.

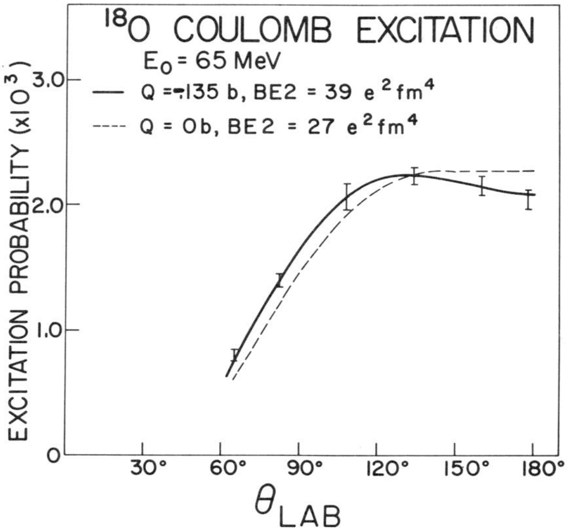

Figure 24
The excitation probability for the 1.98 MeV 2^+ state in ^{18}O fitted without taking effects of the giant dipole resonance into account.

which elastic and inelastic peaks are well separated as a result of careful collimation of the incident beam and of the particles scattered by the target. The peak-to-background ratio was $\approx 10^4{:}1$. The excitation probabilities are shown in Figure 24 together with theoretical fits in which only E2 matrix elements are taken into account. The fitted quadrupole moment, $Q \sim -0.135$ b, is more than a factor of two larger than the rotational value, $Q^{rot} = -0.05$ b. A quadrupole moment much larger than $|Q^{rot}|$ is however not expected on theoretical grounds since Hartree-Fock calculations[39] predict an axially symmetric intrinsic state for ^{18}O.

A possible source of the discrepancy has recently been identified by Aage Winther[41] to be due to virtual excitation of the giant-dipole resonance. This process can be viewed classically as a polarization of the nucleus that is being excited. It has been estimated in perturbation theory for rotational and vibrational nuclei by de Boer and Eichler[42] who concluded that the effect on the observed cross section was small. Winther has repeated their calculation and uncovered (trivial) numerical errors. In particular, their eq. (66) should read

$$\frac{P_{if}^{(12)}(E1)}{P_{if}^{(11)}(E2)} \approx 1.8 \times 10^{-3} \frac{A}{Z^2} \phi(\theta,\xi)E$$

where A,Z refer to the nucleus being excited, E is the bombarding energy in the C.M. system, and the function $\phi(\theta,\xi)$ shown in their Figure 4 should be multiplied by a factor $5\pi\sqrt{6} \approx 38.5$. The dependence of $\phi(\theta,\xi)$ versus θ is much weaker than the angular dependence introduced by the reorientation effect and, therefore, inclusion of the polarization effect modifies both B(E2; $0^+\rightarrow2^+$) and Q_{2^+}. If we correct the previous interpretation of reorientation experiments in the (sd) shell[38] somewhat better agreement with the predictions of the rotational model is obtained.

The most dramatic effect of virtual E1 excitation on measured cross sections has recently been observed[43] with the reaction ^{208}Pb(^{6}Li,αd) at E(^{6}Li) = 24 MeV where the ^{6}Li breakup proceeds through the 3^+ state at 2.18 MeV. The experiment showed that the reaction proceeds by a sequential process, i.e. Coulomb excitation of the 3^+ state followed by breakup of this level into α+d. The cross section observed at backward angles (Figure 25) is $\approx$ 40% lower than expected in first order from B(E2; $1^+\rightarrow3^+$) as determined by various electron scattering experiments. The discrepancy can be explained quantitatively if we take the E1 polarization effect into account. The numerical factor in eq. (15) assumes Migdal's form of the polarizability σ_{-2}, i.e.

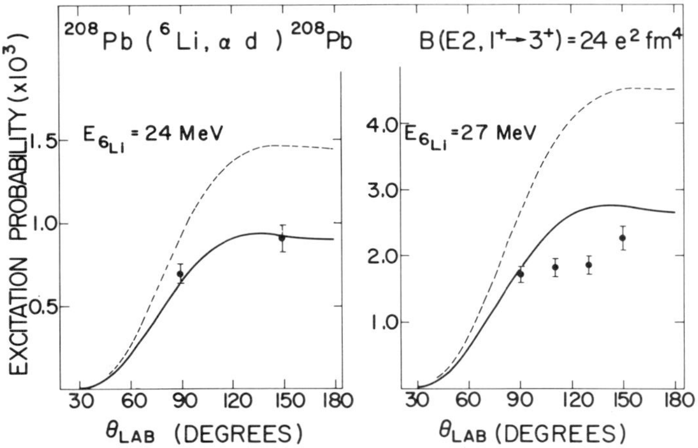

Figure 25
Comparison of experimental excitation probabilities for breakup of ^{6}Li proceeding through the $J^\pi=3^+$ state at 2.184 MeV with Coulomb excitation calculations. Dashed line: E2 excitation only; solid line: projectile polarization taken into account. The elastic scattering cross sections deviate from Rutherford scattering above ~26 MeV. The poor agreement at 27 MeV is probably due to the onset of nuclear reactions.

$$\sigma_{-2} = \int \frac{\sigma(E_\gamma)}{E_\gamma^{\,2}} \, dE_\gamma \approx 3.5 \; A^{5/3} \; \mu b/MeV \, . \qquad (16)$$

Photoabsorption measurements[44] show that the polarizability of ^6Li is about a factor of four larger than calculated from eq. (16). This and the low Z of ^6Li produce the observed large effect on the cross section.

7. CONCLUSIONS

In the preceding sections a variety of experimental techniques have been discussed which are suitable for nuclear lifetimes $\tau < 10$ ps. The unifying aspect of these techniques is that they allow the measurement of electromagnetic static and transition matrix elements and that they greatly benefit from the availability of heavy ion beams.

Static moments can be measured from their interaction either with the electric field gradient produced in Coulomb excitation by a swiftly moving heavy ion, or with the magnetic hyperfine field produced by the strongly ionized atom.

The radiative capture process allows the measurement of radiative widths of unbound states and can be used at high C.M. energies provided heavy ion beams and windowless gas targets are used.

The DSAM and the recoil-distance method make use of the Doppler effect to derive the recoil velocity at the time of gamma-ray emission. These latter methods are quite general since a very large number of nuclear levels can be produced by heavy ion reactions with suitable recoil velocities, especially in heavy nuclei.

The DSAM and the recoil-distance method make use of general kinematic effects associated with the emission of radiation from a moving source. They are therefore not limited to gamma-ray emission, but can be used for nuclei emitting particles with a suitable lifetime (e.g. fission).

REFERENCES

1. Rose, H.J. and Brink, D.M. Rev. Mod. Phys. 39, 306 (1967).
2. Siegert, A.J.F. Phys. Rev. 52, 787 (1937).
3. Chemtob, M. Nucl. Phys. A123, 449 (1969).
4. de Shalit, A. Summary talk in Nuclear Reactions Induced by Heavy Ions, Heidelberg 1969, North-Holland, ed. by Bock, R. and Hering, W.R.
5. Grodzins, L. ibid.
6. Nordhagen, R., Goldring, G., Diamond, R.M., Nakai, K. and Stephens, F.S. Nucl. Phys. A142, 577 (1970).
7. Ben Zvi, I. et al. Nucl. Phys. A121, 592 (1968).
8. Häusser, O., Pelte, D., Alexander, T.K., and Evans, H.C. Nucl. Phys. A150, 417 (1970).

9. Betz, H.D. et al. Phys. Lett. 22, 643 (1966); see also Heckmann, H.H., Hubbard, E.L. and Simon, W.G. Phys. Rev. 129, 1240 (1963).
10. Bethe, H.A. and Salpeter, E.E. Quantum Mechanics of one- and two-electron systems, in "Encyclopedia of Physics" (Flügge, S., ed.) Vol. 35, Springer Verlag Berlin, 1957.
11. Frauenfelder, H. and Steffen, R.M. Angular Correlations, in "Alpha-, Beta- and Gamma-Ray Spectroscopy" (Siegbahn, K., ed.) Vol. 2, North-Holland, Amsterdam, 1965.
12. Faessler, M.A., Povh, B. and Schwalm, D. Annals of Physics 63, 577 (1971).
13. Alexander, T.K. and Allen, K.W. Can. J. Phys. 43, 1653 (1965).
14. Gallant, J.L. Nucl. Instr. and Meth. 81, 27 (1970).
15. Alexander, T.K. and Bell, A. Nucl. Instr. and Meth. 81, 22 (1970).
16. McDonald, A.B., Alexander, T.K., Häusser, O. and Ewan, G.T. Can. J. Phys. in press.
17. Alexander, T.K., Häusser, O., McDonald, A.B. and Ewan, G.T. to be published.
18. Lindhard, J., Scharff, M. and Schiott, H.E. Kgl. Danske Videnskab Selskab Mat.-Fys. Medd 33, 14 (1963).
19. Blaugrund, A.E. Nucl. Phys. 88, 501 (1966).
20. Broude, C. ICPNS, Montreal 1969, 221.
21. Kutschera, W., Pelte, D. and Schrieder, G. Nucl. Phys. A111, 529 (1968).
22. Bister, M., Anttila, A., Piiparinen, M. and Viitasalo, M. Phys. Rev. C3, 1972 (1971).
23. Hvelplund, P., and Fastrup, B. Phys. Rev. 165, 408 (1968).
24. Currie, W.M., Earwaker, L.G. and Martin, J. Nucl. Phys. A135, 325 (1969).
25. Northcliffe, L.C. and Schilling, R.F. Nuclear Data Tables A7, 233 (1970).
26. Graham, R.L., Geiger, J.S. and Ward, D. AECL Report 3742; and Häusser, O., Ward, D., Disdier, D.L. and Gallant, J.L. AECL Report 3912.
27. Hausser, O., Alexander, T.K., McDonald, A.B., Ewan, G.T. and Litherland, A.E. Nucl. Phys. A168, 17 (1971).
28. Pelte, D., Häusser, O., Alexander, T.K. and Evans, H.C. Phys. Lett. 29B, 660 (1969).
29. Häusser, O., Ward, D. and Khanna, F.C. to be published.
30. Gove, H.E. Resonance Reactions, Experiment in Nuclear Reactions, Vol. 1, ed. by North Holland Publishing Co., Amsterdam.
31. Alexander, T.K., Häusser, O., McDonald, A.B., Ferguson, A.J. Diamond, W.T. and Litherland, A.E. Nucl. Phys. to be published; and Diamond, W.T., Alexander, T.K. and Häusser, O. Can. J. Phys. 49, 1589 (1971).
32. Litherland, A.E. et al. Can. J. Phys. 45, 1901 (1967).
33. Häusser, O., Ferguson, A.J., McDonald, A.B., Szoghy, I.M., Alexander, T.K. and Disdier, D.L. Nucl. Phys. to be published.
34. Harvey, M., private communication.

35. Breit, G. and Lazarus, J.P. Phys. Rev. 100, 942 (1955); Breit, G., Guckstern, R.L. and Russell, J.E. Phys. Rev. 103, 727 (1956).
36. Winther, A. and de Boer, J. In Coulomb Excitation, ed. by Alder, K., and Winther, A. (Academic Press Inc., New York) p.303.
37. Häusser, O. and Cusson, R.Y. Can. J. Phys. 48, 240 (1970).
38. Häusser, O., Alexander, T.K., McDonald, A.B. and Diamond, W.T. Nucl. Phys., in press; and references therein.
39. Lee, H.C. and Cusson, R.Y. to be published.
40. Disdier, D.L., Häusser, O., Alexander, T.K. and Ferguson, A.J. to be published.
41. Winther, A., private communication.
42. de Boer, J. and Eichler, J. Advances in Physics Vol. 1, Plenum Press, New York 1969, ed. by Baranger, M. and Vogt, E.
43. Disdier, D.L., Ball, G.C., Häusser, O. and Warner, R.E. to be published.
44. Bazhanov, E.B. et al. Nucl. Phys. 68, 191 (1965).

MEASUREMENT OF ELECTROMAGNETIC TRANSITION PROBABILITIES WITH THE COULOMB EXCITATION METHOD

D. WARD
Atomic Energy of Canada Limited
Chalk River Nuclear Laboratories

SYNOPSIS

In the first part of this lecture I will cover the basic mechanism of Coulomb excitation and show how the probability of exciting a nuclear level can be related to electromagnetic transition matrix elements or B(Eλ)'s. Experiments will be described in which measured cross sections are interpreted to give B(E2) values, firstly in perturbation theory, where only the matrix element connecting the level to the ground state is important, then in more complex cases, where we are obliged to use computer programs. Many matrix elements can be involved in calculating the cross section for a given level. The use of Coulomb excitation in measuring E4 transition probabilities will be described. The emphasis will be on understanding the Coulomb excitation process, experiments with Coulomb excitation, and interpretation of the results to give B(Eλ) values. However in each example the significance of the measurements from the nuclear structure point of view will be mentioned.

In the second part I will cover two other methods of determining transition probabilities by direct lifetime measurement, namely the Doppler Shift Attenuation Method (DSAM) and the recoil distance method. Neither of these techniques is specific to Coulomb excitation; however they can be applied with considerable success to levels populated by Coulomb excitation.

THEORY OF COULOMB EXCITATION

The process by which nuclear levels can be excited in the varying electric field caused by the close passage of a charged particle is called "Coulomb Excitation". Whether or not the target nucleus and projectile come within the range of nuclear forces, Coulomb excitation will occur; however if the initial kinetic energy of the projectile is sufficiently low, so that mutual electrostatic repulsion ensures that nuclear forces play no role, then Coulomb excitation will be the only mechanism for exciting levels in the target nucleus. This condition on the bombarding energy is given by the usual expression for the Coulomb Barrier, E_B,

$$E < E_B = Z_1 Z_2 (1+A_1/A_2)/(A_1^{1/3}+A_2^{1/3}) \qquad (1)$$

where the bombarding energy, E, is given (in the lab system) in MeV, and the suffixes 1 and 2 refer to the projectile and target respectively.* If we wish the surfaces to be separated by a distance, b fermis, we could write:

$$E(b) = Z_1 Z_2 (1+A_1/A_2)/(A_1^{1/3}+A_2^{1/3}+b/1.48)\ . \qquad (2)$$

Below this bombarding energy, the orbit of the projectile will be described by a hyperbola and the cross section for elastic scattering will simply be the Rutherford expression, which in the COM system is:

$$d\sigma/d\Omega = (1/4)\ a^2 \sin^{-4}(\theta/2) \qquad (3)$$

where a is half the distance of closest approach for 180° scattering, which for practical purposes is given in fermis by

$$a \cong 0.72\ (1+A_1/A_2)\ Z_1 Z_2/E \qquad (4)$$

(the bombarding energy will always be in the lab system expressed in MeV).

The most significant approximation in Coulomb excitation is to view the dynamics of the process as a classical Rutherford scattering, i.e. the semiclassical approximation. In this way the cross section for Coulomb excitation for a given level, f, from the ground state can be broken down into the probability for excitation along the orbit, times the differential elastic cross section:

$$(d\sigma/d\Omega)_f = P(g \rightarrow f) \times (d\sigma/d\Omega)_{elastic}\ . \qquad (5)$$

Furthermore, the magnitude and time dependence of the electric field at the target nucleus are easily derived from the dynamics of elastic scattering. The time-dependent Hamiltonian is just:

$$H(t) = \int \rho_n(\vec{r})\phi(\vec{r},t)d\tau \qquad (6)$$

where $\rho_n(\vec{r})$ is the nuclear charge density operator

$$\rho_n(\vec{r}) = \sum_k e\delta(\vec{r} - \vec{r}_k)$$

(the sum extending over the protons) and $\phi(\vec{r},t)$ is the Coulomb

* A radius parameter R_o = 1.48 fermis will be assumed throughout.

potential of the projectile:

$$\phi(\vec{r},t) = Z_1 e/|\vec{r} - \vec{R}(t)| - Z_1 e/R(t) \ . \tag{7}$$

The vectors are referred to an origin at the centre of the nucleus, the position of the projectile is denoted by $\vec{R}(\theta,\phi)$, and a volume element of the target nucleus $d\tau$ by $\vec{r}(\Theta,\Phi)$. The second term in Eq. (7) arises because we must subtract that term in the Hamiltonian which is responsible only for the scattering, since it has already been tacitly included in Eq. (5). We can effect a partial separation between the terms which are concerned only with determining the magnitude and time variation of the electric field, and the terms which are unique to the nuclear levels involved. This is done by noting that $1/|\vec{r}-\vec{R}|$ can be expanded in multipole moments:

$$1/|\vec{r}-\vec{R}| = (1/R(t))\left(1+\sum_{\lambda} (r/R)^{\lambda} P_{\lambda}(\cos\alpha)\right) \tag{8}$$

where α is the angle between the vectors $\vec{r}$ and $\vec{R}$. Using the spherical harmonic addition theorem for $P_{\lambda}(\cos\alpha)$ the Hamiltonian Eq. (6) can be written:

$$H(t) = 4\pi Z_1 e \sum_{\lambda=1}^{\infty} \sum_{\mu=-\lambda}^{\lambda} (2\lambda+1)^{-1} R^{-\lambda-1} Y_{\lambda\mu}(\theta,\phi) M^*(E\lambda,\mu) \tag{9}$$

where

$$M(E\lambda,\mu) = \int r^{\lambda} Y_{\lambda\mu}(\Theta,\Phi)\rho(\vec{r})d\tau \ .$$

This latter quantity is just the nuclear electric multipole operator of order λ which defines the reduced transition probability for an electric transition of order λ between states I_i and I_f through:

$$B(E\lambda, I_i \to I_f) = \sum_{M_i M_f \mu} |\langle I_i M_i | M(E\lambda,\mu) | I_f M_f \rangle|^2 \ . \tag{10}$$

Further progress can now be made by assuming first order perturbation theory, in which case the probability for excitation of the level f from the ground state g is determined by the transition amplitudes $b(g\to f)$, given by

$$b(g\to f) = (i\hbar)^{-1} \int_{-\infty}^{\infty} \langle f|H(t)|g\rangle e^{i\omega t} dt \tag{11}$$

where $\omega = E_f/\hbar$. Since we are interested in the probability for excitation regardless of the orientation of the initial and final states we can write:

$$P(g\rightarrow f) = (2I_g+1)^{-1} \sum_{M_g M_f} |b(g\rightarrow f)|^2 . \quad (12)$$

I will not follow these derivations any further here but refer to the classic paper by Alder et al.;[1] however it can be readily seen that when the Hamiltonian, Eq. (9), is substituted in the perturbation formula, Eq. (11), a term will arise which is just $B(E\lambda;I_i\rightarrow I_f)$. In practice it is not necessary to use perturbation theory since the problem can be dealt with in big computers using a coupled channels formalism as we shall see later, but perturbation theory provides an invaluable insight. For example in Eq. (11), although the structure of the matrix element is fairly complex, one can perhaps intuitively see that the transition amplitude results from integrating the folding of an oscillation, $e^{i\omega t}$, with $\langle H(t)\rangle$ where the time dependence of the $\langle H(t)\rangle$ must look rather like the time dependence of the electric field. The time dependence of the electric field is like a sharp pulse reaching a maximum at R=2a with a half width (in time) equal to $\sim a/v \sim 10^{-21}$sec. Following this line of thought it is evident that, unless the oscillation is longer than the duration of the "interaction pulse", the integral in Eq. (11) will tend to vanish. We could view the perturbation expression for the amplitude as a Fourier transform of the electric field pulse, so that a given level in the nucleus will be excited only if there are large frequency components in the pulse that are equal to those in the nuclear level, i.e. $\omega = E/\hbar$.

The parameter ξ is defined from the ratio of these times

$$\xi = (a/v)/\omega = (a/v)/(E_f/\hbar) . \quad (13)$$

An alternative definition which is often more useful is in terms of the dimensionless parameters η characteristic of the incoming and outgoing projectiles:

$$\eta_i = Z_1 Z_2 e^2/\hbar v_i \quad (14a)$$

$$\eta_f = Z_1 Z_2 e^2/\hbar v_f \quad (14b)$$

$$\xi = \eta_f - \eta_i . \quad (14c)$$

From the preceding arguments we require $\xi \lesssim 1$ and in typical systems ξ will lie in the range 0.05 to 1. In practical terms:

$$\eta_i \cong (Z_1 Z_2/2)(A_1/10.0\ E)^{1/2} . \quad (15)$$

The parameter η is actually the ratio of half the distance of closest approach, a, to the de Broglie wavelength of the projectile, and hence it is a measure of how well the orbit can be described by classical mechanics. For light ions $\eta \cong 3$, whereas for the heaviest ions η may be as high as 100.

The matrix elements $M(E\lambda;g\to f)$ in Eqs. (9) and (10) which govern the response of the nucleus to the electric field are very simply related to the reduced transition probabilities encountered in electromagnetic decays. The processes of Coulomb excitation and electromagnetic decay have an intimate connection. For Coulomb excitation, in 1st order theory, the probability for excitation by multipolarity, λ, is proportional to the reduced transition probability:

$$P(g\to f) \sim B(E\lambda;g\to f) \; . \tag{16}$$

The probability for electromagnetic decay from the same level to the ground state is:

$$1/\tau_\lambda \sim B(E\lambda;f\to g) \; , \tag{17}$$

where τ is the mean lifetime of the level. The relationship between the reduced matrix elements and reduced transition probabilities is:

$$B(E\lambda;i\to f) = (2I_i+1)^{-1} M(E\lambda;i\to f)^2 \; . \tag{18}$$

The matrix elements are symmetrical, i.e.

$$M(i\to f) = (-1)^{I_f-I_i+\lambda} M(f\to i) \tag{19}$$

whereas we can easily see from Eq. (19) that $B(E\lambda)\uparrow$ is related to $B(E\lambda)\downarrow$ by spin factors $(2I_f+1)/(2I_i+1)$. The two processes obey the same selection rules, which for electric multipolarities are:

$$|J_i-J_f| \leq \lambda \leq J_i+J_f \qquad \Delta\pi = (-1)^\lambda . \tag{20}$$

There are two significant differences between excitation and decay. Firstly, magnetic excitations are exceedingly weak, being of order $(v/c)^2$ relative to the corresponding electric multipole. Secondly, the decrease in excitation probability per unit increase in multipolarity is about a factor of 100, whereas in decay the corresponding factor is around 10^6. (We assume here single particle values for $B(E\lambda)$.) For example, in the case of a level sequence 0^+, 2^+, 4^+ the excitation of the 4^+ level can be measurably affected by $B(E4;0\to 4)$, whereas the decay of the same state is not.

Since E2 excitation is by far the most commonly encountered case, we quote here the final expression for the total cross section for E2 Coulomb excitation in first order theory:

$$\sigma(E2;g\to f) \cong 4.819(1+A_1/A_2)^{-2}A_1 Z_2^{-2}(E-\Delta E') \times f(\xi)B(E2;g\to f) \text{ barns} \tag{21}$$

where

$$\Delta E' = (1+A_1/A_2)\Delta E ,$$

with ΔE the excitation energy of the level f in MeV, and the reduced transition probability is in units of e^2barns^2. The function $f(\xi)$ is tabulated in Ref. 1. The probability for E2 excitation by an angle θ in the COM system is:

$$P(E2;g\to f) \cong 3710 \; df(\theta,\xi) \; A_1 E^3 B(E2;g\to f) Z_1^{-2} Z_2^{-4} (1+A_1/A_1)^{-4} . \tag{22}$$

Again, tables for the function $df(\theta,\xi)$ will be found in Ref. 1.

The behaviour of the excitation probability with scattering angle depends on the multipolarity, and on the value of the parameter ξ; for E2 excitations at small values of ξ, say < 0.2, the probability goes very roughly like $\sin^{-4}(\theta/2)$. Hence when we multiply this by the Rutherford cross section, Eqs. (3) and (5), it will turn out that the differential cross section for excitation is relatively flat as a function of scattering angle. As we increase ξ to 2, at which point the probability becomes vanishingly small, the differential cross section peaks at increasingly more backward angles and vanishes at forward angles. This is shown in Fig. 1, (taken from Ref. 1).

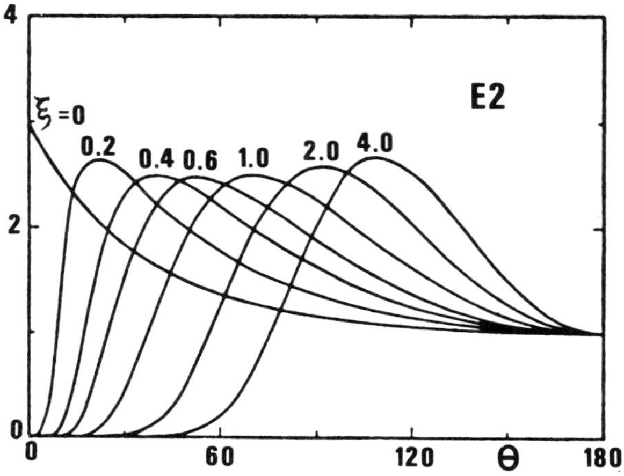

Figure 1
Differential cross sections for E2 excitations in the 1st order perturbation theory (from Ref. 1).

Substituting in Eq. (21) the condition that we bombard at the Coulomb barrier height (Eq. (1)), it will be seen that the cross section for E2 Coulomb excitation in 1st order theory varies approximately as Z_1^2. This shows the great power of heavy ions for inducing Coulomb excitation provided the accelerator is capable of giving the ion at the energy required by Eq. (1). Thus ^{32}S ions give ~250 x (the cross section of protons) and ~15 x (the cross section of ^{4}He ions). For the heaviest nuclei, only the HILAC accelerators (Manchester, Yale, Berkeley) are currently capable of producing ions of say $Z \cong 18$, at the required 190 MeV energy. Tandem accelerators can only manage the feat with ^{16}O ions (80 MeV); however with the upgraded MP Tandem at Chalk River, we should be able to accelerate ^{35}Cl ions to an energy suitable for work on the heaviest nuclei.

In the case of nuclei with well-developed ground state rotational bands, the enhancement of the B(E2)'s may be so great that a perturbation expansion breaks down. For example, putting the numbers into Eq. (22) for ^{16}O ions at 50 MeV on ^{170}Er we find that B(E2;0→2) ~ 1.5 at 180° scattering! In cases like this it is necessary to solve the exact semiclassical equations by numerical methods. Winther and de Boer have developed a method for doing this using a computer program.[2] I shall not describe their formalism in detail but point out that the time-dependent Schrodinger equation, governing the final amplitudes,

$$i\hbar\, db_n/dt = \sum_m \langle n|H(t)|m\rangle \exp\left(i(E_n-E_m)t/\hbar\right) b_m(t) \ , \qquad (23)$$

can be solved numerically to any desired accuracy. The user supplies the matrix elements M(Eλ;n→m) connecting the nuclear levels, and for this set of matrix elements, an essentially exact solution (within the context of the semiclassical approximation) is provided by the program. A few months of working with the program, however, convinces the user what nature knew all along, that the excitation probability for a given level depends to a greater or lesser extent on the matrix elements connecting it to every other level in the nucleus. By using the coupled channels formalism we have lost the explicit relationships between B(E2) values and cross sections, so that we can only say that a given set of matrix elements will produce a certain set of cross sections. To use the Winther-de Boer program effectively it is useful to have a good grasp of perturbation theory, so that one has a guide in deciding which matrix elements are important, and in understanding why changes in the matrix elements affect the cross sections in the way they do.

In perturbation theory the amplitudes, b, are proportional to matrix elements ; thus for a one step process

$$b_{if} \sim \langle i|M|f\rangle \ ,$$

and for a 2-step process, through an intermediate state z,

$$b_{izf} \sim \langle i|M|z\rangle\langle z|M|f\rangle \ .$$

To get the measured probabilities we must add all amplitudes and then square, and in this way cross terms (interference terms) will arise. The cross terms are of course not positive definite quantities, so that the relative signs of matrix elements play a role as soon as we leave 1st order theory. In 2nd order perturbation theory the amplitude for the final state, f, will be

$$b_{izf} = (1/i\hbar^2) \int_{-\infty}^{\infty} dt\langle f|H(t)|z\rangle e^{i\omega_2 t} \int_{-\infty}^{t} dt'\langle z|H(t')|g\rangle e^{i\omega_1 t'} \qquad (24)$$

where ω_1 is related to the excitation energy of the intermediate state $= E_z/\hbar$ and $\omega_2 = (E_f - E_z)/\hbar$. Although the probability for observing the intermediate state may be vanishingly small, its effect on the final state can be considerable. For example, the intermediate state could be at a very high excitation energy, that is $\xi < 1$, in which case the ξ-dependence for observing it goes as $\sim \xi^{-2\pi}$; however the probability for its interfering with a lower state goes only as $\sim \xi^{-1}$.

The dependence of multiple E2 excitation probabilities on the projectile, Z_1, go very roughly as Z_1^{2p} where p is the order of the process, assuming that the probability in order p is much smaller than unity. Thus double excitations such as $0^+ \rightarrow 2^+ \rightarrow 4^+$ go as Z_1^4, whereas $0 \rightarrow 4$ goes as Z_1^2. Furthermore the angular distribution of inelastically scattered particles is quite different from that observed in single excitation; in multiple excitation, the probability is sharply peaked at backward angles.

Summary of the Theory

In summary I would just like to review the approximations used in Coulomb excitation theory. By negligible in this discussion I will mean less than 1% effect on a total cross section or probability for excitation. In other types of measurement, such as gamma-ray angular distributions, or reorientation, these effects may be more important. A list of effects which are usually neglected is given below:

a) relativistic corrections
b) atomic screening
c) quantum electrodynamics
d) structure of the projectile.

Of this list effect a) is negligible;[1] effects b) and c) can be taken into account quite easily; however they tend to cancel and their combined effect is usually negligible.[1] Excitations in the projectile (the fact that the projectile is not a point charge) can be taken into account using a treatment in which combined states of the system are viewed as states of the target nucleus.[3]

The exactness of the semiclassical formalism is greatly improved when the symmetrized velocity $v = (v_i \cdot v_f)^{\frac{1}{2}}$ rather than the incident velocity v_i is used in the calculation.[2] The error introduced will depend on the parameters η and ξ. Provided η is large enough, say 50 or more, which is typical with heavy projectiles, there will be no problems; however if η is down below 10 or so (e.g. ^{4}He on nickel isotopes) then the errors may be significant if ξ is large. Except in the measurement of E4 moments, situations do not arise in which both quantum mechanical corrections and deviations from 1st order perturbation theory occur together; therefore in analysing results with light ions it is best to use perturbation treatments where the quantum mechanical effects can be taken into account properly, rather than use the Winther-de Boer program. Recently Alder and Pauli have presented some results on the deviations between the semiclassical and quantum mechanical treatments.[4]

When second and higher order effects are large it is better to use the Winther-de Boer program. The solution is then essentially exact (but still semiclassical); however it is exact only for the levels and matrix elements specified. The uncertainties concerning the levels and matrix elements which could influence the cross-section of interest are by far the largest uncertainty in most contemporary Coulomb excitation calculations.

APPLICATION OF THE THEORY TO OBTAIN B(Eλ)'s

I will now describe some experiments to illustrate the techniques of extracting reduced transition probabilities from experimental cross sections.

Experimentally, Coulomb excitation can be detected either by direct observation of the scattered particle spectrum, or by the gamma decay of the excited states (or conversion electron spectroscopy). Currently the latter techniques have a far greater sensitivity, and all the experiments I will describe involve gamma-ray spectroscopy. Most of the experiments attributed to the Chalk River group were carried out in collaboration with Drs. R.L. Graham and J.S. Geiger.

1st Order Theory

There are several interesting cases where it is useful to analyse the cross sections in first order theory; these involve nuclei close to closed shells, or nondeformed odd mass nuclei in which the E2 strength is fragmented into many levels. In the applications I shall discuss, the experiment is simply to measure the absolute cross section for production of gamma rays de-exciting the level of interest with ^{4}He ions. An excitation function must be measured in order to establish the validity of the theory, and to distinguish the energy of the parent state from which the gamma ray was emitted. The term $f(\xi)$ in the expression for the cross section introduces a dependence on the state energy ΔE; for example the excitation function for a high-lying level rises much more rapidly than that for a low-lying level.

Because the angular distribution of gamma rays emitted following a single E2 Coulomb excitation has an almost negligible $P_4(\cos\theta)$ coefficient, the gamma detector is set at 54°, with respect to the beam direction, where $P_2(\cos\theta)$ vanishes. A typical experimental arrangement is shown in Fig. 2. In our case we use two detectors at ±54° in order to average out some of the effects of not positioning the standard calibration sources at the beam spot position. Since it is difficult to measure target thicknesses to the required accuracy, we employ targets which are thick enough to stop the beam. In this case the uncertainty in the knowledge of the stopping power, dE/dx, for ^{4}He ions in the material then carries through to the final uncertainty in the B(E2) value. Reliable measurements of dE/dx for ^{4}He ions exist (~±4%) and we have ourselves performed some measurements which we hope will be published shortly. The recent compilation of Northcliffe and Schilling seems to be in excellent agreement with almost all the experimental data for ^{4}He ions known to us.[5] The beam current is integrated in a carefully suppressed Faraday cup so that we have all the ingredients necessary to extract the B(E2) value from Eq. (21), which in the case of a thick target appears as

$$I = (10^6/qe)(N_A/A)\int_{E_i}^{0} \frac{\sigma(E)dE}{dE/dx} , \qquad (26)$$

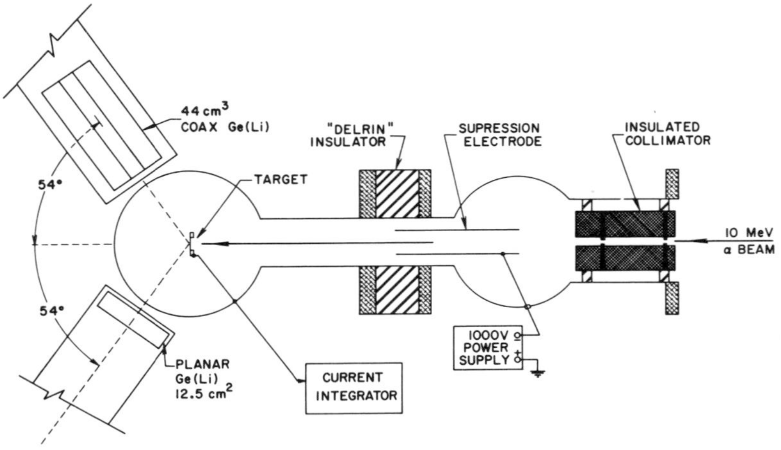

Figure 2
Apparatus for measuring absolute yields of de-excitation gamma rays in Coulomb excitation. The target is thick enough to stop the beam.

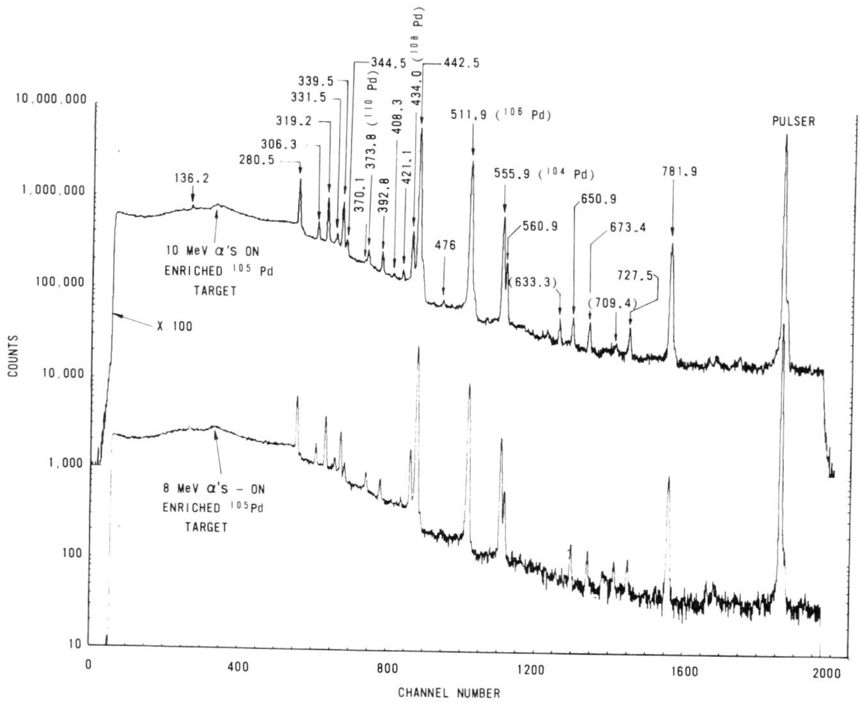

Figure 3
Spectrum of gamma rays from a thick target of ^{105}Pd, enriched to 97%, bombarded with ^{4}He ions. As an example of the sensitivity of the technique it should be noted that γ476 corresponds to an effective average cross section of about 20 μb.

where I is the thick target yield per μcoulomb for the level of interest, deduced from the absolute cross section for de-excitation gamma rays, and must include corrections for conversion electrons, decay by other branches, and for gamma ray feeding from higher lying levels. The factors in front are Avogadro's number N_A and the charge state of the ion = q.

The example I have chosen to illustrate 1st order theory is ^{105}Pd excited by ^{4}He ions. A typical gamma spectrum is shown in Fig. 3. The Ge(Li) detector has vastly increased the sensitivity of gamma-ray measurements over that possible with NaI(Tℓ). For example some of the weak gamma rays seen in Fig. 3 have a cross section of only ~ 20 μb. In this particular case some care has to be taken in obtaining the correct independent level populations, in view of the complexity of the decay scheme. Transition probabilities as a function of bombarding energy are shown in Fig. 4. The constancy of the B(E2) values over the range of bombarding energy (the absolute yield varied by an order of magnitude for

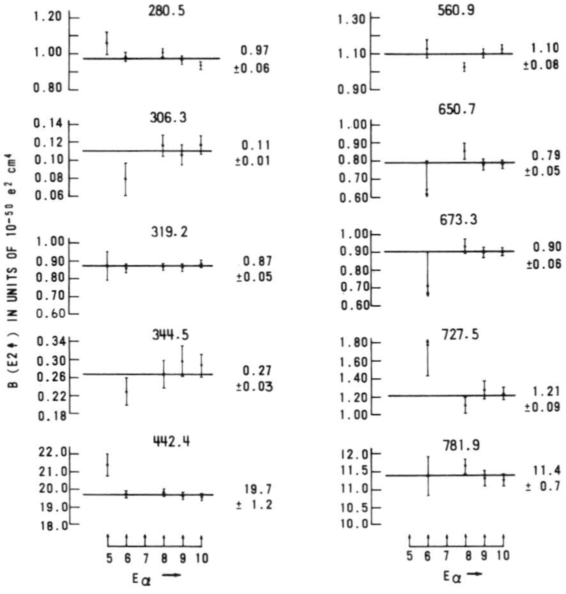

Figure 4
Reduced transition probabilities for levels in ^{105}Pd as a function of ^{4}He bombarding energy.

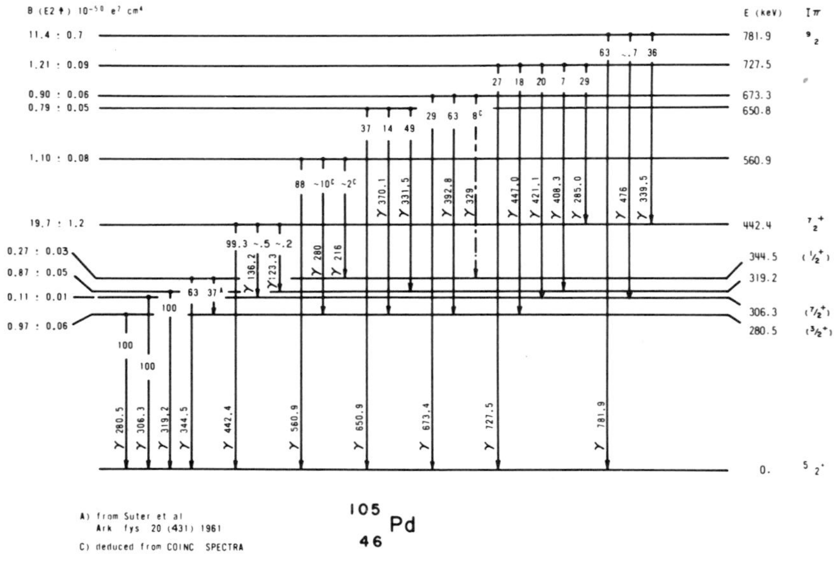

Figure 5
Level scheme and B(E2)'s for ^{105}Pd.

the higher states) is good evidence that the gamma-ray cross feedings have been properly taken into account.

The level scheme from this work (and other experiments) is shown in Fig. 5. As a zeroth order attempt to understand the level structure we can invoke the weak coupling picture;[6] the odd neutron will then be in the $2d_{5/2}$ orbital, and the one phonon excitation of the core coupled to the neutron will give rise to 5 states of spins $1/2^+$ through $9/2^+$. Evidently states of these spins (and more!) can be found in the level scheme. In weak coupling, the B(E2) values should correspond with that of the $2^+ \rightarrow 0^+$ transition in the core. For ^{105}Pd we find that experimentally $B(E2;7/2 \rightarrow 5/2) = 0.148 \pm 0.009\ e^2b^2$ and $B(E2;9/2 \rightarrow 5/2) = 0.069 \pm 0.005\ e^2b^2$ compared with the weak coupling prediction of $0.102 \pm 0.005\ e^2b^2$ for all transitions from the multiplet to the ground state. There are no other levels of anywhere near the right degree of collectivity to be considered as members of the multiplet. The total observed E2 strength, which is $\Sigma B(E2;g \rightarrow f) = 0.374\ e^2b^2$ plus the contribution from the ground state quadrupole moment of ^{105}Pd $\langle Q \rangle_{5/2} = 0.8$ barns, is $0.552\ e^2b^2$. This may be compared with the core strength $B(E2;0 \rightarrow 2) = 0.508\ e^2b^2$. The existence of at least 10 excited states below 1 MeV and the large splitting of the 7/2 and 9/2 levels suggest that an intermediate coupling calculation including perhaps the $g_{7/2}$, $d_{5/2}$, $d_{3/2}$ and $s_{1/2}$ neutron orbitals would be more realistic.[7]

Winther-de Boer Program for Vibrational Nuclei

This example concerns a study of levels in ^{103}Rh, with various ions; the ground state spin is $1/2^-$ in this case. From relative cross sections with ^{4}He, ^{16}O and ^{35}Cl we can establish that levels at 848 and 920 keV are excited by double Coulomb excitation, and gamma ray angular distributions suggest that their spins are almost certainly 7/2 and 9/2 respectively. A level at 881 keV has also been identified as spin 5/2. It is tempting to describe these levels as components of the multiplet $p_{1/2} \otimes 2$ phonons; see Fig. 6. The levels at 295 and 357 keV, spins $3/2^-$ and $5/2^-$ are known to be reasonably well described by the weak coupling model as $p_{1/2} \otimes 1$ phonon. Since there exist almost no data on 2 phonon couplings in odd mass nuclei, we have tried to extract the matrix elements coupling the 1 phonon to the 2 phonon levels. The case of the 9/2 level is relatively simple since it is only excited by one dominant path, namely $1/2 \rightarrow 5/2 \rightarrow 9/2$. On the other hand the 7/2 level is excited by two paths $1/2 \rightarrow 3/2 \rightarrow 7/2$ and $1/2 \rightarrow 5/2 \rightarrow 7/2$. In the lowest order the cross section for excitation for the 7/2 level depends on 4 matrix elements. We know 2 reduced transition probabilities $|M(E2;1/2 \rightarrow 3/2)|^2$ and $|M(E2;1/2 \rightarrow 5/2)|^2$ from work with ^{4}He ions, which could be treated in 1st order theory. If we invoke the gamma-ray branching ratio $\gamma(7/2 \rightarrow 5/2):\gamma(7/2 \rightarrow 3/2)$ we have one further piece of information, although we must bear in mind that the $\gamma(7/2 \rightarrow 5/2)$ is a mixed M1/E2 transition. The way we have applied ourselves to this problem is to invoke 2nd order pertur-

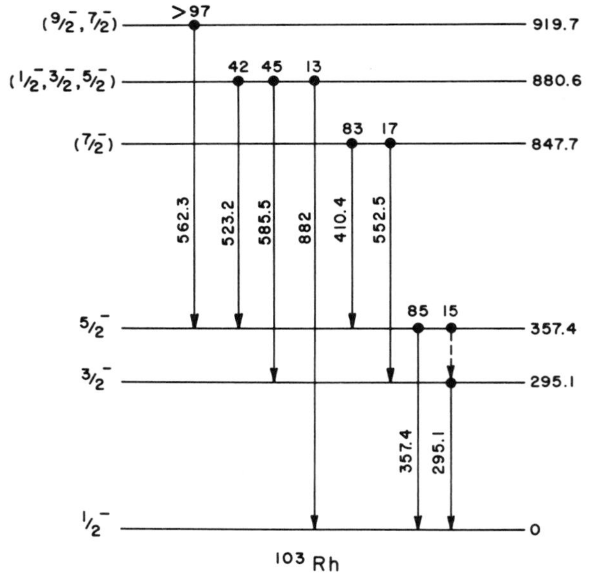

Figure 6
Partial level scheme for ^{103}Rh from Coulomb excitation studies with heavy ions. The three highest levels are believed to correspond with a weak coupling of the $p_{1/2}$ proton to the 2 phonon excitation of the core.

bation theory in which

$$\sigma(7/2) \sim |a\ M(E2;1/2\rightarrow3/2)\ M(E2;3/2\rightarrow7/2) + b\ M(E2;1/2\rightarrow5/2)\ M(E2;5/2\rightarrow7/2)|^2 \ .$$

I will abbreviate the notation in what follows:

$$\sigma(7/2) \sim \alpha M(1/2\rightarrow3/2)^2\ M(3/2\rightarrow7/2)^2 + \beta M(1/2\rightarrow5/2)^2\ M(5/2\rightarrow7/2)^2 + \gamma M(1/2\rightarrow3/2)\ M(3/2\rightarrow7/2)\ M(1/2\rightarrow5/2)\ M(5/2\rightarrow7/2) \ .$$

It can now be seen that we do not require a knowledge of the signs of all the matrix elements, but merely a certain combination of signs. We proceed further by running the Winther-de Boer code for a variety of values of $M(3/2\rightarrow7/2)$ and $M(5/2\rightarrow7/2)$ with the appropriate bombarding conditions. The coefficients α, β and γ were evaluated by fitting the results of these calcu-

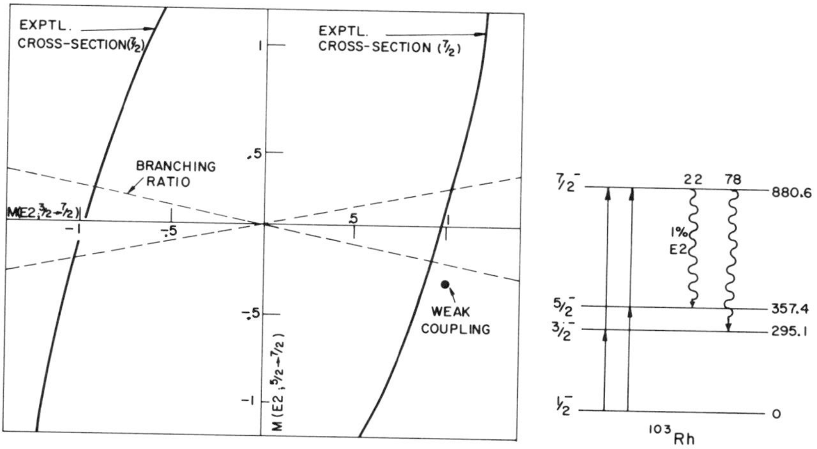

Figure 7
Interpretation of the experimental cross section for the 7/2 level in ^{103}Rh. By using the Winther-de Boer program to fit the coefficients of a 2nd order expansion we can plot contours of constant cross section as functions of the matrix elements. From the gamma-ray branching ratio and E2/M1 admixture in $\gamma(7/2\rightarrow5/2)$ the ratio of the matrix elements is known. The 4 points of intersection then correspond to possible solutions for the matrix elements. (N.B. experimental errors have not been indicated for the purpose of clarity.)

lations. The fit is fairly good and we attribute the small discrepancies to a failure of 2nd order perturbation theory. We now have an analytic expression for the cross section in terms of the unknown matrix elements from which we can plot contours of constant cross section in the plane of M(E2;3/2→7/2) and M(E2;5/2→7/2). This is shown in Fig. 7. The solution will lie along the pair of lines representing the experimental cross section. The gamma branching ratio defines two further lines, passing through the origin, which intersect the experimental cross section contours in 4 places. These intersections define possible pairs of values for the matrix elements.

The weak coupling model gives the following expression for E2 matrix elements between N and N+1 phonons:

$$\langle j_1 j_2 ; j \| E2 \| j_1' j_2' ; j' \rangle = (2j+1)^{\frac{1}{2}} (2j'+1)^{\frac{1}{2}} \times W(j_1 j_2 2 j' ; j j_1') \langle j_1 \| E2 \| j_1' \rangle \, \delta(j_2, j_2') \qquad (27)$$

where we express the matrix element for the coupled system in terms of the core matrix elements; j_1 is a core state coupled with j_2, the particle state, to resultant j. Taking the core values for ^{102}Ru from experiment, we find that $M(E2;9/2\rightarrow7/2)_{exp}$ = 1.32±0.07eb compared with $M(E2;9/2\rightarrow 7/2)_{weak\ coupling}$ = 1.34±0.11 eb. The weak coupling prediction for $M(E2;7/2\rightarrow 5/2)$ and $M(E2;7/2\rightarrow 3/2)$ is shown in Fig. 7. The experimental uncertainty in $M(E2;7/2\rightarrow 5/2)$ is about ±30% on account of the uncertainty in the E2/M1 mixing ratio.

Winther-de Boer Program for Rotational Nuclei

Multiple Coulomb excitation can be used to measure B(E2) values within the ground-state bands of deformed nuclei. With heavy ions such as ^{32}S at 110 MeV or ^{40}A at 140 MeV it is possible to observe excitations up to the 10^+ level in the rare earth region. As an example let us consider how we might extract $B(E2;6\rightarrow 8)$ from an experimental determination of the cross section for excitation of the 8^+ level in ^{152}Sm. In lowest order, the cross section will be proportional to $|M(E2;0\rightarrow 2)M(E2;2\rightarrow 4)M(E2;4\rightarrow 6)M(E2;6\rightarrow 8)|^2$. (The first of these terms is known very accurately from direct lifetime measurement by electronic means.) A possible procedure would be to use successively heavier ions to establish the matrix elements step by step: thus with ^4He we might hope to measure $B(E2;2\rightarrow 4)$ independently of $B(E2;4\rightarrow 6)$ and using this result with ^{16}O ions we might measure $B(E2;4\rightarrow 6)$ independently of $B(E2;6\rightarrow 8)$ etc. However since we are interested in detecting quite small deviations from rotational behaviour (in the rare earths) next order processes should not be neglected. In the next order we will encounter the static moments of the 2^+, 4^+, 6^+ and 8^+ levels, and interference from excitations in the β and γ vibrational bands. Furthermore, there will be interference from octupole bands by E3 excitations, and E4 excitation within the ground band. For example, with typical heavy ions on ^{152}Sm the E4 effect can amount to about a 5% change in the cross section for the 8^+ level. The dependence of the cross sections on static quadrupole moments through the "reorientation effect" is quite interesting since in deformed nuclei the effect comes in a non-linear way, that is the reorientation effect is not proportional to Q_J but includes a $Q_J{}^2$ term (and higher powers). The usual estimate given by de Boer and Eichler[8] that $\Delta P/P \sim Q_2 \Delta E$, where ΔP is the change in the probability of exciting the level on introducing a static quadrupole moment Q_J, greatly underestimates the reorientation effect in deformed nuclei. The combined effect of all the corrections mentioned above is calculated to be typically between 5% and 10% of the 6^+ and 8^+ cross sections, even though a considerable amount of cancellation occurs. The analysis of these experimental cross sections then enters a philosophical phase as to how far we can assume the rotational model to give us unknown matrix elements, such as Q_{8^+}, $B(E2;6_\beta\rightarrow 8_g)$, etc., whilst we look for deviations in the ground band cross sections from the rotational values.

As I have mentioned, the deviations from rotational behaviour in ^{152}Sm are not very marked; $B(E2;6^+\rightarrow 4^+)$ is 1.14 ± 0.04 of rotational and $B(E2;8^+\rightarrow 6^+)$ is 1.23 ± 0.12 of rotational, according

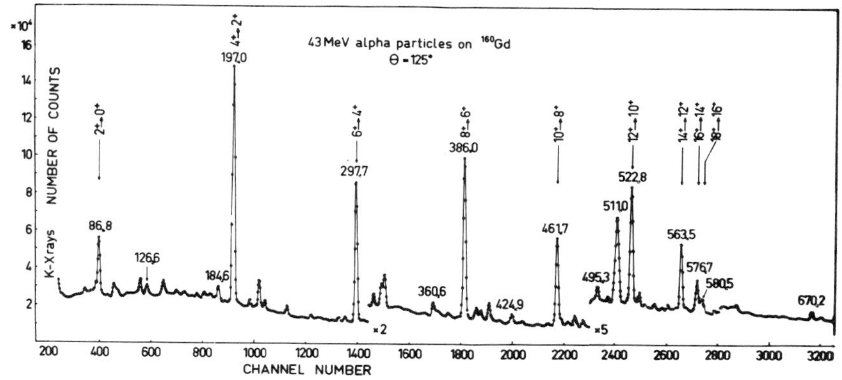

Figure 8
The singles spectrum recorded at 125° relative to the beam direction. The indicated peaks are assigned to the (α,4n) reaction. (After Johnson et al.)

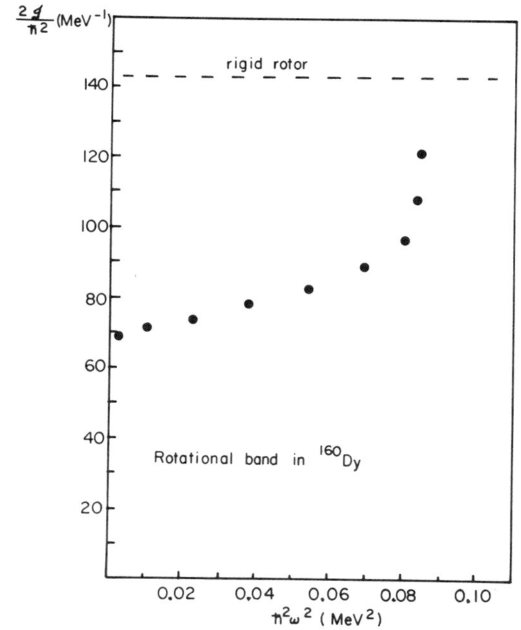

Figure 9
The moment of inertia for ^{160}Dy as a function of the rotational frequency. The dashed curve represents the calculated value for the moment of inertia for a rigid body (after Johnson et al.).

to Doppler Shift Recoil Distance Measurements by Diamond et al.[20] and by Nils Rud and the Chalk River group. I shall be describing this work in more detail later. Similar results have been obtained by Stokstad from cross section measurements;[9] however I think these results are more susceptible to systematic error because of the problems I have already mentioned.

A case where the measurement of cross sections may be extremely useful is in the excitation of very high spin states recently reported by Ryde et al.[10] This group has identified the 16^+ and 18^+ members of the ground state band in ^{160}Dy and in several other rare earth nuclei by (^{4}He,3n) and (^{4}He,4n) reactions (see Fig. 8). The reason why other groups did not assign them several years ago is that the energy spacings are very irregular (for rare earth nuclei). These anomalies have been interpreted in terms of the Coriolis force which has become so large at J = 16 that the neutron pairs are decoupled, and the moment of inertia makes a rigid body value as shown in Fig. 9. This effect was of course predicted by Mottelson and Valatin several years ago.[11] It would be of great interest to measure the transition moment between states on opposite sides of the phase change. To see whether this is feasible or not in Coulomb excitation I have performed a calculation assuming rotational B(E2) values using the Winther-de Boer program, the results of which are shown in Fig. 10. With the upgraded MP tandem at Chalk River I estimate we could have enough counts to do the experiment, although it does look a tough one. In a good rotational nucleus such as ^{160}Dy the problem with interference from the vibrational bands will be much less severe than it was in ^{152}Sm since they lie at a higher energy, and are not so strongly mixed in to the ground band. We would also hope for much larger effects at the phase transition.

The response of the intrinsic state to the forces caused by rotation is one of the most interesting problems in nuclear structure theory and I will discuss this aspect in the second part of this talk.

Measurement of E4 Transition Moments

I have already alluded to the possibility of measuring E4 moments in Coulomb excitation; in fact the possibility was mentioned as early as 1956 in Ref. 1. However only recently have the necessary tools become available; these are the high resolution Ge(Li) detector with very clean in-beam techniques, and the coupled channels computer program taking into account multipolarities λ = 1, 2, 3 and 4. Surprisingly enough the best projectile for observing E4 effects in the rare earths is the ^{4}He ion at low energy, say ~10 MeV. In this case the multiple E2 effect is minimized giving one the best chance of picking out the E4 effect. For example the cross section for exciting a 4^+ level in a deformed rare-earth nucleus with ^{4}He is ~50μb. In general it turns out that the double E2 process is almost incoherent with the E4 process, which means that unlike the familiar reorientation

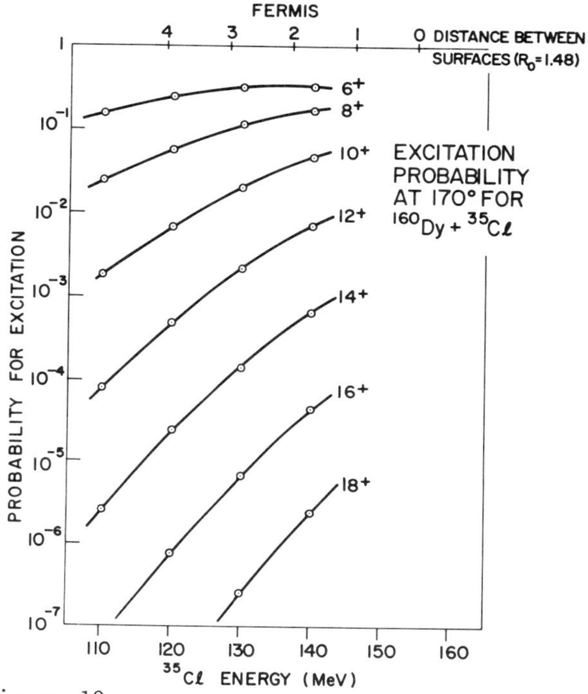

Figure 10
Calculation of the excitation probability for the ground state band in ^{160}Dy with ^{35}Cl ions. Rotational matrix elements were assumed. The separation between the surfaces is also indicated. Observation of the 16^+ level would be at the limit of existing techniques.

effect,[8] the coefficient of the cross term when we square the amplitudes turns out to be small. The effect of the E4 matrix element on the cross section is therefore smaller than one might have hoped and, furthermore, there will be two solutions for M(E4) with opposite signs. So far only one E4 measurement, for the case of ^{152}Sm, has been reported in the literature.[12]

In the case of an even nucleus, knowing the value for M(E4;0→4) does not enable one to distinguish between rotations of the intrinsic shape, or hexa-decupole vibrations. I have therefore been interested in performing measurements in an odd mass nucleus, because in that case E4 matrix elements link two states to the ground state, and one can test to see whether they are in the ratio of Clebsch-Gordon coefficients.* If the hexa-decupole moment resulted from rotations this result would be required.

* In collaboration with B. Ader (Orsay, France).

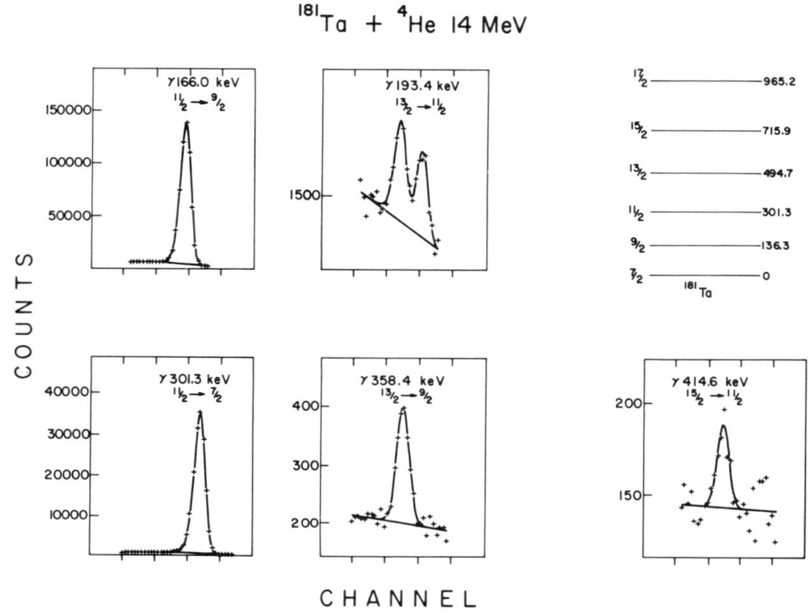

Figure 11
Portions of the gamma-spectrum from a thick target of ^{181}Ta bombarded with 14 MeV ^{4}He ions. The energies of the g.s. band were derived from ^{16}O and ^{35}Cl bombardment.

In a preliminary experiment to test the feasibility of these ideas we have examined the nucleus ^{181}Ta (ground state 7/2). The gamma spectrum for ^{4}He ions on a thick target of ^{181}Ta is shown in Fig. 11. There is no trouble in observing gamma rays from the 13/2 and 15/2 levels in spite of the fact that no special precautions were taken, and the runs were quite short. In principle, the analysis to extract the E4 matrix elements should proceed from an absolute cross section measurement of the 13/2 and 15/2 levels. These cross sections can then be compared with those calculated from known E2 matrix elements in the rotational band. However in this case we do not know the ground band E2 matrix elements with sufficient accuracy, so we have compared the relative cross sections between ^{4}He ions and ^{35}Cl ions assuming rotational E2 matrix elements. With ^{35}Cl projectiles the E4 effect is completely dominated by multiple E2's. This procedure eliminates the need to know conversion coefficients and relative gamma-ray detection efficiencies; however it was necessary to take into account small changes in the gamma-ray angular distributions between ^{35}Cl and ^{4}He ions, and correct for the different feeding from higher levels.

The effect of the E4 matrix elements on the ratios of He/Cl cross sections is shown in Fig. 12. The experiment is consistent with $\beta_4 \cong -0.09$, as would be expected from the work of

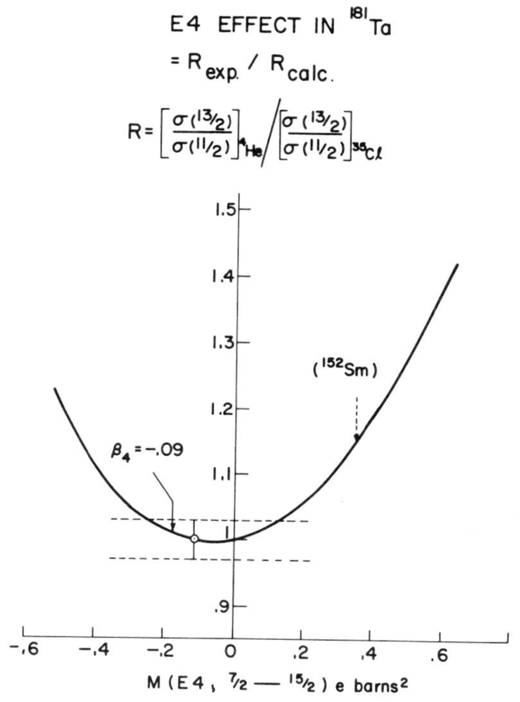

Figure 12
Interpretation of relative cross sections for ^{4}He and ^{35}Cl bombardment in terms of E4 excitations in ^{181}Ta. For $\beta_4 = -0.09$ the E4 transition moments are such that a zero effect would be expected. If the E4 matrix elements had been as large as that encountered in ^{152}Sm ($\beta_4 \cong +0.05$) an effect of about 14% would have been expected.

Hendrie et al.;[13] however no meaningful value can be extracted from these data. The rather small value for the E4 matrix elements, despite the appreciable β_4, is a consequence of the β_2 deformation. There is a contribution of order $\beta_2{}^2$ to the hexadecupole moment (even when $\beta_4 = 0$) which tends to cancel the β_4 term. Although unsuccessful in this particular case, I think this type of experiment should be pursued at the beginning of the rare earth region where β_4 is positive and the effects will be much larger. One should perhaps question the assumption of rotational E2 matrix elements in the ground state band; however all the evidence points to increased stability of the deformation when an odd particle is present. For example in ^{181}Ta there is no measurable deviation (i.e. <0.5 keV) in the energy spacings from the $AJ(J+1) + BJ^2(J+1)^2$ rule up to the 17/2 level. It should also be pointed out that the use of semiclassical theory for the ions may not be adequate for the analysis of these experiments.[12]

Some nuclei in the sd shell are expected to have very large static and transition hexa-decupole moments. Recent Hartree-Fock calculations by Cusson and Lee[14] predict that B(E4;0→4) in ^{28}Si should be ~10 Wu. I have performed some calculations using the Winther-de Boer program to see whether Coulomb excitation measurements are feasible or not. With ^{16}O ions the effect of a B(E4;0→4) = 10 Wu on the 4^+ cross section would be to increase it by about 50% of the double E2 value. As in the previous example, B(E2;2→4) is not known sufficiently accurately for an absolute cross section analysis; therefore one would probably compare cross sections with ^{35}Cl and ^{16}O. The actual yield of the 4^+ level with ^{16}O projectiles is very small and the feasibility of the experiment will depend on just how high one could raise the bombarding energy and still be safely below the barrier. At 23 MeV ^{16}O energy, the ratio of total cross sections for the 4^+ and 2^+ levels is only $1:10^6$. However using a γ-γ coincidence technique with two large NaI detectors in close geometry it may still be feasible to observe the $4^+ \to 2^+$ gamma-ray.

I hope I have shown that the traditional method of Coulomb excitation, in which measured cross sections are interpreted in terms of transition matrix elements, is still an active field of research. I will now go on to describe methods by which lifetimes are measured directly by DSAM and by the recoil distance method where states of interest are populated by Coulomb excitation.

DIRECT LIFETIME MEASUREMENTS FOLLOWING COULOMB EXCITATION

By detecting gamma rays in coincidence with backward scattered particles it is possible to obtain very clean spectra for levels excited by multiple excitation. The backscattered coincidence ensures that the velocity of the recoil is large and well defined in magnitude and direction.

DSAM Following Coulomb Excitation

The Doppler Shift Attenuation Method involves measuring the lineshape (or centroid shift) of gamma rays depopulating an excited state of a nucleus which is recoiling in a medium. The energy of the gamma ray will be shifted by the Doppler effect; however since the recoil population is both decaying and slowing down at the same time, the observed line shape corresponds to folding the exponential decay rate with the velocity history of the recoil. If the stopping power, dE/dx, for the recoils in the medium is known, or assumed, it is a straightforward calculation to extract the time dependence of the velocity, and hence predict the gamma-ray line shape for various values of the nuclear lifetime. Comparison of the calculated lineshapes with that observed will give the nuclear lifetime. If we assume that $dE/dx \sim 1/v$, then

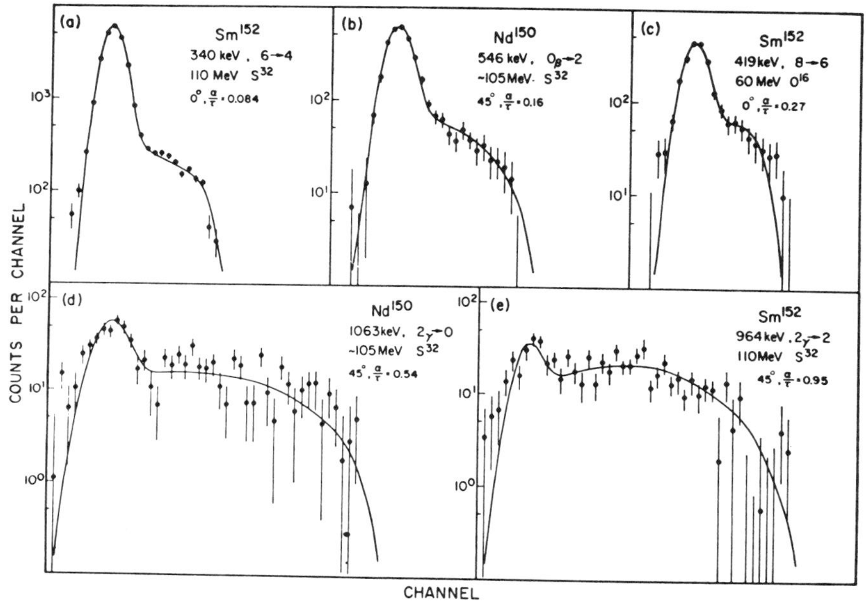

Figure 13
Gamma ray line shapes observed by Stokstad et al., with backscattered coincidence in Coulomb excitation.

v(t) will be an exponentially decaying function $e^{-t/\alpha}$ where α is the characteristic stopping time.

The first DSAM measurements with a backscattered coincidence requirement were reported by Stokstad et al.[15] This paper mainly represented a test of the method, since only levels whose lifetimes were known by other methods were used. Data published by Stokstad et al. is shown in Fig. 13. The parameter α was deduced from the known lifetimes of the levels; however the fact that this is possible over a considerable range of lifetimes (see Fig. 14) indicates that the method is reliable. It is interesting to note that the $6^+ \rightarrow 4^+$ transition in ^{152}Sm ($\tau = 15$ ps) had the best defined lineshape; this is an astonishingly long life time to be measured by DSAM considering that the stopping was relatively fast ($\alpha = 1.2$ ps). It will be seen that the measurement comes entirely from the height of the tail, which is only ~1% of the peak height; however, because of the experimental method, the background is essentially negligible.

The analysis of Stokstad used a treatment first given by Warburton[16] in which it is assumed that

$$(dE/dx) = (dE/dx)_{nuclear} + (dE/dx)_{electronic}$$

with

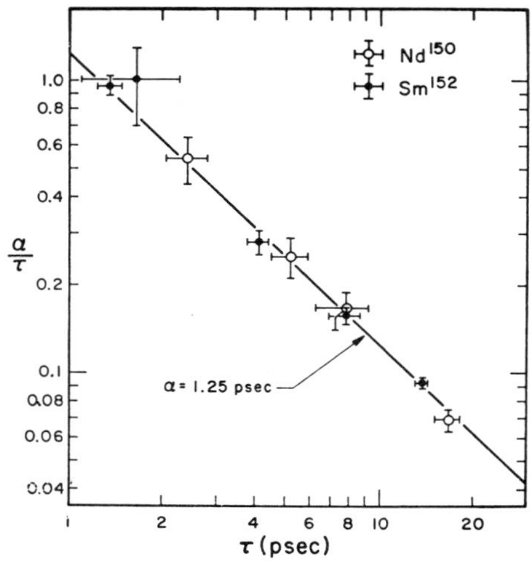

Figure 14
Values of α/τ derived from gamma-ray line shapes versus the known lifetimes of the levels. If the treatment of the Doppler shift attenuation were correct then the data should be fitted by a straight line whose slope is 45 degrees (after Stokstad et al.).

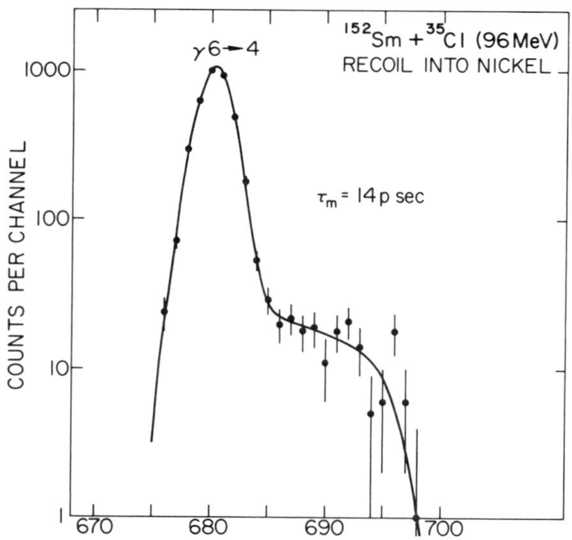

Figure 15
Line shape at 0 degrees for the 6→4 gamma ray in ^{152}Sm observed by the Chalk River group in backscattered coincidence with ^{35}Cl ions. The lifetime is assumed from recoil distance measurements, and there were no free parameters in the treatment of the stopping power (see text).

$$(dE/dx)_{nuclear} \sim K_n(v/v_o)^{-1}$$

$$(dE/dx)_{electronic} \sim K_e(v/v_o)$$

where v is the ion velocity. These assumptions are chosen to make the integration of the equation of motion analytic. In practice, neither of these assumptions is valid, although the results of applying them are clearly quite good. At Chalk River we have treated the problem somewhat differently by using the compilation of Northcliffe and Schilling[5] for $(dE/dx)_{electronic}$, with an empirical fit to the Lindhart results for $(dE/dx)_{nuclear}$. Results from the program have been encouraging, considering that there are no free parameters (there were 2 free parameters in the analysis of Stokstad et al., K_e/K_n and α). An example of this analysis is shown in Fig. 15 which is for the $6^+ \rightarrow 4^+$ transition in ^{152}Sm, but in a nickel stopper. This result came from backscattered coincidence with ^{35}Cl ions at 100 MeV. We have also performed experiments in ^{105}Pd with 100 MeV ^{35}Cl ions.* The lineshape for $\gamma 442$ which de-excites the 7/2 level to the 5/2 ground state is shown in Fig. 16 as fitted by the Chalk River

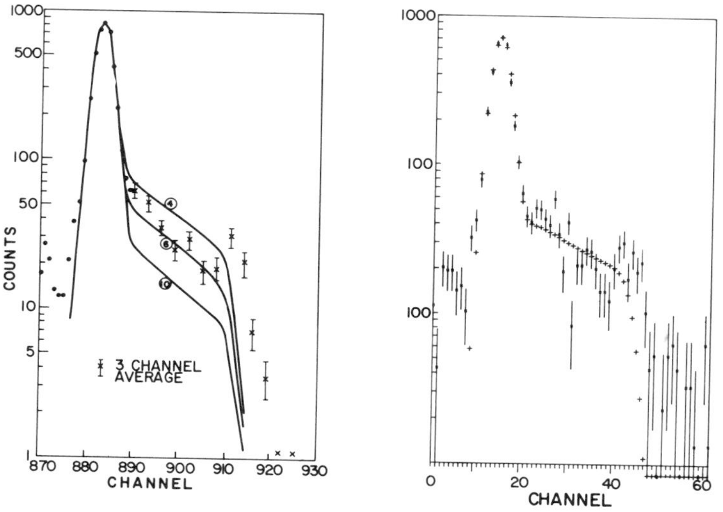

Figure 16
Line shape at 0° for the 442 keV gamma-ray in ^{105}Pd in backscattered coincidence with ^{35}Cl ions. The same data is fitted with the Stokstad program (right), and with the Chalk River program. The sensitivity to the life time is indicated by fits at 4, 6 and 10 ps on the left hand side.

* In collaboration with S. Sie (Chalk River).

program and the same data fitted by the Yale program. The lifetime for the 7/2 level 5.5 ± 1.5 ps is in excellent agreement with that deduced by the recoil distance method.

Recoil Distance Method Following Coulomb Excitation

The recoil distance method was first described by Ian Wright;[18] however most of the early development work was done at Chalk River by Alexander and Allen[19] and others. The first application of the method in Coulomb Excitation with backscattered coincidences was reported by Diamond et al.[20] Briefly, the idea is to place a movable plate behind a thin target so that the recoiling nuclei travel out of the target through the vacuum and come to rest in the plate (or plunger). Lines in the gamma spectrum will be split into two components by the Doppler effect (usually referred to as stopped and moving components); their relative intensities will be determined by the probability that the excited state in the recoil nucleus survived the transit time to the stopper: this depends on the lifetime of the state. The recoil velocity is readily determined from the splitting of the stopped and moving peaks according to the Doppler formula

$$\Delta E_\gamma = E_\gamma v \cos\theta/c \quad . \tag{28}$$

At high recoil velocities (~4% of c) it is necessary to use the relativistic version quoted in Ref. 20. A simple set of numbers to remember is:

for v/c = 2%, transit time ~4ps/mm.

With the many improvements which have been made in "plunger technology" it is possible to measure relative distances to 0.1 mm. and know the absolute zero distance between the plunger and the target to ~0.2 mm. (Although this latter piece of information is frequently not necessary). The target foil is stretched like a drum and is flat to ~0.4 mm. A nice feature of the Chalk River plunger is that the distance between the plunger and the target is continually measured by sampling their capacity, using a pulser and a charge sensitive preamplifier.[21] In this way we always know where the plunger is, which is an important point because thermal expansion from beam heating can cause the separation to drift. Since these experiments are time consuming, we have evolved a system in which an array of 4 Ge(Li) detectors looks at gamma rays in coincidence with the backscattered particles. The data are recorded on magnetic tape as 4 parameter events: gamma pulse height, particle pulse height, time-to-amplitude converter output and target-to-stopper capacity. The experimental set-up is shown in Fig. 17. We have performed several experiments with this arrangement, one of the most successful being the measurement of lifetimes in ^{152}Sm.* Typical

* In collaboration with N. Rud, G.T. Ewan and A. Christy, (Queen's University).

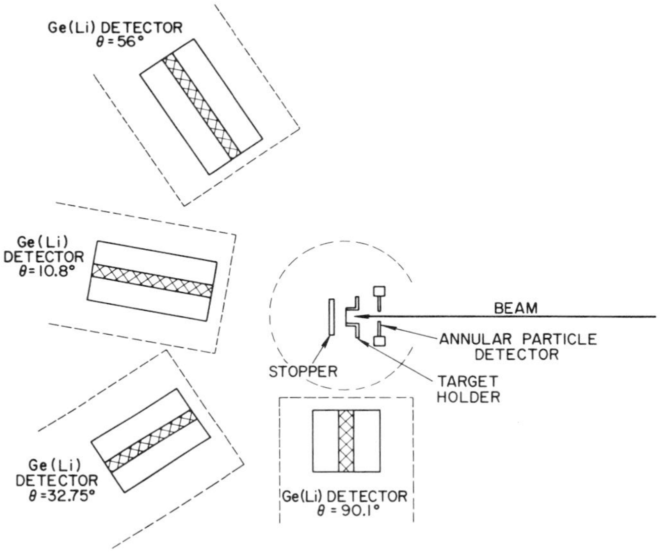

Figure 17
Schematic view of the set-up for recoil distance measurements with backscattered coincidences.

gamma spectra from these measurements are shown in Fig. 18. The stopper in this case was made of nickel. The advantage of nickel is that the backscattered particles from Sm are completely resolved from those from the plunger. Decay curves are shown in Fig. 19. There are a large number of small corrections which must be made before one can extract reliable half-lives. These may be divided into two groups:

Factors which effect the ratio $\gamma_s/(\gamma_s+\gamma_m)$:

(a) Relativistic aberration term in the solid angle for gamma detection $\Omega_m \cong (1 + 2(v/c)\cos\theta)\Omega_s$.
(b) Different efficiencies for γ detection at $E\gamma_s$ and $E\gamma_m$.
(c) Doppler broadening of the stopped line shape, which tends to increase the background on the high energy side and put counts into the moving peak.

Factors which effect the interpretations of the decay curve:

(a) Feeding of the level of interest from higher levels.
(b) Deorientation of the nuclear alignment during the transit across the vacuum[22] and in the stopper itself.

The deorientation effect can be quite spectacular; for example in a 2→0 decay Coulomb excited with backscattered coincidence, the angular distribution of gamma rays is such that the yield at θ=0 is zero. Therefore at zero degrees the yield of the stopped peak will actually increase as the plunger is drawn back, because the deorientation of the alignment in the gap more than compensates for the nuclear decays in the gap. The deorientation mechanism is not fully understood at present; however the Chalk

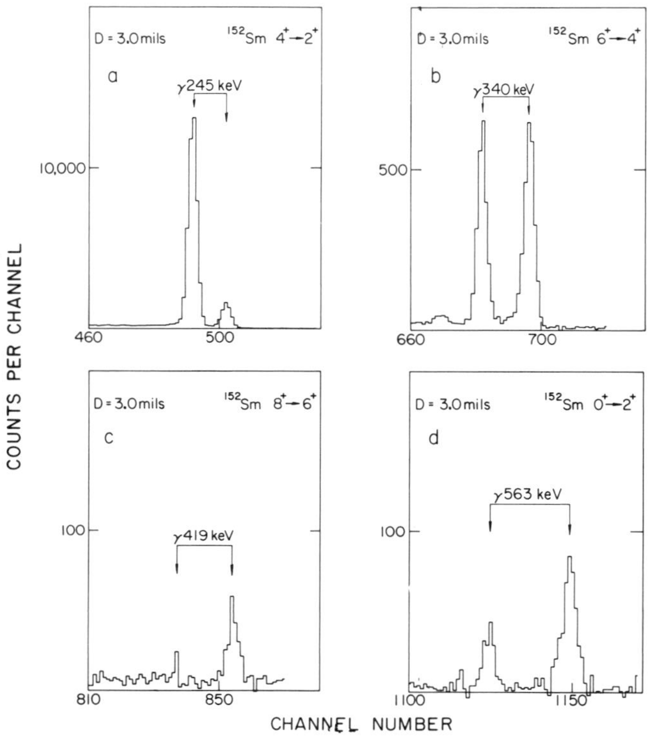

Figure 18
Typical gamma-ray spectra observed at zero degrees for ^{152}Sm excited by ^{35}Cl ions at 100 MeV in backscattered coincidence. The separation between the target foil and the stopper was 3.0 mm. (0.003 ins.) in all cases. The stopped and moving components are indicated.

River group have for sometime been investigating the effect using the recoil distance method.[23] In the present experiment the effect on the decay curves is to change the half life by ~ +5% for the 4→2 and 6→4 transitions; however since the lifetime for vacuum deorientation of ^{152}Sm recoils is fairly well known,[20] the correction can be taken care of with adequate accuracy.

Results for the transition moments in the ground band from this work are in good agreement with values given by Diamond et al.; the uncertainties are comparable. We also determined the lifetime of the β_0 level as $\tau = 8.9 \pm 0.6$ ps; this is in reasonable accord with the value of Fraser et al., $\tau = 7.9 \pm 0.9$ ps, deduced from the excitation cross section with ^{16}O and ^{32}S ions.[24] Deviations from rotational behaviour are apparent in the ground band of ^{152}Sm; these can be well described by the

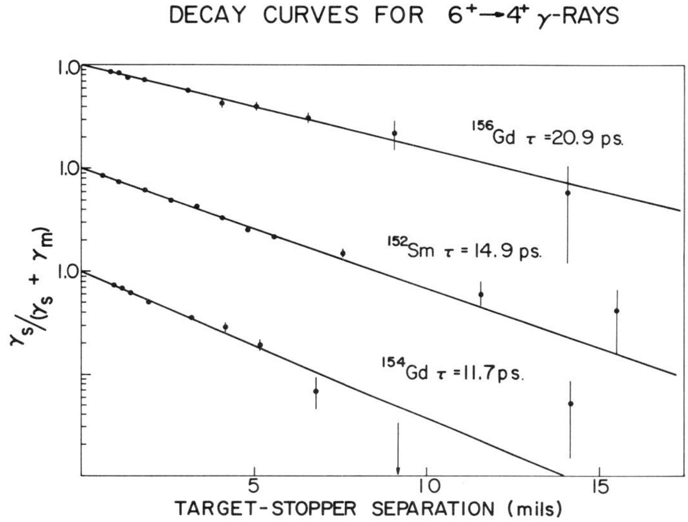

Figure 19
Decay curves for gamma rays de-exciting levels in ^{152}Sm.

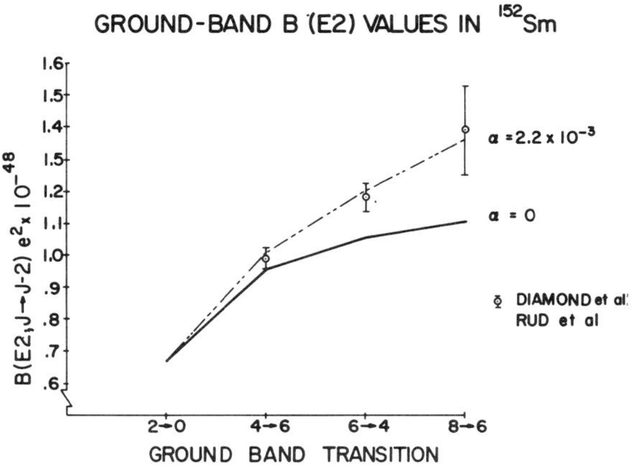

$$B(E2;I \rightarrow I-2) = \frac{5}{16\pi} e^2 Q_0^2 \langle I200|I2,I-2,0\rangle^2$$

$$* \left[1 + \frac{\alpha}{2}\{I(I+1)+(I-2)(I-1)\}\right]^2$$

Figure 20
Comparison of the rotational model predictions with experimental results by Diamond et al. and Rud et al. obtained from the recoil distance method. The form of the correction term involving the stretching parameter α is that expected from mixing the β-vibrational band with the ground band.

stretching parameter α (see Fig. 20). The results for ^{154}Gd of Rud and the Chalk River group can be described by $\alpha = 2.0 \pm 0.9 \times 10^{-3}$. This value is almost identical with the ^{152}Sm value ($\alpha = 2.2 \pm 0.6 \times 10^{-3}$) as would be expected since the level spacings in the two nuclei are very similar.

I will now discuss, in semiquantitative terms, what can be learned from the measured ground band moments in ^{152}Sm. Empirically, the energy spacings in ^{152}Sm can be reproduced by allowing the moment of inertia to increase with spin.[25] This can be associated with changes in the deformation, if we further assume that $\mathcal{J} \sim \beta^2$. Now since the quadrupole moment of the intrinsic state Q_0 is related to β we can write:

$$M(E2;J\rightarrow J-2) \sim (\beta_J + \beta_{J-2})/2 \ . \qquad (29)$$

From fitting the energy levels to give $\mathcal{J}(J)$, hence $\beta(J)$, and using Eq. (29), we find that $B(E2;4\rightarrow 2) = 1.113$, $B(E2;6\rightarrow 4) = 1.269$, and $B(E2;8\rightarrow 6) = 1.431$ times the rotational model prediction. These results are about a factor of 2 greater than those observed experimentally. I think that the right conclusion to be drawn from this is that the increased deformation with spin accounts for only around one half of the total change in the moment of inertia. The other half of the effect has to come from some other cause such as the decoupling of the pairs by the Coriolis force in the rotating intrinsic state. This latter effect, usually referred to as the Coriolis Anti Pairing (CAP) effect, causes the moment of inertia to increase owing to the tendency of the nuclear matter to rotate in a less correlated way, that is the moment of inertia will tend towards the rigid body value as the pairs are broken.

We have another handle on these effects from measurements of the change in mean square radius between the 0^+ and 2^+ levels. Thus $\Delta\langle r^2\rangle/\langle r^2\rangle$ for ^{152}Sm $\cong 5 \times 10^{-4}$ according to Mossbauer and Muonic X-ray measurements.[26,27] In relating these measurements to a change in deformation we can write:

$$\langle r^2\rangle = \frac{3}{5} R_0{}^2(1 + \xi\beta^2)(1 + \frac{5}{4\pi}\beta^2)$$

where the parameter ξ relates to nuclear compressibility under deformation as discussed by Fradkin[28] and others.[29] If we accept the value of ξ which fits the isotope shift data, namely $\xi = -5/8\pi$, then

$$\Delta\langle r^2\rangle/\langle r^2\rangle = (5\beta^2/4\pi)(\Delta\beta/\beta)$$

and the experimental value for $\Delta\langle r^2\rangle$ will correspond to $(\Delta\beta/\beta) =$ 1.5% for the 2^+ level. Again, this is about a factor of two less than we need for the energy spacings; thus the measured ground band B(E2) values should be in agreement with the measured $\Delta\langle r^2\rangle$, and indeed after some trivial manipulations and assuming that the J-dependence of $\Delta\langle r^2\rangle$ and $\Delta\beta$ is like $J(J+1)$, we find that

$$(\Delta\beta/\beta)_{2+} = 6\alpha$$

where α is the previously defined stretching parameter. The B(E2) measurements gave $\alpha = (2.2 \pm 0.6) \times 10^{-3}$ and the $\Delta\langle r^2\rangle$ measurements give $\alpha = 2.5 \times 10^{-3}$. It should be noted that without Fradkin's correction to the relationship between $\langle r^2\rangle$ and β we would have predicted $\alpha = 1.25 \times 10^{-3}$ from the measured $\Delta\langle r^2\rangle$ and a $(\Delta\beta/\beta)_{2+}$ of only 0.75% which would be in even worse agreement with the energy spacings.

Evidently from the ball park agreement of the ground band B(E2) values and the measurements of $\Delta\langle r^2\rangle/\langle r^2\rangle$ it would appear that the Coriolis anti-pairing effect in the lower states of ^{152}Sm does not significantly retard the E2 transitions. However there are no microscopic calculations of these effects as yet.

REFERENCES

1. Alder, K., Bohr, A., Huus, T., Mottelson, B. and Winther, A. Rev.Mod.Phys. 28, 432 (1956).
2. Winther, A. and de Boer, J. "Coulomb Excitation" (Academic Press, New York) 1966.
3. Häusser, O. and Cusson, R.Y. Can.J. Phys. 48, 240 (1970).
4. Alder, K. and Pauli, H.K.A. Nucl.Phys. A129, 193 (1969).
5. Northcliffe, L.C. and Schilling, R.F. Nuclear Data Tables A7. 233 (1970).
6. de Shalit, A. Phys.Rev. 122, 1530 (1961).
7. Choudhury, D.C. Mat.Fys.Medd.Dan.Vid.Selsk. 28, No. 4 (1954).
8. de Boer, J. and Eichler, J. Advances in Physics Vol. 1 (1968).
9. Fraser, I.A., Greenberg, J.S., Sie., S.H., Stokstad, R.G. and Bromley, D.A. Phys.Rev.Lett. 23, 1047 (1969).
10. Johnson, A., Ryde, H., and Sztarkier, J. Phys.Lett. 34B, 605 (1971).
11. Mottelson, B.R. and Valatin, J.G. Phys.Rev.Lett. 5, 511 (1960).
12. Stephens, F.S., Diamond, R.M., Glendenning, N.K., and de Boer, J. Phys.Lett. 24, 1137 (1970).
13. Hendrie, D.L., Glendenning, N.K., Harvey, B.G., Jarvis, O.N., Duhm, H.H., Sudinos, J. and Mahoney, J. Phys.Lett. 26B, 127 (1968).
14. Lee, P. and Cusson, R.Y. Private communication, 1971.
15. Stokstad, R.G., Fraser, I.A., Greenberg, J.S., Sie, S.H. and Bromley, D.A. Nucl.Phys. A156, 145 (1970).
16. Warburton, E.K., Olness, J.W. and Poletti, A.R. Phys.Rev. 160, 938 (1967).
17. Lindhard, J., Scharff, M. and Schiott, H.E. Mat.Fys.Medd. Dan.Vid.Selsk. 33, No. 14 (1963).
18. Wright, I.F. Bull.Am.Phys.Soc. 6, 285 (1961).
19. Alexander, T.K., Allen, K.W. and Healey, D.C. Phys.Lett. 20, 402 (1965).
20. Diamond, R.M., Stephens, F.S., Nakai, K. and Nordhagen, R. Phys.Rev. 3, 344 (1971).

21. Alexander, T.K. and Bell, A. Nucl.Instr. and Meth. 81, 22 (1970).
22. Ben Zvi, I., Gilad, P., Goldberg, M., Goldring, G., Schwarzchild, A., Sprinzak, A. and Uager, Z. Nucl.Phys. A121, 592 (1968).
23. Ward, D., Geiger, J.S. and Graham, R.L. Proc. of 21st Conf. on Nucl. Spectroscopy and Structure of Atomic Nuclei, Moscow, U.S.S.R. 1971.
24. cf. Ref. 9.
25. Diamond, R.M., Stephens, F.S. and Swiatecki, W.J. Phys.Lett. 11, 315 (1964).
26. Yeboah-Amankwah, D., Grodzins, L. and Frankel, R.B. Phys.Rev.Lett. 18, 791 (1967); Kienle, R., Henning, W., Kaindl, G., Korner, H.J., Schaller, H. and Wagner, F. Proc. Int.Conf. on Nuclear Structure, Tokyo, 1967; J.Phys. Soc. of Japan, 24, 207 (1968).
27. Schroder, U., Watter, H.K. and Wien, K. Phys.Lett. 27B, 425 (1968); also Cohe, S.G., Blum, N.M., Chow, Y.W., Frankel, R.B. and Grodzins, I. Phys.Rev.Lett. 16, 322 (1967).
28. Fradkin, E.E. JETP 42, 787 (1962).
29. Cusson, R.Y. and Castel, B. Can.J.Phys. 47, 1189 (1969).

SELECTED TOPICS IN INTERMEDIATE ENERGY PHYSICS

G.E. WALKER
Physics Department
Indiana University

INTRODUCTION

These lectures will briefly outline some aspects of intermediate energy physics which I thought might be of interest from the nuclear physics perspective of this conference. The first two lectures will deal primarily with muon, electron, and proton probes of the nucleus. One purpose of these talks is to introduce you to the overlap and advantage of the detailed probing of the nucleus with different projectiles. The last lecture will be concerned with pion-nucleus interactions and will contain a brief discussion of a technique we are currently using to study pion-nucleus scattering. The method, although somewhat removed from conventional nuclear structure studies, may be instructive.

The recent general interest in the intermediate energy region is a response of the physics community to the exciting new horizons in both the quality and type of data that will become available for interpretation in the medium energy region in the 1970's. The new high quality data will originate from the excellent beam-current and energy-resolution medium energy facilities for accelerating protons and heavy ions scheduled for completion in the early part of this decade.

The Los Alamos Meson Physics Facility (LAMPF),[1-4] a linac, is designed to yield 100-800 MeV protons with channels for pion and muon beams. The intensity will be approximately 10^4 times that previously available from synchrocyclotrons.[4] The beam characteristics depend on the channel but one has typically for the energetic pion channel and spectrometer (EPICS) a high flux with a pion energy range from 50 to 300 MeV with energy resolution better than 50 keV. A pion and particle-physics channel yields pions with up to 600 MeV energy but with poorer energy resolution. The muon channel will have a momentum range of from 0 to 250 MeV/c, greater than 90% purity, polarization > .4, and there will be two simultaneous muon beams. Mid 1972 is the projected date for the first 800 MeV beam.

The Tri-University Meson Facility (TRIUMF), at Vancouver, a negative-ion fixed frequency (isochronous) cyclotron device, is designed to produce H^- ions and after subsequent stripping should yield a high-current proton beam in the energy range 150-500 MeV.[5] This machine should be a source of intense pion beams in the low and medium energy region. Two other facilities (both isochronous cyclotrons), one at the Swiss Institute for Nuclear Research (SIN) and the other at Dubna, will be capable of accelerating high-intensity beams of protons into the intermediate energy region.[6]

The 200 MeV Indiana Cyclotron Facility[7,8] will be an intense and precise source of protons and other light ions in the lower medium-energy region and will also produce energetic beams of heavy ions as well. The facility is composed of three stages: a D.C. preinjector containing the ion source, an injector cyclotron, and the main cyclotron. The two cyclotrons (the smaller being a 1/3 model of the larger) are separated-sector, isochronous, variable-energy machines. The energy range of protons will be from 20 to 200 MeV with an energy resolution of from 20 to 40 keV at 200 MeV. The internal beam current should be 100μA and excellent beam extraction is expected. However, initially one will probably work with $\lesssim$ 10μA beams because of shielding restrictions. The energy range for deuterons is 20-120 MeV; ^{3}He, 30-300 MeV; ^{3}H, 30-80 MeV; ^{4}He, 40-240 MeV, and in general for heavier ions $E \leq 240\ q^2/M$ MeV or $E/A \leq 240\ (q/M)^2$ MeV/nucleon where q is the ionic charge and M is the mass of the nucleus in AMU. The machine is designed to accelerate nuclei whose ionic charge-to-mass ratio, (q/M), is in the range $1/6 \leq q/M \leq 1$. This means, using conventional positive ion sources, one can accelerate heavy ions up to about mass 40. Of course with a better ion source one would be in an excellent position to look for superheavies. Present expectations are for an initial beam through the assembly in 1973.

So, generally speaking, the expectation is for more intense beams of protons, pions, muons, and heavy ions with good energy resolution. It would probably take me several minutes to list the major experimental studies proposed for the Los Alamos or Indiana machines alone. Not only can we expect better data for the kinds of experiments we do now but other experiments with low counting rates can be reconsidered (for example $N + N \rightarrow N + N + \pi$ for studying off-energy-shell effects of the nucleon-nucleon interaction.[6]) As a by-product one has been given the hope of more satisfactory methods for destroying malignant tumor cells using the pion beams from the meson factories.[4]

1. MUON-NUCLEUS INTERACTIONS

Before we discuss muons as a probe of nuclear structure, let me remind you, in passing, that the study of muonium, i.e. the bound $\mu^+ - e^-$ system, is a very interesting aspect of fundamental particle physics.[9]

If we consider muons interacting with nuclei we might study muon-nucleus scattering, much as in electron scattering. In the case of muon-nucleus scattering, the radiative tail problem would be less important because of the larger muon mass.

One could also study muons captured in atomic orbits (μ-mesic atoms).[10] From such investigations one can obtain information (via a detailed study of X-rays resulting from transitions of the muon to lower atomic shells and of muonic atom nuclear γ rays) on nuclear charge distributions, the intrinsic quadrupole moment of deformed nuclei, and the distribution of magnetic dipole moments.

One can also consider nuclear muon capture.[11] The basic process is a weak interaction $\mu^- + p \to \nu_\mu + n$ in the nuclear environment. In the process considerable momentum (~ 70-10 MeV/C) is transferred to the nucleus (compared to the electron capture process) because of the large mass of the annihilated muon. There is one variation, radiative muon capture, $\mu^- + p \to n + \nu_\mu + \gamma$, which is roughly 10^{-4} times rarer than simple capture. It is of importance because the ratio of radiative to normal capture is sensitive to the value assumed for the induced pseudoscalar coupling constant (see below).[12]

If one is able to do capture experiments with polarized muons then the correlation of the momentum of the final emitted neutron with the initial direction of muon spin can be studied and information on nuclear matrix elements extracted.

Today I will discuss in some detail simple muon capture on closed shell (J = T = 0) nuclear targets. In the last few years muon capture has become an important field of study in nuclear structure physics. For our purposes it is sufficient to remember that a muon is a spin ½ lepton with mass $\sim$ 206.8 m_e and half-life $\tau = 2.2 \times 10^{-6}$ sec. It decays via

$$\mu^- \to e^- + \nu_\mu + \overline{\nu}_e \ . \qquad (1.1)$$

Initially the muon is trapped in an atomic orbit and then cascades down rapidly into the 1s muonic atom orbit from which it is captured by the nucleus (or decays). Because of its large mass the Bohr orbit for the muon has a radius approximately 200 times smaller than that of an electron. Nevertheless, to a first rough approximation we could consider the muon wave function ϕ_μ constant over the nucleus, remove it from the nuclear matrix element, and set it equal to its value at the origin $\phi_\mu(o)$. The total capture rate $\Lambda_{\mu c}$ is roughly proportional to $|\phi_\mu(o)|^2 Z$ (Z is the nuclear charge) and since $|\phi_\mu(o)|^2 \propto Z^3$ one obtains a Z^4 dependence of the capture rate (actually a $Z^4_{(\text{effective})}$ must be used in more realistic calculations). As a reference point we note that the capture rate in ^{12}C is roughly equal to the decay rate of a free muon.[13]

The capture of a muon is pictured in weak interaction theory as arising from a local interaction of the lepton V-A weak current with the non-strangeness changing V-A hadronic weak current. Thus we have a current-current interaction

$$J^-_\lambda(\text{lep}) \cdot J^+_\lambda(\text{had}) + \text{Hermitian Conjugate} \qquad (1.2)$$

where

$$J_\lambda = V_\lambda + A_\lambda . \qquad (1.3)$$

Taking matrix elements of the lepton current with the appropriate lepton initial and final states leads to a term of the form

$$\bar{u}_\nu(k')\gamma_\lambda(1 + \gamma_5)\, u_\mu(k) \qquad (1.4)$$

where the u's are solutions of the Dirac equation $(\gamma_\lambda k_\lambda - im)\, u\,(k) = 0$ and k and k' are the momenta for the indicated lepton state. Because of the effect of strong interactions the matrix element (between dressed nucleon states) of the weak hadronic current is more complicated. However, using Lorentz invariance and the Dirac equation one can demonstrate that the matrix element must take the form[14,15]

$$\bar{u}_n(p')\{F_1(q^2)\gamma_\mu + F_2(q^2)\sigma_{\mu\nu}q_\nu + iF_S(q^2)q_\mu$$

$$+ F_A(q^2)\gamma_5\gamma_\mu - iF_P(q^2)q_\mu\gamma_5 - F_T(q^2)\sigma_{\mu\nu}q_\nu\gamma_5\}\tau^{(-)}u_p(p) \qquad (1.5)$$

where q^2 is the four-momentum transfer. By using the universality of weak (Fermi) interactions (U.F.I.) and the conserved vector current hypothesis[16] one can obtain the form factors F_1 and F_2 (the weak magnetism term) from β decay and electron-nucleon scattering and set $F_S = 0$. By using the U.F.I. hypothesis we can take $F_A(o)$ from the β decay of the neutron. The form factor F_P is predicted to be $\sim (7.5/m_\mu)F_A(o)$ by a less certain technique that involves the assumption of a partially conserved axial vector current and the idea that the pseudoscalar term arises mainly from the insertion of single pion line between the (p,n) vertex and the (μ,ν) vertex.[16,17] Terms that do not have appropriate transformation properties under the G-parity operation,[14] charge conjugation followed by a 180° rotation about the second axis in isospin space, are usually set equal to zero. Deletion of these "second class currents" means we can set $F_S = F_T = 0$. I have been overly brief but the point is that some rather global assumptions about the coupling constants go into the theory of muon capture. It is important to test the validity of these assumptions, especially those regarding the value of the induced pseudoscalar coupling constant and the deletion of second class currents.

In order to obtain the usual expression for the capture rate one makes a non-relativistic reduction of the Dirac spinors referring to the nucleons keeping terms through 1/M. After summing over all nucleons in the nucleus the capture rate takes the form[18,19]

$$\Lambda_{\mu c} = \frac{\nu_\mu{}^2(|\phi_\mu|^2)_{av}}{2\pi\hbar^2 c}\,[G_V{}^2M_V{}^2 + 3G_A{}^2M_A{}^2$$

$$+ (G_P{}^2 - 2G_PG_A)M_P{}^2] + \Lambda'_{\mu c}$$

where $\Lambda'_{\mu c}$ represents recoil corrections to the capture rate and is of the order of 10% of $\Lambda_{\mu c}$ in ^{16}O.[15] The terms $M_V{}^2$, $M_P{}^2$ and $M_A{}^2$ contain the nuclear matrix elements and are defined as follows:

$$M^2 = \frac{1}{2J_i+1}\,\overline{\sum}_a\sum_b\left(\frac{\nu_{ab}}{\nu_\mu}\right)^2\int\frac{d\hat{\nu}}{4\pi}\,|\langle b|\sum_{i=1}^{A}\tau_i^{(-)}O(i)e^{-i\vec{\nu}_{ab}\cdot\vec{x}_i}|a\rangle|^2 \qquad (1.7)$$

where for

$$M_V{}^2,\ O(i) = 1,$$

$$M_A{}^2,\ O(i) = \frac{1}{3}\vec{\sigma}_i, \qquad (1.8)$$

$$M_P{}^2,\ O(i) = \hat{\nu}\cdot\vec{\sigma}_i,$$

also

$$m_\mu c^2 = h\nu_{ab}c + E_b - E_a + \text{(small atomic binding energy and n-p mass difference corrections)},$$

and where $|a\rangle$ and $|b\rangle$ are the initial and final nuclear states respectively, $h\nu_{ab}$ is the neutrino momentum, $\hbar\nu_{\mu c} \equiv m_\mu c^2$ is the muon rest mass energy, and $\hat{\nu}$ is a unit vector in the direction of the neutrino momentum. Values adopted previously[19] for the effective coupling constants yield $G_V = 1.03G$, $G_A = 1.36G$, and $G_P = -0.54G$ where $G = 1.01 \times 10^{-5}$ $(\hbar^3/M_p{}^2c)$ is the weak interaction coupling constant. We note that since 105 MeV of energy is liberated when the muon is destroyed and since, for capture on a light closed shell nucleus such as ^{16}O, the important excited states are at roughly 10 to 25 MeV excitation, the neutrino momentum appearing in the matrix elements is ~ 80 to 95 MeV/c.

A few years ago total and (a few) partial muon capture rates on closed shell nuclear targets were measured and one hoped, by doing a theoretical analysis, to test the assumed values for the coupling constants and in particular learn about the validity of the assumptions leading to the value for the induced pseudoscalar coupling constant (which enters only in $G_P{}^2$). But the insensitivity of the rate to this coupling constant and the usual theoretical uncertainties associated with the nuclear wave functions have precluded the obtaining of precise information. (In particular cases progress has been possible by taking some of the nuclear information from other experiments.[20])

One of the most recognizable and well known features of photoabsorption on T = 0 nuclei is the strong excitation of the T = 1 giant dipole resonance (g.d.r.). This excitation proceeds via an E1 multipole; thus if $J_{g.s.}^{\pi}=0^+$, it leads uniquely to $J^{\pi} = 1^-$ excited nuclear final states. Although the momentum transferred to the nucleus in the case of muon capture excitation of the isospin analogue of the g.d.r. ($\approx$ 80 MeV/c in ^{16}O) is considerably more than the $q \simeq E_b - E_a$ = 20 MeV/c delivered in photoabsorption calculations predict strong excitation of the 1^- isospin analogue of the g.d.r. photoabsorption state in muon capture. Suppose we insert a multipole expansion for the plane wave appearing in the nuclear matrix elements. A simple calculation shows that in addition to the strong excitation of a 1^- state via the $Y_{1m}j_1(\nu_{ab}r)$ transition operator appearing in $M_V{}^2$ one expects also important 0^-, 1^-, and 2^- excitations arising from the $[\sigma \times Y_{1m}]^J_M\, j_1(\nu_{ab}r)$ operators appearing in $M_P{}^2$ and $M_A{}^2$.

Actually many of the operators we meet in considering nuclear matrix elements associated with reactions induced by such projectiles as photons, electrons, muons, pions, and protons (in Born approximation) are of the form

$$\sum_{i=1}^{A} \tau_i{}^k f(x_i) \quad , \quad \sum_{i=1}^{A} \sigma_i{}^k f(x_i),$$

$$\sum_{i=1}^{A} \sigma_i{}^k \tau_i{}^{k'} f(x_i), \qquad k,k' = 1,2,3, \tag{1.9}$$

where, for example, if plane-wave Born approximation is adopted, $f(x_i)=\exp(i\vec{q}_{ab}\cdot\vec{x}_i)$, $\vec{q}_{ab}$ being a momentum transfer appropriate to the reaction under consideration. Thus if we are in similar regions of q_{ab} in different processes, the states strongly excited in one process may be, loosely speaking, the spin-isospin analogues of states strongly excited in another process. More precisely if we are considering processes depending only on $\tau_i{}^k \exp(i\vec{q}_{ab}\cdot\vec{x}_i)$ (as in isovector longitudinal transitions in inelastic electron scattering and the $M_V{}^2$ matrix elements appearing in muon capture) we need only assume SU(2) isospin symmetry to relate the matrix elements at a given q. However, to interrelate processes at a given q depending on different operators appearing in eq. (1.9) one may need to assume more approximate symmetries such as Wigner's supermultiplet (SU(4)) symmetry.[21] This symmetry is of course broken by spin-dependent forces in nuclei but can still be advantageously applied. An example is Foldy and Walecka's work[19] relating photoabsorption cross sections to muon capture rates in closed shell nuclei. Clearly just because a symmetry is broken doesn't mean in particular circumstances it cannot give valid and understandably valid results. In fact we deal with approximate models all the time in nuclear physics.

The connection between SU(4) and the operators in (1.9) above arises because the fifteen operators

$$\sum_{i=1}^{A} \tau_i^{\,k}, \quad \sum_{i=1}^{A} \sigma_i^{\,k}, \quad \sum_{i=1}^{A} \sigma_i^{\,k} \tau_i^{\,k'} \qquad (k,k' = 1,2,3) \tag{1.10}$$

have the same mutual commutation properties as the members of the Lie algebra consisting of the infinitesimal generators of SU(4). The operators (1.9) transform like vector operators[19] under SU(4). If for 4n, T = 0, nuclei such as ^{16}O one postulates the ground state is a scalar supermultiplet, then application of a vector operator takes one only to final states that belong to 15-dimensional vector supermultiplets. All members of a given supermultiplet have the same many-particle orbital space wave function and differ only in their spin-isospin wave functions (which must be of the same permutational symmetry type). Of course within the spirit of the model all states belonging to the same supermultiplet are degenerate in energy. These facts plus the trivial application of a generalized Wigner-Eckart theorem for SU(4) and the commutation properties of the operators allow great progress to be made in interrelating nuclear matrix elements for different processes. For example, in muon capture one can show $M_V^2 = M_P^2 = M_A^2$ if the ground state belongs to a scalar supermultiplet and SU(4) symmetry for the excited states is valid.[19] (The analogue to the SU(4) discussion is, for the rotation group, basically that if the ground state has J = 0 and one applies an irreducible tensor operator of rank one to this ground state only J = 1 final states will be reached. For a given J, the different J_z substates are degenerate in energy if the symmetry is exact. Thus one can easily interrelate energy dependent nuclear matrix elements of the "vector" (irreducible tensor of rank one) operators

$$\sum_i j_i^{\,z} f(x_i), \quad \sum_i j_i^{+} f(x_i), \quad \text{and} \ \sum_i j_i^{-} f(x_i)$$

using the usual Wigner-Eckart theorem.)

We know that strong spin-dependent forces exist in nuclei so we carry out the muon capture calculation using p-h model wavefunctions for the excited states with various kinds of spin-dependent forces in the residual interaction and what happens? Typical results[22,23] agree to within a few percent with the SU(4) based conjecture $M_V^2 = M_P^2 = M_A^2$. Actually in ^{16}O and ^{40}Ca one finds M_V^2 is about 10% smaller than M_P^2 and M_A^2 using a Serber-Yukawa residual interaction. If we carry the analysis further some disagreement appears between the SU(4) and p-h predictions (note RPA and Tamm Dancoff p-h calculations yield similar results for transitions involving the high-lying T = 1 states we consider here). The SU(4) model predicts the existence of a giant magnetic quadrupole, 2^- resonance, if a 1^- giant dipole resonance exists. Recent calculations using p-h wave functions predict that in ^{40}Ca the 2^- transition strength should be shared by several states widely separated in energy and it will be interesting to see what experiments will reveal.

There has been some difficulty in the past in obtaining agreement between theory and experiment for total and partial muon capture rates. I am confining my remarks to microscopic calculations where one assumes some shell model wave functions, applies the operators defined earlier for muon capture and calculates the appropriate nuclear transition matrix elements. Consider partial capture rates on ^{16}O where the lowest-lying quartet $(0^-, 1^-, 2^-, 3^-)$ of bound states in ^{16}N are the states of interest in the daughter nucleus. If we use a closed shell g.s. and the p-h amplitudes obtained by several authors we find that the partial capture rates to these lowest states are overestimated in particular, by factors of 2 to 3 for transitions to the 1^- and 2^- states. It was found possible to obtain reasonable agreement with experiment by introducing 2p-2h components in the ground state.[24,25] Since this same technique has been applied in connection with electron scattering induced transition matrix elements, I shall sketch briefly the technique and results.

If we adopt the particle-hole model for the excited states and consider transitions induced from a closed shell g.s. to these excited states by a sum of single nucleon operators, i.e.

$$|c\rangle \rightarrow \sum_k \gamma_k{}^E |p_k - h_k\rangle \text{ via } \sum_{i=1}^{A} O(i) \ , \qquad (1.11)$$

then the matrix elements of interest have the simple form

$$\sum_k \gamma_k{}^E \langle p_k - h_k | \sum_{i=1}^{A} O(i) | c\rangle = \sum_k \gamma_k{}^E (p_k|O|h_k) \qquad (1.12)$$

where the $\gamma_k{}^E$ are the admixture amplitudes for the pure p-h states in a given configuration mixed state as determined by diagonalizing some residual interaction (k stands for a pure p-h configuration and E here refers to the complete set of quantum numbers representing a given configuration mixed p-h state, i.e. the energy, parity J, J_z, T and T_z).

If we now include 2p-2h states in the ground state then we have for the new g.s.

$$|g.s.\rangle = \alpha|c\rangle + \sum_n \beta_n |2p_n - 2h_n\rangle \qquad (1.13)$$

where for normalization $\alpha^2 + \Sigma\, \beta_n{}^2 = 1$. Now we might expect the excited states of interest to be somewhat more complicated than before. However, my impression is that although we have some information on the 2p-2h components in the g.s. (guidance from shell model calculations,[26,27] spectroscopic factors, and 2p-2h components needed to fit other one-body transition rates) we have considerably less information about what to do with the excited states. There are those that feel it is proper to keep the low-lying T = 1 p-h states unchanged. It seems to me an open question. Of course one can always rephrase the problem by saying that we are working in a model space (also, of course, the free transition operator may be renormalized merely because the process takes place in nuclear matter), and the transition operator should be altered. My simple minded approach[24] was to consider several different procedures: 1) keep the p-h states the same, 2) carry out a calculation in a truncated space where p-h states were allowed to mix with 3p-3h states via a residual interaction, and 3) generate the excited states by applying the configuration mixed p-h state operator to the ground state configuration. For the calculation involving muon capture induced transitions to the low-lying states of ^{16}N which we are discussing it was found all three procedures yielded a reduction in the partial capture rates and so I'll proceed by ignoring for the moment any changes in the p-h states.

Then, adopting the more complicated g.s. and keeping the same p-h excited states one obtains an expression of the following form for the square of the transition matrix element:

$$\Big|\alpha \sum_k \gamma_k^E \langle p_k - h_k | \sum_{i=1}^{A} O(i) | c\rangle + \sum_{k,n} \gamma_k^E \beta_n \langle p_k - h_k | \sum_{i=1}^{A} O(i) | 2p_n - 2h_n\rangle \Big|^2 . \qquad (1.14)$$

Of course, the second term reduces to a term that is of the general form of the first term except that the position of particle and hole states is reversed in the matrix element and some additional phases and recoupling coefficients appear. For example, if we assume pairing type 2p-2h states in the g.s. of the form

$$[\{(n\ell_j)_p^2\, J_p = 0,\ T_p = 1\}\{(n\ell_j)_h^{-2} J_h = 0,\ T_h = 1\}]^{J = T = 0} \qquad (1.15)$$

then the general expression for the squared transition matrix element becomes (for a multipole transition operator $O^{j\tilde{\alpha}}$ which transforms like an irreducible tensor of rank j under rotations and an irreducible tensor of rank $\tilde{\alpha}$ under rotations in isospin space)

$$[(2j+1)(2\tilde{\alpha}+1)]^{-1} \left\{ \alpha \sum_k \gamma_k{}^E (j_{pk} \vdots\vdots O^{j\tilde{\alpha}} \vdots\vdots j_{hk}) + \sum_{k,n} \beta_n \gamma k^E \frac{\sqrt{3}(-1)^{j_{pk}+j_{hk}+j}}{(2\tilde{\alpha}+1)[(2j_{pk}+1)(2j_{hk}+1)]^{\frac{1}{2}}} (j_{hk} \vdots\vdots O^{j\tilde{\alpha}} \vdots\vdots j_{pk}) \delta(j_{pk}, j_{pn}) \delta(j_{hk}, j_{hn}) \right\}^2 \qquad (1.16)$$

where the $(\vdots\vdots \;\; \vdots\vdots)$ are single-nucleon double-reduced matrix elements in angular momentum and isospin space and the delta functions result because the single-nucleon states are part of an orthonormal basis and we have only one-body operators linking the p-h and 2p-2h states. For the general type of operator we consider (i.e. those entering in muon capture, electron scattering, and photon absorption, etc.)

$$(j_2 \vdots\vdots O^{j\tilde{\alpha}} \vdots\vdots j_1) = f(j_1, j_2)(j_1 \vdots\vdots O^{j\tilde{\alpha}} \vdots\vdots j_2) \qquad (1.17)$$

where for transverse electric (TE) matrix elements $f(j_1,j_2)=(-1)^{j_1+j_2}$ and for longitudinal coulomb matrix elements, transverse magnetic matrix elements and the matrix elements of the operators appearing in $M_V{}^2$, Mp^2, and $M_A{}^2$, $f(j_1,j_2) = (-1)^{j_1+j_2+1}$. By substituting these relationships into expression (1.16) one obtains a reduction in particular cases from two sources: 1) α is less than one and 2) the second term may have different sign from the first so that destructive interference can occur. Note that the momentum transfer dependence of the m.e. is determined in the reduced matrix elements. After substituting (1.17) into (1.16) the reduced matrix element may often be factored (if for example a particular pure p-h state amplitude dominated the configuration mixed p-h state) out of the expression. This allows one to predict, for certain electron scattering induced transitions, the same momentum transfer dependence as in the closed shell g.s. calculations but with an overall (q-independent) normalization factor.

Of course the simplistic treatment of the excited "p-h" states is worrisome. Even forgetting about possible admixtures of 3p-3h states there is the problem of $(2p\text{-}2h)_{g.s.}$ to (2p-2h) excited state transitions. For muon capture on ^{16}O one calculates, assuming a closed shell g.s. and p-h excited state, a total capture rate roughly sixty percent higher than that observed. [You will understand because of our earlier discussion concerning the similarity of the nuclear matrix elements for different probes that similar problems arise in other processes.] Now the technique I discussed above of putting 2p-2h states in the g.s. and then considering transitions to unaltered p-h states can be used to calculate the total capture rate. Of course, a reduction factor is obtained if only transitions to p-h states are considered. Unfortunately I found[24] that one actually obtained a slight increase in the total rate if transitions to 2p-2h states were included in the calculation (in my calculation the 2p-2h states were all arbitrarily placed at 30 MeV excitation). Of course the important 2p-2h states might be at somewhat higher energy and there is some danger from spurious components so one might still hope to obtain an appropriate reduction of the total capture rate by treating the excited states properly. One way of sidestepping this problem is to eliminate the excited states from the problem by using completeness to reduce the problem to one of calculating the ground state expectation value of a two-body operator. Let me remind you of the basic idea and limitations of this technique.

One is often faced with a calculation of the following type (for example in total capture rates):

$$\sum_b f(q_{ab}) \, |\langle b|\hat{O}(q_{ab})|a\rangle|^2 \tag{1.18}$$

where $|a\rangle$ represents the ground state wave function and the states $|b\rangle$ are final nuclear excited states reached via the reaction. Let $\hat{O}(q_{ab})=\sum_i O_i(q_{ab})$, i.e. be a sum of one-body operators. Rewriting eq. (1.18) we have

$$\sum_b f(q_{ab}) \, \langle a| \sum_j O_j^{\dagger}(q_{ab})|b\rangle\langle b|\sum_i O_i(q_{ab})|a\rangle. \tag{1.19}$$

What we would like to do is bring the sum over b inside and use completeness, i.e. $\Sigma_b |b\rangle\langle b| = 1$, to arrive at the result

$$f(\bar{q}_{ab}) \, \langle a| \sum_{ji} O_j^{\dagger}(\bar{q}_{ab}) O_i(\bar{q}_{ab})|a\rangle \tag{1.20}$$

which is the ground state expectation value of a special kind of two-body operator and $\bar{q}_{ab}$ is some appropriate average value of the momentum transfer. By choosing $\bar{q}_{ab}$ properly we can probably fit a single experiment that depends on eq. (1.20). However, the choice of $\bar{q}_{ab}$ required may be considerably different than we would suspect so that at least we would know that an inconsistency exists. Now when is eq. (1.20) an understandable approximation of eq. (1.19)? For illustrative purposes suppose that only a single state $|b'\rangle$ or group of states closely spaced in energy could be linked to the g.s. via the operator O. We could write equation (1.19) as

$$f(q_{ab'})\langle a|\sum_j O_j^{\dagger}(q_{ab'}) \, |b'\rangle\langle b'|\sum_i O_i(q_{ab'})|a\rangle \tag{1.21}$$

assuming only the state $|b'\rangle$ can be reached.

We can insert matrix elements to all the other states which give no contribution and arrive at the result

$$f(q_{ab'})\langle a|\sum_j O_j^{\dagger}(q_{ab'})\ \sum_b |b\rangle\langle b|\sum_i O_i(q_{ab'})|a\rangle \tag{1.22}$$

$$= f(q_{ab'})\langle a|\sum_{j,i} O_j^{\dagger}(q_{ab'})O_i(q_{ab'})|a\rangle \tag{1.23}$$

which is the desired result if we choose $f(q_{ab'}) = f(\bar{q}_{ab})$. So the idea is that the strength should be concentrated in states closely spaced in energy so that q_{ab} is essentially the same for all states. (How closely depends on two factors, the sensitivity of $f(q_{ab})$ to q_{ab} and the value of the matrix elements in some local region as a function of q_{ab}).

Now we return to muon capture on ^{16}O. From doing particle-hole calculations one expects most of the strength to be located in the giant resonance region — especially concentrated in the giant magnetic quadrupole state. As a theoretical experiment we shall calculate the total capture rate in the closure approximation. As has been previously reported,[15] using closure one finds the simple closed shell approximation to the ground state yields a calculated total rate a factor of two larger than experiment if a value of $\bar{q}_{ab}$ roughly consistent with p-h calculations is adopted. Let us study how the capture rate changes as pairing type 2p-2h admixtures are introduced in the ground state. We take the M_V^2 matrix element as an example. Using closure one obtains

$$\left(\frac{\bar{\nu}_{ab}}{\nu_\mu}\right)^2 \langle a|\sum_{ij} \tau_j^+\tau_i^- e^{i\vec{\nu}_{ab}\cdot\vec{r}_{ij}}|a\rangle \tag{1.24}$$

where

$$\vec{r}_{ij} = \vec{r}_i - \vec{r}_j.$$

There are two types of terms, $i = j$ and $i \neq j$. The $i = j$ terms are particularly simple and yield the number of protons in the ground state — here 8. The $i \neq j$ terms can be calculated using standard techniques. If we adopt harmonic oscillator wavefunctions then the result, combining both kinds of terms and assuming a $1s^4 1p^{12}$ configuration for the ground state, for M_V^2, M_P^2 or M_A^2 is the same and is

$$\left(\frac{\nu_{ab}}{\nu_\mu}\right)^2 [8 - 8e^{-2y}(1+y^2)] \tag{1.25}$$

where $y = (\tfrac{1}{2}\nu_{ab}B)^2$. (1.26)

The symbol B is the oscillator constant ($q_{ab} \equiv \nu_{ab}$ for muon capture). If we take B to be 1.8F then the mean momentum transfer $h\overline{\nu}_{ab}$ required to give agreement between the predicted rate and experiment is ≈ 73 MeV/c which would correspond to an "average" nuclear excitation of about 32 MeV. If one chooses the parameter $h\overline{\nu}_{ab}$ so that the completeness result agrees with a sum over partial transitions using standard p-h wave functions after residual interaction diagonalization, then $h\overline{\nu}_{ab} \approx$ 85 MeV/c which corresponds to a nuclear excitation of 20 MeV. However, the rate predicted in the latter case is much higher than observed. Calculations using finite-well single-particle wave functions yield similar results.[15] Actually in inelastic electron scattering one sees the isospin analogue of the important 2^- state, studied here, at about 20 MeV and in electron scattering one also needs a reduction factor of the same order (1.5 - 2.0) to get agreement between the p-h cross section prediction and experimental results.[28] Let us see if including 2p-2h states in the ground state can make it possible to obtain agreement with experiment without raising $h\overline{\nu}_{ab}$ to what seems like an unreasonable value (based on our prejudice concerning the fact that most of the strength goes to states in the g.r. region and below — as obtained in the p-h model). We consider a ground state of the general form

$$\alpha|c> + \left| \left[\left\{ \beta_{1d}[1d_{\frac{5}{2}}{}^2]+\beta_{2s}[2s_{\frac{1}{2}}{}^2] \right\}^{J_p=0}_{T_p=1} \times \left\{ 1\ p_{\frac{1}{2}}{}^{-2} \right\}^{J_h=0}_{T_h=1} \right]_{J=T=0} \right> . \tag{1.27}$$

As an example the result for $M_V{}^2$ is (compare with eq. (1.25))

$$\left(\frac{\nu_{ab}}{\nu_\mu}\right)^2 \Big\{ 8 - 8e^{-2y}(1+y^2) + e^{-2y}\{\alpha\beta_{2s}[.51y-1.03y^2+.51y^3]$$

$$+ \alpha\beta_{1d} \times .18y^3+\beta_{2s}\beta_{1d}[.82y^2-.82y^3+.21y^4]$$

$$+\beta^2_{2s}[.15y+2.07y^2-1.33y^3+.59y^4]$$

$$+\beta^2_{1d}[1.90y^2-.75y^3+.14y^4]\}\Big\} . \tag{1.28}$$

The results for $M_P{}^2$ and $M_A{}^2$ are somewhat different than for $M_V{}^2$ in that terms independent of y are obtained in the expressions multiplying $\beta_{2s}{}^2$ and $\beta_{1d}{}^2$. These terms originate from $j_0(qr_i)j_0(qr_j)\sigma_j{}^k\sigma_i{}^{-k}\tau_j{}^+\tau_i{}^-$ terms in the two-body operator. One finds for a very wide range of α, β_{2s}, and β_{1d} little effect (a slight increase on the $M_V{}^2$ term for the ν_{ab} range of interest). However, the $M_P{}^2$ and $M_A{}^2$ terms increase when pairing type correlations are included in the g.s. so that the disagreement between theory and experiment is not resolved by this particular mechanism if one still insists that the overwhelming majority of the strength lies in states in the giant resonance region and uses the "unrenormalized" muon capture operators as we have done. Of course some strength probably is transferred above the giant resonance region (look at the photoabsorption data and remember our discussion interrelating the nuclear operators appearing in different probe-induced reactions). Moreover the appropriate transition operator may well be different in the nuclear environment.

Our purpose here was simply to introduce you to the field of muon capture, to remind you of the utility of interrelating weak and electromagnetic probe transition matrix elements, and to illustrate the use of the closure approximation. In the next lecture, using a simplified model for strong probes, we shall discuss the overlap between inelastic proton-nucleus and inelastic electron-nucleus scattering. In the final lecture I will mention a connection between some of the important states studied in muon capture and the states reached via radiative pion capture and pion photoproduction. Clearly for those of you interested in spectroscopy or in testing your models of nuclear wave functions there will be plenty of exciting data from the new accelerators.

2. ELECTRON SCATTERING

Historically, electron scattering from nuclei has been one of the most fertile and accurate sources of information about the nucleus. The fact that the electromagnetic interaction is so well known means that one is really testing the nuclear response without any of the vagaries associated with a lack of knowledge of the reaction mechanism. Moreover, since the interaction is relatively weak there is a range of momentum transfer in light nuclei where first Born approximation is appropriate and thus the theoretical calculations can be greatly simplified. Static moments, r.m.s. charge radii, charge distributions, and transition charge densities are well-known examples of nuclear information obtained from electron scattering. In addition for inelastic electron scattering on closed shell nuclei it has been possible to identify some of the most important nuclear transitions lying in the giant dipole resonance region as the spin-isospin analogues of states reached via, for example, muon capture. As we discussed last time, using such general assumptions as the conserved vector current hypothesis one can easily interrelate various weak (such as β decay and muon capture) and electromagnetic interaction nuclear matrix elements. This global approach allows a more stringent test of proposed nuclear shell model wave functions, and proposed mechanisms when these wave functions fail. From another perspective one can obtain information regarding the relationship between various weak and electromagnetic coupling constants.

The situation for inelastic nucleon-nucleus scattering is more difficult than for inelastic electron-nucleus scattering. When the projectile is a nucleon, the interaction with the nucleus is strong. In addition one has only an approximate knowledge of the form for the required residual nucleon-nucleon "potential" in finite nuclear matter. Thus not only are there computational complexities arising because one must use such techniques as the distorted wave Born approximation or a coupled channel calculation approach but also there are uncertainties with regard to the reaction potential. Further because identical particles are involved, effects due to projectile-target wave function antisymmetrization must be considered.

Faced with such a situation, it is desirable to use inelastic electron scattering results to eliminate some of the uncertainties in the theoretical calculation of nucleon-nucleus scattering. In previous theoretical analyses the combined application of information from experiments using electrons and protons as probes for inelastic scattering on closed shell nuclei has been concerned mainly with nuclear transitions to low-lying (mostly normal parity "collective" T = 0 levels) and low spin states. Our discussion today will review briefly some of the information one can obtain (and previously has obtained) from inelastic electron-nucleus and nucleon-nucleus scattering. Some emphasis will be put on a speculation regarding inelastic nucleon-nucleus information that may be forthcoming from new medium energy high resolution and beam current proton accelerators now under construction (for example LAMPF at Los Alamos and the variable energy Indiana University Cyclotron). If these speculations are realized and combined with recent (and again, perhaps future) inelastic electron scattering results, one will be in a nice position, for example, to study nuclear magnetization densities and the spin-isospin flip strength in the residual nucleon-nucleon interaction in nuclear matter.

The differential cross section for electron scattering in first Born approximation is given by[29]

$$\frac{d^2\sigma}{d\Omega d\omega} = \frac{4\pi\sigma_M(E_o,\theta)}{1+2(\frac{E_o}{M_T})\sin^2\frac{\theta}{2}}\left\{\left(\frac{q_\mu{}^2}{q^2}\right)^2 \sum_{J=0}^{\infty} \frac{|\langle J_f||\hat{M}_J^{coul}(q)||J_i\rangle|^2}{2J_i+1}\,\delta(\omega-E_f+E_i)\right.$$
$$+\left(\frac{q_\mu{}^2}{2q^2}+\tan^2\frac{\theta}{2}\right)\left[\sum_{J=1}^{\infty}\frac{|\langle J_f||\hat{T}_J^{el}(q)||J_i\rangle|^2}{2J_i+1}\right.$$
$$\left.\left.+\frac{|\langle J_f||T_J^{mag}(q)||J_i\rangle|^2}{2J_i+1}\right]\delta(\omega-E_f-E_i)\right\} \tag{2.1}$$

where $q_\mu=(\vec{q},i\omega)$ is the four-momentum transfer, $q=|\vec{q}|$ is the magnitude of the three-momentum transfer, E_o is the incident electron energy, θ is the scattering angle, M_T is the nuclear target mass and $\sigma_M(E_o, \theta)$ is the Mott cross section, which is given by

$$\sigma_M(E_o,\theta) = \left(\frac{\alpha \cos \frac{\theta}{2}}{2E_o \sin^2 \frac{\theta}{2}} \right)^2 . \tag{2.2}$$

We have suppressed in eq. (2.1) correction factors for the finite size of the nucleon and the center-of-mass. The nuclear information is contained in the indicated matrix elements, $J_i\,(J_f)$ represents the spin of the initial (final) nuclear state and the symbols $\hat{O}_J(q)$ denote a sum of single-nucleon operators of multipolarity J. We shall refer to the term

$$\sum_{J=0}^{\infty} \frac{|\langle J_F \| \hat{M}_J^{coul} \| J_i \rangle|^2}{(2J_i+1)} \tag{2.3}$$

as the longitudinal form factor, $F_L{}^2(q)$, for a particular nuclear excitation (in the literature one often finds this expression divided by $(\frac{1}{2}+\tan^2 \frac{\theta}{2})$ defined as $F_L{}^2(q)$) while the term

$$\frac{1}{2J_i+1} \sum_{J=1}^{\infty} \{ |\langle J_f \| \hat{T}_J^{el} \| J_i \rangle|^2 + |\langle J_f \| \hat{T}_J^{mag} \| J_i \rangle|^2 \} \tag{2.4}$$

will be designated the transverse form factor, $F_T{}^2(q)$. The only restriction on q is $q^2 \geq \omega^2$. By energy conservation, ω, the energy lost by the electron, is also the excitation energy of the final nuclear state. For $q^2 >> \omega^2$ and high energy electrons we have essentially

$$q = 2E_o \sin \frac{\theta}{2} ; \tag{2.5}$$

thus by varying either the incident electron energy or the angle of detection (for fixed ω) we can study the momentum transfer dependence of the nuclear transition matrix element leading to a particular final state, i.e. we can map out $F^2(q)$. Note this is not possible in photoabsorption or muon capture since in these processes once ω or $E^f_{nuc} - E^i_{nuc}$ is fixed then so is q.

For illustrative purposes let us consider the longitudinal form factor for elastic scattering from a spin zero target.[29] We take the nucleon charge density to be that of a point particle, i.e.

$$\hat{\rho}_N(\vec{x}) = \delta(\vec{r}_i - \vec{x})e_i \tag{2.6}$$

where $\vec{r}_i$ is the position of the point and e_i is the nucleon charge (i.e. 0 for neutrons and +e for protons). The single nucleon operator $M_{JM}^{coul}(q)$ is given by

$$\hat{M}_{JM}^{coul}(q) = \int d\vec{x} j_J(qx) Y_{JM}(\Omega_x) \hat{\rho}_N(\vec{x}) \tag{2.7}$$

so for our special situation

$$\left.\frac{d\sigma}{d\Omega}\right|_{J=0}^{elastic} = \frac{\sigma_M}{1+2\left(\frac{E_0}{M_T}\right)\sin^2\frac{\theta}{2}} \left|\int \frac{\sin(qx)}{qx} \rho(\vec{x})_{oo} d\vec{x}\right|^2 \tag{2.8}$$

where

$$\rho(\vec{x})_{oo} = \sum_{i=1}^{Z} \int \delta(\vec{r}_i - \vec{x}) |\psi_o(\vec{r}_1 .. \vec{r}_A)|^2 d\vec{r}_1 .. d\vec{r}_A . \tag{2.9}$$

So what we are actually determining when we map out F(q) as a function of q is $\rho(\vec{x})_{oo}$, the distribution of charge in the nucleus. The effect of the delta function is to make an external integration variable the same as a nucleon shell model orbital co-ordinate. I mention this because use of a delta function force in simple inelastic nucleon-nucleus scattering calculations causes a similar simplification, and, as is well known, leads, in plane wave Born approximation, to similar expressions (as those obtained in electron scattering) for the appropriate nuclear transition matrix elements.

For inelastic scattering, in addition to different spins being involved, one would have $\psi_f{}^*\psi_i$ instead of $|\psi_o|^2$ in (2.9) and would be essentially measuring the transition charge density, ρ^J_{fi}, defined by

$$<\psi_f| \sum_{i=1}^{Z} e^{i\vec{q}\cdot\vec{r}_i} |\psi_i> = 4\pi \sum_J \int \rho^J_{fi} j_J(qr) r^2 dr . \tag{2.10}$$

Transition charge densities (and at least one of the analogues in the transverse form factors, the transition magnetization density) can yield extremely useful information for the nucleon-nucleus inelastic scattering problem.

A fairly successful and simple model for describing some of the final states reached via weak and electromagnetic projectile interactions on "closed shell" nuclei is the particle-hole (p-h) model[30] (we have in mind both the Tamm Dancoff and random phase approximations). As mentioned previously, in the Tamm Dancoff approximation one takes linear combinations of single particle excitations of the core (i.e. p-h states) to approximate the final nuclear states reached via electroexcitation. The combination (admixture amplitudes) is obtained by diagonalizing a residual interaction in some truncated Hilbert space of simple p-h states. In this model the nuclear matrix elements in eq. (2.1) take the form (we suppress trivial isospin Clebsch-Gordan coefficients)

$$\left| \sum_k \gamma_k^E (p_k \| O_J \| h_k) \right|^2 \tag{2.11}$$

where the γ's are constants determined in diagonalizing the residual interaction. The symbol k denotes a particular pure p-h configuration such as $1d_{3/2}(1p_{1/2})^{-1}$. The single particle nuclear matrix elements appearing in eq. (2.11) are between the state p_k to which the particle is promoted and the single particle state h_k which has been vacated (the hole state). The operator O_J appearing in the longitudinal matrix elements is simply $M_{coul} \equiv j_J(qx) Y_{JM}(\Omega_x)$; however for the transverse matrix elements the operator is a sum of terms, each term being a product of a Bessel function, a nucleon spin or derivative operator and a vector spherical harmonic. These terms arise from the interaction between the electron and the nuclear convection and magnetization currents. We denote the spin-dependent part of the matrix element by $M_\sigma \equiv j_L(qx) \vec{Y}_{JL1}{}^M(\Omega_x) \cdot \vec{\sigma}$. For this operator one assumes the nucleon magnetization density is that of a point particle so that an expression analogous to eq. (2.6) is adopted with μ_i, the nucleon magnetic moment, replacing e_i. Because of their importance in our later discussion we list explicit results for the single particle matrix elements of M_{coul} and M_σ. One obtains[29]

$$(n_p \ell_p j_p{}^k \| M_J{}^{coul} \| n_h \ell_h j_h{}^k) = (-1)^{j_h + J + \frac{1}{2}} \times \left(\frac{(2\ell_p+1)(2\ell_h+1)(2j_p+1)(2j_h+1)(2J+1)}{4\pi} \right)^{\frac{1}{2}} \begin{Bmatrix} \ell_p & j_p & \frac{1}{2} \\ j_h & \ell_h & J \end{Bmatrix} \times \begin{pmatrix} \ell_p & J & \ell_h \\ 0 & 0 & 0 \end{pmatrix} (n_p \ell_p | j_J(qr) | n_h \ell_h) \tag{2.12}$$

(q-dependence in radial matrix element)

and

$$(n_p \ell_p j_p{}^k \| M_\sigma \| n_h \ell_h j_h{}^k) = (-1)^{\ell_p} \sqrt{6} \times \left(\frac{(2\ell_p+1)(2\ell_h+1)(2L+1)(2j_p+1)(2j_h+1)(2J+1)}{4\pi} \right)^{\frac{1}{2}} \times \begin{Bmatrix} \ell_p & \ell_h & L \\ \frac{1}{2} & \frac{1}{2} & 1 \\ j_p & j_h & J \end{Bmatrix} \begin{pmatrix} \ell_p & L & \ell_h \\ 0 & 0 & 0 \end{pmatrix} (n_p \ell_p | j_L(qr) | n_h \ell_h). \tag{2.13}$$

For the rest of our discussion today we will be confining our remarks to electron and proton scattering on J=T=0 "closed shell" nuclear targets. Most of the electron scattering information previously used in conjunction with proton inelastic scattering calculations has been for scattering leading to T=0 normal parity excited nuclear states where M_{coul} is the important term entering in the electron scattering calculation. We shall consider here an entirely different kind of excited nuclear state, having high spin, non-normal parity, T = 1, and somewhat higher in excitation energy. These "stretched" T = 1 states such as a $1f_{7/2}(1d_{5/2})^{-1}$, 6^- state at $\sim$ 13.88 MeV in ^{28}Si or a $1d_{5/2}$, $(1p_{3/2})^{-1}$, 4^- state at $\sim$ 18.55 MeV in ^{16}O are reached in inelastic electron scattering through the term M_σ. (The derivative terms vanish because the appropriate triangle conditions in a Racah coefficient cannot be satisfied, $J \le \ell_p + \ell_h$.) Recently these T = 1 high spin states have been identified in inelastic electron scattering by working at large angle so that the transverse form factor dominates.[23,28,31] In the M_σ term (for a T = 0 g.s.) $\Delta T = 1$ transitions dominate considerably because the isovector magnetic moment ($\mu_p - \mu_n = 2.79 + 1.91 = 4.70$) is larger than the isoscalar moment ($\mu_p + \mu_n = .88$). Of course these high spin states dominate at large q because their transition matrix elements contain higher order Bessel functions which peak at larger momentum transfer.

3. INELASTIC NUCLEON-NUCLEUS SCATTERING

Now let us turn to direct reaction inelastic nucleon-nucleus scattering. The differential cross section, in distorted wave Born approximation (DWBA), is given by[32]

$$\frac{d\sigma_{if}}{d\Omega_f} = \frac{\mu_i\mu_f}{(2\pi\hbar^2)^2}\frac{k_f}{k_i}\left(\psi^*_{1f}\psi^*_{2f}\chi_f^{(-)},\ V\psi_{1i}\psi_{2i}\chi_i^{(+)}\right) \qquad (3.1)$$

where μ_i and μ_f are appropriate reduced masses for the initial and final partition respectively, k_i and k_f are the magnitudes of the initial and final wave numbers in the center-of-momentum system, the four ψ's are the internal wave functions for the initial and final scattered nucleon and initial and final target nuclear states. The distorted waves, χ, are solutions of the relative co-ordinate Schrodinger equation containing the appropriate optical potential (taken to be that responsible for elastic scattering) and V, the interaction responsible for the inelastic scattering, is the difference between the total projectile-target interaction and the optical potential, i.e. loosely the "residual interaction". Although the DWBA is frequently applied, it is well to keep in mind that, when a particular level is very strongly excited and subsequent transition strength from this level to the excited level being studied by DWBA is comparable (a multiple step process), a more complicated coupled channels calculation must be performed.

For illustrative purposes showing the information obtained by considering transitions to particular kinds of excited levels we shall consider an overly simple, naive, calculation. Subsequently we discuss some of the complications which must be properly treated.

We shall use plane-wave Born approximation (PWBA), a delta force spin- and isospin-dependent interaction, ignore antisymmetrization between the closed shell J = T = 0 target and the projectile, and treat the excited levels as simple p-h states. We take as the zero range interaction responsible for inelastically scattering the nucleon and exciting the target to a p-h state

$$V = \sum_{j=1}^{A'} \left\{ A+B(\vec{\sigma}\cdot\vec{\sigma}_j)+C(\vec{\tau}\cdot\vec{\tau}_j)+D(\vec{\sigma}\cdot\vec{\sigma}_j)(\vec{\tau}\cdot\vec{\tau}_j) \right\} \times \delta(\vec{r}-\vec{r}_j) \quad (3.2)$$

where the unsubscripted spin and isospin operators operate on the projectile and the subscript j is summed over all nucleons in the target nucleus. Constants A through D give the strength for the different spin and isospin dependent components of the simplified potential. Subject to the assumptions mentioned above the differential cross section to a definite level (E, J, T) in PWBA for an initial and final polarized projectile beam is

$$\frac{d\sigma_{if}}{d\Omega_f} = \frac{\mu_i\mu_f}{(2\pi h^2)^2} \frac{k_f}{k_i} \sum_{J_z,T_z} \Big| \sum_{\substack{j_{zp},j_{zh} \\ t_{zp},t_{zh},k}} \gamma_k{}^E (-1)^{j_h^k - j_{zh}} (j_p^k j_{zp} j_h^k - j_{zh} | JJ_z)$$

$$\times (-1)^{\frac{1}{2}-t_{zh}} (\tfrac{1}{2}t_{zp} \tfrac{1}{2}-t_{zh} | TT_z) \int e^{i(\vec{k}_i-\vec{k}_f)\cdot\vec{r}} \eta^+_f(s,t)$$

$$\times \langle j_p^k j_{zp} t_{zp} | \{A+B(\vec{\sigma}\cdot\vec{\sigma}')+C(\vec{\tau}\cdot\vec{\tau}')+D(\vec{\sigma}\cdot\vec{\sigma}')(\vec{\tau}\cdot\vec{\tau}')\}\delta(\vec{r}-\vec{r}')$$

$$| j_h^k j_{zh} t_{zh} \rangle \eta_i(s,t) d\vec{V}_r \Big|^2 \quad (3.3)$$

where the symbol η denotes the projectile spin-isospin functions for the initial (η_i) and final polarization (η_f). Since a delta force and PWBA is being used we expect results similar to those obtained in inelastic electron scattering, and we do in fact trivially obtain the following results for the different kinds of forces. Although most of the results have a wider range of validity we confine our discussion to an initially unpolarized beam, i.e. average over spin orientations and sum over final projectile spin states in eq. (3.3). For the spin-isospin independent potential $\sum_j A\delta(\vec{r}-\vec{r}_j)$

$$\frac{d\sigma_A}{d\Omega_f} = \frac{\mu_i\mu_f}{(2\pi\hbar^2)^2}\frac{k_f}{k_i} 2A^2\delta_{JL} \left| \sum_k \gamma_k^{\ E}(n_p\ell_p j_p{}^k \| M_L^{coul}(q) \| n_h\ell_h j_h{}^k) \right|^2 \tag{3.4a}$$

where $q = |\vec{k}_i - \vec{k}_f|$. For this potential no projectile spin-flip or isospin-flip is allowed and only normal parity $\pi = (-1)^J$, $\Delta T = 0$ nuclear transitions occur. For the isospin dependent potential $\sum_j C(\vec{\tau}\cdot\vec{\tau}_j)\delta(\vec{r}-\vec{r}_j)$ one obtains

$$\frac{d\sigma_C}{d\Omega_f} = \frac{C^2}{A^2}\frac{d\sigma_A}{d\Omega_f} \qquad (\pi=(-1)^J\ ,\quad \Delta T = 1 \text{ only}) \tag{3.4b}$$

where the same result is obtained for (p,p'), (p,n), (n,p) and (n,n') reactions and projectile spin-flip is not allowed. For the spin dependent potential $\sum_j B(\vec{\sigma}\cdot\vec{\sigma}_j)\delta(\vec{r}-\vec{r}_j)$ one finds for an initially unpolarized beam and summing over final projectile spin states

$$\frac{d\sigma_B}{d\Omega_f} = \frac{2}{(2L+1)}\frac{\mu_i\mu_f}{(2\pi\hbar^2)^2}\frac{k_f}{k_i} B^2 \left| \sum_k \gamma_k^E(n_p\ell_p j_p^k \| M_\sigma \| n_h\ell_h j_h^k) \right|^2 \tag{3.4c}$$

where the appropriate single particle matrix element is given by eq. (2.13), and now non-normal parity transitions $\pi = (-1)^{J+1}$ are possible but $\Delta T = 0$. Finally for a spin-isospin dependent potential of the form $\Sigma_j D(\vec{\sigma}\cdot\vec{\sigma}_j)(\vec{\tau}\cdot\vec{\tau}_j)\delta(\vec{r}-\vec{r}_j)$, averaging over initial projectile spins and summing over final spins,

$$\frac{d\sigma_D}{d\Omega_f} = \frac{d\sigma_B}{d\Omega_f}\frac{D^2}{B^2} \qquad \begin{array}{l}\text{(non-normal parity possible; only } \Delta T = 1\\ \text{nuclear transitions are allowed).}\end{array} \tag{3.4d}$$

The same result is obtained for (p,p'), (p,n), n,p) and (n,n') reactions. The $4^-(^{16}O)$ and $6^-(^{28}Si)$, T = 1 non-normal parity states we discussed earlier in connection with electron scattering can only be reached (in the simple model above) in inelastic nucleon scattering via the spin-isospin dependent potential D – see eq. (3.4d). In this case, in both electron and nucleon scattering, only the operator M_σ enters into the single particle matrix elements. Operators of this type also appear in muon capture in the axial vector, $M_A{}^2$, and induced pseudo-scalar, $M_P{}^2$, matrix elements. Note also that at a particular excitation energy one can relate the cross sections (p,n) to (p,p') for the force leading to (3.4d). The charge exchange cross section predicted is a cross check that one is actually looking at a T = 1 state. No charge exchange should come from the symmetry energy induced $\vec{T}\cdot\vec{\tau}$ term in the optical potential because the nuclear g.s. has T = 0.

Before we conclude by discussing briefly what enters into a more realistic calculation, note that since the nucleon is much heavier one obtains at a given scattering angle and incident energy a much higher momentum transfer than is given in electron scattering. For the momentum transfer region [300 to 400 MeV/c], where the high spin states mentioned above are expected to peak, and an incident proton energy of ~ 200 MeV, we are interested in nucleon scattering angles of approximately 30°. In fact, existing lower energy machines are capable of reaching this momentum transfer region. Some advantages of the new variable medium energy machines will be the high beam current and good energy resolution. Hopefully some of the complications discussed below, which make studies of the type we envisage virtually impossible at lower energies, will become less troublesome at higher energies (i.e. presumably the distorted-wave impulse approximation becomes valid).

Now one can easily go beyond the simple calculation we have used for illustrative purposes above. Computer codes exist for treating the problem in DWBA and allowing a finite range "residual interaction". Such calculations are standard. Inclusion of a spin-orbit potential in the optical potential responsible for the distortion means that spin-flip can occur in inelastic scattering even if the residual potential V contains no spin dependence. Thus the spin-orbit contribution must be included and its effect subtracted before conclusions regarding the size of spin dependent terms in V can be made from a comparison with experimental differential cross sections. Earlier work[33] has shown it may be especially important to include the optical spin-orbit potential if conclusions are to be drawn from polarization studies. Previous calculations[33] indicate that the main effect of the distorting optical potential is a reduction of the peak cross section; the shape of the cross section curve, for example, as a function of center of mass scattering angle, being essentially unaffected (exception - electric monopole transitions). The reduction factor of course comes from the imaginary part of the optical potential which produces attenuation of the wave function inside the nuclear volume. This fact also means that using the usual harmonic oscillator (h.o.) wave functions for the nuclear orbitals may be more suspect since the tail region of the nuclear density where the h.o. orbitals are especially inadequate is now being emphasized.

Another feature which must be considered is the effect of projectile target anti-symmetrization. Spin-isospin flip transition can occur from the exchange terms even assuming only a Wigner residual interaction. Clearly more realistic calculations are needed before we can draw definitive conclusions about the utility of studying, and indeed the ability to see, the high spin states under consideration in inelastic proton scattering.

One of the great advantages of the higher energy range is that one can use the distorted-wave impulse approximation (DWIA).[34] The basic idea is that the t matrix one must calculate in nucleon-nucleus scattering (assuming two-body interactions) becomes, at high energy, essentially the t matrix obtained from nucleon-nucleon scattering. The main difference between the integral equations the two types of t matrices must satisfy is that in the nucleon-nucleus case there appears a projection operator eliminating certain intermediate states due to the Pauli exclusion principle (of course the respective propagators are also different). Simple arguments[34] concerning the importance of these unallowed intermediate states for high energy incident projectiles based on, for example, a Fermi gas model of the nucleus give one confidence in replacing the "free" nucleon-bound nucleon t matrix with the free nucleon-nucleon t matrix in the energy region we consider. Since the free nucleon-nucleon t matrices are already properly symmetrized one does not have to worry about calculational problems resulting from anti-symmetrization. The t matrix is usually written in the form[34]

$$t_i = [A + B(\vec{\sigma}\cdot\hat{n})(\vec{\sigma}_i\cdot\hat{n}) + C\,(\vec{\sigma}+\vec{\sigma}_i)\cdot\hat{n} + E(\vec{\sigma}\cdot\hat{q})(\vec{\sigma}_i\cdot\hat{q}) + F(\vec{\sigma}\cdot\hat{p})(\vec{\sigma}_i\cdot\hat{p})] \qquad (3.5)$$

where

$$\hat{q} = \frac{\vec{q}}{|\vec{q}|}, \quad \vec{q} = \vec{k}_f - \vec{k}_i$$

and

$$\hat{n} = \frac{\vec{n}}{|\vec{n}|}, \quad \vec{n} = \vec{k}_i \times \vec{k}_f, \quad \hat{p} = \hat{q} \times \hat{n}.$$

The subscript i on the spin functions refers to the ith target nucleon. The symbols A, B, C, E and F are functions of the isospin state, the square of the local momentum transfer q^2, and the initial and final local C.M. energy, k_i^2 and k_f^2 (of course, for the free nucleon-nucleon case these are the same). As mentioned above the t matrix is a function of the local two-body C.M. energy and local momentum transfer while what one actually uses are the asymptotic (away from the region of nuclear interaction) C.M. energy and momentum transfer. The local and asymptotic variables differ because of refraction and diffraction effects in nucleon-nucleus scattering. The free nucleon-nucleon t matrix varies relatively slowly as a function of bombarding energy so the small difference between local and asymptotic energy is assumed not to introduce appreciable error. However the fact that one is using distorted waves means that the t matrix must be averaged over a range of q. Model calculations by Haybron[35] have shown that for bombarding energies of 100 MeV and above, for the low lying 2^+, T = 0 states in ^{12}C and ^{40}Ca, the differential cross section is altered by a factor of ~ 2 in the forward direction by including such effects.

There are other complications that must be considered (core polarization if the nuclear ground state isn't a closed shell, which it isn't, effect of coupled channels, etc., etc.) and one may finally begin to wonder what is being tested — the p-h model, the importance of multiple-step processes, a particular approximation attendant to the DWIA or the electron scattering nuclear-state identification. In fact, as is often the case, one will probably be "testing" more than one assumption at once. But by using the variable energy and high resolution features of, for example, the new Indiana cyclotron and by employing the electron scattering results one may be able to make definite progress in understanding (and cross-checking our "understanding") the spin and isospin dependence of the residual interaction responsible for the direct process. We may also learn more about the wave functions of these intriguing T = 1 high-spin non-normal parity excitations of "closed shell" nuclei.

For those of you who have more experience dealing with electron scattering experiments it should be pointed out that in nucleon-nucleus scattering life is somewhat easier for the experimentalist because of the absence of the radiative tail associated with electron scattering. Also of course cross sections tend to be much larger. For those of you that favor nucleon-nucleus scattering because "in electron scattering one is only learning about the distribution of protons in the nucleus" I would point out that neutrons, because of their appreciable magnetic moment, play an important role in electron-induced transitions to the non-normal parity states discussed here.

Finally, it could turn out that the particular states considered above just aren't that easy to see or not as important as other new kinds of states (for example, the high spin collective states predicted in the variable moment of inertia model[36] are important states to look for in both electron and proton-nucleus inelastic scattering). Nevertheless use of the combined electron and proton information by theorists and even joint studies by electron scattering facilities and the new proton machines are clearly useful approaches to keep in mind.

4. PION-NUCLEUS SCATTERING

The meson factories will provide us with a fine low and intermediate energy pion microscope. Clearly the uses to which we put this new tool for studying the nucleus depend on the basic properties of the pion so let me briefly remind you of a few basic facts. The pion is a strongly interacting, spin zero, boson that exists in three charge states ($\pi^{\pm}$, charge $\pm e$ and the π°, uncharged). The dominant decay modes and associated mean lives are given below with the rest mass in parenthesis:[37]

$$\pi^{\pm}(139.6\text{ MeV}) \rightarrow \mu^{\pm} + \nu_{\mu}, \quad \tau = 2.6 \times 10^{-8}\text{ sec.}$$

$$\pi^{\circ}(135.0\text{ MeV}) \rightarrow 2\gamma, \quad \tau = .8 \times 10^{-16}\text{ sec.} \qquad (4.1)$$

Note also that

$$\frac{\pi \to e + \nu}{\pi \to \mu + \nu} \approx 10^{-4}. \qquad (4.2)$$

So the pion is an alternative to the nucleon for studying the nucleus via a fundamental strongly interacting probe. Our "understanding" of the pion interaction with nucleons and nuclei is far less than our knowledge of the muon and electron-nucleus interactions discussed earlier. In addition the pion-nucleon interaction is less completely parametrized than the nucleon-nucleon interaction at low and intermediate energy. These statements mainly reflect the fact that, as mentioned earlier, the pion is a strongly interacting probe and that, historically, precise and intense pion fluxes have not been available for experimental use. However it means that our enthusiasm for this new probe as a means of studying nuclear structure must be tempered by our considerable lack of knowledge regarding the pion-nucleus interaction. From a fundamental particle physics perspective this uncertainty is, in fact, exciting, because it means there is something new and "basic" that one may learn regarding the elementary strong pion-nucleon and pion-pion interactions.

By now it is conventional to mention that we have, heretofore, largely ignored the existence of pions in the nucleus. To be sure we are aware of pions as a major mediator of the nucleon-nucleon interaction (especially in the longer range part of the force) and we are careful to speak of "dressed nucleons", i.e. the picture of a cloud of mesons, including pions, surrounding a nucleon. Moreover, we have a healthly respect for the possibility that "meson-exchange currents" in the nucleus may alter the familiar free nucleon weak and electromagnetic probe interactions we most often apply, unchanged, in nuclear matter. However, information and studies have been largely absent regarding the nuclear pion distribution and the pionic degrees of freedom and related modes of excitation in the nucleus. Perhaps by viewing the nucleus with the pion microscope we may learn more about the details of the nuclear pions. However at this stage, in view of the uncertainty in the fundamental pion interaction and the fact that, to some extent, the effects of nuclear pions are included in nuclear model wave functions, this aspect of meson physics, while exciting, appears exceptionally challenging in complex nuclei.

Just as in the case of muon-nucleus interactions discussed earlier, if we consider investigations of pion-nucleus interactions we may study pion-nucleus elastic and inelastic scattering, pions captured in atomic orbits (pi-mesic atoms), and nuclear pion capture. In addition using, for example, a 200 MeV intense proton beam one can study subthreshold pion production in complex nuclei ("threshold" is the 290 MeV lab proton energy required for pion production in free nucleon-nucleon scattering).

Our main emphasis today will be on a technique to be applied to pion-nucleus scattering; however, let me first mention that interactions involving pions can lead to final excited nuclear states belonging to the family of giant resonances (spin and isospin analogues of the photoabsorption giant dipole resonance and the giant quadrupole resonance) studied earlier for weak and electromagnetic probes. Near theshold one has some confidence[38,39] that the effective Hamiltonian appearing in nuclear matrix elements associated with radiative pion capture in nuclei, for example for π^- capture

$$\pi^- + \mathrm{Nuc}(A,Z) \rightarrow \mathrm{Nuc}^*(A,Z-1)+\gamma \ , \tag{4.3}$$

is of the form

$$H_{eff} \propto \sum_j^A \tau_j^{(-)} A\vec{\sigma}_j \cdot \hat{\varepsilon}\ \delta(\vec{r}-\vec{r}_j) \ . \tag{4.4}$$

In eq. (4.4) τ^- is the usual single nucleon isospin lowering operator, σ is nucleon spin operator, $\hat{\varepsilon}$ is a unit vector in the direction of photon polarization. The sum in eq. (4.4) is over all nucleons in the nucleus. The expression for the radiative capture rate contains the following matrix element:

$$\langle f|e^{-i\vec{k}_f\cdot\vec{r}} H_{eff}\phi_{n\ell}(\vec{r})|i\rangle \tag{4.5}$$

where k_f is the final photon wavenumber and $\phi_{n\ell}$ is the pion atomic wave functions. By substituting eq. (4.4) into (4.5), assuming the pion is in the 1s atomic orbit when captured, and noting that approximately 140 MeV becomes available for nuclear excitation and photon energy when the stopped pion is annihilated one predicts that an important class of final nuclear excited states reached via this process consists of the isospin analogues of the giant electric dipole and magnetic quadrupole states seen in electron scattering (in, for example, ^{16}O). Of course if π^- capture results in strong excitation of these L = 1 spin-flip giant resonances this would be exciting because these are exactly the same states that are predicted to be strongly excited in muon capture through the $M_A{}^2$ and $M_p{}^2$ matrix elements. One advantage of using pion-capture to study these states is that one can detect the final γ ray and determine its energy (relatively easily certainly, compared to the detection of the final neutrino that is emitted in muon capture!). If instead of capture of a pion from the lowest atomic orbit (1s) one considers capture from the 1p orbit then one predicts strong excitation of a giant electric quadrupole resonance.[40] One can go further and consider pion photoproduction[41]

$$\gamma + \mathrm{Nuc}(A,Z) \rightarrow \mathrm{Nuc}^*(A,Z+1)+\pi^-$$

or

$$\rightarrow \mathrm{Nuc}^*(A,Z-1)+\pi^+ \quad , \qquad (4.6)$$

where by varying the angle of detection of the final emitted pion one can vary the momentum transfer appearing in the nuclear matrix element. Thus to the extent that the operator (4.4) (or simple extension of it) is appropriate for the region of pion energy studied in pion photoproduction and to the extent that final state interactions can be treated properly, pion photoproduction can be an additional exciting mechanism for studying the nuclear giant resonances.

The pion-nucleon interaction is weak (relative to the nucleon-nucleon interaction) at low energies and at intermediate energies is dominated by the p wave, $J^{\pi} = 3/2^-$, isospin = 3/2 resonance occurring at a pion lab kinetic energy (incident on a nucleon at rest) of ≈ 180 MeV. The dominant feature in pion-nucleus scattering below 500 MeV is the manifestation of the (2T, 2J) = (3,3) resonance mentioned above in the nuclear matter environment. As A increases, the energy at which the peak cross section occurs decreases (interpreted as a shifting of the (3,3) resonance in nuclei). (See ref. 42 for a recent discussion and probable explanation of this shift.)

Most discussions to date, of pion-nucleus scattering, have used techniques that can be reasonably justified in two extremes: high energy ($T_{\pi}^{lab} \gtrsim$ 1-2 GeV) and low energy ($T_{\pi}^{lab} \lesssim$ 100 MeV). Much progess has been made in high energy pion-nucleus scattering through application of the eikonal or Glauber multiple scattering formalism.[43-45] The low energy scattering of pions from nuclear targets has been treated using an optical potential similar to that suggested by Kisslinger[46] with modifications and applications by Ericson and Ericson[47] and others.[48,49] The low energy optical potential utilizes primarily s and p wave pion-nucleon scattering phase shifts (in the various isospin channels) as input information to determine pion-nucleus scattering. A discussion of the low energy optical potential in terms of a multiple scattering series makes it apparent that the correlations in nuclear matter (for example the long range correlations induced by the exclusion principle) are important in altering the effective pion-nucleus potential from a simple sum of pion-nucleon interactions.[50] Application[51] of a modified optical potential (suggested by Kroll[48]) that incorporates the effect of nuclear correlations has been encouraging. The basic pion wave equation in this model is written[51]

$$(-\nabla^2+\mu^2)\Phi(\vec{r}) = \left\{(E_{\pi}-V_c)^2+Ab_0 q_0^2 \rho(\vec{r}) - \vec{\nabla}\cdot\{Ab_1\rho(\vec{r})(1+\tfrac{1}{3}\xi Ab_1\rho(\vec{r}))^{-1}\nabla\}\right\}\cdot \Phi(\vec{r}) \qquad (4.7)$$

where

$E_\pi \rightarrow$ total energy, $q_o \rightarrow$ incident momentum

$\mu \rightarrow$ pion mass

$A \rightarrow$ target atomic number

$\rho(\vec{r}) \rightarrow$ unit normalized nuclear density function

$(1+\frac{1}{3}\xi A b_1 \rho(\vec{r}))^{-1} \rightarrow$ nuclear anti-correlation-induced factor

(4.8)

and where b_0 and b_1 contain the pion-nucleon s wave (b_0) and p wave (b_1) phase shift information. For example (for negative pion scattering)

$$b_o = \frac{4\pi}{q_o^{\,3}} \left(\frac{\mu^2+M_p^2+2E_\pi M_p}{M_p^2}\right) \frac{1}{A} \left[(\tfrac{1}{3}Z+N)\, e^{i\delta^{\ell=0}_{T=3/2}} \sin \delta^{\ell=0}_{T=3/2} + \tfrac{2}{3} Z e^{i\delta^{\ell=0}_{T=1/2}} \sin \delta^{\ell=0}_{T=1/2} \right] \quad (4.9)$$

with a similar expression for b_1 in terms of the p-wave phase shifts. Jones and Eisenberg[51] point out that, while the agreement with experiment is not significantly improved over that obtained adopting a simpler optical potential, one can obtain (using the optical "potential" in (4.7)) reasonable fits to low energy pion nucleus scattering with values of the input parameters (especially b_1) close to those obtained in a multiple scattering theory (as opposed to purely phenomenological optical potential fits). This is important because one wishes to have confidence in the optical potential since, for example, in pion absorption the cross section is quite sensitive to the pion wave function. Even without other applications in mind, one naturally prefers, if possible, to have an accurate theory of pion-nucleus scattering that can be obtained from a model of pion-nucleon scattering supplemented by a multiple scattering theory, incorporating nucleon-nucleon correlations in the nucleus if necessary.

The procedure outlined below for treating pion-nucleus scattering in the intermediate energy region represents work in progress[52] so that I won't be giving any final results. Since this is meant to be only a sketch of another approach to the problem of pion-nucleus scattering, spin and isospin complications will be ignored. One certainly wants to incorporate the experimentally prominent resonances formed in the pion-nucleon system at intermediate energies into the procedure for treating intermediate energy pion-nucleus scattering. Our technique has the following features: A) The basic input data is obtained by utilizing the available pion-nucleon complex phase shifts. Thus several partial waves with phase shifts (and energy dependent inelasticity parameters) up through and including several pion-nucleon resonances are incorporated. B) The form of the parametrization chosen for the pion-nucleon data (a separable "potential" or source function) allows one to obtain relatively easily a pion-nucleus optical potential and to calculate (systematically) corrections to it. Of course one treats the pion relativistically. We can go to the usual high and low energy regions and thus make a comparative study of our results with those obtained using the established formalisms in these asymptotic energy ranges.

We assume the pion-nucleon interaction can be parametrized in terms of an incident pion, described by a Klein-Gordon equation, and a target nucleon whose effect on the pion can be included via a fixed static potential $S(\vec{x})$. (Remember spin and isospin complexities are suppressed.) This allows

$$[\Box - (\frac{mc}{\hbar})^2]\overline{\Psi} = -S(x)\overline{\Psi} \ . \tag{4.10}$$

The stationary state solutions $\overline{\Psi} = \Psi e^{-i\omega t}$ [now with $(\hbar\omega)^2 = (\hbar kc)^2 + m^2c^4$] satisfy the familiar scalar Helmholtz equation,

$$(\nabla^2 + k^2)\Psi = -S(\vec{x})\Psi \ . \tag{4.11}$$

If we now consider non-local separable potentials then

$$S(\vec{x})\Psi(\vec{x}) \rightarrow \int S(\vec{x},\vec{x}\,')\Psi(\vec{x}\,')d\vec{x}\,' \tag{4.12}$$

where

$$S(\vec{x},\vec{x}\,') = \sum_{\ell m} 4\pi\lambda_\ell s_\ell(x)s_\ell(x')Y_{\ell m}(\Omega_x)Y^*_{\ell m}(\Omega_{x'}) \tag{4.13}$$

or written in momentum space

$$S(\vec{k},\vec{k}\,') = \sum_{\ell m} \alpha_\ell s_\ell(k)s_\ell(k')Y_{\ell m}(\Omega_k)Y^*_{\ell m}(\Omega_{k'}) \ . \tag{4.14}$$

This particular technique has been previously suggested for both pion-nucleon[53,54] and nucleon-nucleon scattering[55,56] but has been applied mainly to nucleon-nucleon scattering to obtain a separable potential from the nucleon-nucleon phase shifts[55,56] (the inverse scattering problem with separable potentials).

For the nucleon-nucleon problem one obtains the separable potential from the phase shifts via an expression of the form (assuming for simplicity here no bound state and $\delta(k=0) - \delta(k=\infty)=0$)

$$(s_\ell(k))^2 = \frac{1}{-\beta_\ell} \frac{\sin \delta_\ell(k)}{k} e^{-\Delta_\ell(k)} \tag{4.15}$$

where $\Delta_\ell(k)$ is the principal value integral

$$\Delta_\ell(k) = \frac{2}{\pi} P\int_0^\infty \frac{d'\delta_\ell(k')dk'}{k'^2-k^2} \tag{4.16}$$

and where $\delta_\ell(k)$ is the ℓth partial wave energy-dependent phase shift and β_ℓ is a constant. The results are similar in the case of pion-nucleon scattering except that now $s_\ell(k)$ is complex so that instead of eq. (4.16) one obtains two coupled equations involving the real and imaginary parts of $s_\ell(k)$ (now $\delta_\ell(k)$ is complex because of the possible inelastic channels).

Unfortunately the integral appearing in eq. (4.16) requires a knowledge of the phase shift up to infinite energy. Clearly such information is not obtainable from experiment. However we intend to use the potential s(k) in the region hkc < 500 MeV so that $k'^2 - k^2$ which appears in the denominator in eq. (4.16) will be large as k'^2 goes to ∞. Thus we may hope that the contributions to the integrals at energies where we don't know $\delta_\ell(k')$ will not be important and $s_\ell(k)$ will therefore not be appreciably affected by the "guess" we must make for $\delta_\ell(k')$ when k' is large. Of course one can use several different functional forms for $\delta_\ell(k)$ as $k \to \infty$ and study how $s_\ell(k)$, in the region of planned application, is altered. For high energy applications of a separable potential we note a form has been given that leads to Regge behavior and a finite cross section at infinite energy.[57]

To obtain the effective potential for pion-nucleus elastic scattering one first assumes that the energy transferred to the target, compared to the incident pion energy, is negligible at each step in the possible multiple scattering process. (This allows one to use closure on the target. Within the spirit of a model this assumption can and should be tested for particular examples.) Adopting this assumption and using a separable form for the pion-nucleon "potential" one can proceed to solve exactly for the pion-nucleus many-body scattering amplitude.[58] Except for the lightest nuclei it appears that evaluation of the exact amplitude will be prohibitively long. However one can make a multiple scattering expansion of the many-body amplitude (along with an expansion of the ground state target density in terms of n-particle densities, $1 \leq n \leq A$) and thus generate the effective potential or modified optical potential to a desired degree of complexity.[58] More specifically, we may define the optical potential U(x,y) by writing down the usual integral equation for the Schrodinger equation incorporating outgoing-wave boundary conditions[58]

$$\Psi_k^+(\vec{x}) = e^{i\vec{k}\cdot\vec{x}} - \iiint \frac{d\vec{t}}{(2\pi)^3} \frac{e^{i\vec{t}\cdot(\vec{x}-\vec{y})}}{t^2-k^2-i\eta} U(\vec{y},\vec{z})\Psi_k^{(+)}(\vec{z})d\vec{y}d\vec{z} \tag{4.17}$$

where the scattering amplitude $f(\vec{k}',\vec{k})$ is obtained from

$$f(\vec{k}',\vec{k}) = -\frac{1}{4\pi}\int e^{-i\vec{k}\cdot\vec{x}}U(\vec{x},\vec{y})\Psi_k^{(+)}(\vec{y})d\vec{x}d\vec{y} \; . \qquad (4.18)$$

It has been shown[58] that if 1) the number of terms that must be kept in the iteration of the integral equation for the optical potential is small compared to the number of target particles, 2) a given target particle is not struck more than once in a given term in the multiple scattering series, and 3) the A-nuclear-particle density distribution is written as a simple product of one particle density distributions, then the optical potential $U(\vec{x},\vec{y})$ assumes the simple form

$$U_o(\vec{x},\vec{y}) = \iint \frac{d\vec{p}}{(2\pi)^3}\frac{d\vec{q}}{(2\pi)^3} e^{i\vec{p}\cdot\vec{x}}U_o(\vec{p},\vec{q})e^{-i\vec{q}\cdot\vec{y}} \qquad (4.19)$$

where

$$U_o(\vec{p},\vec{q}) = -4\pi A\rho(\vec{p}-\vec{q})\sum_\ell f_\ell(k)(2\ell+1)P_\ell(\cos\theta_{pq}) \times \frac{s_\ell(p)s_\ell(q)}{[s_\ell(k)]^2} \qquad (4.20)$$

and where

$$\int e^{i(\vec{q}-\vec{p})\cdot\vec{x}}\rho(\vec{x})d\vec{x} \equiv \rho(\vec{q}-\vec{p}) \; . \qquad (4.21)$$

One can extend this result by, for example, relaxing assumption three above and including two-particle correlations in a two-particle density distribution function. As I mentioned, this represents current research so I will not go further except to reiterate that it will be interesting to compare the results obtainable from such a procedure with the pion-nucleus optical potentials presently in use.

In this short discussion of pion-nucleus interactions I have left out several important topics (examples, double charge exchange (π^+,π^-), and information regarding the nuclear wave function high momentum components using pion absorption and production). I would suggest reference 59 for a more comprehensive review. I hope I have communicated to you some of the reasons for the excitement, from a nuclear physics viewpoint, that one has regarding the proton, muon and pion-nucleus data that should be forthcoming soon from the intermediate energy accelerators. Today we talk about bringing pion physics into conventional nuclear physics. It may not be long before we include kaons in our list of available nuclear probes.

REFERENCES

1. LAMPF Users Handbook (Preliminary Version), LA-4586-MS (Lewis E. Agnew, Jr., Liaison Officer, LAMPF Users Group), Jan. 1971.
2. Nagle, D.E. The Los Alamos Meson Physics Facility (LAMPF), Proc. Banff Summer School on Intermediate Energy Nuclear Physics, 1970. Edited by G.C. Neilson, W.C. Olsen, and S. Varma (University of Alberta, Edmonton), p. 284.
3. Search and Discovery. Physics Today 24, 19, 1971.
4. Rosen, L. Physics Today 19, 21, 1966.
5. Warren, J.B. TRIUMF -- Design of Initial Facility and Possible Developments, Proc. Banff Summer School on Intermediate Energy Nuclear Physics, 1970. Edited by G.C. Neilson, W.C. Olsen, and S. Varma (University of Alberta, Edmonton), p.294.
6. Moravcsik, M.J. Physics Today 23, 40, 1970.
7. Rickey, M.E., Sampson, M.B., and Bardin, B.M. IEEE Trans. Nuclear Sci. NS16, 397, 1969.
8. Singh, P.P. Indiana University 200 MeV Cyclotron Facility, Proc. Banff Summer School on Intermediate Energy Nuclear Physics, 1970. Edited by G.C. Neilson, W.C. Olsen, and S. Varma (University of Alberta, Edmonton), p. 315.
9. Hughes, V.W. Physics Today 20, 29, 1967.
10. Engfer, R. Muonic Atoms and Nuclear Structure, Third International Conference on High Energy Physics and Nuclear Structure (Plenum Press, 1970, Edited by S. Devons), p. 104.
11. Fujii, A., and Primakoff, H. Nuovo Cimento 12, 327, 1959.
12. Rood, H.P.C., and Tolhoek, H.A. Nucl. Phys. 70, 658, 1965.
13. Sens, J.C. Phys. Rev. 113, 679, 1959.
14. Weinberg, S. Phys. Rev. 112, 1375, 1958.
15. Luyten, J.R., Rood, H.P.C., and Tolhoek, H.A. Nucl. Phys. 41, 236, 1963.
16. Feynman, R.P. and Gell-Mann, M. Phys. Rev. 109, 193, 1958; Goldberger, M.L. and Treiman, S.B. Phys. Rev. 111, 355, 1958.
17. Wolfenstein, L. Nuovo Cimento 8, 882, 1958.
18. Primakoff, H. Rev. Mod. Phys. 31, 802, 1959.
19. Foldy, L.L. and Walecka, J.D. Nuovo Cimento 34, 1026, 1964.
20. Foldy, L.L. and Walecka, J.D. Phys. Rev. 140, B1339, 1965.
21. Wigner, E. Phys. Rev. 51, 106, 1937.
22. de Forest, T. Phys. Rev. 139, B1217, 1965.
23. Donnelly, T.W., and Walker, G.E. Ann. Phys. 60, 209, 1970.
24. Walker, G.E. Phys. Rev. 174, 1290, 1968.
25. Green, A.M. and Rho, M. Nucl. Phys. A130, 112, 1969.
26. Zucker, A.P., Buck, B., and McGrory, J.B. Phys. Rev. Lett. 21, 39, 1968.
27. Brown, G.E., and Green, A.M. Nucl. Phys. 85, 87, 1966.
28. Sick, I., Hughes, E.B., Donnelly, T.W., Walecka, J.D., and Walker, G.E. Phys. Rev. Lett. 23, 1117, 1969.
29. de Forest, Jr., T., and Walecka, J.D. Advan. Phys. 15, 1, 1966.
30. Brown, G.E., Castillejo, L., and Evans, J.A. Nucl. Phys. 22, 1, 1961.

31. Donnelly, T.W., Walecka, J.D., Walker, G.E., and Sick, I. Phys. Lett. 32B, 545, 1970.
32. Austern, N. Direct Nuclear Reaction Theories, (Wiley, New York, 1970), Chapters 4 and 5.
33. Haybron, R.M., and McManus, H. Phys. Rev. 136B, 1730, 1964.
34. Kerman, A.K., McManus, H.M., and Thaler, R.M. Ann. Phys. 8, 551, 1959.
35. Haybron, R.M. Phys. Rev. 160, 756, 1967.
36. Mariscotti, M.A.J., Scharff-Goldhaber, C.S., and Buck, B. Phys. Rev. 178, 1864, 1969.
37. Rittenberg, A. et al. (Particle Data Group). Review of Particle Properties. Rev. Mod. Phys. 43, 51 (supplement), 1971.
38. Chew, G.F., Goldberger, M.L., Low, F.E., and Nambu, Y. Phys. Rev. 106, 1345, 1957.
39. Delorme, J., and Ericson, T.E.O. Phys. Lett. 21, 98, 1966.
40. Anderson, D.K., and Eisenberg, J.M. Phys. Lett. 22, 164, 1966.
41. Uberhall, H. Analogs of Giant Resonances from Photo-Pion Production and from Muon and Pion Capture, Third International Conference on High Energy Physics and Nuclear Structure (Plenum Press, 1970, Edited by S. Devons), p. 48.
42. Ericson, T.E.O., and Hüfner, J. Phys. Lett. 33B, 601, 1970.
43. Glauber, R.J. Lectures in Theoretical Physics, Vol. I (Interscience, New York, 1959), p. 315.
44. Franco, V., and Coleman, E. Phys. Rev. Lett. 17, 827, 1966.
45. Michael, C., and Wilkin, C. Nucl. Phys. B11, 99, 1969.
46. Kisslinger, L.S. Phys. Rev. 98, 761, 1955.
47. Ericson, M., and Ericson, T. Ann. Phys. 36, 323, 1966.
48. Edelstein, R.M., Baker, W.F., and Rainwater, J. Phys. Rev. 122, 252, 1961.
49. Auerbach, E.H., Fleming, D.M., and Sternheim, M.M. Phys. Rev. 162, 1683, 1967; Phys. Rev. 171, 1781, 1968.
50. Ericson, T.E.O. International School of Physics, Enrico Fermi Course XXXVIII (Academic Press, New York, 1967), p. 253.
51. Jones, W.B., and Eisenberg, J.M. Nucl. Phys. A154, 49, 1970.
52. Work in collaboration with E. Moniz and M. Piepho.
53. Omnes, R. Nuovo Cimento 8, 316, 1958.
54. Gourdin, M., and Martin, A. Nuovo Cimento 8, 699, 1958.
55. Bolsterli, M. and MacKenzie, J. Physics 2, 141, 1965.
56. Tabakin, F. Phys. Rev. 177, 1443, 1969.
57. Bawin, M. Nucl. Phys. B28, 109, 1971.
58. Foldy, L.L., and Walecka, J.D. Ann. Phys. 54, 447, 1969.
59. Ericson, T.E.O. Lectures on Pion-Nucleus Interactions, Proc. Banff Summer School on Intermediate Energy Nuclear Physics, 1970. Edited by G.C. Neilson, W.C. Olsen, and S. Varma (University of Alberta, Edmonton), p. 102.

EXPERIMENTS AT MESON FACTORIES

D.F. MEASDAY
Department of Physics
University of British Columbia

THE ACCELERATORS

The experiments at meson factories are very much governed by the properties of the accelerator and of the available beams, so I would like to start by comparing various accelerators. I hope to illustrate the tremendous improvements that are expected when the new generation is born in a year or two. As the gestation period has been four or five years, many will be familiar with the details, so I intend to keep this section short.

There are three meson factories being built around the world (SIN in Zürich, LAMPF in Los Alamos and TRIUMF in Vancouver). In addition at least two of the older generation of synchrocyclotrons are undergoing major surgery and afterwards will have properties approaching those of the meson factories. The improvement programmes for the CERN and Columbia synchrocyclotrons are well underway, and other machines may follow suit. I shall compare LAMPF and TRIUMF with the CERN synchrocyclotron as it is at present in order to illustrate what superior properties the new machines have. I hope also to show that TRIUMF will have some advantages over LAMPF, although the latter's higher energy and higher current make it the heavyweight among the new machines. SIN will be similar in many ways to TRIUMF. Table 1 lists some of the important parameters.

LAMPF is a proton linear accelerator with a maximum energy of 800 MeV; the machine will come into operation in the middle of next year (1972). It can accelerate protons and H^- ions simultaneously, and the latter can be separated off with a magnet and then easily split into various beams by a stripping foil, followed by a magnet. The energy of the beam can be varied in theory by turning off sections of the accelerator, but at present the proton and H^- beam must have the same energy, and the people who want to use the protons to produce secondary meson beams will want as high a proton energy as possible most of the time. One idea may change this conclusion: it has been suggested that one strip the H^- ions part way down the machine and hope the resulting protons focus sufficiently as they coast (or slightly decelerate) to the end of the machine.

Table 1

	TRIUMF	CERN SC (present)	LAMPF
Proton Energy (MeV)	150 - 500 (540?) (3 independent beams)	600 (fixed energy)	200 - 800 in steps (H^+ and H^- with same energy)
Energy Resolution for Proton Beam	1 MeV initially ≃100 keV eventually	≃3 MeV	Beam ≃9 MeV Energy loss mode ≃30 keV
Neutron Beam (sec^{-1} over 10 cm^2)	10^7	10^5	10^8
Duty Cycle $\Delta\tau \simeq 1$ μsec.	100%	50%	6%
$\Delta\tau \simeq 3$ nanosec.	10 to 20%	≃20%	≃4%
Pion Energies (MeV)	20 - 300	70 - 300	20 - 550
Pion Fluxes π^+ (sec^{-1})	10^7 to 10^8	10^4 to 10^5	10^8 to 10^9
π^- (sec^{-1})	10^6 to 10^7	$\simeq 10^5$	10^7 to 10^8
Pion Energy Resolution (energy loss mode)	≃30 keV	≃1 MeV	≃30 keV
Stopped π^- ($sec^{-1}g^{-1}cm^2$)	$\simeq 10^7$	$\simeq 10^4$	$\simeq 10^8$
Stopped μ^- ($sec^{-1}g^{-1}cm^2$)	$\simeq 10^6$	$\simeq 10^3$	$\simeq 10^7$

TRIUMF will come into operation a year or so after LAMPF (late 1973). It will have a maximum energy of only 500 MeV, but because it accelerates H^- ions, it will be possible eventually to have three external beams which are continuously and independently variable in energy from 150 to 500 MeV. Probably only two beams will be installed in the first year, but the third beam which is intended for high resolution work will be added in 1974. This feature of independent variation in energy of the beams will be of immense importance for nucleon-nucleon experiments and for studies of the proton-nucleus interaction.

For existing high-energy synchrocyclotrons, the energy resolution of the proton beam has been typically a few MeV, and this has severely hampered studies of nuclear structure. Resolutions of from 30 to 100 keV can be expected in the future, and that is why there is such a strong effort going toward nuclear (as opposed to particle) physics. The 200 MeV cyclotron at Indiana University will also be an excellent tool in this energy region. One word of warning though: the accelerators at Orsay (150 MeV) and at Uppsala (185 MeV) have both achieved proton beam resolutions of about 200 keV, which is quite adequate for light nuclei, and a lot of work has already been done on nuclear studies. Their results must not be ignored as there will often not be much difference between reactions induced by protons at either 150 or 185 MeV and 500 or 800 MeV. There is a monograph by Clegg entitled "High-Energy Nuclear Reactions",[1] and there are numerous reviews, so that some aspects of this subject are well established.

A major part of the experimental programme at the meson factories will utilize secondary beams, i.e., beams of particles which are produced when the primary proton beam strikes a target. Such beams are seldom used at low energy accelerator laboratories and the ogre of intensity limitation is rarely a problem. Now LAMPF will accelerate 1 mA of protons whilst SIN and TRIUMF will accelerate only 100 μA, and so LAMPF will produce secondary beams at least ten times more intense than the other meson factories. Remember that at existing machines, most experiments with pion, muon or neutron beams have been limited by the intensity of the beam.

For the first example of secondary beams let us take neutrons, produced using the reaction D(p,n)2p as a source. This produces a flux of neutrons of about the same energy as the incident proton and with an energy width of about 1.4 MeV at a production angle of 0° (or greater for thick production targets). At CERN the attainable flux over 10 cm^2 is only 10^5 neutrons sec^{-1}, whilst at TRIUMF we expect 10^7 neutrons sec^{-1}, and LAMPF will achieve 10^8 neutrons sec^{-1}. (These fluxes as quoted are limited by considerations other than incident proton flux.) We can now seriously consider using neutrons of a few hundred MeV for nuclear physics studies, whereas up till now only exploratory experiments have been made. For studies where a better energy resolution is needed, the reaction $^7Li(p,n)^7Be$ has proved suitable as a source,[2,3] but the neutron intensity is down by at least a factor of ten. The lithium reaction also has a higher background of lower energy neutrons, but not much is known about this tail for either reaction at energies of a few hundred MeV. For both reactions the neutron energy is approximately the same as the energy of the incident protons, and so the variable energy of the TRIUMF proton beams is thus of great advantage in many neutron studies.

Pion beams are one of the prime examples where LAMPF has a distinct advantage. The mesons will have more variety and be of greater abundance; this is because the proton beam has a higher energy and can therefore produce higher energy mesons, and in addition the meson flux will reflect the higher proton flux available at LAMPF. However, it is essential to remember that all new machines will have pion beams far superior to those extant, and anyway not all experiments in the future will be intensity limited. The resolutions which are being planned are over an order of magnitude better than anything that has so far been available. The pion has now come of age as a nuclear probe.

Muons are obtained from the decay of pions in flight. The muons are retained by magnetic focussing devices which have large angular acceptances: LAMPF is building a classical quadrupole channel; Zürich is building a superconducting solenoid; whilst TRIUMF is planning an inexpensive quadrupole channel which looks more like an ordinary pion channel. The new machines will produce muon fluxes over 1000 times more intense than exist at present, and will rejuvenate studies of muonic atoms.

I wish to make some comments about duty cycle, i.e., the fraction of time the machine produces particles. The great advantage of Van de Graaff accelerators is that they produce a continuous flux of particles. However, high energy accelerators are pulsed and this pulsing can be classified into two categories: first, one with a fast time constant (a few nanosecs.) where the cause is the radio-frequency which performs the acceleration; second, a pulsing with a slower time constant (tens or hundreds of microseconds) where the cause is the batch-processing nature of the accelerator (synchrotrons or synchrocyclotrons), or the fact that one deliberately turns off and on linear accelerators to allow the electronic equipment to cool off.

Now, counters are affected differently according to intrinsic time constants in their mode of operation. We can simplify the discussion by dividing detectors into two categories: those with time constants of about a microsecond, e.g., Si(Li), Ge(Li), NaI, CsI, Charpak proportional counters, spark chambers, etc., and those with time constants of a few nanoseconds, e.g., plastic scintillators and Cerenkov counters.

For slow counters a cyclotron is effectively on 100% of the time, and so TRIUMF and SIN will be at an advantage with such equipment. A synchrocyclotron with slow excitation is moderately good (duty cycle ≃ 50%), whilst LAMPF is rather poor (6 to 12%). The worst machines from this point of view are electron linacs for which even the new "good duty cycle" machines in the few hundred MeV range at Saclay and MIT have duty cycles around 1 or 2%.

For fast counters there is much less difference between the various machines. This is because plastic scintillators can count faster than the intrinsic bunching in cyclotrons, whilst the bunching in linear accelerators is still a little too fast. This subject is quite complicated, so I have arbitrarily taken a time constant of three nanosecs. and estimated that TRIUMF will eventually have a duty cycle of about 15%, whilst LAMPF will have about 4%. Depending on the numbers one chooses and the assumptions one makes, these duty cycles can vary by at least a factor of two.

Up till now, experiments on intermediate energy machines have normally been limited by the beam intensity. Now the fluxes will be so much higher that one may have too much beam, and the experiments will be limited by the maximum instantaneous counting rate that can be tolerated in the detectors; if this becomes the case, it will be possible to take data faster on the good duty cycle machines.

Let me note in passing that there is one type of experiment which prefers a very poor duty cycle. These are experiments where the event rate is very low, and so it is easy to eliminate background by demanding that real events occur on the rare occasion that the machine is on. Neutrino experiments come into such a category, and it seems that LAMPF will have so many advantages for such studies (high intensity and low duty cycle) that it will not be worthwhile duplicating a neutrino facility elsewhere.

EXPERIMENTS AT MESON FACTORIES

I have just explained that the era we are entering is one of beam intensities 1000 or 10,000 times greater, energy resolutions 10 or 100 times better. How can we take advantage of such improvements? Let me divide the experiments into two parts: first particle physics and second nuclear physics. The choice of subject matter is quite personal and reflects my own interests.

PARTICLE PHYSICS EXPERIMENTS

i) The Pion-Proton Interaction

Although pion beams have been available for a relatively short time, the pion-nucleon interaction is better understood than the nucleon-nucleon interaction. The advantage of the pion-nucleon situation is first that the pion has no spin; this simplifies the phase-shift analysis and reduces the number of experiments which need be performed at each energy. Secondly, both systems have two isotopic spin states ($T = 0$ and $T = 1$ for the nucleon-nucleon case and $T = 1/2$ and $T = 3/2$ for the pion-nucleon case); however, positive and negative pions are conveniently available whilst neutron beams have been of low quality, and this has meant that the nucleon-nucleon $T = 0$ phase-shifts are poorly known.

Pion-nucleon phase-shifts are available up to 2 GeV [4,5] and a little higher. There is still some controversy over the details, but the basic features are firmly established. The energy region from 70 to 290 MeV has been recently remeasured at CERN. The total cross sections are already available,[6] and the differential cross sections are being analysed. It seems to me to be unnecessary to repeat these measurements in the immediate future. However, the energy region below 70 MeV is very poorly known, and some new measurements are essential. The difficulty is that present pion beams are very poor at low energies. There are two reasons: first, the production of such pions is lower than that of higher energy pions and, second, the low energy pions decay more quickly. The meson factory beams will enable experiments to be performed down to about 20 MeV or so.

The low energy pion-nucleon interaction is interesting because at present there is much disagreement over the values of the pion-proton scattering lengths.[7,8] Various analyses disagree, and a direct measurement is difficult at present because of the dominance of p-wave scattering to energies as low as 50 MeV. The existing data could be improved by at least a factor of 5 with the new machines.

ii) The Nucleon-Nucleon Interaction

Many nuclear physicists assume that our knowledge of free nucleon-nucleon scattering is satisfactory. This is not even true for the case of proton-proton scattering; the Livermore phase-shift group made an excellent analysis up to 450 MeV [9] and the continuation up to 3 GeV is possible [10] as long as there is more theory than experiment, but even as low as 735 MeV there are not enough data for a purely phenomenological analysis to be completed.[11] There is still a need therefore for proton-proton experiments at the higher energies available at LAMPF.

The neutron-proton interaction is very poorly understood. The Livermore analysis published in 1969 showed that there were enormous holes in our knowledge and that a larger variety of neutron-proton scattering experiments were needed. If anything the situation has deteriorated since then.

An excellent neutron beam was recently developed at the Princeton-Pennsylvania Accelerator. The 3 GeV proton beam struck a heavy target and the energy of the neutrons thus produced was measured using a time-of-flight technique. This enables experiments to be carried out simultaneously at neutron energies over a broad range of approximately 200 to 2000 MeV. Differential [12] and total [13] cross sections have been measured for free neutron-proton scattering. However, when an attempt was made to include these new results into a phase-shift analysis, it was found that there was a fundamental inconsistency between the new data and previous data.[14] This is illustrated in Fig. 1 where a comparison is made between the PPA data at 343 MeV and earlier data from Liverpool at 350 MeV.[15] The PPA data exhibit a much steeper

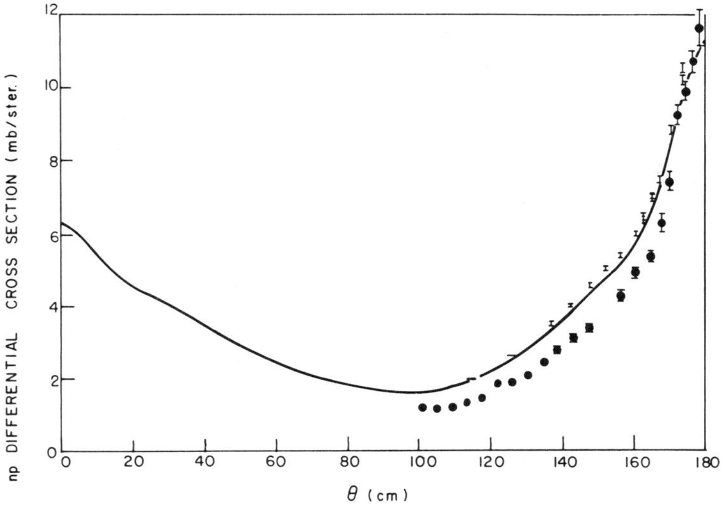

Figure 1
Neutron-proton differential cross-sections; a comparison between the new PPA data at 343 MeV [12], dots, and the older Livermore data at 350 MeV [15], crosses. The curve is from the Livermore phase-shift solution X at 347 MeV. (From Arndt and Roper.[14])

rise in the backward direction; I think it is not unfair to point out that the PPA apparatus was not ideal for a precision measurement of a differential cross section as the solid angle acceptance of their spectrometer changed rapidly in this angular region However, this does not alter the obvious conclusion that a completely new remeasurement of the differential cross section is essential and this cannot be done at the PPA as it is being closed down. The other two basic parameters, the total cross section and the polarization, should also be remeasured, but in addition more sophisticated measurements are required on the triple scattering parameters and spin correlation coefficients.

How can these phase-shifts be used? Many low energy nuclear physicists like to work with a potential. To obtain the nucleon-nucleon potential from the phase-shifts used to be a game of hide and seek. One guessed a form for the potential and then calculated the phase-shifts from the potential. A significant improvement was made recently by Benn and Scharf [16] who used the Marchenko theory to derive the potential from the phase-shifts, and thereby they took all the fun out of the game! If we had a perfect knowledge of the phase-shifts, it would now be possible to derive a unique version of the nucleon-nucleon potential for use in the non-relativistic Schrödinger equation.

Above 300 MeV pion production becomes possible in nucleon-nucleon collisions. Not much is known about the inelastic

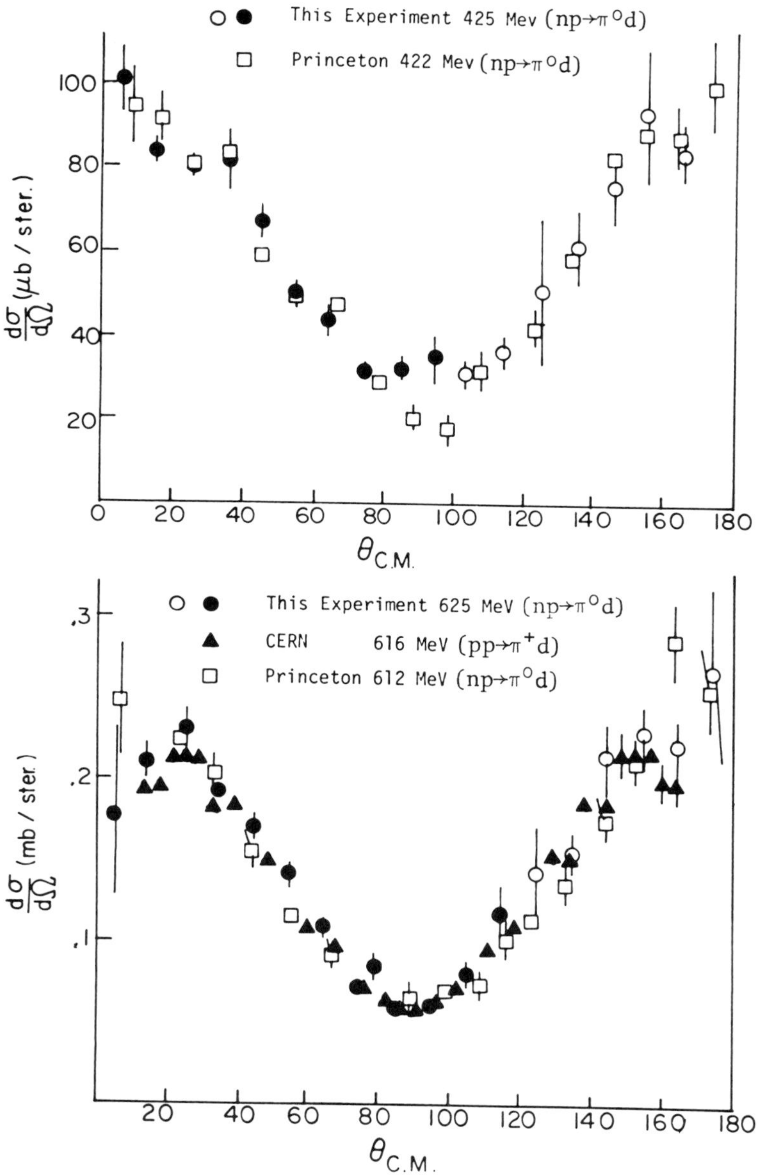

Figure 2
Differential cross-sections for the reaction $np \to \pi^0 d$ compared to the CERN data for the reaction $pp \to \pi^+ d$. No asymmetry about 90^0 is observed for the $np \to \pi^0 d$ reaction. (From S.S. Wilson et al.[20])

nucleon-nucleon interaction. Our knowledge of the differential cross-section for the reaction ($p + p \rightarrow \pi^+ + d$) is adequate up to 1000 MeV,[17] and there are several asymmetry measurements using polarized protons,[18,19] but little else is known about this reaction and there are many experiments to be done before it will be possible to enumerate scattering amplitudes.

There are many other pion production reactions, e.g.,

$$p + p \rightarrow p + n + \pi^+$$
$$p + p \rightarrow p + p + \pi^0 \quad .$$

Even the total cross sections for such reactions are poorly known in some energy ranges.

It is relevant to point out two recent studies of the reaction $n + p \rightarrow \pi^0 + d$. This reaction is important as a test of the conservation of isotopic spin. The deuteron has isotopic spin $T = 0$, and the pion has $T = 1$, so the conservation of the isotopic spin requires that only $T = 1$ states contribute to this reaction. The (np) system is half $T = 0$ and half $T = 1$; thus the cross section should be half of the pure $T = 1$ transition, which is the reaction $p + p \rightarrow \pi^+ + d$. This factor of two is difficult to measure, but an easier test to remember is that the differential cross section for the $p + p \rightarrow \pi^+ + d$ reaction must be symmetric about 90° because of the identical nature of the protons in the initial state. There would be no similar restriction on the $n + p \rightarrow \pi^0 + d$ reaction if isotopic spin were not conserved. Two experiments have been reported recently and no asymmetry was observed[20,21]. This is illustrated in Fig. 2 where the reactions $n + p \rightarrow \pi^0 + d$ and $p + p \rightarrow \pi^+ + d$ are compared. The results could also be due to chance, but this lack of asymmetry is a necessary consequence of isotopic spin conservation.

iii) Time Reversal

Many experiments have been performed looking for failures of time-reversal invariance, and yet the only clear evidence for such behaviour is the now well established decay of the K_L^0 into $\pi^+ \; \pi^-$ and $2\pi^0$.[22,23] The decay breaks CP invariance, and using the basic assumption of CPT invariance, one deduces that time-reversal invariance is broken as well.

I shall restrict myself to the experiments on time-reversal which can be performed at intermediate energies. One of the tests has been the verification of detailed balance in the reactions $n + p \rightarrow \gamma + d$ and $\gamma + d \rightarrow n + p$. It was postulated that at a neutron energy of about 600 MeV, there might be a breakdown of time-reversal invariance due to an interference between two modes for the reaction, one being "direct" and the other having an intermediate $\Delta(1236)$. It was suggested that the differential cross sections for the two reactions might be different if performed at the same centre-of-mass energy. This test is quite difficult as neither reaction is easy to measure. The particular

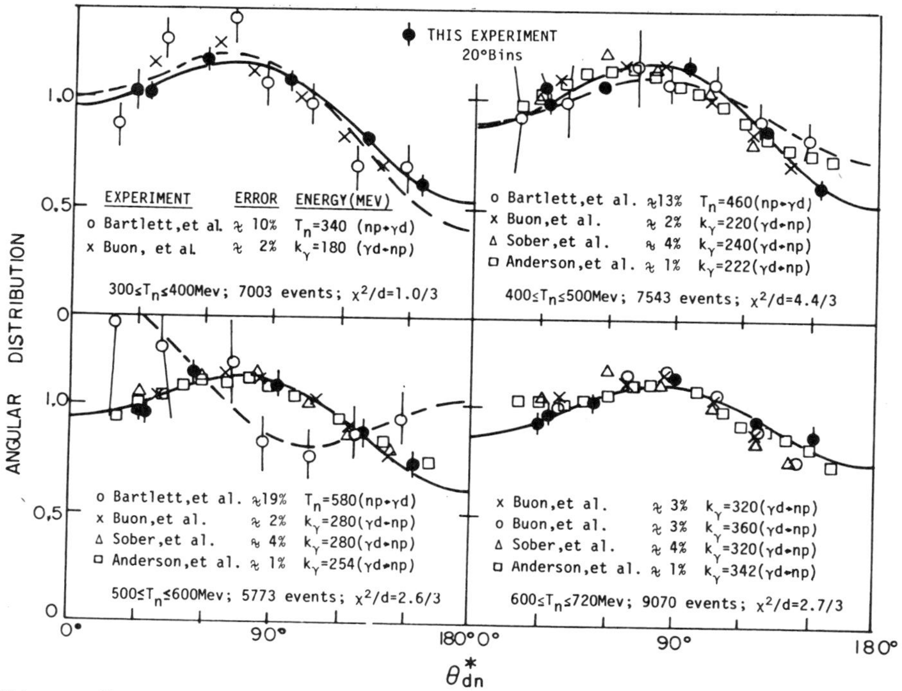

Figure 3
A comparison of the differential cross sections of the reactions ($np \rightarrow \gamma d$) and ($\gamma d \rightarrow np$) at the same total energy in the centre-of-mass. (From Schrock et al.[24])

difficulty of the neutron-proton capture is the intense background from the reaction $n + p \rightarrow \pi^0 + d \rightarrow \gamma + \gamma + d$, which is about 30 times more intense. The difficulty of the measurement can be assessed from the fact that although Barshay first suggested this test in 1966, it was only this year (1971) that dependable results were published, and no difference was observed between the two reactions.[24] The comparison is illustrated in Fig. 3.

Another similar test is also being pursued on the reactions $p + d \leftrightarrow {}^3He + \gamma$ Some information is already available on the photodisintegration reaction,[25] but none yet on the capture reaction.

A further comparison which has been recommended is between the reactions $\pi^- + p \leftrightarrow \gamma + n$ for pion energies between 100 and 300 MeV.[26,27,47] It was suggested that T-reversal invariance (in this case related to C-reversal invariance) might be violated at the first resonance. Again both reactions are experimentally difficult to observe. The pion capture has to compete against a background of charge exchange which can be as much as seventy times more intense, whilst the photoproduction study suffers from the lack of free neutron targets. The deuteron must be used with subsequent unfortunate corrections for the spectator proton. There is some evidence for the failure of time reversal invariance, but further experiments are needed.

The last test I want to mention is a comparison between measurements of polarization and asymmetry. By polarization I mean experiments where a particle with spin is scattered off a target, and one measures the spin alignment after the scattering. By asymmetry I mean using a beam of particles with their spins aligned, and then scattering them from a target and measuring the asymmetry of particles scattered to one side and the other. Now, if the target has zero spin then parity invariance alone requires the equality of the polarization and asymmetry measurements for the same scattering angle; thus for example in p - ^{4}He or p - ^{12}C scattering, the polarization and asymmetry would be the same as the violation of parity invariance in nuclear forces, which is known to be extremely small and to be due to the weak interactions.[28]

The obvious target to choose is the proton, and several experiments have been performed with negative results. Sudarshan has suggested a model for a strong interaction force which does not conserve time reversal invariance, and Bryan and Gersten [29] have recently made some calculations in which they have shown the effect is much larger for neutron-proton scattering than for proton-proton scattering. Again, however, the experiments are difficult, and we must not expect a speedy solution to this problem.

NUCLEAR PHYSICS EXPERIMENTS

One valid query made by low energy physicists is whether it is worth so much money and effort to study nuclear physics at meson factories. Although the new accelerators appear to be costly, if one added up the budgets of all the Van de Graaff laboratories, one would obtain a far larger sum. The meson factory budget is concentrated on a particular facility, but many universities will participate in the programme, so a naive comparison is very unfair to the meson factories. Their great advantage is that they give us a completely new way of obtaining knowledge about the nucleus with tools which have already indicated their usefulness, and that is why so many of us are confident that much exciting information will be gleaned on the winnowing floors of our factories.

I have chosen a few topics to illustrate the sorts of experiments which will be carried out; the list is by no means exhaustive, but I have tried to select the subject matter to exhibit the variety of experiments that will be possible.

i) Muonic Atoms

Muonic atoms have been studied now for nearly twenty years.

When a μ^- slows down in a material, it stops and is captured by the Coulomb field of a nucleus. The energy levels are similar to those of the electronic hydrogen atom, and because of the relatively long life of the muon (2.2 μsec.), it has time to reach the lowest orbit. Since the muon is much heavier than the electron, the orbits are closer to the nucleus, and for the heavier nuclei the lowest orbit is completely within the nucleus. The energies of the levels are thus quite different from those which would be calculated for a point nucleus, and the measurement of these energies gives a very accurate determination of the size of the nucleus. There have been several recent reviews,[30,31] so I shall not give any more details.

The energy levels can be measured precisely by determining the energies of the X-rays given off in the cascade. These X-rays are in the energy range of 100 keV to 10 MeV or so, and can be detected conveniently with Ge(Li) counters. A single measurement gives only one piece of information, and the most accurate piece of information which comes from muonic atoms is the root-mean-square radius of the nucleus; this can be obtained to better than 1% for nuclei heavier than magnesium. However, we should also like to know the shape of the charge distribution of the nucleus, and to do this several transitions need be measured. The normal way is to postulate that the charge distribution of the nucleus has a Saxon-Woods shape, i.e.,

$$\rho = \frac{\rho_0}{1 + \exp\,[(r - c)/a]}$$

where c is the radius at which $\rho = 0.5\ \rho_0$, and a is a measure of the surface thickness. Many authors prefer the parameter t, the distance in which ρ drops from $0.9\ \rho_0$ to $0.1\ \rho_0$. These are related by t = 4.4a.

Fig. 4 gives the energy level scheme for muonic neodymium (Z = 60). The full lines are the unperturbed levels for a point nucleus, and the dashed lines give the experimentally observed position of the levels as deduced from the energies of the X-rays. Now, the normal decay route for the muonic atom is the main series 4f - 3d - 2p - 1s. Measurements of the 3d - 2p and 2p - 1s transitions give very accurate information on the root-mean-square radius of the nucleus, but unfortunately give similar information such that the shape of the charge distribution cannot be inferred. If one plots the Saxon-Woods parameters c and t, the experiments determine an allowed band of values. It was pointed out by Anderson et al. [32] that the energy of the $2s_{1/2}$ - $2p_{1/2}$ transition has quite a different dependence on c and t, and therefore is a valuable tool for determining these parameters more precisely. This is illustrated in Fig. 5 for ^{203}Tl (Z=81), where the banded areas represent the allowed values from the measurements on the relevant transitions, made by Pearce et al.[33] Unfortunately, this ≃ 1 MeV X-ray is of low intensity, about 0.5% of muon captures, so the measurements are difficult but very worthwhile.

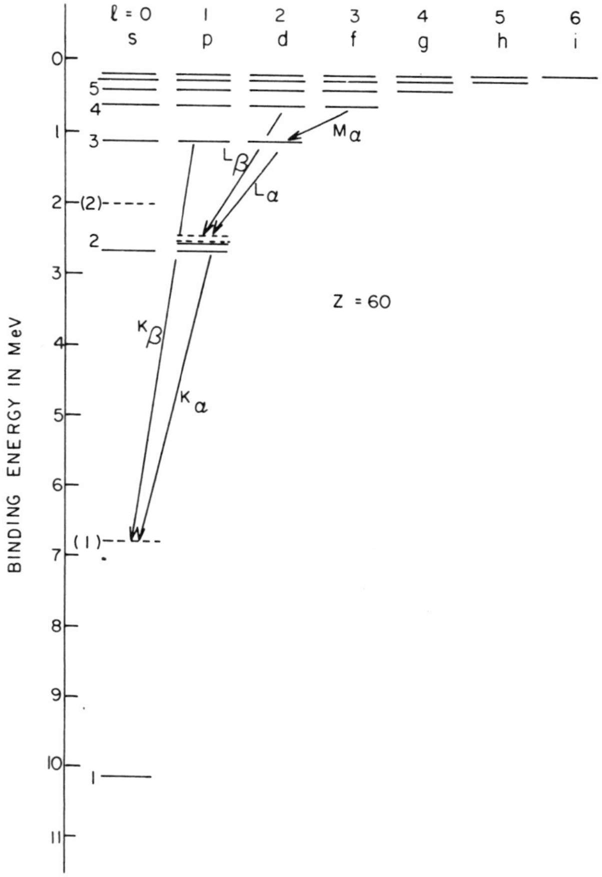

Figure 4
The energy levels for muonic neodymium (Z = 60). The full lines are for a point nucleus and the dashed lines for a finite nuclear charge distribution. Note that the $2s_{1/2}$ and $2p_{1/2}$ states are no longer degenerate. (From Devons and Duerdoth.[31])

This sort of work obviously competes fairly directly with electron scattering measurements. The latter experiments give a better idea of the shape of the nucleus, and are also better for very light nuclei; however, with careful determinations of the $2s_{1/2}$ - $2p_{1/2}$ X-ray the tables may turn more in favour of the muonic atom work. Certainly one great danger of past results has been that the electron scattering and muonic atom measurements have been analysed independently of one another, whereas really they complement each other and should be analysed together.

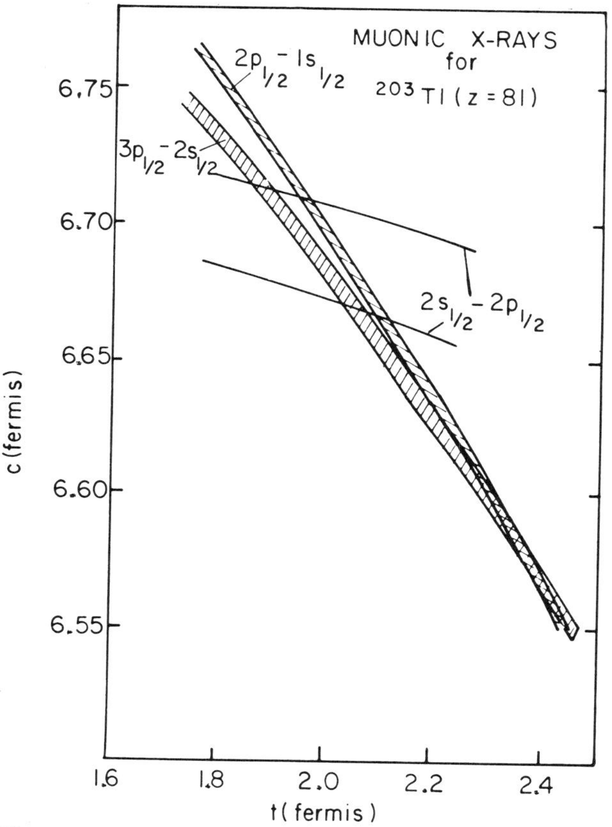

Figure 5
The banded areas are the allowed values for c and t as determined from measurements of the $3p_{1/2}-2s_{1/2}$ and $2p_{1/2}-1s_{1/2}$ muonic X-rays. The other two lines limit the values as determined by the $2s_{1/2}-2p_{1/2}$ transition. (The only QED correction made for this figure is vacuum polarization).[33]

ii) Pionic Atoms

Negative pions act in a similar way when stopped in materials; they form pionic atoms and cascade down from the higher to the lower orbits, giving out X-rays.[34] The main differences are that the strong interaction of the pion with the nucleus causes larger shifts of the energy levels and that there is also absorption by the nucleus which is so strong that it can compete with the X-ray transition for the heavier elements. This absorption gives the levels a finite life-time and so an energy width which can sometimes be measured directly from the width of the X-rays given out in the transition to that level. For example in ^{23}Na the 1s level width is about 10 keV. The 2p level width is 35 eV, and that is too small to measure directly, as a typical Ge(Li) detector would have a resolution of about 1 keV for this 3d-2p

X-ray of 70 keV. However, the width can be deduced from the relative intensities of the 3d - 2p and 2p - 1s lines which indicate that about 96% of pions which reach the 2p level are absorbed before they make a radiative transition.

The absorption of pions in the 2p levels means that the intensity of the 2p - 1s X-ray becomes so small for elements heavier than sodium that it is impossible to distinguish it from the background. This difficulty is compounded by the width of the line which camouflages it rather effectively. One idea which does not seem to have been pursued is to investigate the 3p - 1s, or even the 4p - 1s, transitions. These lines of the Lyman series for muonic titanium are illustrated in Fig. 6 taken from reference 35. We can see that these higher transitions have a fairly low intensity, 9% and 4% respectively, of the 2p - 1s transition. However, to revert to pionic atoms which should be similar, these higher transitions have two advantages: first, the absorption in the 3p - 4p levels may not compete so strongly with the radiative transition probability so the intensity of the transitions to the 1s state will not be reduced as much; and secondly, the X-ray is of higher energy and will stand out more easily from the background. It may even be necessary to reduce the background by requiring coincidences between transitions to the 3p or 4p level and the 3p - 1s or 4p - 1s transition. Such sophisticated experiments are very difficult with present intensities but will soon be feasible.

The final topic on pionic atoms that I shall discuss is the case of hydrogen. Pionic hydrogen is unusual as the π-p bound system is neutral and very small by atomic standards. As this small object bounces around it meets other protons, and the Coulomb fields of the other protons cause Stark mixing of the pionic hydrogen's atomic levels. The system would normally cas-

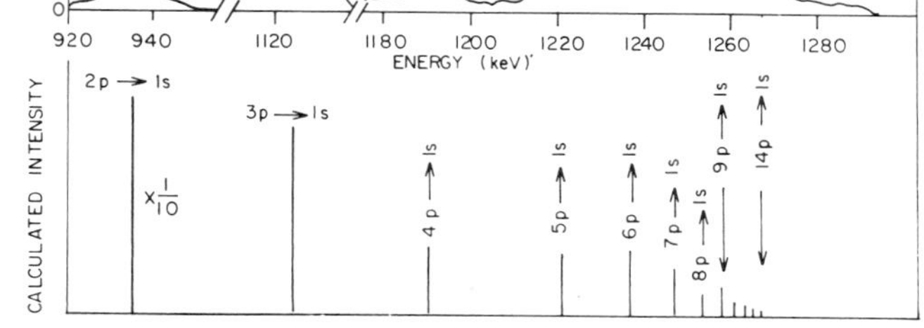

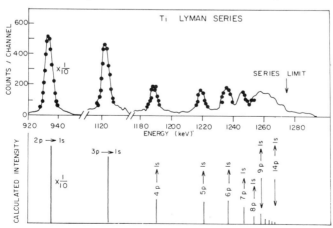

Figure 6
The relative intensities of the Lyman series of muonic X-rays in titanium. (From Kessler et al.[35])

cade through the levels 6h - 5g - 4f - 2p - 1s, and when the 1s level was reached nuclear absorption would occur via the two reactions ($\pi^- + p \rightarrow \pi^0 + n$) and ($\pi^- + p \rightarrow \gamma + n$). However, the collisions of the pionic hydrogen mix, for example, the 4f level with the 4d, 4p and 4s levels; now the wave function of the 4s pion is much larger at the proton than the 4f and so nuclear absorption becomes more likely. It has been calculated that in a liquid the pion is absorbed before it reaches the n = 4 levels. However, Zavattini's group at CERN has recently observed the 2p - 1s pionic hydrogen X-ray in gaseous hydrogen[36,37] for which the Stark mixing is much reduced. A precision measurement of this 2.43 keV (≃ 5000 X.U.) line would be of great importance as the change in energy of the 1s level from the point nucleus value would yield a value for the pion-proton scattering length.

iii) The (π^+,p) and (p,π^+) Reactions

I shall now discuss some nuclear reactions which are interesting because of their unusual properties. The (π^+,p) reaction and its time reverse (p,π^+) have recently been studied in three laboratories, at CERN (E_p = 600 MeV), at Uppsala (E_p = 185 MeV),[39] and at SREL (E_π = 68 MeV).[40] Both reactions are difficult to study as the cross section is extremely small, and data at only one or two angles are available at present.

For an example let us consider the Uppsala data of Dahlgren et al. for which sufficient energy resolution was achieved that levels in the final nucleus were clearly separated. The reactions $^{12}C(p,\pi^+)\,^{13}C$ and $^{9}Be\,(p,\pi^+)^{10}Be$ were studied with an energy resolution of 1.5 MeV, using the 185 MeV proton beam of the Uppsala synchrocyclotron. Fig. 7 shows these results for carbon at 45° and 90° (the only two angles studied). The ground-state transition is clearly observed at 45° (100 nb/sr) but is indistinct at 90° (>7 nb/sr).

The unusual properties of the (p,π^+) are in the kinematics. As an example let us take the Uppsala case; the proton brings in about 600 MeV/c momentum, but the 40 MeV pion takes away only 100 MeV/c, and so the nucleus suffers a momentum shock of 500 MeV/c If the proton proceeds by direct capture to become a $1p_{1/2}$ neutron in ^{13}C, then one is probing very-high-momentum components of the wave function. Normally we consider 200 MeV/c to be the highest momentum of the nucleons in a nucleus. Of course the nucleus will try to absorb the momentum transfer as efficiently as possible and two-step processes are quite likely. Attempts at calculating the cross section have been made [41] but enormous changes in the cross section can be achieved by slight variation of the parameters. With the current interest in high-momentum components it is worth continuing studies on these reactions but it is essential to have much more information,i.e., complete differential cross sections at several energies. Only then will sufficient constraints be put on the calculations that one might have

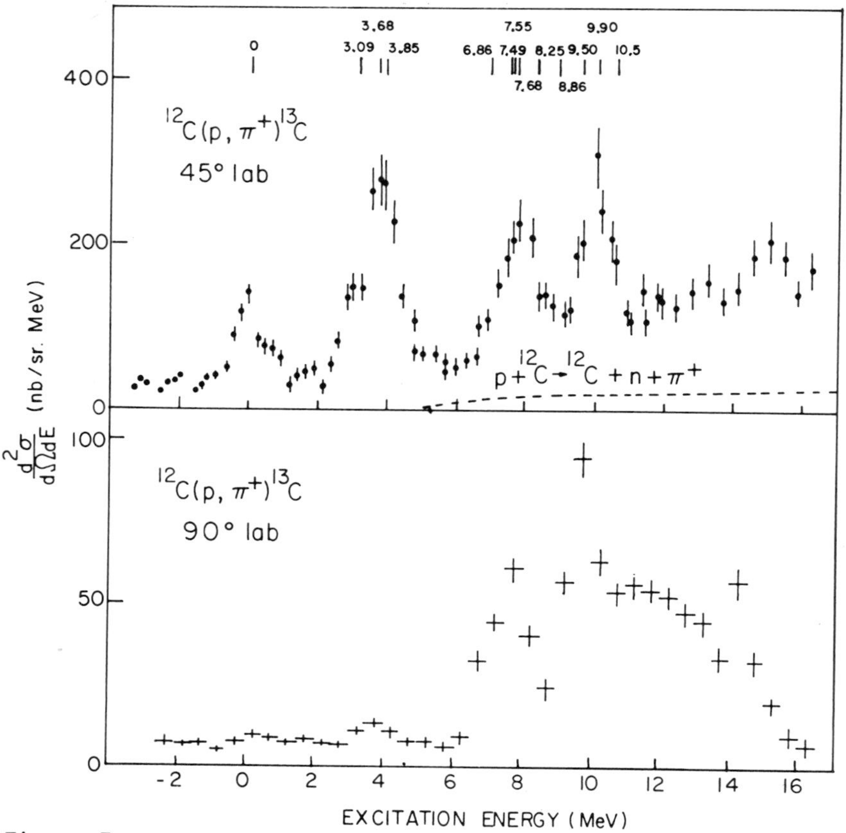

Figure 7
Energy spectra for the reaction $^{12}C(p,\pi^+)^{13}C$; obtained at the Uppsala synchrocyclotron (E_p = 185 MeV). (From Dahlgren et al.[39])

confidence in extracting information about the nucleus rather than about the reaction mechanism.

iv) The (n,p) and (p,n) Reactions

Charge exchange reactions have been studied in various guises. The (^3He,t) and (t,^3He) reactions have been quite thoroughly studied. The (π^+, π^0) reactions would have the great advantage that the pion interaction is well known, but it is very difficult to detect with good energy resolution the two decay gamma rays of the π^0. Some information on the total cross section of such reactions is available at $E_\pi \simeq 180$ MeV,[42] but data on the differential cross section are desirable.

I would therefore like to point out some of the advantages of the (n,p) and (p,n) reactions. First, it is important to recall that free neutron-proton scattering is not completely documented and that a concurrent programme on that subject would be necessary. Several experiments have been completed on the (p,n) reaction, for example at E_p = 30 and 50 MeV,[43] and at E_p = 94 MeV.[44] The neutrons are detected by a time-of-flight technique and resolutions of about 2 MeV were achieved at the higher energy. Neutron detectors can be calibrated quite accurately and so absolute cross-sections can be obtained.

Only one very crude measurement has been made on the (n,p) reaction at synchrocyclotron energies.[45] The experiment had a poor energy resolution (≃ 10 MeV) and was severly limited by the intensity of the 150 MeV neutron beam at Harvard. In the future all these problems will disappear and excellent data will be within our reach.

Of what interest are such experiments? One interest would be a precision comparison of the (n,p) and (p,n) reactions on a nucleus such as ^{12}C. This might help to add more confusion to the problems connected with the β-decay of ^{12}B and ^{12}N.[46] We might obtain some information on the isotopic spin impurity of various levels. This could be applied to many nuclei. Of course this comparison can be done with the (^3He,t) and (t,^3He) reactions, but one has the added complexity of three nucleon projectiles. It seems that the nucleon-nucleon direct reaction has the advantage of theoretical simplicity even if the problems are shifted over to the experimenter who must deal with the neutron.

There are other interesting reactions as well. A precision comparison of (n,d) and (p,d) reactions would help elucidate the problems surrounding neutron and proton mass distributions in nuclei. It is also possible to obtain polarized neutrons of reasonable energy resolution, so that measurements of polarization effects in neutron-induced reactions will become possible.

I hope that I have been able to give a slight view of the vistas ahead. Of course there are many more topics I could have covered; the LAMPF Program Advisory Committee has already received over 70 proposals for experiments and 65% of these proposals are nuclear physics experiments. The possibilities are enormous and I trust the TRIUMF programme will be as interesting and as varied.

REFERENCES

1. A.B. Clegg, "High Energy Nuclear Reactions", Clarendon Press (1965).
2. C.J. Batty et al., Nucl. Inst. Meth. 68, 273 (1969).
3. J.A. Jungerman and F.P. Brady, Nucl. Inst. Meth. 89, 167 (1970).
4. D.J. Herndon, A. Barbaro-Galtieri and A.H. Rosenfeld, UCRL 20030 πN (1970).
5. S.L. Baker et al., Nucl. Phys. B30, 116 (1971).
6. A.A. Carter et al., Nucl. Phys. B26, 445 (1971).
7. R.G. Moorhouse, Ann. Rev. Nucl. Sci. 19, 301 (1969).
8. J. Hamilton and J. Lyng Petersen, Nucl. Phys. B29, 2951 (1971).

9. M.H. McGregor, R.A. Arndt, and R.M. Wright, Phys. Rev. 182, 1714 (1969).
10. T. Ueda, Phys. Rev. Letters, 26, 588 (1971).
11. S.I. Bilenkaya et al., Nucl. Phys. B13, 375 (1969).
12. R.E. Mischke et al., Phys. Rev. Letters 23, 542 (1969).
13. R.E. Mischke et al., Phys. Rev. Letters 25, 1724 (1970).
14. R.A. Arndt and L.D. Roper, Phys. Rev. Letters 26, 1260 (1971).
15. A. Ashmore et al., Nucl. Phys. 36, 258 (1962).
16. J. Benn and G. Scharf, Nucl. Phys. A134, 481 (1969).
17. C. Richard-Serre et al., Nucl. Phys. B20, 413 (1970).
18. C. Dolnick, Nucl. Phys. B22, 461 (1970).
19. M.G. Albrow et al., Phys. Letters 34B, 337 (1971).
20. S.S. Wilson et al., Phys. Letters 35B, 83 (1971).
21. D.F. Bartlett et al., Phys. Rev. D1, 1984 (1970).
22. W.S.C. Williams, Chapter 14 of "An Introduction to Elementary Particles", 2nd edition, Academic Press (1971).
23. E.G.P. Rowe and E.J. Squires, Reports on Progress in Physics 32, 273 (1969).
24. B.L. Schrock et al., Phys. Rev. Letters 26, 1659 (1971).
25. P. Picozza et al., Nucl. Phys. A157, 190 (1970).
26. A.I. Sanda and G. Shaw, Phys.Rev. Letters, 26, 1057 (1971).
27. A. Donnachie and G. Shaw, Phys.Rev. Letters, 35B, 419 (1971).
28. K.S. Krane et al., Phys. Rev. Letters 26, 1579 (1971).
29. R. Bryan and A. Gersten, Phys. Rev. Letters 26, 1000 (1971).
30. C.S. Wu and L. Wilets, Ann. Rev. Nucl. Sci. 19, 527 (1969).
31. S. Devons and I Duerdoth, Advances in Physics 2, 295 (1969).
32. H.C. Anderson et al., Phys. Rev. 187, 1565 (1969).
33. R.M. Pearce et al., Swiss Physical Society, Lausanne Meeting (1971), and private communication.
34. E.H.S. Burhop, High Energy Physics, Vol. III, Academic Press (1969).
35. D. Kessler et al., Phys. Rev. Letters 18, 1179 (1967).
36. J. Bailey et al., Phys. Letters 33B, 369 (1970).
37. M. Leon, Phys. Letters 35B, 413 (1971).
38. J.J. Domingo et al., Phys. Letters 32B, 309 (1970).
39. S. Dahlgren et al., Phys. Letters 35B, 219 (1971).
40. K. Gotow, p. 374 of High Energy Physics and Nuclear Structure, Ed. S. Devons, Plenum Press (1970).
41. A. Reitan, Nucl. Phys. B29, 525 (1971).
42. D.T. Chivers et al., Nucl. Phys. A126, 129 (1969).
43. A.S. Clough and C.J. Batty, Nucl. Phys. A143, 385 (1970).
44. A. Langsford et al., Nucl. Phys. A113, 433 (1968).
45. D.F. Measday and J.N. Palmieri, Phys. Rev. 161, 1071 (1967).
46. J. Blomqvist, Phys. Letters 35B, 375 (1971).
47. P. Noelle and W. Pfeil, Nucl. Phys. B31, 1 (1971).

ON THE FOUNDATIONS OF THE VMI MODELS

J.N. URBANO AND J. DA PROVIDENCIA
Laboratorio de Fisica
Universidade de Coimbra

We briefly report a derivation of the VMI models for rotations and vibrations in the framework of the Generator Coordinate Method (GCM). Here we consider a two dimensional version for rotations only.

Let $\Phi(x,\beta,\phi)$ be an intrinsic wave function for a well deformed nucleus, labelled by two parameters: an orientation angle ϕ and a quadrupole deformation parameter β. The GCM consists of solving the equation

$$\int\{(\beta\phi|H|\beta'\phi') - E_m(\beta\phi|\beta'\phi')\}e^{im\phi'}f(\beta')d\beta'd\phi' = 0 \qquad (1)$$

where E_m is the energy of the states with angular momentum $\underline{m}$ about the axis of rotation, $f(\beta)$ is a weight function and the overlap and energy kernels are defined as usual. If these are well represented in a gaussian and quadratic approximation for $\phi-\phi'$,

$$(\beta\phi|\beta'\phi') = \exp\{-\tfrac{1}{2}K(\beta,\beta')(\phi-\phi')^2 - G(\beta,\beta')\}$$

$$(\beta\phi|H|\beta'\phi')/(\beta\phi|\beta'\phi') = \mathcal{H}_0(\beta,\beta') - \tfrac{1}{2}\mathcal{H}_2(\beta,\beta')(\phi-\phi')^2,$$

the integration over ϕ' in (1) can be carried out. It yields

$$\int\{A(\beta,\beta') + \frac{m^2}{2\mathcal{J}(\beta,\beta')} - E_m\}e^{-G(\beta,\beta')}f(\beta')\,d\beta' = 0 \qquad (2)$$

where $A(\beta,\beta')$ and $\mathcal{J}(\beta,\beta')$ can be easily related to $K(\beta,\beta')$, $\mathcal{H}_0(\beta,\beta')$ and $\mathcal{H}_2(\beta,\beta')$.

In order to solve eq. (2) one expands its kernel in the variables β and β' around the value β_m which minimizes

$$E_m(\beta) = A(\beta,\beta) + \frac{m^2}{2\mathcal{J}(\beta,\beta)} \quad . \qquad (3)$$

Eq. (2) describes then a vibrational motion of the variable β the frequencies of which, because a well-deformed nucleus is quite stable against β vibrations, are expected to come much higher than the rotational ones obtained from eq.(3). The VMI formula then results immediately assuming that β_m does not change much with $\underline{m}$, so that $A(\beta_m,\beta_m)$ is well reproduced by a quadratic function of $(\beta_m - \beta_0)$.

A NEW CRANKING MODEL FOR THE SOLUTION OF THE LARGE AMPLITUDE, NON-ADIABATIC TDHF EQUATIONS

R. BASSERMAN AND D.J. ROWE
Department of Physics,
University of Toronto

In the past, the TDHF equations have not been solved without an adiabatic approximation except in the small amplitude (RPA) limit. We avoid this approximation by reducing the difficult TDHF problem to a simpler constrained but static HF problem. Unlike previous cranking models our constraint is defined such that it induces (within an approximation) the same deformation at a particular instant in time as would occur in the freely vibrating nucleus. To the extent that this objective is achieved, our cranking model provides exact solutions to the TDHF equations.

The TDHF Hamiltonian for a Q.Q force can be expressed in terms of a constrained static Hamiltonian $G(\alpha,\beta)$:

$$H(t) = H_0 - Xq(t)Q = G(\alpha,\beta) - X[q(t) - \alpha]Q + \Lambda(\alpha,\beta)$$

$$G(\alpha,\beta) = H_0 - X\alpha Q - \Lambda(\alpha,\beta)$$

where the constraining field $\Lambda(\alpha,\beta)$ is <u>a priori</u> an unknown function of the instantaneous quadrupole moment α and its instantaneous time derivative β. For given α and β, the difference between the solutions for $H(t)$ and $G(\alpha,\beta)$ can be calculated in time-dependent perturbation theory. $\Lambda(\alpha,\beta)$ is then defined such that this difference vanishes.

To obtain a practical expression for $\Lambda(\alpha,\beta)$ it is unfortunately necessary to assume that departures from harmonic motion in $q(t)$ occur slowly in time. However, this is clearly very much better than the adiabatic approximation that $q(t)$ itself varies slowly in time. With the self-consistency equations

$$\alpha = \langle Q\rangle, \qquad i\hbar\beta = \langle [Q,H]\rangle ,$$

$\Lambda(\alpha,\beta)$ is completely defined.

In the small amplitude limit the above cranking model reproduces the usual RPA expressions for a Q.Q interaction. It can be applied to other Hamiltonians by retaining the constraining field appropriate to the Q.Q interaction and performing standard constrained HF calculations.

PROPERTIES OF TWO-BODY RANDOM HAMILTONIAN ENSEMBLES

O. BOHIGAS
Institut Physique Nucléaire
Orsay, France

Recently new ensembles of random matrices have been introduced and treated numerically.[1,2] They take into account the correlations between the elements of the NxN hamiltonian matrix induced by the Pauli principle and the two-body character of the interaction, when dealing with a system $(a,b,\ldots)^n_J$ of n identical fermions distributed among a set a,b,... of degenerate single particle orbits and of total angular momentum J. The independent random variables are not, as in the case of the Gaussian orthogonal ensemble (GOE), the $N(N+1)/2$ elements of an NxN real symmetric hamiltonian matrix, but the m two-body matrix elements defining the two-body interaction ($m \ll N(N+1)/2$). By standard shell model techniques, the NxN hamiltonian matrix corresponding to n particles and total angular momentum J is calculated in terms of the m independent two-body matrix elements, which are selected at random from a given distribution.

It has been found that:

i) The level density of two-body random hamiltonian ensembles (TBRE) is nearly normal (Gaussian),[1,2] in contrast to the well known Wigner's semi-circle law, that holds for GOE in the limit of large N.

ii) The k-th neighbour spacing distributions of GOE and TBRE are identical for $k=0$,[3,4] but definite departures of TBRE results from GOE exist for $k \gtrsim 3$. The effect of the correlations included in TBRE is to broaden the GOE distribution.[5] Comparison of the different theoretical predictions with the experimental data is attempted.

REFERENCES

1. J.B. French and S.S.M. Wong, Phys. Lett. 33B (1970) 449.
2. O. Bohigas and J. Flores, Phys. Lett. 34B (1971) 261.
3. J.B. French and S.S.M. Wong, Phys. Lett. 35B (1971) 5.
4. O. Bohigas and J. Flores, Phys. Lett. 35B (1971) 383.
5. O. Bohigas and J. Flores, Contribution to International Conference on Statistical Properties of Nuclei, Albany, 1971.

APPLICATION OF SPECTRAL DISTRIBUTIONS METHOD TO QUASI-PARTICLE SUBSPACES

H. NISSIMOV†
Institut de Physique Nucléaire
Division de Physique Théorique
Orsay, France

The full BCS Hamiltonian is given, as is well known, by a quasi-particle conserving part $H_c \equiv H_{11}+H_{22}$ and q.p. non-conserving ones: $H_3 \equiv H_{31}+H_{13}$ and $H_4 \equiv H_{40}+H_{04}$. The decomposition to these terms is an orthogonal one, i.e.: $\langle\langle H_c \cdot H_3 \rangle\rangle^k = \langle\langle H_c \cdot H_4 \rangle\rangle^k = \langle\langle H_3 \cdot H_4 \rangle\rangle^k = 0$ where $\langle\langle\ \rangle\rangle^k$ stands for the trace calculated in the subspace of states with k q.p. The moments of H_c can be calculated and propagated to different K-subspaces by the usual French methods. In this contribution we give the propagation formulae for the second moments of the H_3 and H_4 operators in terms of a small number of basic quantities. The Spectral Distributions Method, applied so far to particle systems, can be therefore transcribed to quasi-particle ones. A useful application of this generalization is the study of the goodness of the BCS approximation: From the above-mentioned formulae we derive expressions for the partial widths: $\sigma^2_{k\to k\pm 2}$ and $\sigma^2_{k\to k\pm 4}$. The intensities $\sigma^2_{k\to k'}/(\bar{E}_k-\bar{E}_{k'})^2$ (where $\bar{E}_k$ is the centroid energy of the K-subspace) measure the extent to which different K-subspaces mix and can serve as a criterion for possible truncation of the q.p. space. The norms of H_3 and H_4, the width of the BCS ground state and different mixing intensities have been calculated for nuclei in the Tin region using different two-body interactions.

Numerical results and conclusions will be published soon.

I am grateful to Prof. R. Arvieu for suggesting the problem and for many useful discussions and to Dr. O. Bohigas for invaluable help in performing the numerical calculations.

†Permanent address: Department of Theoretical Physics,
The Hebrew University of Jerusalem,
Israel.

ANHARMONIC VIBRATIONS

JOÃO DA PROVIDENCIA
Laboratorio de Fisica
Universidade de Coimbra

Consider a wave function $\Phi(\gamma)$ dependent on some complex parameter $\gamma = \alpha + i\beta$, with α and β real. The real part α is the collective coordinate describing the deformation of the system. The imaginary part β must be introduced in order to allow the system to acquire momentum in the α degree of freedom. We define α in such a way that $\langle\Phi'|\Phi'\rangle = 1$, where $\Phi' = \partial\Phi/\partial\alpha$. Then the time dependent variational principle

$$i[\langle\delta\Phi|\dot{\Phi}\rangle - \langle\dot{\Phi}|\delta\Phi\rangle] - \delta[\langle\Phi|H|\Phi\rangle/\langle\Phi|\Phi\rangle] = 0$$

leads to the classical equations describing the collective motion:

$$\dot{\alpha} = \partial\mathcal{H}/\partial p_\alpha, \qquad \dot{p}_\alpha = -\partial\mathcal{H}/\partial\alpha.$$

The collective momentum p_α is given by $p_\alpha = 2\beta$ and the collective Hamiltonian $\mathcal{H}$ is given by

$$\mathcal{H} = F(\alpha) + p_\alpha^2\, M(\alpha),$$

where

$$F(\alpha) = \langle\Phi|H|\Phi\rangle_{\beta=0}$$

and

$$M(\alpha) = \tfrac{1}{4}\{\langle\Phi'|H|\Phi'\rangle - \langle\Phi''|H|\Phi\rangle - 2F(\alpha)\}_{\beta=0} .$$

Here $\Phi'' = \partial^2\Phi/\partial\alpha^2$. A derivation of the quantal collective Hamiltonian, based on the generator coordinate method, may also be given. The two approaches lead formally to the same Hamiltonian simply because $\Phi(\gamma)$ is nothing else but the coherent state considered by Biedenharn.

For a complete account of the method see: Lidia Ferreira and Helena Caldeira, Nuclear Physics, to be published. Professor F.M.H. Villars has, independently, derived the same quantal collective Hamiltonian (private communication).

STRUTINSKY CALCULATION FOR LIGHT NUCLEI

H.J. KRAPPE
Hahn-Meitner Institut
Berlin, W. Germany

The ground state binding energies of s-d shell nuclei are calculated with the Strutinsky method. For the liquid drop part Myers and Swiatecki's parametrization was used without pairing and Wigner terms. To account for the Hartree-Fock gap in the single particle spectrum, the shell model energies were shifted in a schematic way according to a prescription given by Levinson and Bar Touv. These modified energies were used to construct the shell correction to the binding energies. Though it is possible to reproduce the odd-even mass differences and the extra binding energy of α-nuclei in this way, the shell correction becomes insensitive to the changes in the uncorrected shell model level densities from magic to mid-shell nuclei. Therefore Levinson's strength of the residual interaction was reduced by 35%. The remaining discrepancies to the experimental binding energies of all nuclei with $10 \leq Z \leq 20$ were smaller than 2.5 MeV compared to 5.5 MeV for the pure liquid drop model. The deformation energy of ^{32}S was calculated including β_2 and β_4 deformations. Assuming a deformation independent Hartree-Fock gap there is no second minimum for large β_2 deformations. Without Hartree-Fock gap it shows up even after inclusion of pairing forces in the calculation of the shell correction.

ISOSPIN SPLITTING OF THE GIANT RESONANCE

RENZO LEONARDI
Istituto di Fisica dell'
Universita Bologna

We can define the giant resonance energy in the channel T^1 by

$$E_{T^1} = \int \sigma(E,T\rightarrow T^1)\, dE / \int E^{-1}\sigma(E,T\rightarrow T^1)\, dE,$$

where $T^1 = T+1$, T, $T-1$. We are able to give a sum rule for each of the six integrated cross sections entering our definition of E_{T^1}. Hence we can investigate the isospin structure of the giant resonance and, in particular, estimate the isospin splitting

$$\Delta E = E_{T+1} - E_T.$$

We found

$$\Delta E = \frac{T+1}{A} U$$

where U is 50-70 MeV for nuclei with A between 30 and 80. For heavier nuclei U is greater (≃120 MeV). The most favourable situation for the channel T+1 is isodoublets ($T = \frac{1}{2}$) in the region of light nuclei where the strengths of the T channel and the T+1 channel are of the same order of magnitude. Detailed calculations are given in ref. 1,2,3.

REFERENCES

1) R. Leonardi and M. Rosa-Clot, Phys. Rev. Lett. 23, 874 (1969).
2) R. Leonardi and M. Rosa-Clot, Rivista del Nuovo Cimento, 1, 1, 1971.
3) R. Leonardi, Bologna pre-print, July 1971.

SHORT RANGE CORRELATIONS AND (γ,p)/(e,e'p) REACTIONS AT INTERMEDIATE ENERGIES

W. WEISE
Institute of Theoretical Physics
University of Erlangen-Nuremberg
Erlangen, Germany

(γ,p) and (e,e'p) reactions at real/virtual photon energies $\omega \gtrsim 100$ MeV are known to provide a useful tool for testing the momentum distribution of single bound protons in the region of high momenta ($p \gtrsim 300$ MeV/c), where short range nucleon-nucleon correlations (SRC) are expected to be of some relevance. A systematic study of the effects of SRC's on proton angular distributions from (γ,p) and (e,e'p) reactions has been carried out on the basis of an independent particle model (IPM) modified by two-nucleon correlation factors (Jastrow's method†). The results are to be summarized as follows:

1) (γ,p)-data at ω=97 MeV could by no means be reproduced within the frame of the pure IPM. Agreement could be obtained, however, after the inclusion of SRC's. The order of magnitude as well as the shape of the proton angular distributions turns out to depend sensitively upon the details of SRC's.
2) Pronounced effects of SRC's on (e,e'p) coincidence cross sections are observed in the range of high recoil momenta ($q_R \gtrsim 200$ MeV/c) of the residual nucleus.

The results obtained are consistent with previous calculations on (γ,pn), (γ,n) and (π,NN) processes††. There is some evidence that the concept of SRC's provides a useful basis for our understanding of a variety of particle-nucleus reactions at intermediate energies.

†R. Jastrow, Phys. Rev. 98 (1955) 1479.

††W. Weise et al., Nucl.Phys. A162 (1971) 330, and ref. therein.

GAS SCINTILLATION COUNTERS

C.A.N. CONDE AND A.J.P. POLICARPO
Laboratorio de Fisica
Universidade de Coimbra

Several gases, especially the noble gases (A,Xe,Kr,He,A-N_2,etc.), are scintillators. They have rather poor energy resolution but their scintillation time is fast (10^{-7} - 10^{-9}s).

Application of an electric field doesn't affect the primary scintillation but introduces secondary scintillation due to the excitation of gas molecules by electrons on their way to the anode. This secondary component increases with the field and has a rise time dependent on the electron transit time (a few μs). Its initially poor energy resolution, due to the uncertain position and orientation of the nuclear particle track, is much improved if the secondary light emission is confined to a small volume of the gas. Such a "Gas Proportional Scintillation Counter"[1] can then achieve energy resolutions comparable to those of an ionization proportional counter but has an advantage for fast timing applications. X-ray studies[2] show that if the light output is increased by increasing the anode voltage, the photomultiplier resolution becomes negligible. Resolutions similar to those for solid state detectors are then obtainable. So far a resolution of 11% has been obtained for the 5.9 keV X-radiation from a ^{55}Fe source.

The characteristics of the counter depend on the gas. Gases like Xe, A-2.5%N_2, Xe-He may be useful for γ-ray, charged particle and neutron detection[1]. Furthermore the possibility of observing the correlated standard charge pulse and the scintillation pulse simultaneously may have special applications.

Various geometries which confine the light production to a small gas volume and improve the light collection efficiency are being studied; e.g. a high pressure Xe γ-counter, which is the gas scintillation counter analog of the Frish ionization chamber.

REFERENCES

(1) Nuclear Instrum. & Meth. 53, 5 (1967); 55, 105 (1967); 58, 151 (1968); IEEE Trans. on Nuclear Science, Vol. NS-15, No.3, 84 (1968).
(2) Nuclear Instrum. & Meth. (to be published).

CONFERENCE HIGHLIGHTS

L.E.H. TRAINOR
Physics Department
University of Toronto

I think that the most important thing we learned is why physicists like so much using logarithmic scales. This point is illustrated in Figure 1 where the time is plotted on a linear scale horizontally and on a logarithmic scale vertically. The horizontal scale shows the time allotted and the vertical scale the time taken. It is actually generous on the part of the speaker because when he is granted 10 minutes he actually uses only 9. Having shown his virtuousness he then requests a 5 minute extension and goes on talking for another 22 minutes. Of course this procedure is scientifically based and so everyone who presented "short" papers used it.

Fig. 1 Value of logarithmic scales

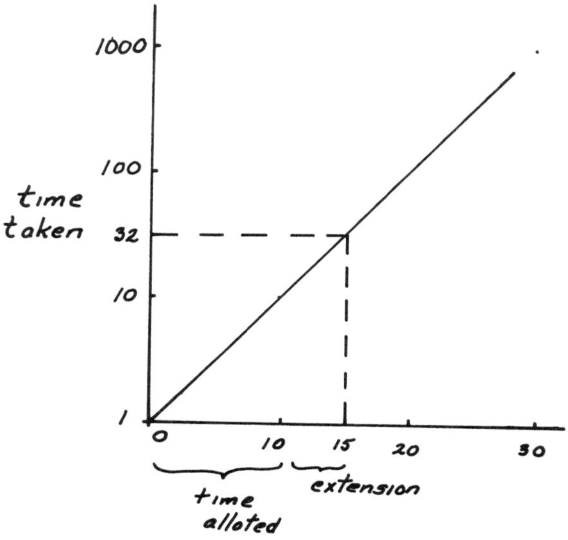

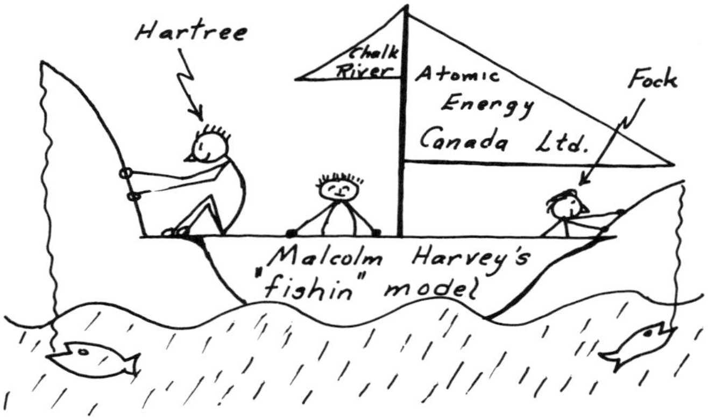

Fig. 2 Harvey's Fission Model

As many of you know, Canada is very proud of its nuclear energy program and we were very fortunate to have Dr. Malcolm Harvey here from Chalk River to tell us about his latest fission model. I have made a rough sketch of it in Figure 2. Dr. Harvey claims he got the idea on a fishing trip with two friends called Hartree and Fock.

Fig. 3 Niagara Falls ≡ 3.6×10^9 watts

For the most part there was a lot of good will at the summer school, but we had some anxious moments during the bitter feud between the Myers and Harvey camps. It all started when Dr. Bill Myers introduced his "liquid drop model". He didn't bring his pictures along, so I have taken the liberty of sketching his model in Figure 3.

But Malcolm Harvey is a hard man to put down and sure enough he came up with an alternative to the liquid drop model which he called the "solid drip model". In a stunning display of arithmetical prowess, he showed that his model, which I have illustrated in Figure 4, could produce just as much power by dripping in 1 cc. of natural uranium per second. Besides you could save all that water spilt over the Niagara Gorge and send it to needy countries like Arabia.

Perhaps the most important discovery of the summer school was made on Sunday afternoon before the lectures got started. Scarcely two hours after he arrived at Mont Tremblant Lodge, Professor Villars discovered the self-energy problem climbing up the mountain, as shown in Figure 5. When he returned to the Lodge he started some important calculations. I have reproduced his final equation in diagrammatic form in Figure 6. It's too bad he didn't wait 2 or 3 days for Dr. Kuo to show him how to

Fig. 4 AECL's Solid Drip Model

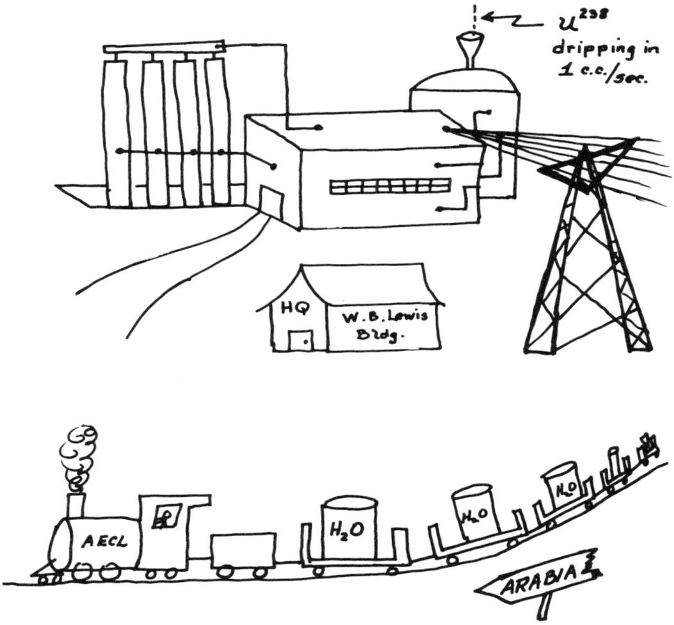

Fig. 5 Self-Energy Problem at Mont Tremblant

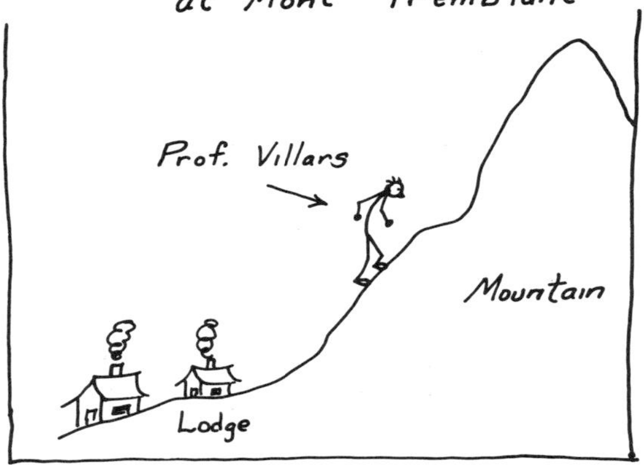

Fig. 6 Villars' Equation (pre-Kuo)

fold up these diagrams because he could have saved a lot of time and wouldn't have had to subtract so many infinities.

At any rate, Dr. Villars' calculations led him to propose a basic new model which he called the "cranking model". Because

Fig. 7 Solution to Self-Energy Problem

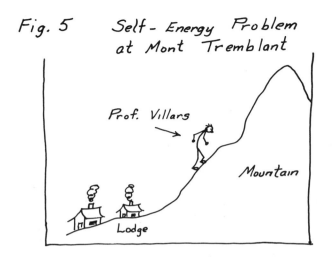

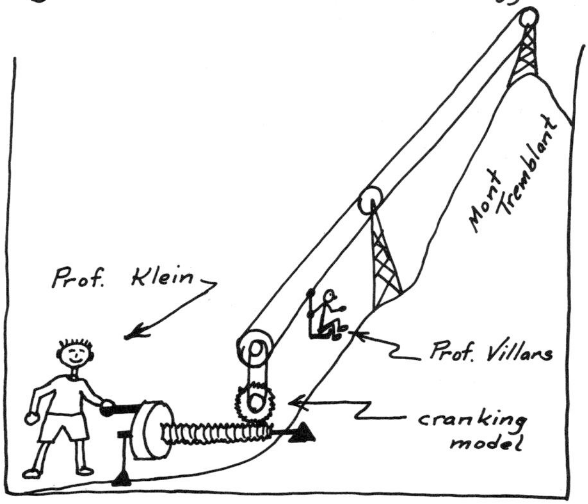

Fig. 8 Tom Kuo's Folding Machine

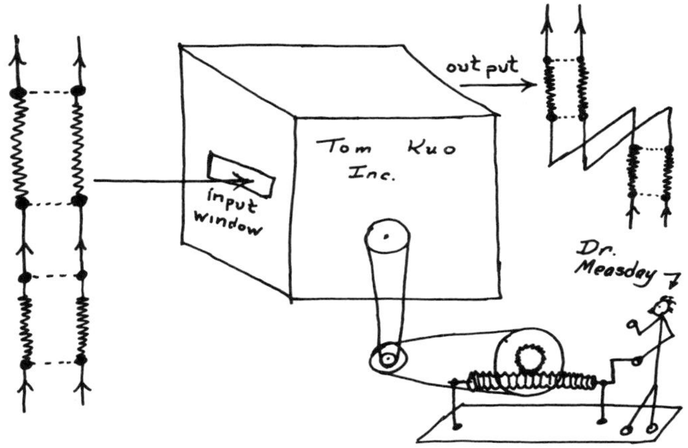

it was new and unique, I have reproduced it for you in Figure 7. Professor Klein got so excited about it all that he wanted to do most of the cranking, and that was one of the most important contributions to the conference.

Professor Kuo was so impressed with the "cranking model" as a powerful tool in nuclear physics that he invented a folding machine for his diagrams. It wasn't too hard to turn, and Dr. Measday said he was interested in intermediate energy problems, so Professor Klein let him turn the crank this time so he could make a contribution to the conference too. As you can see from Figure 8, the complicated diagrams going into the machine are folded into simple ones that any school boy can understand.

Actually I don't think many people realized during Kuo's lecture that his diagrams were road maps of the area around Mont Tremblant. I discovered this when my family drove through the park on Sunday afternoon. Since I like to share some of my bitter experiences with others, I have reproduced a map of Mont Tremblant Park in Figure 9. The straight sections of a Kuo graph represent "model" roads entering and leaving the park and the wiggly line represents the gravelled road through the Park, which is out of this world, just like Tom told us it would be.

One of the puzzling things about Kuo diagrams was the presence of all those Q-boxes. Obviously Q stood for Quebec, but we didn't know what was inside the boxes until Dr. Lo Iudice and some of his Italian friends sampled night life in the neighbouring communities.

Professor Arima seemed anxious to tell us something about Japanese culture, so he developed for us the diagrammatic

Fig. 9. Road Map of Mont Tremblant Park using Kuo diagrams.

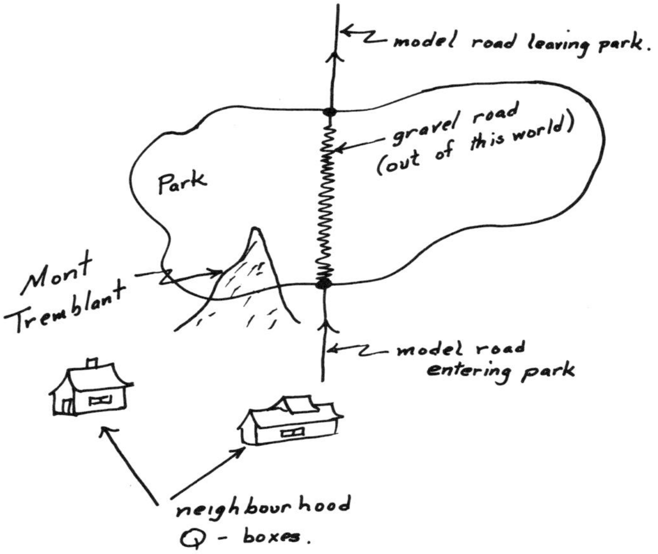

Fig. 10 Arima's Diagrams of Japanese Culture

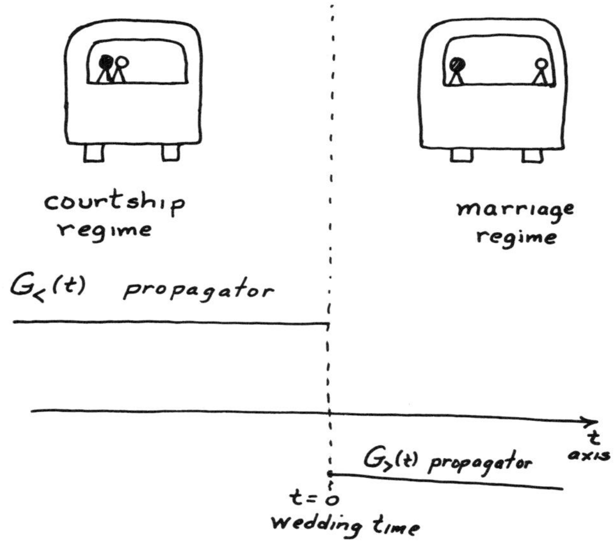

analysis shown in Figure 10. My important contribution to the conference, which you are hearing tonight for the first time, is to recognize that these are time-ordered diagrams so that we can introduce Greens function propagators in a space-time representation to illustrate the marriage discontinuity at t=0.

But the most interesting part of Professor Arima's presentation was his discussion of pairing effects, 4-particle clusters and exotic states. I hope the illustrations in Figure 11 will help you to understand what his lectures were all about. I don't know what all this has to do with shells, unless he was just trying to humour Sam Wong and Andres Zuker.

I suppose it has something to do with his Newfoundland background, but I was impressed that Professor French only took 3 lectures to prove that a Gaussian was symmetrical. That gave him a full 2 lectures to give examples. I thought his best example was the one shown in Figure 12 and so I have called it French's theorem in his honour. As you can see it shows that dN/dt is a Gaussian where N is the number of people swimming in the pool and T is the ambient temperature. Bruce spent a lot of time on fluctuations. He was particularly unhappy about the curve not fitting at the low temperature end because Sam Wong was always in the pool. In any case people were pretty happy with his statement of the Central Limit Theorem which showed that the number of people in the pool at saturation was given by the ratio:

$$N_{sat} = \frac{\text{Volume of Pool}}{\text{Average volume per person}}$$

Fig. 11 Pairing Effects and Exotic States in Arima's Model

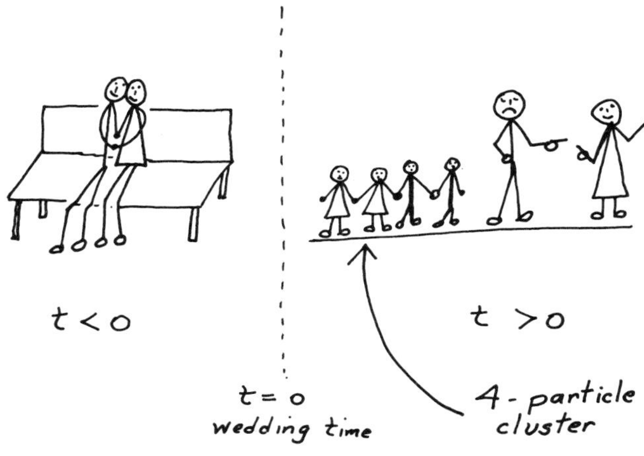

Fig 12 French's Theorem

Let $\dot{N}$ = no. of people swimming

T = outside Temperature

Then dN/dT is Gaussian for physicists

Proof:

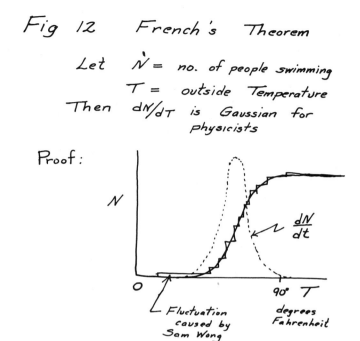

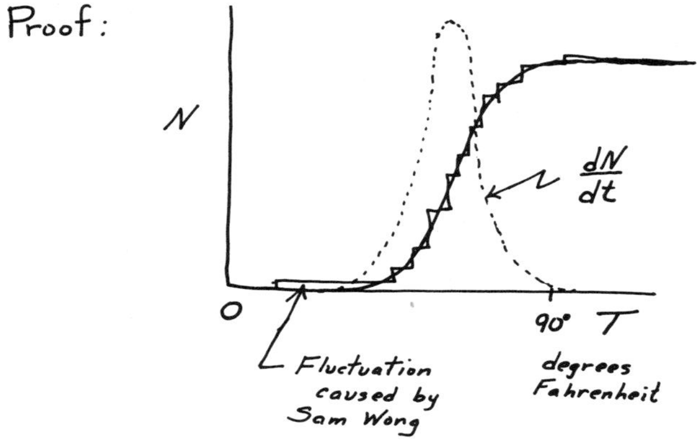

Somehow you can't have a conference on any subject in nuclear physics without all these scattering people turning up, and this summer school was no exception. Dr. George Walker talked about ee', pp', ππ' and so forth up to 10^3 MeV or so and he called it intermediate energy physics. Obviously he had not been reading the recent literature ("New linear accelerator to take wing", Physics Today, April 1971, p.9) on chicken-chicken scattering at about 10^{33} MeV. I would say intermediate energy physics is probably sand experiments in wind tunnels, and Dr. Walker's physics is low-energy - just peanuts in this big-league game.

We might have had a good discussion on chicken-chicken scattering if Dr. Tom Drake hadn't turned up and raised 2 fundamental objections:

1. experiment in this field is for the birds.
2. what have we got against ducks?
   (one thing: ducking tends to reduce cross-sections and counting rates).

He suggested duck-duck scattering, because one could use an audio detector to count the quacks. His slides were a bit much, but I have tried to summarize his results in Figure 13, where the total d-d (duck-duck) cross-section is plotted against E_d, the incident duck energy. The most interesting feature is the abrupt rise at E_e, the threshold for egg production. And that is how Dr. Zuker got into the discussion. He showed how a

Fig. 13 Cross-Section for duck-duck Scattering (according to Drake).

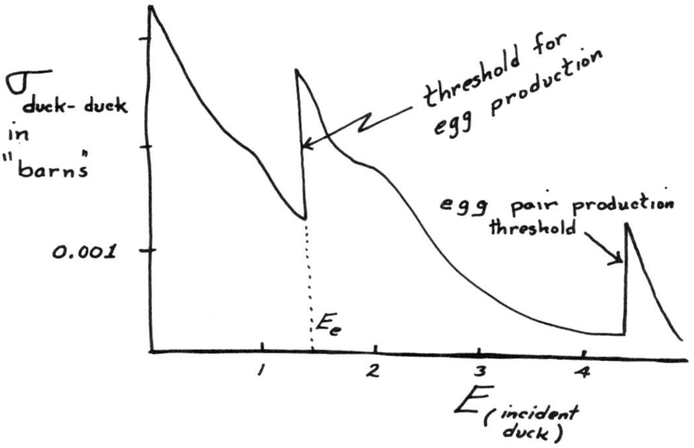

1 barn = 10^3 square duck feet

shell model could introduce bound states into the duck problem. He also explained how one could understand these results by doing a coupled channel analysis on the egg-egg scattering problem.

Well, no conference summary would be complete without a historical sketch of how it came about in the first place. It goes like this.

Once upon a time there was an English farm boy who used to watch the ducks running around the farmyard and scattering off

Fig. 14 RPA = random path approach

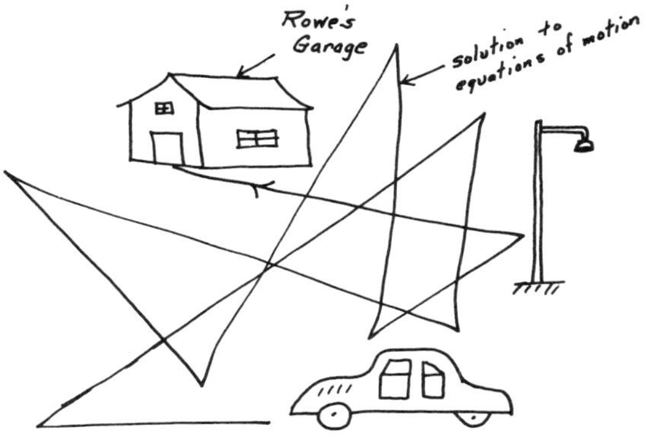

RPA + Open Space = OSRPA

each other. So he became a physicist and by some good fortune married a gracious woman named Una and moved to the University of Toronto. His name was David Rowe.

The Rowes had one little problem - objects kept getting in the way of Una's car, like lamp posts, for example. Now David is not the kind of chap to leave a lady with a problem. Being a shrewd observer of all natural phenomena, a trait common to all farm boys educated at Cambridge and Oxford, David soon recognized the source of difficulty in the RPA method of driving (Random Path Approach - see Figure 14). Using the equations of motion technique he developed an improved method which he called OSRPA (Open Space RPA). It was tailor-made for Canadian driving conditions and geography.

Professor Rowe was so excited about his discovery that he wanted to tell it to all of the important people in the world. That meant organizing an international summer school in nuclear physics at Mont Tremblant, Quebec.

PARTICIPANTS

BRAZIL

L.C.M. do Amaral, Department of Physics, University of Rio de Janeiro, C/RJ, Gb, Zc-20, Rio de Janeiro

*J. Goldemberg, Department of Physics, University of São Paulo, Caixa Postal 8.105, São Paulo and Department of Physics, University of Toronto, Toronto, Ontario, Canada

CANADA

J. Asai, Department of Physics, University of Windsor, Windsor, Ontario

A. Ayad, Department of Physics, University of Toronto, Toronto, Ontario

R. Basserman, Department of Physics, University of Toronto, Toronto, Ontario

R. Beukens, Department of Physics, University of Toronto, Toronto, Ontario

K.L. Bhatia, Institute of Theoretical Physics, McGill University, Montreal, P.Q.

A.H. Chung, Department of Physics, University of Toronto, Toronto, Ontario

Mme B. Cujec, Département de Physique, Université Laval, Cité Universitaire, Québec, P.Q.

R. Dayras, Département de Physique, Université Laval, Cité Universitaire, Québec, P.Q.

A. Degré, Laboratoire de Physique Nucléaire, Université de Montréal, Montréal, P.Q.

W.R. Dixon, Division of Physics, National Research Council, Ottawa, Ontario

T.W. Donnelly, Department of Physics, University of Toronto, Toronto, Ontario (now at Stanford University, Stanford, California)

*T.E. Drake, Department of Physics, University of Toronto, Toronto, Ontario

Miss P. Dworkin, Department of Physics, University of Toronto, Toronto, Ontario

R. Esch, Department of Physics, University of British Columbia, Vancouver, B.C.
G.F.H. Graebner, Department of Physics, University of Saskatchewan, Saskatoon, Sasckatchewan
P.W. Green, Nuclear Research Centre, Department of Physics, University of Alberta, Edmonton, Alberta
M.W. Greene, Tandem Accelerator Laboratory, McMaster University, Hamilton, Ontario
V.K. Gupta, Nuclear Research Centre, Department of Physics, University of Alberta, Edmonton, Alberta
*M. Harvey, Atomic Energy of Canada Ltd., Chalk River Nuclear Laboratories, Chalk River, Ontario
*O. Haüsser, Atomic Energy of Canada Ltd., Chalk River Nuclear Laboratories, Chalk River, Ontario
N. Lo Iudice, Department of Physics, University of Toronto, Toronto, Ontario
A. Jaccarini, Département de Physique, Université Laval, Cité Universitaire, Québec, P.Q.
K.P. Jackson, Department of Physics, University of Toronto, Toronto, Ontario
A. Jopko, Department of Physics, McMaster University, Hamilton, Ontario
W.E. Kieser, Department of Physics, University of Toronto, Toronto, Ontario
*W. Knowles, Atomic Energy of Canada Ltd., Chalk River Nuclear Laboratories, Chalk River, Ontario
P.H.C. Lee, Atomic Energy of Canada Ltd., Chalk River Nuclear Laboratories, Chalk River, Ontario
*A.E. Litherland, Department of Physics, University of Toronto, Toronto, Ontario
G. Lougheed, Department of Physics, University of Toronto, Toronto, Ontario
J.H. Matthews, Department of Physics, Mount Allison University, Sackville, New Brunswick
*D.F. Measday, Department of Physics, University of British Columbia, Vancouver, B.C.
S. Morris, Department of Physics, Sir George Williams University, Montreal, P.Q.
W.Y. Ng, Department of Physics, University of Toronto, Toronto, Ontario
C. Ngo-Trong, Department of Physics, University of Toronto, Toronto, Ontario
A. Pilt, Tandem Accelerator Laboratory, McMaster University, Hamilton, Ontario
J.-M. Poutissou, Laboratoire de Physique Nucléaire, Université de Montréal, Montréal, P.Q.
C. Rolfs, Department of Physics, University of Toronto, Toronto, Ontario
*D.J. Rowe, Department of Physics, University of Toronto, Toronto, Ontario
R.C. Sharma, Department ofPhysics, Sir George Williams University, Montreal, P.Q.

H.S. Sherif, Nuclear Research Centre, Department of Physics, University of Alberta, Edmonton, Alberta
N.K. Sherman, Division of Physics, National Research Council, Ottawa, Ontario
J.A. Stronach, Department of Physics, University of Saskatchewan, Saskatoon, Saskatchewan
*L.E.H. Trainor, Department of Physics, University of Toronto, Toronto, Ontario
H.P. Trautvetter, Department of Physics, University of Toronto, Toronto, Ontario
R. Volders, Laboratoire de Physique Nucléaire, Université de Montréal, Montréal, P.Q.
T. Vo Van, Department of Physics, University of Toronto, Toronto, Ontario
*D. Ward, Atomic Energy of Canada Ltd., Chalk River Nuclear Laboratories, Chalk River, Ontario
D.C.S. White, Nuclear Research Centre, Department of Physics, University of Alberta, Edmonton, Alberta
S.S.M. Wong, Department of Physics, University of Toronto, Toronto, Ontario

ENGLAND

B. Underwood, Nuclear Physics Laboratory, University of Oxford, Keble Road, Oxford
M.R. Wormald, Nuclear Physics Laboratory, University of Oxford, Keble Road, Oxford

FRANCE

O. Bohigas, Division de Physique Théorique, Institut de Physique Nucléaire, B.P. No. 1, 91-Orsay
H. Nissimov, Division de Physique Théorique, Institut de Physique Nucléaire, B.P. No. 1, 91-Orsay
S.K.M. Wong, C.E.N. Saclay, Service de Physique Théorique, B.P. No. 2, Saclay, 91-Gif-sur-Yvette
A. Zuker, C.E.N. Saclay, Service de Physique Théorique, B.P. No. 2, Saclay, 91-Gif-sur-Yvette

GERMANY

R. Fraser, Institut für Theoretische Physik der Universität Frankfurt, 8/10 Robert Mayer Strasse, 6 Frankfurt/Maine 1 (now at University of Toronto)
H.M. Hofmann, Institut für Theoretische Physik der Universität Erlangen/Nürnberg, 852 Erlangen
H.J. Krappe, Hahn-Meitner Institut, 100 Glienickerstrasse, 1 Berlin 39
A. Sevgen, Max Planck Institut für Kernphysik, Postfach 1248, 69 Heidelberg 1
W. Weisse, Institut für Theoretische Physik der Universität Erlangen/Nürnberg, 852 Erlangen

INDIA

J.C. Parikh, Physical Research Laboratory, Navrangpura, Ahmedabad 9

ITALY

A. d'Andrea, Istituto di Fisica "G. Marconi", Universita di Roma, Rome
S. Frullani, Istituto Superiore di Sanita, Viale Regina Elena 299, Rome
Sgna. M. Guidetti, Istituto di Fisica, Politecnico, Torino 10129.
R. Leonardi, Istituto di Fisica "A. Righi", Universita di Bologna, Via Irneria 46, Bologna
B. Mosconi, Istituto di Fisica Teorica, Universita di Firenze, Firenze 50125
E. Olivieri, Istituto Matematico, Universita di Roma, Rome.
A. Pompei, Istituto di Fisica, Universita di Cagliari, Cagliari
O. Ragnisco, Istituto di Fisica "G. Marconi", Universita di Roma, Rome
M. Scalia, Istituto Matematico, Universita di Roma, Rome

PORTUGAL

C.A.N. Conde, Laboratorio de Fisica da Universidade Coimbra, Coimbra
A. Policarpo, Laboratorio de Fisica da Universidade Coimbra, Coimbra
J. da Providencia, Laboratorio de Fisica da Universidade Coimbra, Coimbra
J.N. Urbano, Laboratorio de Fisica da Universidade Coimbra, Coimbra

U.S.A.

*A. Arima, Institute for Theoretical Physics, S.U.N.Y., Stony Brook, N.Y. 11790
R. Braley, Lewis Research Center, N.A.S.A., Cleveland, Ohio 44135
*A.G.W. Cameron, Department of Physics, Yeshiva University, New York 10033
W. Ford, Lewis Research Center, N.A.S.A., Cleveland, Ohio 44135.
*J.B. French, Department of Physics and Astronomy, University of Rochester, Rochester, N.Y. 14627
C.C. Fu, Physics Department, S.U.N.Y., Buffalo, N.Y. 14214.
S. Jena, Department of Physics, Carnegie Mellon University, Pittsburgh, Pa. 15212
V. Joshi, Department of Physics, Duke University, Durham, N. Carolina 27706
*A. Klein, Department of Physics, University of Pennsylvania, Philadelphia, Pa. 19104

*T.T.S. Kuo, The Physical Laboratory, S.U.N.Y., Stony Brook, N.Y. 11790

S.-Y. Li, Department of Physics, University of Rochester, Rochester, N.Y. 14627

*W.D. Myers, Lawrence Radiation Laboratory, Berkeley, Calif. 94720

P.D. Parker, Physics, Department, Yale University, New Haven Conn. 06520

I. Reichstein, Department of Physics, University of Indiana, Bloomington, Ind. 47401 (now at University of Toronto)

J.K. Tuli, Department of Physics, University of Indiana, Bloomington, Ind. 47401

*F.M.H. Villars, Department of Physics, Massachusetts Institute of Technology, Cambridge, Mass. 02139

*G.E. Walker, Department of Physics, University of Indiana, Bloomington, Ind. 47401

E.F. Zganjar, Physics Department, Louisiana State University, Baton Rouge, La. 70803

* Speakers.

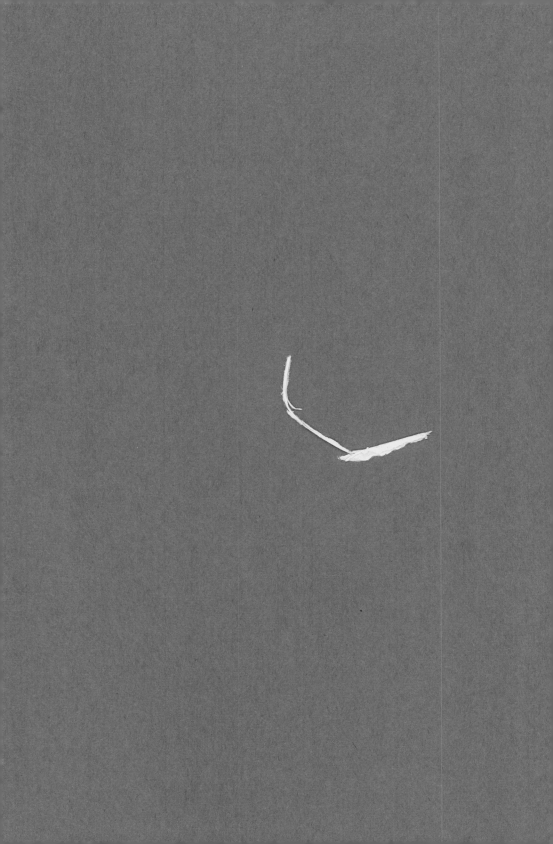